Ergodic Theory, Dynamical Systems, and the Continuing Influence of John C. Oxtoby

CONTEMPORARY
MATHEMATICS

678

Ergodic Theory, Dynamical Systems, and the Continuing Influence of John C. Oxtoby

Oxtoby Centennial Conference
October 30–31, 2010
Bryn Mawr College, Bryn Mawr, PA

Williams Ergodic Theory Conference
July 27–29, 2012
Williams College, Williamstown, MA

AMS Special Session
Ergodic Theory and Symbolic Dynamics
January 17–18, 2014
Baltimore, MD

Joseph Auslander
Aimee Johnson
Cesar E. Silva
Editors

American Mathematical Society
Providence, Rhode Island

2010 *Mathematics Subject Classification.* Primary 37A05, 37B05, 37A40, 37P40, 37B50, 37B10, 37A30, 37A20, 01A70.

Library of Congress Cataloging-in-Publication Data

Names: Auslander, Joseph, 1930- editor. | Johnson, Aimee, 1961- editor. | Silva, Cesar Ernesto, 1955- editor. | Oxtoby, John C.
Title: Ergodic theory, dynamical systems, and the continuing influence of John C. Oxtoby/Joseph Auslander, Aimee Johnson, Cesar E. Silva, editors.
Description: Providence, Rhode Island : American Mathematical Society, [2016] | Series: Contemporary mathematics ; volume 678 | "Oxtoby Centennial Conference, October 30–31, 2010, Bryn Mawr College, Bryn Mawr, PA." | "Williams Ergodic Theory Conference, July 27–29, 2012, Williams College, Williamstown, MA." | "AMS Special Session on Ergodic Theory and Symbolic Dynamics, January 17–18, 2014, Baltimore, MD." | Includes bibliographical references and index.
Identifiers: LCCN 2016020577 | ISBN 9781470422998 (alk. paper)
Subjects: LCSH: Measure theory–Congresses. | Ergodic theory–Congresses. | Topological spaces–Congresses. | AMS: Dynamical systems and ergodic theory – Ergodic theory – Measure-preserving transformations. msc | Dynamical systems and ergodic theory – Topological dynamics – Transformations and group actions with special properties (minimality, distality, proximality, etc.). msc | Dynamical systems and ergodic theory – Ergodic theory – Nonsingular (and infinite-measure preserving) transformations. msc | Dynamical systems and ergodic theory – Arithmetic and non-Archimedean dynamical systems – Non-Archimedean Fatou and Julia sets. msc | Dynamical systems and ergodic theory – Topological dynamics – Multi-dimensional shifts of finite type, tiling dynamics. msc | Dynamical systems and ergodic theory – Topological dynamics – Symbolic dynamics. msc | Dynamical systems and ergodic theory – Ergodic theory – Ergodic theorems, spectral theory, Markov operators. msc | Dynamical systems and ergodic theory – Ergodic theory – Orbit equivalence, cocycles, ergodic equivalence relations. msc | History and biography – History of mathematics and mathematicians – Biographies, obituaries, personalia, bibliographies. msc
Classification: LCC QA312.E74 2016 | DDC 515/.48–dc23
LC record available at http://lccn.loc.gov/2016020577
Contemporary Mathematics ISSN: 0271-4132 (print); ISSN: 1098-3627 (online)

DOI: `http://dx.doi.org/10.1090/conm/678`

∞ The paper used in this book is acid-free and falls within the guidelines
established to ensure permanence and durability.
Visit the AMS home page at `http://www.ams.org/`

10 9 8 7 6 5 4 3 2 1 21 20 19 18 17 16

Contents

Preface

The work of John C. Oxtoby was fundamental for the fields of measure theory and ergodic theory, and paved the way for many advances in these fields. His contributions were numerous and include his classic monograph, "Measure and Category," along with the oft-quoted Oxtoby-Ulam theorem. His life was celebrated at the Oxtoby Centennial Conference held at Bryn Mawr College on October 30–31, 2010. This conference, held at the institution where Oxtoby spent most of his professional life, featured 14 speakers and over 30 participants and included recollections on Oxtoby's life by his son David and his daughter Margaret. The webpage for this conference can be found at

http://web.williams.edu/Mathematics/csilva/Oxtoby_Centennial_Conf.htm

These proceedings serve as a tangible record of that celebration along with two subsequent conferences: the Williams Ergodic Theory Conference, held at Williams College on July 27-29, 2012, and the AMS Special Session on Ergodic Theory and Symbolic Dynamics at the joint AMS/MAA meetings in Baltimore on January 17-18, 2014. Webpages for these events can be found at:

http://web.williams.edu/Mathematics/csilva/Williams_ErgodicTheory2012.html

http://jointmathematicsmeetings.org/meetings/national
/jmm20142160_program_ss31.html#title

The articles contained in this volume are from the participants of these three conferences and comprise a variety of topics in both measurable and symbolic dynamics. These include a survey article on the life and work of Oxtoby, providing a source of information about the many ways Oxtoby's work influenced mathematical thought in this and other fields.

We thank the co-organizers of the Oxtoby Centennial Conference, Leslie Cheng and Paul Melvin, for their invaluable help in organizing such an endeavor. We gratefully acknowledge the National Science Foundation, grant DMS-1058120, and the Bryn Mawr College 1902 Lecture Fund for their financial support. The National Science Foundation grant DMS-1058120 also supported the Williams Ergodic Theory Conference, with additional financial support provided by Williams College. Finally, we thank all of the speakers for their wonderful talks and the participants for making these three conferences such a success.

Joseph Auslander
Aimee Johnson
Cesar E. Silva

Oxtoby Centennial Conference, organized by Joseph Auslander, Leslie Cheng, Aimee Johnson, Paul Melvin, and Cesar E. Silva. Bryn Mawr College, October 30–31, 2010

Ethan Akin (CCNY), *Good measures on the Cantor set*

Tim Austin (Brown University), *The quantitative ergodic theorem and embeddings of groups into Banach spaces*

Robert Devaney (Boston University), *The fractal geometry of the Mandelbrot set*

Jane Hawkins (UNC Chapel Hill), *Families of type III_0 ergodic transformations in distinct orbit equivalent classes*

Fern Hunt (NIST), *A model of routing in computer networks*

Mike Keane (Wesleyan University), *Understanding ergodicity*

Kathryn Lindsey (Cornell University), *Tight upper bounds on the number of invariant components on translation surfaces by connected component of moduli space*

Nelson Markley (University of Maryland and Lehigh University), *Completing the Weil-Hedlund-Anosov program*

Dan Mauldin (University of North Texas), *Some problems and techniques inspired by John C. Oxtoby*

Kevin McGoff (University of Maryland), *Random subshifts of finite type*

V.S. Prasad (University of Massachusetts, Lowell), *Continuous versions of the homeomorphic measures theorem*

Bethany Springer (Colorado State), *Some results in almost continuous dynamics*

Susan Williams (University of South Alabama), *Oxtoby and Toeplitz minimal systems: An excursion*

Kelly Yancey (University of Illinois), *Continuous and smooth realizations of ergodic transformations of finite entropy*

Participants in addition to speakers

Joseph Auslander (University of Maryland)

Karl Backs (University of North Texas)

Andres del Junco (University of Toronto)

Joanna Furno (University of North Carolina, Chapel Hill)

Geoff Goodson (Towson University)

Gernot Greschonig (University of Vienna and University of Maryland)

Aimee Johnson (Swarthmore College)

Jonathan King (University of Florida)

Lorelei Koss (Dickinson College)

Kathleen Madden (Drew University)

Nelson Markley (University of Maryland and Lehigh University)

Dan Mauldin (University of North Texas)

David McClendon (Swarthmore College)

Kevin McGoff (University of Maryland)

Paul Melvin (Bryn Mawr College)

Joseph Moser (West Chester University)

Anima Nagar (Indian Institute of Technology Delhi, India)

Mahesh Nerurkar (Rutgers University Camden)

David Oxtoby (Pomona College)

Vincent Peck (IBM)

Vallorie J. Peridier (Temple University)

Karl Petersen (University of North Carolina, Chapel Hill)

Amit Priyadarshi (Rutgers University)

Joe Rosenblatt (University of Illinois, Urbana-Champaign)

Niketa Salvi (Colorado State University and DePaul University)

Michael Sears (University of the Witwatersrand)

Nicholas Sharpe (University of Florida)

Cesar E. Silva (Williams College)

Bethany Springer (Colorado State University and University of Toronto)

Lisa Traynor (Bryn Mawr College)

Conan Wu (Princeton University)

Andy Yingst (University of South Carolina)

Figure 1. Oxtoby Centennial Conference Participants.
Courtesy of Cesar E. Silva

Figure 2. Oxtoby Centennial. Undergraduate Lecture Participants.
Courtesy of Cesar E. Silva

Figure 3. Williams Ergodic Theory Conference Participants.
Courtesy of Cesar E. Silva

Williams Ergodic Theory Conference, organized by Joseph Auslander and Cesar E. Silva. Williams College, July 27–29, 2012

Jon Aaronson (Tel Aviv University, Israel), *Rational weak mixing in infinite measure spaces*

Ethan Akin (The City College, New York, USA), *Homeomorphisms of the Cantor set*

Rachel Bayless (University of North Carolina, Chapel Hill, USA), *Negative generalized Boole transformations and Krengel entropy*

Darren Creutz (Vanderbilt University, USA), *Stabilizers of ergodic actions of lattices and commensurators*

Alexandre Danilenko (Institute for Low Temperature Physics, Ukraine), *Odometer actions of Heisenberg group*

Jared Hallett (Williams College), *On Li-Yorke measurable sensitivity*

Jane Hawkins (University of North Carolina, Chapel Hill, USA), *Complex dynamics and infinite ergodic theory*

Yuri Karlovich (Universidad Autonoma del Estado de Morelos, Mexico), *Ergodic number theory and AP factorization of almost periodic matrix functions*

Zemer Kosloff (Tel Aviv University, Israel), *Maharam extensions of nonsingular Bernoulli shifts*

Mariusz Lemanczyk (Nicolaus Copernicus University, Poland), *Self-joinings of higher order in the problem of non-reversibility of ergodic flows*

Tudor Pădurariu (University of California, Los Angeles), *On rationally ergodic and rationally weakly mixing rank-one transformations*

Kyewon Park (Ajou University, South Korea), *Entropy dimension: examples*

V.S. Prasad (University of Massachusetts Lowell, USA), *Tilings of the integers arising from a class of infinite measure preserving transformations*

Emmanuel Roy (University of Paris 13, France), *Prime Poisson suspensions*

Ilya Vinogradov (Princeton University), *Effective bisector estimate for PSL(2,C) with applications to circle packings*

Kelly Funk Yancey (University of Illinois) *Generic homeomorphisms*

Participants in addition to speakers

Joseph Auslander (University of Maryland)

Rachel Bayless, (University of North Carolina, Chapel Hill)

Darren Creutz (Vanderbilt University)

Irving Dai (Harvard University)

Andres del Junco (University of Toronto)

Stanley Eigen (Northeastern University)

Xavier Garcia (University of Minnesota)

Arshag Hajian (Northeastern University)

Shelby Heinecke (MIT)

Aimee Johnson (Swarthmore College)

Yuri Karlovich (Universidad Autonoma del Estado de Morelos, Mexico)

Zemer Kosloff (Tel Aviv University, Israel)

David McClendon (Swarthmore College)

Frank Morgan (Williams College)

Tudor Padurariu (UCLA)

Karin Reinhold (SUNY Albany)

Cesar E. Silva (Williams College)

Emily Wickstrom (Williams College)

Ben Wilson (University of North Carolina, Chapel Hill)

Evangelie Zachos (Princeton University)

AMS Special Session on Ergodic Theory and Symbolic Dynamics, organized by Aimee Johnson and Cesar E. Silva. Baltimore, MD, January 17–18, 2014

Tim Austin (Courant Institute, New York University), *Partial difference equations over compact Abelian groups*

Jon Chaika (University of Utah), *Topological mixing for residual sets of interval exchange transformations*

Joanna Marie Furno (Dickinson College), *Orbit equivalence classes of p-adic transformations*

Jane M. Hawkins (NSF/University of N. Carolina at Chapel Hill), *Bernoulli properties and Julia sets for maps of the real projective plane*

Joseph L Herning (The George Washington University), *Bijective substitutions without topological subshift factors isomorphic to their maximal equicontinuous factors*

Aaron Hill (University of North Texas), *The isomorphism problem for rank-1 systems*

Steve Kass (Drew University), *A sufficient condition for non-soficness of higher-dimensional subshifts*

Konstantin Medynets (United States Naval Academy), *Second order ergodic theorem for substitution tiling systems*

David M. McClendon (Ferris State University), *Speedup equivalence of ergodic* $\mathbb{Z}^d-$*actions*

Kevin McGoff (Duke University), *Random* $\mathbb{Z}^d$ *subshifts of finite type*

Francis C. Motta (Colorado State University), *Optimally topologically transitive orbits of the Bernoulli shift map*

Nicholas Ormes (University of Denver), *Extender sets and multidimensional subshifts*

Ronnie Pavlov (University of Denver), *A characterization of topologically completely positive entropy for shifts of finite type*

Karl Petersen (University of North Carolina at Chapel Hill), *Measuring complexity and interconnectivity in dynamical systems*

David Ralston (SUNY College at Old Westbury), *Symbolic dynamics and the infinite staircase*

E. Arthur Robinson, Jr. (George Washington University), *Parry's topological transitivity and* f*-expansions*

Joseph Rosenblatt (University of Illinois at Urbana-Champaign), *Coboundaries and ergodic sums*

Ayşe A. Şahin (DePaul University), *Recent developments in directional dynamics*

Jane Wang (Princeton University), *Weak rational ergodicity and rank-one transformations*

Kelly B Yancey (University of Maryland), *Weakly mixing vs. rigid in the infinite setting*

Contemporary Mathematics
Volume **678**, 2016
http://dx.doi.org/10.1090/conm/678/13639

Conjugacy in the Cantor set automorphism group

Ethan Akin

ABSTRACT. We survey, and extend, results on the adjoint action of the homeomorphism group $H(X)$ on the space of surjective continuous maps, $C_s(X)$, where X is a Cantor set. We look also at the restriction of the action to various dynamically defined subsets of $C_s(X)$, e. g. the sets of topologically transitive maps, chain transitive maps, chain mixing maps, etc. In each case, we consider whether there exist elements with a dense conjugacy class and if so, what the generic elements look like.

Introduction

When a group acts on itself by the adjoint action, the orbit of an element is its conjugacy class. In the case of the group $H(X)$ of homeomorphisms on a compact metric space X, the adjoint action extends to $C(X)$, the space of all continuous maps on X. For a continuous map f on X we refer to its $H(X)$ orbit as its conjugacy class and denote it by $H(X) \cdot f$.

The case of $H(X)$ is of special interest because of its dynamic interpretation. We can regard a map $f \in C(X)$ as a discrete time dynamical system on the state space X, describing the evolution by iteration: $x_{n+1} = f(x_n)$. The maps f and g are conjugate precisely when there exists $h \in H(X)$ such that $g = h \circ f \circ h^{-1}$. This says that h is an isomorphism from f to g in the category of dynamical systems.

It is convenient to restrict attention to surjective maps, which form a closed subset $C_s(X) \subset C(X)$. A dynamical system is a pair (X, f) with X a nonempty, compact metric space and $f \in C_s(X)$. It is invertible when f is injective, so that $f \in H(X)$.

Considerable work has been done analyzing the adjoint action of $H(X)$ for the special case when X is a Cantor space, i.e. a metric space homeomorphic to the usual Cantor set. The basis of clopen sets makes everything more tractable in that case. Furthermore, the Cantor set plays a central role in the theory of dynamical systems. In various guises it appears as the state space, e.g. for coding-related systems like subshifts and for algebraic systems like adding machines (also called odometers). Furthermore, every system on a state space with no isolated points has an almost one-to-one lift to a system on the Cantor set.

For a Polish topological group G like $H(X)$, it is of special interest when the adjoint action of G on itself is topologically transitive, i.e. when there exists g whose conjugacy class is dense in G. Such a $g \in G$ is called a *transitive element*. A group

2010 *Mathematics Subject Classification.* Primary 37B05, 37B10, 37E99, 22F50.

which admits such transitive elements is said to have the *Rohlin Property.* The name is motivated by an ergodic theory result of Rohlin concerning the automorphism group of the Lebesgue space. When G satisfies the Rohlin Property then the set of transitive elements is a dense G_δ subset of G. It sometimes happens that a single conjugacy class is residual, i.e. it contains a dense G_δ set. Since distinct conjugacy classes are disjoint but any two residual subsets meet, it follows that there is at most one residual conjugacy class. When such a class exists we call its members *transitive elements of residual type* and we say that the group has the *Strong Rohlin Property.* Finally, when the diagonal action of G on any finite product G^n admits a residual conjugacy class then we say that G has *ample generics.* For a survey, see [**10**].

For X a Cantor space it was shown by Glasner and Weiss in [**9**] and, independently, in [**3**] that the automorphism group $H(X)$ has the Rohlin Property. For a certain class of good measures μ on X it was shown in [**2**] that $H_\mu(X)$, the closed subgroup of automorphisms which preserve μ, has the Strong Rohlin Property. It was shown that the product of the universal adding machine (see below) with the identity on a Cantor space provides a transitive element of residual type. Using Fraïssé theory, Kechris and Rosendal showed in [**12**] that $H(X)$ has the Strong Rohlin Property. An explicit description of a transitive element of residual type was then given in [**4**]. With μ the Haar measure on $X = 2^{\mathbb{N}}$ it was shown in [**12**] that $H_\mu(X)$ has ample generics and in [**13**] that $H(X)$ itself has ample generics. Transitivity results with different topologies were proved in [**8**].

A conjugacy invariant subset of $C(X)$ defines a dynamic property, i.e. one which is an invariant under topological conjugacy. Conversely, we will consider the subsets defined by various dynamic concepts associated with recurrence, transitivity and mixing. We recall that the following subsets of $C_s(X)$ are closed.

- $C_s(X;1)$, the surjective maps which admit fixed points.
- $CT(X)$, the chain transitive maps, which contains $TT(X)$, the topologically transitive maps, as a dense G_δ subset.
- $CR(X)$, the chain recurrent maps, which contains the maps with dense recurrent points as a dense G_δ subset.
- $CM(X)$, the chain mixing maps, which contains $WM(X)$, the weak mixing maps, as a dense G_δ subset.

So $CR(X;1) = CR(X) \cap C_s(X;1)$ and $CM(X;1) = CM(X) \cap C_s(X;1)$ are closed as well. If a chain transitive map admits a fixed point then it is chain mixing and so we need not consider $CT(X;1)$. Note that the density results assume that X is a Cantor space.

For such a conjugacy invariant subset S of $C_s(X)$ with X a Cantor space, we wish to consider whether the restriction of the adjoint action to S admits transitive elements or transitive elements of residual type. For example, Hochman [**11**] proved that the universal adding machine is a transitive element of residual type for $CT(X)$. Shimomura in [**15**], [**16**], [**17**] has extended these results, as we will describe below. If f is a transitive element for S, it means that every element of S can be uniformly approximated by a map conjugate to f, i.e. it is f up to a change of variables.

Notice that when we say that $f \in S \subset C_s(X)$ is a transitive element for S we are referring to the adjoint action of $H(X)$ on S. Meanwhile, f itself defines a dynamical system on X which may or may not be topologically transitive.

The analysis proceeds by using finite approximations for $f \in C_s(X)$ as in [**3**].

In general, if $\mathcal{A}$ is an open cover of a compact metric space X, we can regard the elements of $\mathcal{A}$ as providing a finite approximation to X. Think of them as pixels covering X. We can represent $f \in C(X)$ by using $f^{\mathcal{A}}$, the set of pairs $(U_1, U_2) \in \mathcal{A} \times \mathcal{A}$ such that $f(U_1)$ meets U_2. Here the relationship is usually not that of a function, $f(U_1)$ usually meets several members of $\mathcal{A}$. When X is a Cantor space, we restrict attention to the case when $\mathcal{A}$ is a decomposition of X by clopen subsets.

There are different ways to think about this finite setup. Most authors, including Bernardes and Darji [**7**] and Shimomura in his papers, regard the elements of $\mathcal{A}$ as vertices of a directed graph with the oriented edges the pairs in $f^{\mathcal{A}}$. The original function f is conjugate to an inverse limit constructed from such graphs. The projective Fraïssé constructions of Kwiatkowska in [**13**] lead to similar inverse limit constructions.

I prefer the equivalent approach of regarding the set $f^{\mathcal{A}}$ as a relation on $\mathcal{A}$, see, e.g. [**1**] Chapter 5. A relation on a set A is just a subset $R \subset A \times A$. It is called a *surjective relation* when each of the two coordinate projections maps it onto A. A function f on A is a special case of a relation such that $f(x) = \{y \in A : (x, y) \in f\}$ is a singleton for every x and, abusively, we write $f(x)$ for both the singleton set and its unique element. We can iterate relations, and various dynamics concepts extend to relations. For us, a *system* will be a pair (A, R) with A a compact metric space and R a closed, surjective relation on A. In Section 1 we review these relation concepts from [**1**] and recall various dynamic constructions like inverse limits and the sample path system of a relation. For example, given a relation on a finite set, the sample path system is the subshift of finite type associated with the relation. For a continuous surjective map on a compact metric space, the sample path system is the natural homeomorphism lift. After the introductory general remarks of this section, all our spaces are assumed to be either Cantor spaces or finite.

In Section 2 we describe the elementary properties of the representation procedure. Instead of finite decompositions $\mathcal{A}$ on a Cantor space X, our tool for building representations will be an *indexed partition* $\alpha : X \twoheadrightarrow I$, a continuous surjective map to a nonempty, finite discrete space. Since such a function is locally constant, a partition is just a decomposition $\mathcal{A}^{\alpha}$ whose members have been indexed by the set I.

We pick out a collection of finite sets to serve as the index sets. Let $\mathcal{I}$ be the countable set of all nonempty, finite subsets of finite products of of $\mathbb{N}$, the set of positive integers. This ensures that if ϕ is a nonempty relation on $I \in \mathcal{I}$ then ϕ itself is a member of $\mathcal{I}$.

If $f \in C_s(X)$ and $\alpha : X \twoheadrightarrow I$ is a partition then

$$f^{\alpha} \quad =_{def} \quad (\alpha \times \alpha)(f)$$

is a surjective relation on I. If $\phi = f^{\alpha}$, then we will say that ϕ *represents* f *via* α.

If $\mathcal{C}(X, \mathcal{I})$ is the countable set of indexed partitions, equipped with the discrete topology, then then $\Gamma_0 : C_s(X) \times \mathcal{C}(X, \mathcal{I}) \to \mathcal{I}$ by $(f, \alpha) \mapsto f^{\alpha}$ is a locally constant map. Projecting away from the second coordinate we obtain the *representation relation*

$$\Gamma \quad = \quad \{ \ (f, \phi) \in C_s(X) \times \mathcal{I} \ : \ \phi = f^{\alpha} \ \text{ for some } \ \alpha \in \mathcal{C}(X, \mathcal{I}) \ \}.$$

That is, $\Gamma(f)$ is the set of surjective relations which represent f.

If $f, g \in C_s(X)$ and $g = h^{-1} \circ f \circ h$ with $h \in H(X)$ then as subsets of $X \times X$, $f = (h \times h)(g)$. Thus, if ϕ represents f via α, then ϕ represents g via $\alpha h =_{def} \alpha \circ h$. It follows that $\Gamma^{-1}(\phi) \subset C_s(X)$ is open and conjugacy invariant.

For $f, g \in C_s(X)$ we write $f \sim_\alpha g$ if $\alpha f = \alpha g$, or, equivalently, if $f(x)$ and $g(x)$ lie in the same element of $\mathcal{A}^\alpha$ for all x. If $f \sim_\alpha g$ then $f^\alpha = g^\alpha$. On the other hand, if $f^\alpha = g^\alpha$ then there exists $h \in H(X)$ such that $h \sim_\alpha 1_X$ and $h \circ f \circ h^{-1} \sim_\alpha g$.

In Section 3 we obtain the results about representations. For example, Theorem 3.2:

Theorem: *If $\beta : X \twoheadrightarrow J$ is a partition, ϕ is a surjective relation on I, and $f \in C_s(X)$ then there exist $g \in H(X)$ and $\alpha : X \twoheadrightarrow I$ such that*

$$g \sim_\beta f \qquad \text{and} \qquad g^\alpha = \phi.$$

This implies that $\Gamma^{-1}(\phi)$ is dense in $C_s(X)$ for every surjective relation ϕ on a set in $\mathcal{I}$. It follows that the functions $f \in C_s(X)$ which can be represented by every surjective relation form a conjugacy invariant, dense, G_δ subset. This subset is exactly the set of transitive elements for $C_s(X)$.

In general, a closed, conjugacy invariant subset K of $C_s(X)$ is characterized by the set $\Gamma(K)$ of relations in $\mathcal{I}$ which represent its members. Furthermore, the closure of the conjugacy class of $f \in C_s(X)$ consists of exactly those $g \in C_s(X)$ such that $\Gamma(g) \subset \Gamma(f)$.

We call a closed, conjugacy invariant set K a *conjugacy transitive set* when it is the closure of a single conjugacy class. That is, the restriction of the action of the group $H(X)$ to K is topologically transitive. In that case, $Trans(K) = \{f : \overline{H(X) \cdot f} = K\}$, the set of *conjugacy transitive elements* of K, form a G_δ subset of K which is, of course, dense.

We conclude the section by describing a lifting property introduced in [**4**] and extended in [**17**]. This property and an equivalent factoring property provide a sufficient condition that the conjugacy class of an element $f \in C_s(X)$ be a G_δ set and show, in particular, that f is then a homeomorphism. We say that such a homeomorphism is of *residual type*. If K is a closed, conjugacy transitive subset, then $Trans(K)$ contains at most one conjugacy class whose elements are of residual type.

Finally, in Section 4 we collect the results built using the tools of the preceding sections. We show that $C_s(X), C_s(X;1), CR(X), CR(X;1), CT(X)$ are conjugacy transitive subsets each of which admits a conjugacy transitive element of residual type. On the other hand, $CM(X)$ and $CM(X;1)$ are conjugacy transitive subsets for which I can construct no conjugacy transitive element of residual type and I conjecture that they do not exist.

The set $CM(X;1)$ is of special interest. It is, in fact, conjugacy minimal. This follows from a theorem of Shimomura that if $f \in C_s(X)$ is not a periodic function then the closure of the conjugacy class of f contains $CM(X;1)$. This also shows that the only other conjugacy minimal subset of $C_s(X)$ is the singleton $\{1_X\}$. There is a simple inverse limit construction which yields elements $CM(X;1)$ in a G_δ subset which we label $H(X;1!)$. $f \in H(X;1!)$ if it is a homeomorphism with a unique fixed point and for every $x \in X$, other than the fixed point, the $\pm$ orbit $\{f^k(x) : k \in \mathbb{Z}\}$ is dense in X. The construction is associated with a little semigroup, the analysis of which shows that some members of $H(X;1!)$ are topologically mixing while others are not even weak mixing. The weak mixing elements of $H(X;1!)$ form a dense G_δ subset of $CM(X;1)$.

1. Relations and Maps

Our spaces are metric spaces with metrics (all labeled d) bounded by 1. On a finite product $A_1 \times \cdots \times A_n$ we use the metric $d = max_{i=1}^n d \circ (\pi_i \times \pi_i)$. On a countably infinite product $\Pi_{i=1}^\infty A_i$ we use $d = max_{i=1}^\infty \frac{1}{2^{i-1}} d \circ (\pi_i \times \pi_i)$. Here π_i is the ith coordinate projection. On a finite set or a discrete space like $\mathbb{Z}$, the set of integers, or like $\mathbb{N}$, the set of positive integers, we use the zero-one metric.

We will use the relation notation following [**1**] and so we briefly review it.

For sets A, B a relation $R : A \to B$ is a subset of $A \times B$. R is a *relation on* A when $B = A$. A map is a relation such that $R(x) = \{y : (x, y) \in R\}$ is a singleton set for every $x \in A$. For $A_0 \subset A$, the image $R(A_0) = \bigcup_{x \in A_0} R(x)$. $R(A_0)$ is the projection to B of $R \cap (A_0 \times B) \subset A \times B$. $R^{-1} : B \to A$ is defined to be $\{(y, x) : (x, y) \in R\}$. For $B_0 \subset B$ we let $R^*(B_0) = \{x \in A : R(x) \subset B_0\} = A \setminus R^{-1}(B \setminus B_0)$. So $R^*(B_0) \subset R^{-1}(B_0) \cup R^*(\emptyset)$. If R is a map, then $R^*(B_0) = R^{-1}(B_0)$. The relation is called *surjective* when $R(A) = B$ and $R^{-1}(B) = A$, or, equivalently, $R(x) \neq \emptyset$ and $R^{-1}(y) \neq \emptyset$ for all $x \in A$, $y \in B$.

If $R : A \to B$ and $S : B \to C$ then the composition $S \circ R : A \to C$ is the image under the projection to $A \times C$ of the set $(R \times C) \cap (A \times S) \subset A \times B \times C$. Composition is associative.

For a relation R on A we let $R^{n+1} = R^n \circ R$ for $n = 1, 2, ..$ and let $R^0 = 1_A$ and we define $|R| = \{x : (x, x) \in R\}$. We let $Per(R) = \{n \in \mathbb{N} : |R^n| \neq \emptyset\}$. If $A_0 \subset A$ then $R \cap (A_0 \times A_0)$ is the *restriction* of R to A_0.

A relation F on A is *reflexive* when $1_X \subset F$, where 1_X is the identity map on X. The relation is *symmetric* when $F = F^{-1}$. We will say that F*satisfies transitivity* when $F \circ F \subset F$. Then $F \cap F^{-1}$ satisfies symmetry and transitivity and restricts to an equivalence relation on $|F|$. We call the equivalence classes in $|F|$ the *basic sets* of F. We don't call F a transitive relation because we give the latter term a different, dynamic, meaning, see below.

N. B. From now on we will assume that our spaces A are metric spaces which are either compact or discrete.

A closed relation R is a closed subset of $A \times B$. From compactness various properties of closed relations follow. A map is continuous iff it is a closed relation. Furthermore, the composition of closed relations is closed and the image of a closed set by a closed relation is closed. So if B_0 is open in B then $R^*(B_0)$ is open in A. If R is a closed relation on A then $|R|$ is a closed subset of A. If F is a closed relation which satisfies transitivity, then the basic sets $\{F(x) \cap F^{-1}(x) : x \in |F|\}$ are closed sets.

For a relation R on A, we define the following relations associated with R:

- The *orbit relation* is $\mathcal{O}R = \bigcup_{n=1}^\infty R^n$.
- The *wandering relation* is $\mathcal{N}R = \overline{\mathcal{O}R}$
- The *chain relation* $\mathcal{C}R = \bigcap_{\epsilon > 0} \mathcal{O}(V_\epsilon \circ R \circ V_\epsilon)$ where $V_\epsilon = \{(x, y) : d(x, y) < \epsilon\}$.

The relations $\mathcal{N}R$ and $\mathcal{C}R$ are closed and the relations $\mathcal{O}R$ and $\mathcal{C}R$ satisfy transitivity. Clearly, $R \subset \mathcal{O}R \subset \mathcal{N}R \subset \mathcal{C}R$ and if A is discrete, then $\mathcal{O}R = \mathcal{N}R = \mathcal{C}R$.

For a closed relation R on A, we will say that R is *transitive* when $\mathcal{O}R = A \times A$, *topologically transitive* when $\mathcal{N}R = A \times A$, and *chain transitive* when $\mathcal{C}R = A \times A$. We call R *periodic* when $|R^n| = A$ for some $n \in \mathbb{N}$. We call R *recurrent* when $|\mathcal{O}R| = A$, *topologically recurrent* when $|\mathcal{N}R| = A$ and *chain recurrent* when $|\mathcal{C}R| = A$. We will refer to these as the three *transitivity properties* and the three *recurrence properties*.

For subsets $U, V \subset A$ the *hitting time set* is

$$N_R(U,V) = \{n \in \mathbb{N} : R^n(U) \cap V \neq \emptyset\} = \{n \in \mathbb{N} : R^n \cap (U \times V) \neq \emptyset\}.$$

A closed relation R is topologically transitive iff for all nonempty open $U, V \subset A$, $N_R(U,V) \neq \emptyset$. It is topologically recurrent iff for all nonempty open $U \subset A$, $N_R(U,U) \neq \emptyset$.

If A is discrete, then for a surjective relation R the three recurrence properties coincide and the three transitivity properties coincide.

A relation R on A is called *mixing* if there exists a positive integer N such that $R^N = A \times A$. It then follows, by induction, that $R^n = A \times A$ for every $n \geq N$. To see this, observe that if R is mixing, then it is surjective and so, given $x_1, x_2 \in A$, there exists $x_3 \in A$ such that $x_3 \in R(x_1)$. By inductive hypothesis $x_2 \in R^n(x_3)$ and so $x_2 \in R^{n+1}(x_1)$. R is called *chain mixing* if for every $\epsilon > 0$ the relation $V_\epsilon \circ R \circ V_\epsilon$ is mixing. R is called *topologically mixing* if for every $\epsilon > 0$ there exists a positive integer N such that $n \geq N$ implies $V_\epsilon \circ R^n \circ V_\epsilon = A \times A$. Of course, topological mixing implies chain mixing. Again, if A is discrete, these three concepts coincide. We refer to these as the three *mixing properties.* By the nine *dynamics properties* we mean the three recurrence, the three transitive and the three mixing properties.

If R is a closed, surjective relation on A and S is a closed relation on A with $R \subset S$ then S is surjective and it satisfies any of the nine dynamics properties when R satisfies the corresponding property. Notice that if S is a map then $R \subset S$ implies $R = S$ since R is assumed surjective.

If R is a closed, surjective relation on A with A compact, we let and $A_R \subset A^{\mathbb{Z}}$ be the *sample path space* with shift homeomorphism S_R on A_R and the projection map $p_0 : A_R \to A$.

$$\begin{aligned} A_R &= \{z \in A^{\mathbb{Z}} : (z_n, z_{n+1}) \in R \text{ for all } n \in \mathbb{Z}\}, \\ S_R(z)_n &= z_{n+1}, \qquad p_0(z) = z_0. \end{aligned} \tag{1.1}$$

The projection p_0 is a continuous surjection with $(p_0 \times p_0)(S_R) = R$. Also, $Per(S_R) = Per(R)$.

Proposition 1.1. *Let R be a closed surjective relation on a finite set A.*

R	*recurrent*	$\Longrightarrow$	(A_R, S_R)	*has dense periodic points.*
R	*transitive*	$\Longrightarrow$	(A_R, S_R)	*is topologically transitive.*
R	*mixing*	$\Longrightarrow$	(A_R, S_R)	*is topologically mixing.*

Proof. If R is recurrent then A is a union of basic sets for R and if $z \in A_R$ then all $z_i \in A$ lie in the same basic set. For any positive integer N there is a sequence $z_N = u_0, ..., u_{k+1} = z_{-N}$ with $(u_i, u_{i+1}) \in R$ for $0 \leq i \leq k$. The word $z_{-N}....z_N u_1...u_k$ on $[-N, N+k]$ extends to a periodic point in A_R which agrees with z on $[-N, N]$.

If R is transitive then A consists of a single basic set and so A_R contains a point z in which every finite word of A_R occurs, i.e. z is a transitive point.

If $R^M = A \times A$ and $z, w \in A_R$ then for any positive integer N there is a sequence $z_N = u_0, ..., u_M = w_{-N}$. So there is an element q of A_R which agrees with z on $(-\infty, N]$ and $S^{M+N}(q)$ agrees with w on $[-N, \infty)$. Thus, (A_R, S_R) is topologically mixing. $\square$

We will write (A, R) for a pair consisting of a compact metric space A and closed relation R on A. We will call the pair a *dynamical system* or just a system, when the relation R is surjective.

We will say that p is a map of systems, and write $p : (A_1, R_1) \to (A_2, R_2)$ when $p : A_1 \to A_2$ is a continuous map with $(p \times p)(R_1) \subset R_2$, or, equivalently $R_1 \circ p \subset p \circ R_2$. Since R_1 is surjective, the latter inclusion is an equality if R_2 is a map. If R_1 and R_2 are maps, then $p : (A_1, R_1) \to (A_2, R_2)$ iff the following diagram commutes:

$$\begin{array}{ccc} A_1 & \xrightarrow{R_1} & A_1 \\ {\scriptstyle p}\downarrow & & \downarrow{\scriptstyle p} \\ A_2 & \xrightarrow{R_2} & A_2. \end{array}$$

In general, $p : (A_1, R_1) \to (A_2, R_2)$ implies $p : (A_1, R_1^n) \to (A_2, R_2^n)$ for $n \in \mathbb{Z}$, $p : (A_1, \mathcal{A}R_1) \to (A_2, \mathcal{A}R_2)$ for $\mathcal{A} = \mathcal{O}, \mathcal{N}, \mathcal{C}$ and $p(|R_1|) \subset |R_2|$. It follows that $Per(R_1) \subset Per(R_2)$.

We will write $p : (A_1, R_1) \twoheadrightarrow (A_2, R_2)$ when $p : A_1 \twoheadrightarrow A_2$ is a continuous surjection with $(p \times p)(R_1) = R_2$. We will then say that p is surjective or that p maps R_1 onto R_2 or that (A_2, R_2) is a factor of (A_1, R_1). In general,

$$\begin{aligned} (p \times p)(R_1) &= p \circ R_1 \circ p^{-1}, \qquad \text{and so} \\ (p \times p)(R_1) = R_2 \quad &\Longleftrightarrow \quad p \circ R_1 \circ p^{-1} = R_2. \end{aligned} \tag{1.2}$$

For example, for every $i \in \mathbb{Z}$, the projection $p_i : (A_R, S_R) \twoheadrightarrow (A, R)$ maps the sample space homeomorphism S_R back onto the relation R itself.

In the special case when $R = f$ is a continuous map, we label the sample path space pair (X_f, S_f) as $(\hat{X}, \hat{f})$ and call it the *natural lift* of f to a homeomorphism. If $z \in \hat{X}$ then $z_{i+n} = f^n(z_i)$ for all $i \in \mathbb{Z}$ and $n \in \mathbb{N}$. That is, the ith coordinate determines all of the later coordinates.

Lemma 1.2. *(a) Let (A_i, R_i) be systems for $i = 1, 2, 3$. If $p : (A_1, R_1) \twoheadrightarrow (A_2, R_2)$ and $q : A_2 \to A_3$ then $q \circ p : (A_1, R_1) \twoheadrightarrow (A_3, R_3)$ iff $q : (A_2, R_2) \twoheadrightarrow (A_3, R_3)$.*

(b)Let $p : (A_1, R_1) \to (A_2, R_2)$ with $p(A_1) = A_2$, i.e. p is surjective between the underlying spaces. If for each $(a, b) \in R_2$ either $R_2(a) = \{b\}$ or $R_2^{-1}(b) = \{a\}$ then $(p \times p)(R_1) = R_2$, i.e. $p : (A_1, R_1) \twoheadrightarrow (A_2, R_2)$.

Proof. (a): $qp(A_1) = q[p(A_1)] = q(A_2)$. Hence, $qp(A_1) = A_3$ iff $q(A_2) = A_3$. Similarly, $(qp \times qp)(R_1) = (q \times q)[(p \times p)(R_1)] = (q \times q)(R_2)$. Hence, $(qp \times qp)(R_1) = R_3$ iff $(q \times q)(R_2) = R_3$.

(b): Assume $(a,b) \in R_2$ and $R(a) = \{b\}$. There exists $a_1 \in A_1$ such that $p(a_1) = a$ and there exists b_1 such that $(a_1, b_1) \in R_1$ because R_1 is surjective. Since $p \times p$ maps R_1 into R_2, $(a, p(b_1)) \in R_2$. Since $R(a)$ is a singleton, $p(b_1) = b$ and so $(a,b) = (p \times p)(a_1, b_1)$. If, instead $R^{-1}(b) = \{a\}$ then first lift b and proceed as before. □

If C is a closed subset of A and R is a closed relation on A then the *restriction* $R|C$ of R to C is given by $R \cap (C \times C)$. When (A,R) is a system, i. e. R is surjective, then $(C, R|C)$ is a *subsystem* when $R|C$ is a surjective relation on C, i.e. when $C \subset R^{-1}(C) \cap R(C)$. For a system (A,R) the closed set $C \subset A$ is called an transitive, topologically transitive, or chain transitive subset when the restriction $(C, R|C)$ is a subsystem which satisfies the corresponding property. Similarly, for the three recurrence properties and the three mixing properties.

If $p : (A,R) \to (B,S)$ and $C \subset A$ and $(C, R|C)$ is a subsystem of (A,R) then $(p(C), (p \times p)(R|C))$ is a system which satisfies any of the dynamics properties that $(C, R|C)$ does. Since $(p \times p)(R|C) \subset S|p(C)$ it follows that $(p(C), S|p(C))$ is a subsystem satisfying the same dynamics properties.

For compact spaces, an *inverse sequence of spaces* is a sequence of continuous surjections $\{p_{n+1,n} : A_{n+1} \twoheadrightarrow A_n : n \in \mathbb{N}\}$. For $m > n$ we define $p_{m,n} : A_m \twoheadrightarrow A_n$ to be the composition $p_{n+1,n} \circ \cdots \circ p_{m,m-1}$. The *inverse limit* A_∞ is defined by

$$A_\infty \ = \ \{ \ z \in \Pi_{n \in \mathbb{N}} \ A_n \ : \ z_n = p_{n+1,n}(z_{n+1}) \quad \text{for all } n \in \mathbb{N} \ \}. \tag{1.3}$$

The surjection $p_n : A_\infty \twoheadrightarrow A_n$ is the restriction of the projection to the nth coordinate.

If f is a surjective map on X then we can let $A_n = X, p_{n+1,n} = f$ for all n. The resulting inverse limit space is just a relabeling of $\hat{X}$, the natural lift of f. To be precise, if $w \in X_\infty$ we define $z \in \hat{X}$ by $z_{-n} = w_n$ for all $n \in \mathbb{N}$ and $z_k = f^{k+1}(w_1)$ for $k = 0, 1, \ldots$.

We will say that the sequence *bifurcates* when for every $n \in \mathbb{N}$ and $x \in A_n$ there exist $m > n$ and $y_1 \neq y_2 \in A_m$ such that $p_{m,n}(y_1) = x = p_{m,n}(y_2)$. It is clear that if the sequence bifurcates then the limit space A_∞ is perfect, i.e. it has no isolated points. Conversely, if all the A_n's are finite and A_∞ is perfect (and so is a Cantor set), then the sequence bifurcates.

If all of the spaces A_n are perfect then A_∞ is perfect whether the sequence bifurcates or not. In particular, if f is a surjective continuous map on a Cantor set X, then the natural homeomorphic lift $\hat{X}$ is perfect and so is a Cantor set as well.

A map from $\{p_{n+1,n} : A_{n+1} \twoheadrightarrow A_n\}$ to $\{q_{n+1,n} : B_{n+1} \twoheadrightarrow B_n\}$ is a sequence $\{h_n : A_n \to B_n\}$ of continuous maps such that

$$q_{n+1,n} \circ h_{n+1} = h_n \circ p_{n+1,n}$$

for all $n \in \mathbb{N}$. By restricting $\Pi_n \ h_n : \Pi_n \ A_n \to \Pi_n B_n$ we obtain a continuous map $h_\infty : A_\infty \to B_\infty$ with $q_n \circ h_\infty = h_n \circ p_n$ for all $n \in \mathbb{N}$. If each h_n is surjective then for any $z \in B_\infty$, $\{(q_n \circ h_\infty)^{-1}(z)\}$ is a decreasing sequence of nonempty compact sets and the nonempty intersection is $(h_\infty)^{-1}(z)$. Hence, h_∞ is surjective when the h_n's are.

By identifying $A_1 \times A_1 \times A_2 \times A_2$ with $A_1 \times A_2 \times A_1 \times A_2$ we define $(A_1 \times A_2, R_1 \times R_2)$. Then $Per(R_1 \times R_2) = Per(R_1) \cap Per(R_2)$. The similar identification of $(A_1 \times A_2)^{\mathbb{Z}}$ with $A_1^{\mathbb{Z}} \times A_2^{\mathbb{Z}}$ identifies $(A_{R_1 \times R_2}, S_{R_1 \times R_2})$ with $(A_{R_1} \times A_{R_2}, S_{R_1} \times S_{R_2})$.

Similarly, for inverse sequences $\{p_{n+1,n} : A_{n+1} \twoheadrightarrow A_n\}$ and $\{q_{n+1,n} : B_{n+1} \twoheadrightarrow B_n\}$ we can naturally identify $(A \times B)_\infty$, the inverse limit of $\{p_{n+1,n} \times q_{n+1,n} : A_{n+1} \times B_{n+1} \twoheadrightarrow A_n \times B_n\}$, with $A_\infty \times B_\infty$.

An *inverse sequence of systems* is a sequence of continuous surjections $\{p_{n+1,n} : (A_{n+1}, R_{n+1}) \twoheadrightarrow (A_n, R_n) : n \in \mathbb{N}\}$. With the above identifications, R_∞, the inverse limit of $\{p_{n+1,n} \times p_{n+1,n} : R_{n+1} \twoheadrightarrow R_n\}$, is a closed surjective relation on A_∞. So we say that the system (A_∞, R_∞) is the limit of the inverse sequence of systems. The *reverse system* $\{p_{n+1,n} : (A_{n+1}, R_{n+1}^{-1}) \twoheadrightarrow (A_n, R_n^{-1}) : n \in \mathbb{N}\}$ has limit $(A_\infty, R_\infty^{-1})$.

Definition 1.3. *An inverse sequence of systems* $\{p_{n+1,n} : (A_{n+1}, R_{n+1}) \twoheadrightarrow (A_n, R_n) : n \in \mathbb{N}\}$ *satisfies the* Shimomura Condition *if for every* $n \in \mathbb{N}$ *there exists* $m > n$ *so that* $p_{m,n} \circ R_m : A_m \twoheadrightarrow A_n$ *is a mapping.*

Proposition 1.4. *Let* $\{p_{n+1,n} : (A_{n+1}, R_{n+1}) \twoheadrightarrow (A_n, R_n) : n \in \mathbb{N}\}$ *be an inverse sequence of systems.*

(a) *If the sequence satisfies the Shimomura Condition then* R_∞ *is a surjective continuous map on* A_∞.
(b) *If the sequence and its reverse both satisfy the Shimomura Condition then* R_∞ *is a homeomorphism on* A_∞.
(c) *If for every* $n \in \mathbb{N}$, $(a,b),(a,c) \in R_{n+1}$ *implies* $p_{n+1,n}(b) = p_{n+1,n}(c)$ *then the sequence satisfies the Shimomura Condition.*
(d) *If each* A_n *is finite and* R_∞ *is a mapping, then the sequence satisfies the Shimomura Condition.*

Proof. (a): Assume $(x,u),(x,v) \in R_\infty$. To show that $u = v$ it suffices to show that $u_n = v_n$ for every $n \in \mathbb{N}$. Let $m > n$ be such that $p_{m,n} \circ R_m$ is a map. Since $(x_m, u_m),(x_m, v_m) \in R_m$, it follows that $u_n = p_{m,n}(u_m) = p_{m,n}(v_m) = v_n$. Hence, the surjective closed relation R_∞ is a surjective continuous map.

(b): If the the sequence and its reverse satisfy the condition, then R_∞ and R_∞^{-1} are mappings. This implies that R_∞ is a homeomorphism with inverse R_∞^{-1}.

(c): The assumption says that $p_{n+1,n} \circ R_{n+1}$ is a mapping.

(d): Call $(a,b),(a,c) \in R_n$ with $b \neq c$ a V in R_n. If the Shimomura Condition fails then there exists $n \in \mathbb{N}$ so that for every $m > n$ there exists a V in R_m which projects to a V in R_n. If A_n is finite, there are only finitely many V's in R_n. Hence, there exists a V in R_n which is an image of V's in R_m for m arbitrarily large. By compactness there is a V $(z,u),(z,v) \in R_\infty$ which projects to a V in R_n. Hence, R_∞ is not a mapping. □

If f is a surjective map on X then with $A_n = X, p_{n+1,n} = f$, we obtain the inverse sequence of systems with $(A_n, R_n) = (X, f)$ for all n whose inverse limit is the natural lift $(\hat{X}, \hat{f})$, as described above. Notice that $f \circ f$ and $f \circ f^{-1} = 1_X$ are maps and so the sequence and its reverse satisfy the Shimomura Condition.

If (A, R) is a system and N is a positive integer we define the *N-fold discrete suspension* (A_N, R_N) with $A_N = A \times \{1, \dots, N\}$ and

$$R_N = \{ ((a,i),(a,i+1)) : a \in A, i < N \} \cup \{ ((a,N),(b,0)) : (a,b) \in R \}. \tag{1.4}$$

Clearly, R_N is a map, or a homeomorphism, when R is. Notice that

$$((a,i),(b,j)) \in (R_N)^N \quad \Longleftrightarrow \quad (a,b) \in R \quad \text{and} \quad j = i. \tag{1.5}$$

We can identify the successive suspensions $((X_M)_N, (R_M)_N)$ with (X_{MN}, R_{MN}) by

$$((x,i),j) \mapsto (x, j + N(i-1)). \tag{1.6}$$

The construction is functorial. If $p : (A,R) \to (B,S)$ then $p_N : (A_N, R_N) \to (B_N, S_N)$ where $p_N = p \times 1_{\{1,\dots,N\}}$. If p is surjective then p_N is. Hence, if $\{p_{n+1,n} : (A_{n+1}, R_{n+1}) \twoheadrightarrow (A_n, R_n)\}$ is an inverse system then so is $\{(p_{n+1,n})_N : ((A_{n+1})_N, (R_{n+1})_N) \twoheadrightarrow ((A_n)_N, (R_n)_N)\}$. We leave to the reader the easy proof of the following.

Lemma 1.5. *With the obvious identifications, the inverse limit of* $\{(p_{n+1,n})_N : ((A_{n+1})_N, (R_{n+1})_N) \twoheadrightarrow ((A_n)_N, (R_n)_N)\}$ *becomes* $((A_\infty)_N, (R_\infty)_N)$. *If the original sequence bifurcates or satisfies the Shimomura Condition, then so does the* N*-fold suspension sequence.*

□

For a continuous map f on A with A compact, there are different definitions for topological transitivity, see [**5**]. As defined above, we say a map f is topologically transitive when $\mathcal{N}f = A \times A$, or, equivalently, the *hitting time set* $N_f(U,V) \neq \emptyset$ for all open,nonempty $U, V \subset A$. It is equivalent to $Trans_f \neq \emptyset$ when $x \in Trans_f$ if $\{f^i(x) : i \in \mathbb{N}\}$ is dense in X. A topologically transitive map is surjective.

The map is *minimal* when $Trans_f = A$. Equivalently, for every nonempty open $U \subset A$ the sequence of open sets $\{f^{-n}(U) : n \in \mathbb{N}\}$ covers A (and so has a finite subcover).

We write $C_s(A,B)$ for the space of continuous surjections from A to B, $H(A)$ for the homeomorphism group and $C_s(A)$ for the space of continuous surjections on A. All are equipped with the sup metrics. We let $CR(A), CT(A), CM(A)$ denote the subsets of chain recurrent, chain transitive and chain mixing surjective continuous mappings on A. We let $TT(A)$ and $MM(A)$ denote the subsets of topologically transitive and minimal maps on A. A map $f \in C_s(A)$ is *weak mixing* when $f \times f \in TT(A \times A)$. We let $WM(A)$ denote the weak mixing maps on A.

Proposition 1.6. *Let* A *be a compact metric space with* $C(A)$ *the space of continuous maps on* A.

(a) The sets $C_s(A), CR(A), CM(A)$ *and* $C(A; Per \supset Q) = \{f : |f^n| \neq \emptyset$ *for all* $n \in Q\}$ *are closed subsets of* $C(A)$ *for* Q *any subset of* $\mathbb{N}$.

(b) The sets $H(A), TT(A), MM(A), WM(A)$ *and* $\{f \in H(X) : f(x) = x$ *or* $\overline{\{f^i(x) : i \in \mathbb{Z}\}} = X$ *for all* $x \in A\}$ *are* G_δ *subsets of* $C(A)$. *The set of* $f \in C(A)$ *which admit exactly one fixed point is a* G_δ *subset of* $C(A)$.

Proof. (a) If U is a proper subset of A then the condition $f(A) \subset U$ is an open condition by compactness. Hence, $C_s(A)$ is closed. Similarly, if U is an open subset of $A \times A$ then $\mathcal{C}f \subset U$ implies that $\overline{\mathcal{O}(V_\epsilon \circ f \circ V_\epsilon)} \subset U$ for some $\epsilon > 0$. It thus follows that $\mathcal{C}f \subset U$ is an open condition. Hence, $CR(A) = \{f : 1_A \subset \mathcal{C}f\}$ and $CT(A) = \{f : A \times A \subset \mathcal{C}f\}$ are closed in $C_s(A)$, see also [**1**] Chapter 7. By [**1**] Exercise 8.22 it follows that a chain transitive f is not chain mixing iff it factors over a nontrivial periodic orbit, see also [**14**]. This last is an open condition and so $CM(A)$ is closed as well. The condition $|f^n| = \emptyset$ is equivalent to $inf_{x \in A}\ d(f^n(x), x) > 0$ and this is an open condition. It follows that $C(A; Per \supset Q)$ is closed.

(b) For every $\epsilon > 0$ the condition $(f \times f)[(X \times X) \setminus V_\epsilon \subset (X \times X) \setminus 1_X$ is an open condition. Intersecting over rational ϵ we see that the condition that f be injective is G_δ.

For fixed open sets U, V the condition $N_f(U, V) \neq \emptyset$ is an open condition. Intersecting over U, V in a countable basis we see that $TT(A)$ is G_δ.

The map $q : C_s(A) \to C_s(A \times A)$ given by $f \mapsto f \times f$ is continuous. Hence, $WM(A) = q^{-1}(TT(A \times A))$ is G_δ.

A map f is minimal iff for every open $U \subset$ nonempty, there exists $L \in \mathbb{N}$ such that $X \subset \bigcup_{k=0}^{L} f^{-k}(U)$. For each open set U, this is an open condition because it is equivalent to finding a closed finite cover $\{A_0, \dots, A_N\}$ of X such that for $i = 0, \dots, N$, $f^{k_i}(A_i) \subset U$, for some $0 \leq k_i \leq L$. Intersecting over a countable basis we see that f minimal is a G_δ condition.

Given $\epsilon > 0$, let $G(f, \epsilon) = \{x : d(f(x), x)) < \epsilon\}$. It is easy to check that for a homeomorphism f, the condition that every point x is either fixed by f or has a dense $\pm$ orbit is equivalent to the condition that for every $\epsilon > 0$ and every nonempty open $U \subset A$ there exists $L \in \mathbb{N}$ such that $X \subset G(f, \epsilon) \cup \bigcup_{k=-L}^{L} f^{-k}(U)$. For each open U and $\epsilon > 0$ this is an open condition as above. Notice that if A is a closed set then $A \subset G(f, \epsilon)$ iff $sup_{x \in A}\ d(f(x), x) < \epsilon$. Intersecting over rational ϵ and U in a countable basis we obtain a G_δ condition.

Given $\epsilon \geq 0$, let $K(f, \epsilon) = \{x : d(f(x), x)) \leq \epsilon\}$ so that $K(f, 0) = |f|$. Observe that if U is any open set which contains $K(f, \epsilon)$ for some $\epsilon \geq 0$ then there exists $\epsilon_1 > \epsilon$ such that $K(f, \epsilon_1) \subset U$. A mapping f has at most one fixed point iff for every $\epsilon > 0$ there exists $\delta > 0$ so that $x_1, x_2 \in K(f, \delta)$ implies $d(x_1, x_2) < \epsilon$. That is, if $K(f, \delta) \times K(f, \delta) \subset V_\epsilon$. For each $\epsilon > 0$ and $\delta > 0$ fixed this is an open condition on f. Taking the union over $\delta > 0$ and then the intersection over rational $\epsilon > 0$ we obtain the G_δ condition. Intersect with the closed set of f such that $|f| \neq \emptyset$ and we obtain the G_δ set of maps which admit a unique fixed point. □

Remark: There are different definitions for topological transitivity, see [**5**]. We call a map f topologically transitive when $\mathcal{N}f = A \times A$ which is equivalent to nonempty hitting times sets $N(U, V)$ as described above. It is equivalent to $Trans_f \neq \emptyset$ when $x \in Trans_f$ if $\{f^i(x) : i \in \mathbb{N}\}$ is dense in X. On the other hand, if (A, f) is obtained by compactifying $\mathbb{Z}$ so that the translation map $n \mapsto n + 1$ extends to the homeomorphism f on A which is not topologically transitive in this sense although it does have a dense $\pm$ orbit. Happily, when the space is perfect then all definitions agree. For example, if A is perfect and the $\pm$ orbit of x is dense then either $x \in Trans_f$ or $x \in Trans_{f^{-1}}$. To see this observe that x is either in the closure of $\{f^i(x) : i \in \mathbb{N}\}$ or $\{f^{-i}(x) : i \in \mathbb{N}\}$. Suppose the first, then there is a sequence $i_n \to \infty$ such that $f^{i_n}(x) \to x$. So for every $k \in \mathbb{N}$, $f^{i_n - k}(x) \to f^{-k}(x)$. Thus, the entire $\pm$ orbit is in the closure of the forward orbit and so $x \in Trans_f$. Otherwise, $x \in Trans_{f^{-1}}$ which implies that f^{-1} is topologically transitive. Since the inverse of a topologically transitive homeomorphism is topologically transitive (e.g. $\mathcal{N}(f^{-1}) = (\mathcal{N}f)^{-1}$), it follows from either case that f is topologically transitive.

If $1 \in Per(f)$ then $Per(f) = \mathbb{N}$. We denote will write $C_s(X; 1)$ for $C_s(X; Per \supset \{1\})$, the set of surjective maps which admit a fixed point. Similarly, we will write

$H(X;1)$ for $H(X) \cap C_s(X;1)$ and $CM(X;1)$ for $CM(X) \cap C_s(X;1)$ etc. We will write $H(X;1!)$ for the set of homeomorphisms f such that f has a unique fixed point and if $x \in X$ is not the fixed point then the $\pm$ orbit $\{f^i(x) : i \in \mathbb{Z}\}$ is dense in X.

We will apply all this to two special cases: $A = X$ a *Cantor space*, a Cantor set equipped with an ultrametric d, i.e.

$$d(x,y) \quad \leq \quad max(d(x,z), d(z,y))$$

or to A a finite set with the zero-one metric, also an ultrametric. Notice that if A is a finite set then the metrics $d = max_{i=1}^N \, d \circ (\pi_i \times \pi_i)$ on A^N and $d = max_{i \in \mathbb{Z}} \, \frac{1}{2^{|i|}} d \circ (\pi_i \times \pi_i)$ on $A^{\mathbb{Z}}$ are ultrametrics. With d an ultrametric $V_\epsilon = \{(x,y) : d(x,y) < \epsilon\}$ is a clopen equivalence relation.

2. Representations of Mappings via Indexed Partitions

Our spaces X, X_1, X_2, etc are all Cantor spaces, i.e. nonempty, zero-dimensional, perfect, compact metric spaces equipped with ultra-metrics. The maps are assumed to be continuous. We repeatedly use *The Uniqueness of Cantor*, the observation that all Cantor spaces are homeomorphic and, in particular, as nonempty clopen subsets of a Cantor space are Cantor spaces they are all homeomorphic to one another. A *decomposition* $\mathcal{A}$ of X is a finite, pairwise disjoint cover of X by nonempty clopen subsets. Since the metric d on X is an ultrametric, the set of balls $\{V_\epsilon(x) : x \in X\}$ is a decomposition of X for any $\epsilon > 0$.

Let $\mathfrak{I}$ denote the countable set of all nonempty, finite subsets of $\mathbb{N}^n$ for $n = 1, 2, \ldots$. Notice that if $\phi : I_1 \to I_2$ is a nonempty relation between elements of $\mathfrak{I}$ then $\phi \in \mathfrak{I}$. We regard $\mathfrak{I}$ as a discrete set and the elements of $\mathfrak{I}$ as finite discrete spaces. Recall that a discrete space uses the zero-one metric d.

An *indexed partition* (hereafter, just a *partition*) is a continuous surjection $\alpha : X \twoheadrightarrow I$ with $I \in \mathfrak{I}$. We define $\mathcal{A}^\alpha = \{A^i = \alpha^{-1}(i) : i \in I\}$ to be the *associated decomposition* of X. If $\mathcal{A}_1, \mathcal{A}_2$ are decompositions then $\mathcal{A}_1$ *refines* $\mathcal{A}_2$ iff for every $A_1 \in \mathcal{A}_1$ there exists a -necessarily unique- $A_2 \in \mathcal{A}_2$ such that $A_1 \subset A_2$. We will say of partitions $\alpha_1 : X \twoheadrightarrow I_1, \alpha_2 : X \twoheadrightarrow I_2$ that α_1 refines α_2 when $\mathcal{A}^{\alpha_1}$ refines $\mathcal{A}^{\alpha_2}$.

If $\mathcal{A}$ is any decomposition of X and $I \in \mathfrak{I}$ has the same cardinality as $\mathcal{A}$ then there is a partition $\alpha : X \twoheadrightarrow I$ such that $\mathcal{A}^\alpha = \mathcal{A}$. Since there are decompositions of any positive finite cardinality, it follows that for any $I \in \mathfrak{I}$ there exist partitions $\alpha : X \twoheadrightarrow I$.

Let $\mathfrak{C}(X, \mathfrak{I})$ denote the set of all indexed partitions. Each clopen subset of X is a finite union of basic sets and so there are only countably many clopen sets. Hence,for each $I \in \mathfrak{I}$ there are only countably many partitions $\alpha : X \twoheadrightarrow I$. Since $\mathfrak{I}$ is countable, it follows that $\mathfrak{C}(X, \mathfrak{I})$ is countable. We give it the discrete topology.

If $\alpha_1 : X_1 \twoheadrightarrow I_1$ and $\alpha_2 : X_2 \twoheadrightarrow I_2$ are partitions then $\alpha_1 \times \alpha_2 : X_1 \times X_2 \twoheadrightarrow I_1 \times I_2$ is a partition. If $X_1 = X_2 = X$ then we write $\alpha_1 \otimes \alpha_2 : X \to I_1 \times I_2$ by $\alpha_1 \otimes \alpha_2(x) = (\alpha_1(x), \alpha_2(x))$. This is usually not surjective and so is only an indexed partition when we restrict the range to $(\alpha_1 \otimes \alpha_2)(X) \subset I_1 \times I_2$.

The *mesh* of a finite collection $\mathcal{A}$ of sets is $max \; diam\{A : A \in \mathcal{A}\}$. For a partition $\alpha : X \to I$ the mesh of $\mathcal{A}^\alpha$, also called the *mesh* of α, is

$$max\{d(x,y) : \alpha(x) = \alpha(y)\} \quad = \quad inf\{\epsilon : (\alpha \times \alpha)^{-1}(1_I) \subset V_\epsilon\}. \tag{2.1}$$

The *thickness* of α is the minimum of the diameters, i. e.

$$min\ \{\ max\{d(x,y) : \alpha(x) = \alpha(y) = i\}\ \ :\ i \in I\ \}.$$

The *Lebesgue number* of α is

$$min\{d(x,y) : \alpha(x) \neq \alpha(y)\} \quad = \quad max\ \{\epsilon : V_\epsilon \subset (\alpha \times \alpha)^{-1}(1_I)\}. \tag{2.2}$$

Note that $(\alpha \times \alpha)^{-1}(1_I) = \bigcup_{i \in I} A^i \times A^i$ is a clopen neighborhood of the diagonal. If a set has diameter less than the Lebesgue number of α then α is constant on it and so it is contained in an element of $\mathcal{A}^\alpha$.

Proposition 2.1. *Let $\alpha : X \twoheadrightarrow I$ and $\beta : X \twoheadrightarrow J$ be partitions.*

(a) *The following are equivalent.*
  (i) *α refines β.*
  (ii) *There exists $\pi : I \twoheadrightarrow J$ such that $\pi \circ \alpha = \beta$.*
  (iii) *The relation $\beta \circ \alpha^{-1} : I \to J$ is a map.*
  (iv) *β is constant on every element of $\mathcal{A}^\alpha$.*
  When these conditions hold, $\pi = \beta \circ \alpha^{-1}$ is the unique map such that $\pi \circ \alpha = \beta$.

(b) *If the mesh of α is less than Lebesgue number of β, then α refines β. If, in addition, mesh α < thickness β then each element of $\mathcal{A}^\beta$ contains at least two elements of $\mathcal{A}^\alpha$, or, equivalently, α is not constant on any element of A^β.*

PROOF. (a) (i) $\Rightarrow$ (ii): Define $\pi(i) = j$ if $\alpha^{-1}(i) \subset \beta^{-1}(j)$.

(ii) $\Rightarrow$ (i): If $\pi(i) = j$ and $\pi \circ \alpha = \beta$ then $\alpha^{-1}(i) \subset (\pi \circ \alpha)^{-1}(j) = \beta^{-1}(j)$.

(ii) $\Rightarrow$ (iii): Because α is a surjective map, $\alpha \circ \alpha^{-1} = 1_I$ and so $\pi \circ \alpha = \beta$ implies $\pi = \pi \circ \alpha \circ \alpha^{-1} = \beta \circ \alpha^{-1}$.

(iii) $\Rightarrow$ (ii): Because α is a map $1_X \subset \alpha^{-1} \circ \alpha$. Hence, $\beta \subset \beta \circ \alpha^{-1} \circ \alpha$. As the composition of maps, the latter is a map and inclusion between maps implies equality. Hence, $\pi \circ \alpha = \beta$ with $\pi = \beta \circ \alpha^{-1}$.

(iii) $\Leftrightarrow$ (iv): Obvious.

(b) The first part is clear from the definition of the Lebesgue number.

If $B \in \mathcal{A}^\beta$ contains a unique $A \in \mathcal{A}^\alpha$ then since $\mathcal{A}^\alpha$ refines $\mathcal{A}^\beta$, $A = B$. Then *mesh* $\alpha \geq$ *diam* A = *diam* $B \geq$ *thickness* β. □

Proposition 2.2. *Let $I, J \in \mathcal{I}$.*

(a) If $\alpha : X \twoheadrightarrow I$ is a partition and $\pi : J \twoheadrightarrow I$, then there exists $\beta : X \twoheadrightarrow J$ such that $\pi \circ \beta = \alpha$.

(b) If $\alpha_1 : X_1 \twoheadrightarrow I$ and $\alpha_2 : X_2 \twoheadrightarrow I$ are partitions then there exists a homeomorphism $h : X_1 \to X_2$ such that $\alpha_1 = \alpha_2 \circ h$.

PROOF. (a): For each $i \in I$, $\alpha^{-1}(i)$ is a nonempty clopen subset of X and so is a Cantor space. Since $\pi^{-1}(i)$ is a nonempty subset of J, it is an element of $\mathcal{I}$. We can choose $\beta : \alpha^{-1}(i) \twoheadrightarrow \pi^{-1}(i)$ for all $i \in I$ and concatenate.

(b): Choose a homeomorphism between the Cantor spaces $h : \alpha_1^{-1}(i) \to \alpha_2^{-1}(i)$ for all $i \in I$ and concatenate. □

If $\alpha : X_2 \twoheadrightarrow I$ is a partition and $f : X_1 \to X_2$ is a continuous map then we write $\alpha f : X_1 \to I$ for $\alpha \circ f$. If f is surjective, i.e. $f \in C_s(X)$, then αf is a partition.

If $\alpha : X_2 \twoheadrightarrow I$ is a partition then for continuous maps $f, g : X_1 \to X_2$ we write $f \sim_\alpha g$ when $\alpha f = \alpha g$, or, equivalently, if $f(x)$ and $g(x)$ lie in the same member of the decomposition $\mathcal{A}^\alpha$ of X_2 for all $x \in X_1$. This defines a clopen equivalence relation on $C_s(X_1, X_2)$. Clearly, $f \sim_\alpha g$ implies $d(f,g) \leq mesh\ \alpha$ and so we can use these equivalence relations to measure closeness of approximations between maps in $C(X)$.

Corollary 2.3. *If $\alpha : X_2 \twoheadrightarrow I$ is a partition and $p : X_1 \twoheadrightarrow X_2$, then there exists a homeomorphism $h : X_1 \to X_2$ such that $p \sim_\alpha h$.*

PROOF. Apply Proposition 2.2(b) with $\alpha_2 = \alpha$ and $\alpha_1 = \alpha \circ p$. □

Remark: With $X_1 = X_2 = X$ it follows that $H(X)$ is dense in $C_s(X)$.

For f a closed relation on X and a partition $\alpha : X \twoheadrightarrow I$, we let $f^\alpha = (\alpha \times \alpha)(f) \subset I \times I$. So $(i,j) \in f^\alpha$ iff there exists $(x,y) \in f$ such that $\alpha(x) = i$ and $\alpha(y) = j$ and so iff $A^i \cap f^{-1}(A^j) \neq \emptyset$. Clearly, if the relation f is a surjective relation on X then f^α is a surjective relation on I and $\alpha : (X,f) \twoheadrightarrow (I, f^\alpha)$ is a surjective system map.

If f is a surjective map, i.e. $f \in C_s(X)$, then

$$(\alpha \otimes \alpha f)(X) \ = \ f^\alpha, \tag{2.3}$$

and so $\alpha \otimes \alpha f : X \twoheadrightarrow f^\alpha$ is a partition.

Proposition 2.4. *If $\alpha : X \twoheadrightarrow I$ is a partition and $\epsilon > 0$ is less than the Lebesgue number of α, then $f^\alpha = (V_\epsilon \circ f \circ V_\epsilon)^\alpha$ for any closed surjective relation f on X.*

PROOF. If ϵ is less than the Lebesgue number then the clopen equivalence relation V_ϵ is contained in $(\alpha \times \alpha)^{-1}(1_I)$.

If $(x,y) \in f$ and $(x,x_1),(y,y_1) \in V_\epsilon$ then $\alpha(x) = \alpha(x_1)$ and $\alpha(y) = \alpha(y_1)$. Hence, $(\alpha \times \alpha)(f) = (\alpha \times \alpha)(V_\epsilon \circ f \circ V_\epsilon)$. □

For any closed surjective relation f on X, we have the surjective map of systems $\alpha : (X,f) \twoheadrightarrow (I, f^\alpha)$ and so $\alpha : (X, f^n) \twoheadrightarrow (I, (f^\alpha)^n)$ for $n \in \mathbb{Z}$, $\alpha : (X, \mathcal{C}f) \twoheadrightarrow (I, \mathcal{O}f^\alpha)$, and $Per(f) \subset Per(f^\alpha)$. From this and Proposition 2.4 we obtain

Proposition 2.5. *If f is a surjective closed relation on X and $\alpha : X \twoheadrightarrow I$ is a partition then*

$$\begin{array}{lllll} f & \textit{chain recurrent} & \Longrightarrow & f^\alpha & \textit{is recurrent.} \\ f & \textit{chain transitive} & \Longrightarrow & f^\alpha & \textit{is transitive.} \\ f & \textit{chain mixing} & \Longrightarrow & f^\alpha & \textit{is mixing.} \end{array} \tag{2.4}$$

□

For partitions $\alpha_1 : X_1 \twoheadrightarrow I_1$ and $\alpha_2 : X_2 \twoheadrightarrow I_2$, we write $(p, \pi) : \alpha_1 \twoheadrightarrow \alpha_2$ when $p : X_1 \twoheadrightarrow X_2$ and $\pi : I_1 \twoheadrightarrow I_2$ such that the following diagram commutes:

$$\begin{array}{ccc} X_1 & \xrightarrow{\alpha_1} & I_1 \\ {\scriptstyle p}\downarrow & & \downarrow{\scriptstyle \pi} \\ X_2 & \xrightarrow{\alpha_2} & I_2 \end{array}$$

i.e. $\pi \circ \alpha_1 = \alpha_2 \circ p$.

Proposition 2.6. *With $f_1 \in C_s(X_1)$ and $f_2 \in C_s(X_2)$ assume that $p : (X_1, f_1) \twoheadrightarrow (X_2, f_2)$ is a surjective system map. If $\alpha_1 : X_1 \twoheadrightarrow I_1$ and $\alpha_2 : X_2 \twoheadrightarrow I_2$ are partitions and $(p, \pi) : \alpha_1 \twoheadrightarrow \alpha_2$. then*

$$\begin{aligned} &(\pi \times \pi) \circ (\alpha_1 \otimes \alpha_1 f_1) = (\pi \circ \alpha_1 \otimes \pi \circ \alpha_1 f_1) = (\alpha_2 \otimes \alpha_2 f_2) \circ p, \\ &\textit{and so} \qquad (\pi \times \pi)(f_1^{\alpha_1}) \quad = \quad f_1^{\pi \circ \alpha_1} \quad = \quad f_2^{\alpha_2}. \end{aligned} \tag{2.5}$$

Furthermore, the following diagram of systems commutes:

$$\begin{array}{ccc} (X_1, f_1) & \xrightarrow{\alpha_1} & (I_1, f_1^{\alpha_1}) \\ {\scriptstyle p}\downarrow & & \downarrow{\scriptstyle \pi} \\ (X_2, f_2) & \xrightarrow{\alpha_2} & (I_2, f_1^{\alpha_2}). \end{array}$$

PROOF. Because (p, π) maps α_1 to α_2 we have $\pi \circ \alpha_1 = \alpha_2 \circ p$ and $\pi \circ \alpha_1 f_1 = \alpha_2 \circ p \circ f_1 = \alpha_2 f_2 \circ p$. This and (2.3) imply (2.5). The commutative diagram of surjective system maps is then clear. $\square$

Proposition 2.7. *(a) Assume $f_1 \in C_s(X_1)$ and $\alpha_1 : X_1 \twoheadrightarrow I$ is a partition. Let $h : X_1 \to X_2$ be a homeomorphism. If $f_2 = h \circ f_1 \circ h^{-1} \in C_s(X_2)$ and $\alpha_2 = \alpha_1 \circ h^{-1} : X_2 \twoheadrightarrow I$, then*

$$\begin{aligned} \alpha_2 \otimes \alpha_2 f_2 \quad &= \quad (\alpha_1 \otimes \alpha_1 f_1) \circ h^{-1}. \\ \textit{and so} \qquad f_2^{\alpha_2} \quad &= \quad f_1^{\alpha_1}. \end{aligned} \tag{2.6}$$

(b) Assume $\alpha_1 : X_1 \twoheadrightarrow I$ and $\alpha_2 : X_2 \twoheadrightarrow I$ are partitions and that $f_1 \in C_s(X_1), f_2 \in C_s(X_2)$. If $f_1^{\alpha_1} = f_2^{\alpha_2}$ then there exists a homeomorphism $h : X_1 \to X_2$ such that $\alpha_2 \circ h = \alpha_1$ and $h^{-1} f_2 h \sim_{\alpha_1} f_1$.

PROOF. (a) We have $h : (X_1, f_1) \twoheadrightarrow (X_2, f_2)$ and $(h, 1_I) : \alpha_1 \twoheadrightarrow \alpha_2$. So (2.6) follows from (2.5).

(b) Let $\phi = f_1^{\alpha_1} = f_2^{\alpha_2} \in \mathfrak{I}$. $\alpha_1 \otimes \alpha_1 f_1 : X_1 \twoheadrightarrow \phi$ and $\alpha_2 \otimes \alpha_2 f_2 : X_2 \twoheadrightarrow \phi$ are partitions and so by Proposition 2.2 (b) there is a homeomorphism $h : X_1 \to X_2$ such that $(\alpha_1 \otimes \alpha_1 f_1) = (\alpha_2 \otimes \alpha_2 f_2) \circ h$. That is, $\alpha_1 = \alpha_2 \circ h$ and $(\alpha_2 f_2) \circ h = \alpha_1 f_1$. Hence, $\alpha_1 \circ (h^{-1} f_2 h) = (\alpha_2 h) \circ (h^{-1} f_2 h) = (\alpha_2 f_2) \circ h = \alpha_1 f_1$ and so $h^{-1} f_2 h \sim_{\alpha_1} f_1$. $\square$

The following estimate was proved by Bermudez & Darji [**7**] and by Shimomura [**17**].

Proposition 2.8. *Let $f, g \in C_s(X)$. If $g^\alpha \subset f^\alpha$ then $d(f, g) \leq$ mesh $\mathcal{A}^\alpha +$ mesh $f\mathcal{A}^\alpha$ with $f\mathcal{A}^\alpha = \{f(A) : A \in \mathcal{A}^\alpha\}$.*

PROOF. For all $x \in X$, $(\alpha(x), \alpha(g(x))) \in g^\alpha \subset f^\alpha$ and so there exists $y \in X$ such that $(\alpha(x), \alpha(g(x))) = (\alpha(y), \alpha(f(y))$. Since, $d(f(x), f(y)) \leq mesh\ fA^\alpha$ and $d(f(y), g(x)) \leq mesh\ \mathcal{A}^\alpha$ the result follows from the triangle inequality. $\square$

If $f \sim_\alpha g$ for $f, g \in C_s(X)$ then $\alpha \otimes \alpha f = \alpha \otimes \alpha g$ and so $f^\alpha = g^\alpha$ by (2.3). Hence, with the discrete topology on $\mathcal{C}(X, \mathcal{I})$ the map $\Gamma_0 : C_s(X) \times \mathcal{C}(X, \mathcal{I}) \to \mathcal{I}$ by $(f, \alpha) \mapsto f^\alpha$ is locally constant. We can regard Γ_0 as the set of triples $\{(f, \alpha, \phi) : f^\alpha = \phi\}$ and project away from the second coordinate to define

$$\Gamma =_{def} \{(f, \phi) : f^\alpha = \phi \text{ for some } \alpha \in \mathcal{C}(X, \mathcal{I})\} \subset C_s(X) \times \mathcal{I}. \tag{2.7}$$

We will say that ϕ, a surjective relation on $I \in \mathcal{I}$, *represents* $f \in C_s(X)$ if there exists a partition α such that $f^\alpha = \phi$, i.e. if $\phi \in \Gamma(f)$. Equivalently, ϕ represents f if there exists a system surjection $\alpha : (X, f) \twoheadrightarrow (I, \phi)$.

From Proposition 2.6 we obtain the following.

Corollary 2.9. *Let $f \in C_s(X)$. (a) If $\pi : (I, \phi) \twoheadrightarrow (I_1, \phi_1)$ and $\phi \in \Gamma(f)$, then $\phi_1 \in \Gamma(f)$.*

(b) If $p : (X, f) \twoheadrightarrow (X_1, f_1)$ then $\Gamma(f_1) \subset \Gamma(f)$.

(c) If $(\hat{X}, \hat{f})$ is the natural lift of f to a homeomorphism then $\Gamma(f) = \Gamma(\hat{f})$.

PROOF. (a): If $\alpha : X \twoheadrightarrow I$ is a partition with $f^\alpha = \phi$ and $\alpha_1 = \pi \circ \alpha$ then $(1_X, \pi) : \alpha \twoheadrightarrow \alpha_1$ and so Proposition 2.6 implies $f^{\alpha_1} = \phi_1$.

(b): If $\alpha_1 : X_1 \twoheadrightarrow I$ is a partition with $f_1^{\alpha_1} = \phi$ and $\alpha = \alpha_1 \circ p$, then $(p, 1_I) : \alpha \twoheadrightarrow \alpha_1$, and so Proposition 2.6 implies $f^\alpha = \phi$.

(c): Since p_0 maps $(\hat{X}, \hat{f})$ onto (X, f), it follows from (b) that $\Gamma(f) \subset \Gamma(\hat{f})$.

Now let $\alpha : \hat{X} \twoheadrightarrow I$ be a partition and $\phi = \hat{f}^\alpha$. Because every clopen set is a finite union of basic sets it follows that there is a finite list of coordinates so that $\alpha(z)$ depends only on the value of z at each of these coordinates. Furthermore, if i is the smallest index in this set of coordinates then $z_n = f^{n-i}(z_i)$ if $n \geq i$. This means that $\alpha(z)$ is a function of $z_i = p_i(z)$. That is, there exists a partition $\beta : X \twoheadrightarrow I$ such that $\alpha = \beta \circ p_i$. Because $p_i : (\hat{X}, \hat{f}) \twoheadrightarrow (X, f)$ and $(p_i, 1_I) : \alpha \twoheadrightarrow \beta$, Proposition 2.6 implies that $\phi = \hat{f}^\alpha = f^\beta$. Hence, $\phi \in \Gamma(f)$. $\square$

Because it is of interest to know whether a system (X, f) factors over a non-trivial periodic orbit, we observe the following.

Proposition 2.10. *For $f \in C_s(X)$ and a permutation γ on $J \in \mathcal{I}$ the following are equivalent.*

(i) *f factors over the permutation γ, i.e. there is a system surjection $\beta : (X, f) \twoheadrightarrow (J, \gamma)$.*
(ii) *$\gamma \in \Gamma(f)$.*
(iii) *There exists $\phi \in \Gamma(f)$ which factors over γ.*

PROOF. (i) $\Leftrightarrow$ (ii) and (ii) $\Rightarrow$ (iii) are obvious.

If $\alpha : X \twoheadrightarrow I$ with $f^\alpha = \phi$ and $\pi : (I, \phi) \twoheadrightarrow (J, \gamma)$ then $\alpha : (X, f) \twoheadrightarrow (I, \phi)$ and $\pi \circ \alpha : (X, f) \twoheadrightarrow (J, \gamma)$ are surjective system maps. $\square$

3. Representation Characterizations

In this section we justify our emphasis on the use of surjective relations on finite sets to study $C_s(X)$ and $H(X)$ for X a Cantor space. We will show that a closed, conjugacy invariant subset of $C_s(X)$ is characterized by the set of relations which represent it, i.e. by $\Gamma(K)$. We develop the inverse limit construction which will be used in studying the examples in the next section. Finally, we describe a useful sufficient when the limit of such a construction has a G_δ conjugacy class and so is a residual element of the closure of its conjugacy class.

We will use homeomorphisms to identify the various Cantor spaces which turn up in our constructions so that we regard all our partitions as lying in $\mathcal{C}(X, \mathcal{I})$. If $f_1 \in C_s(X_1), f_2 \in C_s(X_2)$ we will say they are *conjugate* if there exists a homeomorphism $h : X_1 \to X_2$ such that $h \circ f_1 \circ h^{-1}$ is in the orbit of f_2 with respect to the $H(X_2)$ adjoint action on $C_s(X_2)$. This definition is, of course, independent of the choice of homeomorphism h which is used to identify the two spaces. Equivalently, f_1 and f_2 are conjugate if there exists a homeomorphism h such that $h \circ f_1 \circ h^{-1} = f_2$.

We first use the sample path space construction to show that every surjective relation ϕ on an element I of $\mathcal{I}$ can be represented by an element of $H(X)$. Usually, the sample path system of the finite system (I, ϕ) will do, but a bit of extra work is needed to assure that the associated space is Cantor.

Theorem 3.1. *If $\alpha : X \twoheadrightarrow I$ is a partition and ϕ is a surjective relation on I then there exists $f \in H(X)$ such that $f^\alpha = \phi$ and $Per(f) = Per(\phi)$.*

If ϕ is recurrent then f can be chosen with dense periodic points.

If ϕ is transitive then f can be chosen topologically transitive with dense periodic points.

If ϕ is mixing then f can be chosen topologically mixing with dense periodic points.

Proof. On $[2] = \{0, 1\}$ let $U = [2] \times [2]$ so that $([2]_U, S_U)$ is the full shift with $[2]_U$ the Cantor space $[2]^{\mathbb{Z}}$. Let $\phi_1 = \phi \times U$ on $I \times [2] = I_1$. Let $(X_1, f_1) = ((I_1)_{\phi_1}, S_{\phi_1})$, the sample path system for ϕ_1. Observe that $X_1 = I_\phi \times [2]_U$ is a Cantor space and that $(f_1)^{p_0} = \phi_1$. Furthermore, $Per(f_1) = Per(\phi) \cap Per(U) = Per(\phi)$. See the description associated with Proposition 1.1.

With $\pi : I_1 \twoheadrightarrow I$ the first coordinate projection, we define $\alpha_1 : \pi \circ p_0 : X_1 \twoheadrightarrow I$. Since $(1_{X_1}, \pi) : p_0 \twoheadrightarrow \alpha_1$, (2.5) implies $f_1^{\alpha_1} = (\pi \times \pi)((f_1)^{p_0}) = (\pi \times \pi)(\phi_1) = \phi$.

By Proposition 2.2 (b) there exists a homeomorphism $h : X_1 \to X$ so that $\alpha \circ h = \alpha_1$. Let $f = hf_1h^{-1}$. Since h maps f_1 to f, (2.6) implies that $f^\alpha = \phi$. Clearly $Per(f) = Per(f_1)$.

The transitivity and mixing results follow from Proposition 1.1. □

Next, we show that any element of $C_s(X)$ can be approximated by a homeomorphism which is represented by ϕ.

Theorem 3.2. *If $\beta : X \twoheadrightarrow J$ is a partition, ϕ is a surjective relation on I, and $f \in C_s(X)$ then there exists $g \in H(X)$ such that*

- *$g \sim_\beta f$.*
- *There exists a partition $\alpha : X \twoheadrightarrow I$ such that $g^\alpha = \phi$.*
- *$Per(f) \cap Per(\phi) \subset Per(g)$.*

- *If f is chain recurrent and ϕ is recurrent then g can be chosen with dense periodic points.*
- *If f is chain mixing and ϕ is mixing then g can be chosen topologically mixing and with dense periodic points.*

PROOF. Apply Proposition 2.2(a) to get a partition $\alpha_1 : X \twoheadrightarrow I \times J$ such that $\pi_2 \circ \alpha_1 = \beta$. Define $\alpha_2 = \pi_1 \circ \alpha_1 : X \twoheadrightarrow I$ and $\phi_1 = \phi \times f^\beta$. By Theorem 3.1 there exists a homeomorphism g_1 on X such that $g_1^{\alpha_1} = \phi_1$. By (2.5)

$$\begin{aligned} g_1^\beta &= (\pi_2 \times \pi_2)(\phi_1) = f^\beta, \\ g_1^{\alpha_2} &= (\pi_1 \times \pi_1)(\phi_1) = \phi, \end{aligned} \tag{3.1}$$

By Proposition 2.7(b) there exists $h \in H(X)$ such that $h^{-1} \circ g_1 \circ h \sim_\beta f$. Let $g = h^{-1} \circ g_1 \circ h$ and let $\alpha = \alpha_2 \circ h$. By (2.6), $g^\alpha = g_1^{\alpha_2} = \phi$.

The homeomorphism g_1 can be chosen so that $Per(g_1) = Per(\phi_1) = Per(f^\beta) \cap Per(\phi) \supset Per(f) \cap Per(\phi)$.

If f is chain recurrent then f^β is recurrent. If, in addition, ϕ is recurrent then the product ϕ_1 is recurrent.

If f is chain mixing then f^β is mixing. If, in addition, ϕ is mixing then the product ϕ_1 is mixing.

Hence, by Theorem 3.1 we can choose g_1 as required and so the conjugate g satisfies the additional properties as well. □

Corollary 3.3. *If $I \in \mathfrak{I}$ and ϕ a surjective relation on I, then $\Gamma^{-1}(\phi)$ is an open, conjugacy invariant, dense subset of $C_s(X)$. Furthermore,*

(i) *$H(X) \cap \Gamma^{-1}(\phi)$ is dense in $H(X)$ and open relative to $H(X)$.*
(ii) *If ϕ is mixing then $\{f \in H(X) : f \in \Gamma^{-1}(\phi)$ and f is topologically mixing with dense periodic points $\}$ is dense in $CM(X)$.*
(iii) *If ϕ is recurrent then $\{f \in H(X) : f \in \Gamma^{-1}(\phi)$ and f has dense periodic points $\}$ is dense in $CR(X)$.*
(iv) *$\{f \in H(X) : f \in \Gamma^{-1}(\phi)$ and $Per(\phi) \subset Per(f)\ \}$ is dense in $\{f \in C_s(X) : Per(\phi) \subset Per(f)\ \}$.*

PROOF. $\Gamma^{-1}(\phi)$ is open because the map $\Gamma_0 : C_s(X) \times \mathcal{C}(X, \mathfrak{I}) \to \mathfrak{I}$ is locally constant. It is conjugacy invariant by (2.6). The rest then follows from Theorem 3.2. □

We turn now the the inverse limit constructions.

Definition 3.4. *We say that an inverse sequence of systems*

$$\{\ p_{n+1,n} : (I_{n+1}, \phi_{n+1}) \twoheadrightarrow (I_n, \phi_n)\ :\ n \in \mathbb{N}\ \}$$

with $I_n \in \mathfrak{I}$ for all n, is a Shimomura Sequence *when it is bifurcating and satisfies the Shimomura Condition. It is an* invertible Shimomura Sequence *when, in addition, $\{\ p_{n+1,n} : (I_{n+1}, \phi_{n+1}^{-1}) \twoheadrightarrow (I_n, \phi_n^{-1})\}$ satisfies the Shimomura Condition as well.*

Theorem 3.5. *If (X, f) is the inverse limit of a Shimomura Sequence $\{ p_{n+1,n} : (I_{n+1}, \phi_{n+1}) \twoheadrightarrow (I_n, \phi_n) \}$, then X is a Cantor space with $f \in C_s(X)$, and if the sequence is an invertible Shimomura Sequence, then $f \in H(X)$.*

$\Gamma(f)$ consists of the factors of $\{\phi_n\}$. That is, a surjective relation ϕ on $I \in \mathfrak{I}$ lies in $\Gamma(f)$ iff there exists $n \in \mathbb{N}$ and $\pi : I_n \twoheadrightarrow I$ with $\phi = (\pi \times \pi)(\phi_n)$.

PROOF. X is a Cantor space because the sequence bifurcates. The surjective relation f on X is a map by Proposition 1.4 which also says that f is a homeomorphism if the sequence is invertible.

If $p_n : X \twoheadrightarrow I_n$ is the projection from the limit then $f^{p_n} = (p_n \times p_n)(f) = \phi_n$. Hence, $\phi_n \in \Gamma(f)$.

If $\beta : X \twoheadrightarrow I$ is a partition, then because X is the inverse limit, and I is finite, β factors through $p_n : X \twoheadrightarrow I_n$ for sufficiently large n. That is, there exists $n \in \mathbb{N}$ and $\pi : I_n \twoheadrightarrow I$ such that $\beta = \pi \circ p_n$. By (2.5) $f^\beta = (\pi \times \pi)(\phi_n)$ and so f^β is a factor of ϕ_n.

For the converse, Corollary 2.9 (a) implies that any factor of an element of $\Gamma(f)$ lies in $\Gamma(f)$. □

Definition 3.6. *If $(I, \phi), (I_1, \phi_1)$ are surjective relations on elements of $\mathfrak{I}$, we call ϕ_1 a* +directional lift *of ϕ if there exists $\pi : I_1 \twoheadrightarrow I$ such that*

(i) $(\pi \times \pi)(\phi_1) = \phi$.
(ii) $\pi \circ \phi_1 : I_1 \twoheadrightarrow I$ *is a map.*
(iii) $\pi^{-1}(i)$ *contains more than one element for every $i \in I$.*

We then say that π induces the lift *or the lift occurs* via π.

It is a ±directional lift *if π satisfies, in addition,*

(iv) $\pi \circ \phi_1^{-1} : I_1 \twoheadrightarrow I$ *is a map,*

or, equivalently, if, in addition, ϕ_1^{-1} a +directional lift of ϕ^{-1} via π.

Notice that if $\pi : (I_1, \phi_1) \twoheadrightarrow (I, \phi)$ induces a + directional lift and $\pi_2 : (I_2, \phi_2) \twoheadrightarrow (I_1, \phi_1)$ and $\pi_3 : (I, \phi) \twoheadrightarrow (I_3, \phi_3)$ are surjections with $I_2, I_3 \in \mathfrak{I}$, then $\pi_3 \circ \pi \circ \pi_2$ induces a + directional lift (I_2, ϕ_2) of (I_3, ϕ_3).

Thus, an inverse sequence $\{ p_{n+1,n} : (I_{n+1}, \phi_{n+1}) \twoheadrightarrow (I_n, \phi_n)\}$ is a Shimomura sequence exactly when $I_n \in \mathfrak{I}$ for all n and for every n there exists $m > n$ such that $p_{m,n}$ induces a +directional lift, from ϕ_n to ϕ_m.

Lemma 3.7. *Let $\alpha : X \twoheadrightarrow I, \alpha_1 : X \twoheadrightarrow I_1$ be partitions and $f \in C_s(X)$. If mesh α_1 is less than the thickness of α and α_1 refines both α and αf then f^{α_1} is a + directional lift of f^α induced by the unique surjective map $\pi = \alpha \circ \alpha_1^{-1}$ such that $\alpha = \pi \circ \alpha_1$.*

If $f \in H(X)$ and, in addition, α_1 refines $\alpha(f^{-1})$ then f^{α_1} is a ± directional lift of f^α induced by π.

PROOF. If α_1 refines α, then Proposition 2.1(a) implies that $\pi = \alpha \circ \alpha_1^{-1}$ is the unique surjective map such that $\alpha = \pi \circ \alpha_1$ and (2.5) implies $(\pi \times \pi)(f^{\alpha_1}) = f^\alpha$.

If *mesh* α_1 is less than the thickness of α, Proposition 2.1(b) implies that each member of $\mathcal{A}^\alpha$ contains more than one element of $\mathcal{A}^{\alpha_1}$ and so no $\pi^{-1}(i)$ is a singleton.

$f^{\alpha_1} = (\alpha_1 \times \alpha_1)(f) = \alpha_1 \circ f \circ \alpha_1^{-1}$ (see (1.2)) and so $\pi \circ f^{\alpha_1} = (\alpha f) \circ \alpha_1^{-1}$. By Proposition 2.1(a) again this is a map if α_1 refines αf.

Thus, from the hypotheses it follows that π induces a + directional lift from f^{α} to f^{α_1}.

If $f \in H(X)$ we apply the result to f^{-1} to obtain a $\pm$ directional lift when, in addition, α_1 refines $\alpha(f^{-1})$. □

We call $\{ \alpha_n : X \twoheadrightarrow I_n \ : \ n \in \mathbb{N} \}$ a *basic sequence of partitions* when

- α_{n+1} refines α_n for all $n \in \mathbb{N}$.
- $mesh \ \alpha_n \ \to \ 0$ as $n \ \to \ \infty$.

For a basic sequence, let $p_{n+1,n} : I_{n+1} \twoheadrightarrow I_n$ be $\alpha_n \circ \alpha_{n+1}^{-1}$, the unique surjection such that $\alpha_n = p_{n+1,n} \circ a_{n+1}$.

Theorem 3.8. *Let $\{ \alpha_n : X \twoheadrightarrow I_n \}$ be a basic sequence of partitions. If $f \in C_s(X)$ then $\{ p_{n+1,n} : (I_{n+1}, f^{\alpha_{n+1}}) \twoheadrightarrow (I_n, f^{\alpha_n})\}$ is a Shimomura sequence with inverse limit (X, f). If $f \in H(X)$ then it is a invertible Shimomura Sequence.*

Proof. For every n, if m is sufficiently large, then $mesh \ \alpha_m$ is less than the minimum of the Lebesgue number of α_n, the Lebesgue number of $\alpha_n f$ and the thickness of α_n. By Lemma 3.7 $p_{m,n} : I_m \twoheadrightarrow I_n$ induces a +directional lift from f^{α_n} to f^{α_m}. It follows that the sequence is a Shimomura Sequence.

The surjections $\alpha_n : (X, f) \twoheadrightarrow (I_n, f^{\alpha_n})$ induce a surjection from (X, f) onto the inverse limit. because the $mesh \ \alpha_n \to 0$, the map is injective and so is an isomorphism from (X, f) to the inverse limit.

If $f \in H(X)$ we can choose m large enough that α_m refines $\alpha_n(f^{-1})$ as well showing that the sequence is a invertible Shimomura Sequence. □

Corollary 3.9. *If $\{ \alpha_n : X \twoheadrightarrow I_n \}$ is a basic sequence of partitions, and $f \in C_s(X)$, then $\Gamma(f)$ consists of the factors of $\{f^{\alpha_n}\}$.*

Proof. This is immediate from Theorems 3.8 and 3.5. □

If $\mathcal{Q}$ is a collection of surjective relations on elements of $\mathcal{I}$, we let

$$\Gamma^*(\mathcal{Q}) \ = \ \{f \in C_s(X) : \Gamma(f) \subset \mathcal{Q}\}. \tag{3.2}$$

That is, $f \in \Gamma^*(\mathcal{Q})$ when all the relations which represent f are in $\mathcal{Q}$.

Definition 3.10. *Let $\mathcal{Q}$ be a nonempty collection of surjective relations on elements of $\mathcal{I}$. We say that $\mathcal{Q}$* satisfies Condition Γ *if*

- *$\mathcal{Q}$ is closed under factors. That is, if ψ is a factor of $\phi \in \mathcal{Q}$ then $\psi \in \mathcal{Q}$.*
- *$\mathcal{Q}$ admits + directional lifts. That is, if $\phi \in \mathcal{Q}$ then there exists $\psi \in \mathcal{Q}$ which is a + directional lift of ϕ.*

Proposition 3.11. *If $f \in C_s(X)$ then $\Gamma(f)$ is closed under factors and if $\phi \in \Gamma(f)$ then there exists $\psi \in \Gamma(f)$ which is a $\pm$ directional lift of ϕ. In particular, $\Gamma(f)$ satisfies Condition Γ.*

PROOF. $\Gamma(f)$ is closed under factors by Corollary 2.9 (a).

If $\phi \in \Gamma(f)$ then there exists a partition $\alpha : X \twoheadrightarrow I$ such that $\phi = f^{\alpha}$. Construct a basic sequence of partitions $\{\alpha_n\}$ with $\alpha_1 = \alpha$. By Theorem 3.8 $\{(I_n, f^{\alpha_n})\}$ is a Shimomura sequence and it is invertible if $f \in H(X)$. Hence, for sufficiently large n, the map $p_{n,1} : I_n \to I$ induces a + directional lift f^{α_n} of $f^{\alpha_1} = \phi$ and if $f \in H(X)$ then n large enough implies that the lift is $\pm$ directional.

For general $f \in C_s(X)$, we apply the $\pm$ result to the natural lift $\hat{f} \in H(\hat{X})$. By Corollary 2.9 (c) $\Gamma(f) = \Gamma(\hat{f})$ and so every $\phi \in \Gamma(f)$ admits a $\pm$ directional lift in $\Gamma(f)$. □

Now we obtain our main characterization result.

Theorem 3.12. *Let $K_0 \subset K \subset C_s(X)$ be nonempty. Let $\mathcal{Q}$ be a collection of surjective relations on elements of $\mathcal{I}$.*

(a) $\Gamma(\Gamma^*(\mathcal{Q})) \subset \mathcal{Q}$, $K \subset \Gamma^*(\Gamma(K))$ *and*

$$\Gamma^*(\Gamma(\Gamma^*(\mathcal{Q}))) = \Gamma^*(\mathcal{Q}),\ \Gamma(K) = \Gamma(\Gamma^*(\Gamma(K))).$$

(b) $\Gamma^*(\mathcal{Q}) \subset C_s(X)$ *is closed and conjugacy invariant.*

(c) $\Gamma(K)$ *satisfies condition* Γ.

(d) *If K_0 is dense in K, then $\Gamma(K_0) = \Gamma(K)$.*

(e) *If K_0 and K are conjugacy invariant and $\Gamma(K_0) = \Gamma(K)$ then K_0 is dense in K.*

(f) *If K is closed in $C_s(X)$ and is conjugacy invariant then $K = \Gamma^*(\Gamma(K))$. If K is closed in $H(X)$ and is conjugacy invariant then $K = H(X) \cap \Gamma^*(\Gamma(K))$.*

(g) *If $\mathcal{Q}$ satisfies condition Γ then*

$$\mathcal{Q} = \Gamma(\Gamma^*(\mathcal{Q})) = \Gamma(H(X) \cap \Gamma^*(\mathcal{Q})).$$

PROOF. (a) The two inclusions are obvious and then imply the equations by monotonicity of Γ and Γ^*.

(b) If $\alpha : X \twoheadrightarrow I$ is a partition and $\{f_n\}$ is a sequence in $C_s(x)$ converging to f then αf_n is eventually constant at αf and so eventually f_n^{α} is eventually constant at f^{α}. If $f_n \in \Gamma^*(\mathcal{Q})$ for all n then $f^{\alpha} \in \mathcal{Q}$. As α was arbitrary $f \in \Gamma^*(\mathcal{Q})$. $\Gamma^*(\mathcal{Q})$ is conjugacy invariant by (2.6).

(c) Clearly, $\Gamma(K) = \bigcup_{f \in K} \Gamma(f)$ and Condition Γ is preserved by arbitrary unions. So the result follows from Proposition 3.11.

(d) Clearly, $\Gamma(K_0) \subset \Gamma(K)$. If $\phi \in \Gamma(K)$ then $f^{\alpha} = (\alpha \otimes \alpha f)(X)$ for some $f \in K$ and $\alpha : X \twoheadrightarrow I$. Because K_0 is dense in K there exists $f_0 \in K_0$ such that $\alpha f_0 = \alpha f$. Hence, $f_0^{\alpha} = f^{\alpha} = \phi$ and so $\phi \in \Gamma(K_0)$.

(e) If $f \in K$ and $\alpha : X \twoheadrightarrow I$ is a partition then $f^{\alpha} \in \Gamma(K) = \Gamma(K_0)$ and so there exists $\beta : X \twoheadrightarrow I$ and $f_0 \in K_0$ such that $f_0^{\beta} = f^{\alpha}$. By Proposition 2.7 (b) there exists a homeomorphism $h \in H(X)$ such that $f \sim_{\alpha} h^{-1} f_0 h$. Since K_0 is conjugacy invariant $h^{-1} f_0 h \in K_0$. Since α was arbitrary, K_0 is dense in K.

(f) $K \subset \Gamma^*(\Gamma(K)) = K_1$. By (b) K_1 is closed and conjugacy invariant. By (a) $\Gamma(K) = \Gamma(K_1)$. By (d), K is dense in K_1. Since K is closed, it equals K_1.

Now assume that $K \subset H(X)$ is closed in the relative topology. So if $\overline{K}$ is the closure in $C_s(X)$ then K is dense in $\overline{K}$ and $K = \overline{K} \cap H(X)$. Since $\overline{K}$ is closed and conjugacy invariant it follows that $\overline{K} = \Gamma^*(\Gamma(\overline{K}))$ which equals $\Gamma^*(\Gamma(K))$ by (c). Hence, $K = \overline{K} \cap H(X) = \Gamma^*(\Gamma(K)) \cap H(X)$.

(g) Let $\phi \in \mathcal{Q}$. By Condition Γ we can inductively build a Shimomura sequence $\{ p_{n+1,n} : (I_{n+1}, \phi_{n+1}) \twoheadrightarrow (I_n, \phi_n) \}$ with $\phi_1 = \phi$ and $\phi_n \in \mathcal{Q}$ for all $n \in \mathbb{N}$. Let (X, f) be the inverse limit of the sequence. By Theorem 3.5 X is a Cantor set with $f \in C_s(X)$ and every element of $\Gamma(f)$ is a factor of some ϕ_n. Because $\mathcal{Q}$ is closed under factors, we have $\Gamma(f) \subset \mathcal{Q}$, i. e. $f \in \Gamma^*(\mathcal{Q})$. If $(\hat{X}, \hat{f})$ is the natural lift of f to a homeomorphism then by Corollary 2.9(c) $\Gamma(\hat{f}) = \Gamma(f)$ and so $\hat{f} \in \Gamma^*(\mathcal{Q})$. After identification of $\hat{X}$ with X via a homeomorphism, we have that $\hat{f} \in H(X) \cap \Gamma^*(\mathcal{Q})$ with $\phi \in \Gamma(\hat{f})$.

The reverse inclusions follow from (a) and monotonicity of $\Gamma(\cdot)$. □

Thus, $K \mapsto \Gamma(K)$ and $\mathcal{Q} \mapsto \Gamma^*(\mathcal{Q})$ are inverse bijections between the collection of closed, conjugacy invariant subsets of $C_s(X)$ and the collection of those sets of surjective relations on elements of $\mathcal{I}$ which satisfy condition Γ.

Corollary 3.13. *If $K \subset C_s(X)$ is closed and conjugacy invariant then $H(X) \cap K$ is a dense G_δ subset of K.*

Proof. $\mathcal{Q} = \Gamma(K)$ satisfies Condition Γ by (c) and by (f) $K = \Gamma^*(\mathcal{Q})$. Hence, (g) implies that $\Gamma(K) = \mathcal{Q} = \Gamma(H(X) \cap K)$. So by (e) $H(X) \cap K$ is dense in K. Since $H(X)$ is a G_δ subset of $C_s(X)$, its intersection with K is a G_δ subset of K. □

Theorem 3.14. *If K, K_0 are conjugacy invariant subsets of $C_s(X)$ and K is closed, then $K_0 \subset K$ iff $\Gamma(K_0) \subset \Gamma(K)$. In particular, for $f \in C_s(X)$ the closure of the conjugacy class of f is $\{g \in C_s(X) : \Gamma(g) \subset \Gamma(f)\}$.*

Proof. Clearly, $K_0 \subset K$ implies $\Gamma(K_0) \subset \Gamma(K)$. By Theorem 3.12 (f) $K = \Gamma^*(\Gamma(K))$ and by (d) and (f) the closure $\overline{K_0} = \Gamma^*(\Gamma(K_0))$. Hence, $\Gamma(K_0) \subset \Gamma(K)$ implies $\overline{K_0} \subset K$.

If K is the closure of the conjugacy class of f and K_0 is the conjugacy class of g then $g \in K$ iff $K_0 \subset K$ and so iff $\Gamma(g) = \Gamma(K_0) \subset \Gamma(K) = \Gamma(f)$. □

From Corollary 2.9 we obtain the following result from **[15]**.

Corollary 3.15. *(a) If $f, f_1 \in C_s(X)$ and (X, f_1) is a factor of (X, f) then f_1 is in the closure of the conjugacy class of f.*

(b) If $f \in C_s(X)$ and $(\hat{X}, \hat{f})$ is the natural lift of f to a homeomorphism then, (after identifying $\hat{X}$ with X) the closure of the conjugacy classes of f and of $\hat{f}$ agree.

□

Discussed by most of the authors whose work I am describing is the important question of when the conjugacy class of a map f is a G_δ in $C_s(X)$. In that case, the class is a residual subset of its closure. Here we will describe sufficient conditions which are convenient to apply. They come from Shimomura **[17]** and from **[4]**.

We will say that a Shimomura Sequence $\{ p_{n+1,n} : (I_{n+1}, \phi_{n+1}) \twoheadrightarrow (I_n, \phi_n) \}$ is *pointed* when I_1 is a singleton and so (I_1, ϕ_1) is the trivial system. Any Shimomura sequence can be adjusted to become pointed by inserting the trivial system at level 1 and shifting the other index numbers up by one. We will call this the *pointed extension* of the original sequence.

Definition 3.16. *Let $\{ p_{n+1,n} : (I_{n+1}, \phi_{n+1}) \twoheadrightarrow (I_n, \phi_n) \}$ be a pointed Shimomura sequence. If $f \in C_s(X)$ then f satisfies* the lifting property *with respect to the sequence if for every $\epsilon > 0, n \in \mathbb{N}$ and $\alpha : (X, f) \twoheadrightarrow (I_n, \phi_n)$ there exists $m > n$ and $\beta : (X, f) \twoheadrightarrow (I_m, \phi_m)$ such that $\alpha = p_{m,n} \circ \beta$ and with mesh $\beta < \epsilon$.*

Theorem 3.17. *Let $\{ p_{n+1,n} : (I_{n+1}, \phi_{n+1}) \twoheadrightarrow (I_n, \phi_n) \}$ be a pointed Shimomura sequence. The set of $f \in C_s(X)$ which satisfy the lifting property with respect to the sequence is a G_δ subset of $C_s(X)$. If the set is nonempty then it is exactly the set of $f \in C_s(X)$ which are conjugate to the inverse limit of $\{ p_{n+1,n} \times p_{n+1,n} : \phi_{n+1} \twoheadrightarrow \phi_n \}$.*

PROOF. Given $\epsilon > 0$ and $\alpha : X \twoheadrightarrow I_n$ we let $G(\alpha, \epsilon)$ be the set of $f \in C_s(X)$ such that $f^\alpha = \phi_n$ and there exists $\beta : X \twoheadrightarrow I_m$ with *mesh* $\beta < \epsilon$ and such that $f^\beta = \phi_m$ and $p_{m,n} \circ \beta = \alpha$. $G(\alpha, \epsilon) \subset C_s(X)$ is open, because the map Γ_C is locally constant. That is, for any α, β the relations f^α, f^β are unaltered as f varies in a small enough open neighborhood. The intersection over the countable set of α's and rational $\epsilon > 0$ is the G_δ set of maps with the lifting property.

If f satisfies the lifting property then using (2.6) it is easy to check that any conjugate of f satisfies the lifting property.

Now assume that f satisfies the lifting property. Since (I_1, ϕ_1) is trivial, there is a unique $\alpha_1 : (X, f) \twoheadrightarrow (I_1, \phi_1)$. Let $n_1 = 1$. Inductively, we use the lifting property to define an increasing sequence $\{n_i\}$ and $\alpha_i : (X, f) \twoheadrightarrow (I_{n_i}, \phi_{n_i})$ with *mesh* $\alpha_i < 1/i$ for $i > 1$ and with $\alpha_i = p_{n_{i+1},n_i} \circ \alpha_{i+1}$.

Now define $q_n : (X, f) \twoheadrightarrow (I_n, \phi_n)$ for all $n \in \mathbb{N}$ by $q_n = \alpha_i$ if $n = n_i$, and by $q_n = p_{n_{i+1},n} \circ \alpha_{i+1}$ if $n_i \leq n < n_{i+1}$. These define a map from (X, f) onto the inverse limit (I_∞, ϕ_∞). Since the *mesh* $\alpha_i \to 0$ as $i \to \infty$ it follows that the map from X to I_∞ is injective and so is a homeomorphism. This shows that f is conjugate to the inverse limit map. $\square$

We show that for homeomorphisms, the lifting property is preserved by discrete suspension. This requires a little construction.

Lemma 3.18. *Assume $f \in H(X)$, R is a closed relation on A and N is a positive integer. If $p : (X_N, f_N) \to (A_N, R_N)$ is a continuous map between the N-fold suspensions, then there exist a homeomorphism $h : (X_N, f_N) \to (X_N, f_N)$ and a continuous map $\tilde{p} : (X, f) \to (A, R)$ such that $p \circ h = (\tilde{p})_N$.*

PROOF. Let $p(x, k) = (p_1(x, k), n(x, k))$ for $(x, k) \in X \times [1, N]$. Since p maps f_N to R_N, it maps f_N^N to R_N^N. By (1.5) this says that $p(f(x), k) = (p_1(f(x), k), n(f(x), k))$ has $n(f(x), k) = n(x, k)$ and $(p_1(x, k), p_1(f(x), k)) \in R$.

Since $n : X_N \to [1, N]$ is continuous, it follows that $X_j = \{x \in X : n(x, 1) = j\}$ is a clopen, f invariant subset of X for $j = 1, \ldots, N$. Define h to equal $(f_N)^{1-j}$ on the clopen invariant $X_j \times [1, N] \subset X_N$. Since X_N is thus decomposed into invariant pieces, h maps (X_N, f_N) to itself. Since $n(f_N(x,k)) = n(x,k)+1 \ mod\ N$, we have that $n(h(x,1)) = 1$ for all x. Thus, if we let $\hat{p}(x) = p_1(h(x,1))$ we see that $p(h(x,1)) = (\tilde{p}(x), 1)$ and so for $k = 1, \dots, N$, $p(h(x,k)) = (\tilde{p}(x), k)$. That is, $p \circ h = (\tilde{p})_N$.

If $x \in X_j$ then $h(x,1) = (x,1)$ if $j = 1$ and $\tilde{p}(f(x)) = p_1(f(x),1)$. Otherwise $h(x,1) = (f^{-1}(x), N-j+2)$, $\tilde{p}(x),1) = p_1(f^{-1}(x), N-j+2)$ $\tilde{p}(f(x)) = p_1(h((f_N)^N(x,1)) = p_1((f_N)^N(h(x,1)) = p_1(x, N-j+2)$. In either case, we have $(\tilde{p}(x), \tilde{p}(f(x))) \in R$. That is, $\tilde{p} : (X,f) \to (A,R)$. □

Theorem 3.19. *If $f \in H(X)$ satisfies the lifting property for the pointed Shimomura sequence $\{ p_{n+1,n} : (I_{n+1}, \phi_{n+1}) \twoheadrightarrow (I_n, \phi_n) \}$ and N is a positive integer then the N-fold suspension f_N on X_N satisfies the lifting property for the pointed extension of $\{ (p_{n+1,n})_N : ((I_{n+1})_N, (\phi_{n+1})_N) \twoheadrightarrow ((I_n)_N, (\phi_n)_N) \}$.*

PROOF. If $\alpha : (X,f) \twoheadrightarrow (I_n, \phi_n)$ then $(\alpha)_N : (X_N, f_N) \twoheadrightarrow ((I_n)_N, (\phi_n)_N)$ is a lift of the map to the trivial system. That is, we can lift the bottom level of the pointed extension of the suspension.

Given $\alpha : (X_N, f_N) \twoheadrightarrow ((I_n)_N, (\phi_n)_N)$ and $\epsilon > 0$, apply Lemma 3.18 to get $h : (X_N, f_N) \to (X_N, f_N)$ a homeomorphism and $\gamma : (X,f) \twoheadrightarrow (I_n, \phi_n)$ so that $\alpha \circ h = (\gamma)_N$. Let $\delta > 0$ be an ϵ modulus of uniform continuity for h and choose $\beta : (X,f) \twoheadrightarrow (I_{n+k}, \phi_{n+k})$ so that $p_{n+k,n} \circ \beta = \gamma$ and the mesh of β is less than δ. Then the mesh of $(\beta)_N$ is less than δ and so $\tilde{\beta} = ()_{\underline{\ }N} \circ h^{-1}$ has mesh less than ϵ. Furthermore,

$$(p_{n+k,n})_N \circ \tilde{\beta} \ = \ (\gamma)_N \circ h^{-1} \ = \ \alpha,$$

as required. □

We describe an alternative to the lifting property which is a bit easier to use.

Definition 3.20. *If $\{ p_{n+1,n} : (I_{n+1}, \phi_{n+1}) \twoheadrightarrow (I_n, \phi_n) \}$ is a pointed Shimomura sequence, then we says that the the sequence has the* factoring property *if whenever $q_1 : (I_k, \phi_k) \twoheadrightarrow (I_n, \phi_n)$ with $k > n$, there exists $m > k$ and $q_2 : (I_m, \phi_m) \twoheadrightarrow (I_k, \phi_k)$ such that $q_1 \circ q_2 = p_{m,n}$.*

Theorem 3.21. *Let $\{ p_{n+1,n} : (I_{n+1}, \phi_{n+1}) \twoheadrightarrow (I_n, \phi_n) \}$ be a pointed Shimomura sequence with limit (X,f). The sequence has the factoring property iff f has the lifting property with respect to the sequence.*

PROOF. Write $\alpha_n = p_n : (X,f) \twoheadrightarrow (I_n, \phi_n)$, the projection from the inverse limit.

Assume f has the lifting property and we are given $q_1 : (I_k, \phi_k) \twoheadrightarrow (I_n, \phi_n)$.

Let $\epsilon > 0$ be smaller than the Lebesgue number of α_k. By the lifting property applied to $q_1 \circ \alpha_k$, there exists $q : (X,f) \twoheadrightarrow (I_m, \phi_m)$ such that $p_{m,n} \circ q = q_1 \circ \alpha_k$ and with *mesh* $q < \epsilon$. Because ϵ is smaller than the Lebesgue number of α_k, $q_2 = \alpha_k \circ q^{-1} : I_m \twoheadrightarrow I_k$ is a map with $q_1 \circ q_2 = p_{m,n}$. We can always replace q_2 by $q_2 \circ p_{m',m}$ with $m' > m$ arbitrarily large and so we can get $m > k$ if it was not already. This is the factoring property.

Now assume the factoring property and suppose we are given $\epsilon > 0, n \in \mathbb{N}$ and $\alpha : (X, f) \twoheadrightarrow (I_n, \phi_n)$.

Because (X, f) is the inverse limit, α factors through $\alpha_k : X \twoheadrightarrow I_k$ for k sufficiently large. This is because any clopen set is a finite union of basic sets and $\mathcal{A}^{\alpha}$ consists of finitely many clopen sets. We choose k large enough that $k > n$ and $mesh\ \alpha_k < \epsilon$. Thus, there exists $q_1 : (I_k, \phi_k) \twoheadrightarrow (I_n, \phi_n)$ so that $\alpha = q_1 \circ \alpha_k$. By the factoring property, there exist $m > k$ and $q_2 : (I_m, \phi_m) \twoheadrightarrow (I_k, \phi_k)$ so that $q_1 \circ q_2 = p_{m,n}$.

Now use the factoring property again to get $q_3 : (I_r, \phi_r) \twoheadrightarrow (I_m, \phi_m)$ so that $q_2 \circ q_3 = p_{r,k}$. Let $\beta = q_3 \circ \alpha_r : (X, f) \twoheadrightarrow (I_m, \phi_m)$. Since, $q_2 \circ \beta = p_{r,k} \circ \alpha_r = \alpha_k$ it follows that $mesh\ \beta \leq mesh\ \alpha_k < \epsilon$. Furthermore,

$$p_{m,n} \circ \beta \ = \ q_1 \circ q_2 \circ q_3 \circ a_r \ = \ q_1 \circ p_{r,k} \circ a_r \ = \ q_1 \circ \alpha_k \ = \ \alpha.$$

This proves the lifting property. □

We will say that $f \in C_s(X)$ is *of residual type* when its conjugacy class is a G_δ subset of $C_s(X)$ and so is a residual subset in its closure.

Theorem 3.22. *If $f \in C_s(X)$ is of residual type then $f \in H(X)$.*

PROOF. Let K be the closure in $C_s(X)$ of the conjugacy class of f. Hence, K is a closed, conjugacy invariant subset of $C_s(X)$. By Corollary 3.13 $H(X) \cap K$ is a dense, G_δ subset of K. By definition, the conjugacy class of f is dense in K. If the conjugacy class is a G_δ then these two residual subsets intersect. That is, there exists $g \in H(X)$ which is conjugate to f. Since $H(X)$ is itself conjugacy invariant, $f \in H(X)$. □

We will use the lifting property and the factoring property to construct examples of residual type. However, I do not know whether the property is necessary, i.e. whether any f of residual type admits a Shimomura sequence with respect to which it has the lifting property. So, for example, I do not know if the discrete suspensions of a homeomorphism of residual type are necessarily of residual type. Nonetheless, Theorem 3.19 will suffice for construction purposes.

4. Examples

Following Bernardes and Darji [**7**], we define for integers $N, M \geq 1, L \geq 0$, an *Nloop*, or a loop of length N, to be a relation isomorphic to ϕ_N on $[1, N] = \{1, ..., N\} \in \mathfrak{I}$ given by $\{(i, i+1) : i = 1, ..., N-1\} \cup \{(N, 1)\}$, i.e. translation by 1 on the the group $\mathbb{Z}/N\mathbb{Z}$. An $N - L - M$ *dumbbell* is a relation isomorphic to $\phi_{N,L,M}$ on $[1, N+L+M-1] = \{1, ..., N+L+M-1\} \in \mathfrak{I}$ given by $\{(i, i+1) : i = 1, ..., N+L+M-2\} \cup \{(N, 1), (N+L+M-1, N+L)\}$. Notice that a 1 loop and a $1-0-1$ dumbbell are both the trivial surjective relation on a singleton.

For the $N - L - M$ dumbbell, $\phi_{N,L,M}$ restricts to a loop on $[1, N]$. It is called the *in-loop* of the dumbbell. The restriction to $[N+L, N+L+M-1]$ is a loop called the *out-loop* of the dumbbell. We call the restriction to $[N, N+L]$ the *connecting path* of the dumbbell. We extend the language via the -unique- isomorphism to any $N-L-M$ dumbbell.

If $L \geq 1$ then the inloop points precede the outloop points with respect to the partial order given by $\mathcal{O}\phi_{N,L,M}$. On the other hand, if $L=0$, then the connecting path is trivial and we call the $N-0-M$ dumbell an $N-M$ *wedge.* In that case, $\phi_{N,0,M}$ is isomorphic to $\phi_{M,0,N}$ by the map which sends i to $M+i$ for $i=1,..,N-1$ and to $i-N+1$ for $i=N,...,N+M-1$.

If $L \geq 1$ and K is a positive integer, then the K-fold suspension $(\phi_{N,L,M})_K$ is isomorphic to $\phi_{NK,(L-1)K+1,MK}$ by $(i,j) \mapsto K(i-1)+j$. If $L=0, K>1$, then $(\phi_{N,0,M})_K$ is not a dumbell as the two loops share a common path of length K

Proposition 4.1. *Every surjective relation ϕ on a finite set I is a factor of a finite disjoint union of dumbbells.*

Every transitive ϕ on I is a factor of a single loop.

Every recurrent ϕ on I is a factor of a finite disjoint union of loops which may taken to be of the same length.

Proof. We can regard a surjective relation $\phi \subset I \times I$ as describing a directed graph with vertices I and with edges given by the pairs in ϕ.

If ϕ is transitive, i.e. I is a single basic set then we can choose a path along the graph which begins and ends at the same vertex and which passes through every edge. This expresses ϕ as a factor of a loop. If I is recurrent and so is the union of basic sets then ϕ is the factor of a finite number of loops. Taking the least common multiple N of the lengths, we can lift each loop to an N loop and so express ϕ as a factor of a disjoint union of N loops.

We can include any edge for a surjective relation in a path and extend it forward and backward until on each side a repeat of a vertex occurs. This exhibits a dumbbell which maps into ϕ and which hits the given edge. As there are only finitely many edges we can express ϕ as a factor of a finite union of dumbbells. $\square$

Lemma 4.2. *If $f \in H(X)$ and ϕ_N is the N-loop on $[1,N]$ then the following are equivalent:*

(i) *There exists $p : (X,f) \to ([1,N], \phi_N)$.*
(ii) *There exists a partition $\alpha : X \twoheadrightarrow [1,N]$ such that $f^\alpha = \phi_N$.*
(iii) $\phi_N \in \Gamma(f)$.
(iv) *(X,f) is an N-fold suspension of some system (Y,g).*

If $p : (X,f) \to ([1,N], \phi_N)$ exists, it is necessarily a system surjection.

Proof. If $p : (X,f) \to ([1,N], \phi_N)$ then $p(f^k(x))$ is congruent to $p(x) + k \ mod \ N$. Thus, $p(X) = [1,N]$ and so $p : (X,f) \twoheadrightarrow ([1,N], \phi_N)$ since ϕ_N is a map (see Lemma 1.2(b). Since such a system map p is necessarily a surjection and ϕ_N is a permutation, the equivalence of (i), (ii) and (iii) follows from Proposition 2.10.

(iv) $\Rightarrow$ (i): If $(X,f) = (Y_N, g_N)$ and so $X = Y \times [1,N$ then the second coordinate projection from X to $[1,N]$ maps X onto $[1,N]$ and f onto ϕ_N.

(i) $\Rightarrow$ (iv): If $p : (X,f) \to ([1,N], \phi_N)$ then we let $Y = p^{-1}(1)$ which is a clopen $g = f^N$ invariant subset of X. Define the homeomorphism $h : Y_N \to X$ by $h(x,i) = f^{i-1}(x)$ for $i = 1,...,N$. Clearly, for $i < N$, $f(h(x,i)) = f^i(x) = h(x,i+1) = h(g_N(x,i))$, while $f(h(x,N)) = f^N(x) = h(f^N(x),1)) = h(g_N(x,N))$. $\square$

We will repeatedly use the rigidity of maps between loops and dumbbells as described in the following Lemma.

Lemma 4.3. *(a) There exists a map* $p : ([1, M], \phi_M) \to ([1, K], \phi_K)$ *iff* K *is a divisor of* M. *In that case, given* $i \in [1, M]$ *and* $j \in [1, K]$ *there is a unique such* p, *necessarily surjective, with* $p(i) = j$.

(b) There exists a map $p : ([1, N + L + M - 1], \phi_{N,L,M}) \to ([1, K], \phi_K)$ *iff* K *is a divisor of* N *and of* M. *In that case, given* $i \in [1, N + L + M - 1]$ *and* $j \in [1, K]$ *there is a unique such* p, *necessarily surjective, with* $p(i) = j$.

(c) If $p : ([1, N + L + M - 1], \phi_{N,L,M}) \to ([1, N_1 + L_1 + M_1], \phi_{N_1,L_1,M_1})$ *and* p *is not surjective, then either* $p([1, N + L + M - 1])$ *is the in-loop or the out-loop of* ϕ_{N_1,L_1,M_1}.

(d) Assume $L_1 \geq 1$. *There exists a map* $p : ([1, N + L + M - 1], \phi_{N,L,M}) \twoheadrightarrow ([1, N_1 + L_1 + M_1], \phi_{N_1,L_1,M_1})$ *iff* N_1 *is a divisor of* N, M_1 *is a divisor of* M *and* $L_1 \leq L$. *If* $N \leq i \leq N + L$ *and* $N_1 \leq j \leq N_1 + L_1$ *then such a* p *exists with* $p(i) = j$ *iff* $i - N \geq j - N_1$ *and* $N + L - i \geq N_1 + L_1 - j$. *In that case,* p *is unique.*

PROOF. (a): ϕ_M is a map with $(\phi_M)^M = 1_{[1,M]}$ and so if p maps ϕ_M to ϕ_K then $(\phi_K)^M = 1_{[1,K]}$ and so $K|M$. Since $[1, M]$ is the ϕ_M orbit of any of its points, the map p is uniquely determined by its value on any point. By rotating in the ϕ_K loop we see that the value of $p(i)$ can be arbitrarily chosen.

(b): By (a) applied to the restriction of p to the inloop and the outloop, we see from (a) that for p to exist we must have $K|N$ and $K|M$. In that case, the map from $[1, N+L+M-1]$ to $[1, K]$ obtained by sending i to its congruence class *mod* K is a map from $\phi_{N,L,M}$ to ϕ_K. Again we can rotate the ϕ_K loop to get an arbitrary value of $p(i)$.

If $N \leq i \leq N + L$ then by moving forward and backward we see that p is uniquely determined by the value $p(i)$. If i is on the inloop then (a) implies that the restriction to the inloop is uniquely determined by $p(i)$. Thus, $p(N)$ is determined and since N is on the connecting path, p is determined. Similarly, if i is on the outloop.

(c): The only proper subsystems of the dumbbell are the inloop, the outloop and their disjoint union. The image of p is a subsystem in which any two elements can be connected by a chain. It follows that the image is either the inloop our the outloop when p is not onto. When $p([1, N + L + M - 1]) = [1, N_1 + L_1 + M_1 - 1]$ then the map of systems is surjective by Lemma 1.2(b).

(d): When $L_1 > 0$, the dumbbell ϕ_{N_1,L_1,M_1} is not transitive. The only transitive subsets are the inloop and the outloop. The image of the inloop and the outloop of $\phi_{N,L,M}$ are transitive subsets. Furthermore, if $L = 0$ then $\phi_{N,L,M}$ would be transitive and with a transitive image. Hence, $L > 0$. Since p maps $\mathcal{C}\phi_{N,L,M}$ to $\mathcal{O}\phi_{N_1,L_1,M_1}$ we see that p takes inloop to inloop and outloop to outloop. Note that this also uses $L_1, L > 0$. It thus follows from (a) that $N_1|N$ and $M_1|M$ are required.

If $N_1 < j < N_1 + L_1$ then j is not in one of the endloops and so any pre-image is not in an endloop. Also, $N_1 + L_1 - 1$ is not in the outloop and $N_1 + L_1 \in phi_{N_1,L_1,M_1}(N_1 + L_1 - 1)$. Similarly, $N_1 \in \phi^{-1}_{N_1,L_1,M_1}(N_1 + 1)$. Thus, the preimage of the points of the connecting path in ϕ_{N_1,L_1,M_1} must lie in the connecting path of $\phi_{N,L,M}$. Hence, $L \geq L_1$. Furthermore, if $N_1 \leq j \leq N_1 + L_1$, i.e. j lies on the connecting path in the image, and $p(i) = j$ then $p(i - (j - N_1)) = N_1$ and $p(i + (N_1 + L_1 - j)) = N_1 + L_1$. Hence, $i - (j - N_1)$ and $i + (N_1 + L_1 - j)$ lie in the connecting path of $\phi_{N,L,M}$. Thus, $N \leq i - (j - N_1)$ and $i + (N_1 + L_1 - j) \leq N + L$.

Finally, given i and j satisfying the above conditions, p is uniquely defined by taking $i+k$ to $j+k$ for $N_1 - j \leq k \leq N_1 + L_1 - j$. At the endpoints we are in the inloop on the left and the outloop on the right. We then move around the loops. □

Remark: Notice that when $N = N_1, L = L_1, M = M_1$ and $L_1 \geq 1$ then the only surjection between the dumbbells has $i = j$. That is, the identity map is the unique surjection from a dumbbell to itself when the connecting path is nontrivial.

We will call a closed, conjugacy invariant subset K of $C_s(X)$ *conjugacy transitive* when it is the closure of the conjugacy class of some $f \in K$, i.e. $K = \overline{H(X) \cdot f}$. Such elements f are called *conjugacy transitive elements* and the set of such elements is denoted $Trans(K)$. Thus, K is conjugacy transitive when $Trans(K)$ is nonempty, in which case it is a dense G_δ subset of K. We call K *conjugacy minimal* when $Trans(K) = K$. By Theorem 3.12(e)

$$f \in Trans(K) \quad \Leftrightarrow \quad \Gamma(f) \; = \; \Gamma(K). \tag{4.1}$$

A conjugacy transitive point f is of residual type when the conjugacy class $H(X) \cdot f$ is a G_δ subset of K. Since distinct conjugacy classes are disjoint and two dense G_δ subsets meet, it follows that there is at most one dense G_δ conjugacy class. Theorem 3.22 implies that such residual type transitive points are contained in $H(X)$.

If $\mathcal{Q}$ is a collection of surjective relations on members of $\mathcal{I}$ and $K \subset C_s(X)$ we will say that $\mathcal{Q}$ *generates* $\Gamma(K)$ if $\mathcal{Q} \subset \Gamma(K)$ and every element of $\Gamma(K)$ is a factor of some relation in $\mathcal{Q}$.

Example 1 - $\mathbf{C_s(X)}$:

By Theorem 3.1 $\Gamma(C_s(X))$ is the set of all surjective relations on members of $\mathcal{I}$. Since the set of surjective relations on elements of $\mathcal{I}$ is countable, Corollary 3.3 and the Baire Category theorem imply that the set of $f \in C_s(X)$ with $\Gamma(f) = \Gamma(C_s(X))$ is a dense G_δ subset of $C_s(X)$. Thus, $C_s(X)$ is conjugacy transitive and this set is $Trans(C_s(X))$ by (4.1).

Let $f, g \in C_s(X)$ with g a conjugacy transitive point of $C_s(X)$. If (X, g) is a factor of (X, f) then f is a conjugacy transitive point of $C_s(X)$ because

$$\Gamma(C_s(X)) \; = \; \Gamma(g) \; \subset \; \Gamma(f) \; \subset \; \Gamma(C_s(X)).$$

In particular, if $g \in H(X)$ is a conjugacy transitive homeomorphism and g_1 is an arbitrary member of $C_s(X)$ then by using a homeomorphism from $X \times X$ to X, we can regard $f = g \times g_1$ as a member of $C_s(X)$ which has g as a factor and so is conjugacy transitive. On the other hand, if g_1 is not injective then f is not and so its conjugacy class is a subset of $C_s(X) \setminus H(X)$ which is dense in $C_s(X)$.

We describe a Shimomura sequence with the factoring property whose limit is a conjugacy transitive element of $C_s(X)$. For $n \geq 2, i = 1, ..., 4^{n-1}$ let $\{(I_{i;n}, \phi_{i;n})\}$ be a collection of $n! - 2^{n+1} - n!$ dumbbells with the sets $\{I_{i;n}\}$ pairwise disjoint. Label by $a_{i;n}, e_{i;n}, b_{i;n}$ the vertices in $I_{i;n}$ at positions $n!, n! + 2^n, n! + 2^{n+1}$ respectively. That is, these are the vertices at the left end, the mid-point, and the right end of the connecting path for $\phi_{i;n}$. Let (J_1, Φ_1) be a trivial system and for $n \geq 2$ let (J_n, Φ_n) be the disjoint union of the systems $\{(I_{i;n}, \phi_{i;n}) : i = 1, \ldots, 4^{n-1}\}$. $P_{2,1}$ is the unique map to the trivial system. For $n \geq 2$, $P_{n+1,n}$ on $(I_{i;n+1}, \phi_{i;n+1})$ is the unique map to $(I_{j;n}, \phi_{j;n})$ taking $e_{i;n+1}$ to $e_{j;n}$ when $i = 4(j-1)+2$ or $i = 4(j-1)+3$. When $i = 4(j-1)+1$ it is the unique map onto the inloop which takes $e_{i;n+1}$ to $a_{j;n}$. When $i = 4(j-1)+3$ it is the unique map onto the outloop taking $e_{i;n+1}$ to $b_{j;n}$. Thus, each level n dumbbell is the image of two level $n+1$ dumbbells and each endloop at level n is the image of a single level $n+1$ dumbbell. Hence, for $k \geq 1$ each level n dumbbell is the image under $P_{n+k,n}$ of 2^k dumbbells and each endloop is the image of 2^{k-1} dumbbells.

Theorem 4.4. *$\{P_{n+1,n} : (J_{n+1}, \Phi_{n+1}) \twoheadrightarrow (J_n, \Phi_n)\}$ is an invertible pointed Shimomura sequence which satisfies the factoring property. If (X, f) is the limit, then $f \in H(X)$ is a conjugacy transitive point for $C_s(X)$ of residual type.*

PROOF. The map $P_{n+1,n}$ takes the first 2^n vertices of the connecting path for $\phi_{i;n+1}$ into the same endloop as the initial vertex and the last 2^n vertices to the same endloop as the last vertex, it follows that $P_{n+1,n}$ realizes a $\pm$ directional lift and so the sequence is an invertible Shimomura sequence.

If (X, f) is the limit then $f \in H(X)$ and by Theorem 3.5 $\Gamma(f)$ is generated by the $\{\Phi_n\}$. Every finite union of dumbbells is clearly a factor of (J_n, Φ_n) for sufficiently large n and by Proposition 4.1 every surjective relation is a factor of a finite union of dumbbells. Hence, $\Gamma(f) = \Gamma(C_s(X))$ and so f is a conjugacy transitive point for $C_s(X)$.

It remains to check the factoring property which will imply that f is of residual type.

If $p : (J_k, \Phi_k) \twoheadrightarrow (J_1, \Phi_1)$ then $P_{k+1,k}$ provides the required factoring since Φ_1 is trivial. Let $p : (J_k, \Phi_k) \twoheadrightarrow (J_n, \Phi_n)$ with $k > n > 1$. For each $i = 1, ..., 4^{n-1}$ the dumbbell $(I_{i;n}, \varphi_{i;n})$ is hit by at least one and at most 4^{k-1} dumbbells in Φ_k. In addition, each endloop may be the image of some dumbbells in Φ_k, again at most 4^{k-1} of them. Choose m large enough that $2^{m-n-1} > 4^{k-1}$.

First we allocate each the dumbbells of Φ_m to a dumbbell of Φ_k. For each $\phi_{i;n}$ we distribute the dumbbells which hit it via $P_{m,n}$ among those which hit it via p so that each of the latter is allocated at least one. The dumbbells which are mapped onto the inloop of $\phi_{i;n}$ are distributed among the dumbbells, if any, which are mapped onto the inloop via p again so that each of the latter receives at least one. If there are none then these level m dumbbells are allocated to map onto the inloop of some dumbbell of Φ_k which maps onto $\phi_{i;n}$ via p. Similarly, we allocate for the outloops.

Having made these allocations the maps are determined as follows. If p maps $\phi_{r;k}$ onto $\phi_{i;k}$ then the vertex $e_{i;k}$ is the image of a vertex e in the connecting path for $\phi_{r;k}$. If $\phi_{s;m}$ has been allocated to $\phi_{r;k}$ then we choose the unique map $(I_{s;m}, \phi_{s;m}) \twoheadrightarrow (I_{r;k}, \phi_{r;k})$ which maps $e_{s;m}$ to e. Because $m > k$ the inequalities given in Lemma 4.3 (d) are satisfied. If $P_{m,n}$ maps $\phi_{s;m}$ onto the inloop of $\phi_{i;k}$ then $\phi_{s;m}$ has been allocated to $\phi_{r;k}$ which either maps onto $\phi_{i;r}$ or onto its inloop. In either case, we map $\phi_{s;m}$ onto the inloop of $\phi_{r;k}$ in such a way that $e_{s;m}$ is mapped onto a vertex of the inloop of $\phi_{r;k}$ which is mapped by p onto $a_{i;k}$. There may be several of these and any one will do. We use a similar procedure for the outloops. □

An explicit description is given in [**4**] of this map f, unique up to conjugacy, which is conjugacy transitive for $C_s(X)$ and is of residual type. The existence of such an element had earlier been proved in [**12**].

Call $f \in H(X)$ *simple* if there are only finitely many chain recurrent points. This is equivalent to saying that there are finitely many periodic points and the chain recurrent points are periodic. Call $f \in C_s(X)$ *very simple* if there are only finitely many chain recurrent points and every point of X is eventually mapped to a periodic point. We now reprove a result of Batista et al [**6**] showing that the simple homeomorphisms and the very simple maps are each dense in $C_s(X)$.

Let T be the translation homeomorphism on $\mathbb{Z}$, $T(n) = n+1$. With C a Cantor set let $\tilde{X} = \{e_-, e_+\} \cup \mathbb{Z} \times C$ be the two-point compactification of $\mathbb{Z} \times C$ so that $\{n\} \times C \to e_{\pm}$ as $n \to \pm\infty$. Let $\tilde{f} \in H(\tilde{X})$ be the homeomorphism which extends $T \times 1_C$ on $\mathbb{Z} \times C$.

Proposition 4.5. *The homeomorphism $\tilde{f}$ is of residual type in $C_s(\tilde{X})$.*

Proof. We will sketch the proof leaving the final details to the interested reader.

We use as our Cantor set $C = \{1,2\}^{\mathbb{N}}$, with $C_N = \{1,2\}^N$ the set of words of length N. We define an inverse sequence. Let (I_1, ϕ_1) be the trivial system. For $n \geq 2$, let

$$\begin{aligned} I_n &= [1, 2^{n+1}] \times C_n \cup \{(1,1), (2^{n+1}, 1)\}, \\ \phi_n &= \{((i,w),(i+1,w)) : 1 \leq i < 2^{n+1}, w \in C_n\} \cup \\ &\{((1,1),(1,w)) : w \in C_n \cup \{1\}\} \cup \{((2^{n+1},w),(2^{n+1},1) : w \in C_n \cup \{1\}\}. \end{aligned} \tag{4.2}$$

and

$$p_{n+1,n}(i, wa) = \begin{cases} (1,1) \quad \text{for } 1 \leq i \leq 2^n, wa \in C_{n+1} \cup \{1\} \\ (i - 2^n, w) \quad \text{for } 2^n < i \leq 2^n + 2^{n+1}, wa \in C_{n+1}, \\ (2^{n+1}, 1) \quad \text{for } 2^n + 2^{n+1} < i \leq 2^{n+2}, wa \in C_{n+1} \cup \{1\}. \end{cases} \tag{4.3}$$

Thus, we think of I_n as a set of parallel threads indexed by C_n and gathered at the endpoints $(1,1)$ and $(2^{n+1}, 1)$.

Define the map $h_n : \tilde{X} \to I_n$ by mapping e_- to $(1,1)$, e_+ to $(2^{n+1}, 1)$ and

$$(i, z) \mapsto \begin{cases} (1,1) \quad \text{for } i \leq -2^n, \\ (i + 2^n, \pi_n(z)) \quad \text{for } -2^n < i \leq 2^n, \\ (2^{n+1}, 1) \quad \text{for } 2^n < i. \end{cases} \tag{4.4}$$

It is easy to check that $\{h_n\}$ induces an isomorphism from $(\tilde{X}, \tilde{f})$ to the inverse limit. Also it is clear that the sequence is an invertible Shimomura sequence.

Now suppose $p : (I_k, \phi_k) \twoheadrightarrow (I_n, \phi_n)$ with $k > n > 1$. The factoring property is obtained from the following observations.

- Each endpoint of ϕ_k is mapped to an endpoint of ϕ_n.
- If the two endpoints of ϕ_k were mapped to the same endpoint of ϕ_n then the image, which is ϕ_n would be a transitive relation which it is not.

- The orderings given by $\mathcal{O}\phi_k$ and $\mathcal{O}\phi_n$ then imply that the left endpoint is mapped to the left and the right to the right.
- Each thread of ϕ_k is mapped by p onto a thread of ϕ_n.
- If (i,w) is a vertex on a thread in ϕ_n and it is the image of the w' thread in ϕ_k then there is a unique j such that $p(j,w') = (i,w)$ and this equation uniquely determines p on the w' thread.
- If $m > k > n$ then each thread in I_n is the image of 2^{m-n} threads in I_m.

It follows that with m large enough we can use the image under p of each I_k thread to allocate at least one I_m thread with the same image under $p_{m,n}$. Then we pick a point in each I_n thread and pull back to define the map uniquely given the allocations. □

If $\alpha_L : \tilde{X} \to [1,L]$ by

$$\alpha_L(x) = \begin{cases} 1 & \text{for } x = e_-, \text{ and } x = (i,c) \text{ with } i \leq 1, \\ i & \text{for } x = (i,c) \text{ with } 1 \leq i \leq L, \\ L & \text{for } x = e_+, \text{ and } x = (i,c) \text{ with } i \geq L, \end{cases} \tag{4.5}$$

then $\tilde{f}^{\alpha_L} = \phi_{1,L\,1}$. Hence, for every positive L the $1 - L - 1$ dumbbell is a factor of $(\tilde{X}, \tilde{f})$.

For N a positive number, we define the factor map $\pi_N : (\tilde{X}_N, \tilde{f}_N) \twoheadrightarrow (\tilde{X}, \tilde{f})$ by

$$\pi_N(x,n) = \begin{cases} e_\pm & \text{for } x = e_\pm, \quad 1 \leq n \leq N, \\ (N(i-1)+n, c) & \text{for } x = (i,c), \ i \in \mathbb{Z}, \quad 1 \leq n \leq N. \end{cases} \tag{4.6}$$

The map is one-to-one except over the endpoints which are lifted to periodic orbits of period N. By using the identification defined by (1.6) and the functoriality of the suspension operation we obtain a natural factor map $\pi_{N,M} : (\tilde{X}_N, \tilde{f}_N) \twoheadrightarrow (\tilde{X}_M, \tilde{f}_M)$ whenever, M is a divisor of N. That is, $\pi_{N,M}$ is the M-fold suspension of the map $\pi_{N/M}$.

In the language of [**4**] the threads become *spirals*.

Let (X_n, f_n) be the disjoint union of $n!$ disjoint copies of the $n!$-fold suspension of $(\tilde{X}, \tilde{f})$. By decomposing the collection of copies in f_{n+1} by sets of size $n+1$ and then using $\pi_{(n+1)!,n!}$ with a common range on each set of copies, we obtain a an inverse sequence of maps $p_{n+1,n} : (X_{n+1}, f_{n+1}) \twoheadrightarrow (X_n, f_n)$. Let (X_∞, f_∞) be the inverse limit. By choosing homeomorphisms to a common space X, we can regard $f_n \in H(X)$ for $1 \leq n \leq \infty$.

Theorem 4.6. *If $K_0 \subset H(X)$ is the union of the conjugacy classes of f_n for $1 \leq n < \infty$ then K_0 is a conjugacy invariant collection of simple homeomorphisms which is dense in $C_s(X)$. The homeomorphism f_∞ is a conjugacy transitive point for $C_s(X)$.*

Proof. By Proposition 4.1 every surjective relation on an element of $\mathcal{J}$ is a factor of a finite union of dumbbells and every finite union of dumbbells is a factor of f_n for n sufficiently large. Hence, $\Gamma(K_0) = \Gamma(C_s(X))$. By Theorem 3.12 (e) K_0 is dense in K.

Since every f_n is a factor of f_∞ it follows that $\Gamma(C_s(X)) = \Gamma(K_0) \subset \Gamma(f_\infty) \subset \Gamma(C_s(X))$. So f_∞ is a conjugacy transitive homeomorphism. □

Let $\tilde{X}^+ = \{e_-\} \cup \{(i,c) \in \tilde{X} : i \leq 0\}$. Choose $c_0 \in C$ and let $e_0 = (0, c_0)$ and define $\tilde{f}^+ \in C_s(\tilde{X}^+)$ by

$$\tilde{f}^+(x) \quad = \quad \begin{cases} \tilde{f}(x) & \text{for } x = e_-, \text{ and } x = (i,c), i < 0, \\ e_0 = (0, c_0) & \text{for } x = (0,c). \end{cases} \tag{4.7}$$

Thus, $\tilde{f}^+$ is a surjective map and that every point except the fixed point e_- is eventually mapped to the fixed point e_0.

Let (X_n^+, f_n^+) be the disjoint union of $n!$ disjoint copies of the $n!$-fold suspension of $(\tilde{X}^+, \tilde{f}^+)$. Again we can choose homeomorphisms to a common space X and regard $f_n \in C_s(X)$ for $1 \leq n < \infty$.

Theorem 4.7. *If $K_0 \subset C_s(X)$ is the union of the conjugacy classes of f_n^+ for $1 \leq n < \infty$ then K_0 is a conjugacy invariant collection of very simple mappings which is dense in $C_s(X)$.*

PROOF. Just as in Theorem 4.6. □

Recall that for $f \in C_s(X)$, $Per(f) = \{n : |f^n| \neq \emptyset\}$. So, for example, f has a fixed point iff $1 \in Per(f)$, in which case $Per(f) = \mathbb{N}$. By Proposition 1.6 the set

$$C_s(X; Per \supset Q) \quad = \quad \{\ f \in C_s(X)\ :\ Q \subset Per(f)\ \}$$

is a closed, conjugacy invariant subset of $C_s(X)$ for any $Q \subset \mathbb{N}$. Recall that we write $C_s(X;1)$ for $C_s(X; Per \supset \{1\})$. So $C_s(X;1)$ is the closed set of maps which admit a fixed point.

Theorem 4.8. *For any $Q \subset \mathbb{N}$, the set $C_s(X; Per \supset Q)$ is a conjugacy transitive subset of $C_s(X)$. With $1 \in Q$ the set $C_s(X;1)$ has a conjugacy transitive element of residual type.*

PROOF. Recall that if R_2 is a factor of R_1 then $Per(R_1) \subset Per(R_2)$. Hence, if $\phi \in \Gamma(C_s(X; Per \supset Q))$ then $Q \subset Per(\phi)$. Conversely, if $Q \subset Per(\phi)$ then Theorem 3.2 implies that $\Gamma^{-1}(\phi) \cap C_s(X; Per \supset Q)$ is dense in $C_s(X; Per \supset Q)$. It follows that $\Gamma(C_s(X; Per \supset Q))$ consists exactly of the surjective relations ϕ on members of $\mathfrak{I}$ such that $Q \subset Per(\phi)$. Since $\Gamma^{-1}(\phi)$ is open, the Baire Category Theorem implies that $C_s(X; Per \supset Q)$ is a conjugacy transitive set.

Now consider the fixed point case. It is clear from Proposition 4.1 that every surjective relation ϕ with $|\phi| \neq \emptyset$ is a factor of a disjoint union of dumbbells together with a single extra vertex which is related only to itself. Now we adjust the Shimomura sequence of Theorem 4.4 to obtain $P^*_{n+1,n} : (I^*_{n+1}, \Phi^*_{n+1}) \twoheadrightarrow (I^*_n, \Phi^*_n)\}$ as follows: At every level $n \geq 2$ we obtain (J^*_n, Φ^*_n) we adjoin a single new pair $\{1\}, \{(1,1)\}$. We alter $P_{n+1,n}$ so that $P^*_{n+1,n}$ it maps 1 at level $n+1$ and the dumbbell $I_{2,n+1}$ both to 1 at level n. Above 1 at level $m > n$ there is 1 and 2^{m-n-1} dumbbells. So if p is any map from level k onto level n we can allocate to the k level dumbbells mapped to 1 and to 1 at level k, all of the dumbbells at level m which are mapped by $P_{m,n}$ down to 1 at n.

Again we leave the details to the reader.

The result is an invertible Shimomura sequence satisfying the factoring property and with limit f have a unique fixed point over the 1's and with $\Gamma(f)$ containing arbitrarily large unions of large dumbbells. It follows that f is a conjugacy transitive point for $C_s(X;1)$ which is of residual type. □

Example 2 - CR(X) and CT(X):

Recall that $CR(X)$ and $CT(X)$ are the closed, conjugacy invariant subsets of chain recurrent elements of $C_s(X)$ and of chain transitive elements of $C_s(X)$, respectively. By Proposition 4.1 $\Gamma(CR(X))$ is generated by the finite disjoint unions of loops and $\Gamma(CT(X))$ is generated by the loops.

Theorem 4.9. *(a) $CR(X)$ is a conjugacy transitive subset of $C_s(X)$ with a transitive element of residual type.*

(b) For $g \in C_s(X)$ chain recurrent, let $g_n \in C_s(X)$ be isomorphic to n disjoint copies of the $n!$-fold suspension of g. If K_0 is the union of the conjugacy classes of $g_1, g_2, ...$, then K_0 is a dense subset of $CR(X)$.

(c) If K_1 is the set of $f \in H(X)$ with a dense set of periodic points, K_1 is a dense subset of $CR(X)$.

(d) For any $Q \subset \mathbb{N}$, the set $C_s(X; Per \supset Q) \cap CR(X)$ is a conjugacy transitive subset of $C_s(X)$. The set $CR(X;1)$ of chain recurrent maps admitting a fixed point has a conjugacy transitive point of residual type.

Proof. (a) Theorem 3.2 implies that for ϕ any recurrent relation on an element of $\mathcal{I}$, the set $\Gamma^{-1}(\phi) \cap CR(X)$ is open and dense. As there are countably many such relations ϕ, it follows from the Baire Category Theorem that $CR(X)$ is a transitive set.

Let $\{k_n\}$ be an increasing sequence with $k_1 = 1$ and such that $k_n | k_{n+1}$. For $n \geq 2$, let $\{(I_n, \phi_{i,n} : i = 1, \dots, 2^{n-1}\}$ be disjoint k_n-loops with $e_{i,n}$ a chosen point of I_n, and let (J_n, Φ_n) be their disjoint union. Let $(J_1, \Phi_1) = (\{1\}, \{(1,1)\})$ with $e_{11} = 1$. Define $P_{n+1,n} : (J_{n+1}, \Phi_{n+1}) \to (J_n, \Phi_n)$ be the map uniquely defined by

$$P_{n+1,n}(e_{i,n+1}) = e_{i,n} = P_{n+1,n}(e_{i+2^{n-1},n+1}) \quad \text{for} \quad i = 1, \dots, 2^{n-1}.$$

Using the now familiar allocation argument and Lemma 4.3(a), it is easy to check that this defines an invertible Shimomura sequence which satisfies the factoring property. The inverse limit is clearly chain recurrent. If we choose $k_n = n!$ then every finite disjoint union of loops is a factor of Φ_n for n sufficiently large, and so the resulting inverse limit is a transitive element of $CR(X)$ of residual type.

(b) If g is chain recurrent then every g_n is chain recurrent and any finite disjoint union of loops is a factor of g_n for n sufficiently large. By Theorem 3.12 (e) K_0 is dense in $CR(X)$.

(c) In (b) use $g \in H(X)$ with dense periodic points, e. g. the shift homeomorphism on $\{0,1\}^{\mathbb{Z}}$. Then every element of K_0 has dense periodic points, i.e. $K_0 \subset K_1$.

(d) Again Theorem 3.2 implies that $C_s(X; Per \supset Q) \cap CR(X)$ is a conjugacy transitive subset of $C_s(X)$.

In the fixed point case we proceed exactly as in Theorem 4.8. We define $P^*_{n+1,n}$: $(I^*_{n+1}, \Phi^*_{n+1}) \twoheadrightarrow (I^*_n, \Phi^*_n)\}$ by adjoining $(\{1\}, \{(1,1)\})$ at every level n with $n \geq 2$. The we alter the map $P_{n+1,n}$ so that

$$P^*_{n+1,n}(1) = 1 = P^*_{n+1,n}(e_{1,n+1}),$$

but is otherwise unchanged. Again it is easy to check that the sequence satisfies the factoring property. It is clear that $\Gamma(CR(X;1))$ is generated by finite unions of loops together with a single disjoint fixed point. Hence, when $k_n = n!$ the limit is a conjugacy transitive element of $CR(X;1)$. □

Call $\{k_n\}$ a *divisibility sequence* when it is an increasing sequence of positive integers such that $k_1 = 1, k_n | k_{n+1}$ for all $n \in \mathbb{N}$., For such a sequence the sequence of group homomorphisms

$$\{p_{n+1,n} : (\mathbb{Z}/k_{n+1}\mathbb{Z}, \phi_{n+1}) \twoheadrightarrow (\mathbb{Z}/k_n\mathbb{Z}, \phi_n)\}$$

defines an inverse sequence of loops where each ϕ_n is translation by 1. Let $P^* = \{k \in \mathbb{N} : k | k_n \text{ for sufficiently large } n\}$. The inverse limit (X, f) is a minimal system which maps onto a k-loop iff $k \in P^*$. It is called the *adding machine* or *odometer* associated with the sequence $\{k_n\}$ although it really depends only on the set P^*. When $P^* = \mathbb{N}$, e.g. when $k_n = n!$, then the system is called the *universal adding machine.* It factors onto every loop.

It is easy to see that the homeomorphism constructed in part (a) above is the product of the identity on a Cantor set with the adding machine associated with the sequence $\{k_n\}$. In particular, as noted by Shimomura, [**17**], the product of the identity on a Cantor set with the universal adding machine is a transitive element of residual type for $CR(X)$.

Theorem 4.10. *(a) If $\{k_n\}$ is a divisibility sequence then*

$$\{ \, p_{n+1,n} : (\mathbb{Z}/k_{n+1}\mathbb{Z}, \phi_{n+1}) \twoheadrightarrow (\mathbb{Z}/k_n\mathbb{Z}, \phi_n) \, \}$$

is an invertible Shimomura sequence satisfying the factoring property and so the associated adding machine is of residual type.

(b) $CT(X)$ is a conjugacy transitive subset of $C_s(X)$ with a the universal adding machine transitive element of residual type.

(c) For $g \in C_s(X)$ chain transitive, let $g_n \in C_s(X)$ be isomorphic to the $n!$-fold suspension of g. If K_0 is the union of the conjugacy classes of $g_1, g_2, \dots$, then K_0 is a dense subset of $CT(X)$.

PROOF. It is easy to check the factoring property for the inverse sequence associated with a divisibility sequence. For the universal adding machine every loop is a factor and so it is a transitive element for $CT(X)$. This proves (a) and (b). Finally, (c) is proved just like (b) of Theorem 4.9. □

Part (b) was first shown by Hochman in [**11**].

Corollary 4.11. *If $f \in C_s(X)$ is a conjugacy transitive element of $C_s(X)$, $CR(X)$ or $CT(X)$ then f admits no periodic points, i. e. $|f^n| = \emptyset$ for all $n \in \mathbb{N}$.*

PROOF. If $n \in Per(f)$ is a closed, conjugacy invariant condition. So if it is true for f, then it is true for every element of the closure of the conjugacy class of f. In particular, since the adding machines have no periodic points, they cannot be the in the closure of such a conjugacy class. □

Example 3 - One Point Compactification of Incomparable Adding Machines:

It can happen that an element $f \in H(X)$ of residual type is the inverse limit of a sequence which does not satisfy the factoring condition. This shows that the factoring property is a property of the sequence itself and not just of the inverse limit.

Let p_i be the ith prime number in $\mathbb{N}$ counted in increasing order. Observe that for $n, k \geq 1$ a p_i^n loop cannot map to a p_j^k if $i \neq j$.Fix a positive number K. Let $(J_K, 1_K) = (\{1, \dots, K\}, \{(1,1), \dots, (K,K)\})$.

Now let $(I_1, \varphi_1) = (J_1, 1_1)$. For $n > 1$ let (I_n, ϕ_n) be the disjoint union of $(J_K, 1_K)$ and a p_i^n loop for $i = 1, \dots, Kn$. The map $p_{n+1,n} : (I_{n+1}\ \phi_{n+1}) \to (I_n, \phi_n)$ maps as follows:

- $(J_K, 1_K)$ is mapped to the trivial loop $(J_1, 1_1)\})$ in $(J_K, 1_K)$.
- The p_i^{n-1} loop is mapped to $(\{j\}, \{(j,j)\})$ for $j = 1, \dots, K$ and $i = Kn+j$.
- The p_i^{n+1} loop is mapped to the p_i^n loop for $i = 1, \dots, Kn$.

For every K the inverse sequence is an invertible Shimomura sequence and the inverse limits are all the same. The common limit is the one-point compactification of the disjoint union of $\{p_i^n\}$ adding machines.

For $K = 1$ it is easy to check that the inverse sequence satisfies the factoring property. Suppose $k > n > 1$ and p maps (I_k, ϕ_k) to (I_n, ϕ_n). Then p must map the p_i^k loop to the p_i^n loop for $i = 1, \dots, Kn = n$ and the remaining loops and the point $\{1\}$ must map to $(J_1, 1_1)$. From this, the factoring is easy to obtain. It follows that the inverse limit is of residual type.

On the other hand, for any $K > 1$ the sequence does not satisfy the factoring property. The map p can only factor when for $i = Kn + (k-n)K + j$ with $j = 1, \dots, K$, the p_i^k loop is mapped to $(\{j\}, \{(j,j)\})$. But if p does this then we can compose with a nontrivial permutation of $\{1, \dots, K\}$ to obtain a map which does not factor. Similarly, no subsequence satisfies the factoring property.

Example 4 - $\mathbf{CM(X)}$ and $\mathbf{CM(X;1)}$:

In general, a continuous surjective map on a compact metric space is chain mixing iff it is chain transitive and in addition it has no nontrivial loop as a factor. See [**1**] Chapter 3, Exercise 22, as well as [**14**]. In particular, if f is chain transitive and has a fixed point then it is chain mixing. Thus, $CM(X;1) = CM(X) \cap C_s(X;1) = CT(X) \cap C_s(X;1)$ is the set of chain mixing maps which admit a fixed point.

We recall a standard numerical lemma.

Lemma 4.12. *Given a pair of relatively prime, positive integers, every sufficiently large positive integer is a positive mixture of these two. In detail, for* $M, N \in \mathbb{N}$ *if* $1 = xM - yN$ *with* $x, y \geq 0$ *and* $K \geq M + (yN+1)N$ *then there exist* $a, b \in \mathbb{N}$ *such that* $K = aM + bN$.

Proof. For an integer $j \geq 0$ there are unique integers j_1, j_2 such that $j_1 \geq 0$, $N > j_2 \geq 0$ and $j = j_1 N + j_2$. Then $(M + (yN+1)N) + j = (1 + j_2 x)M + (1 + j_1 + (N - j_2)y)N$. □

Recall that an $N - M$ *wedge* is a system isomorphic to a $N - 0 - M$ dumbbell. We extend Proposition 4.1.

Proposition 4.13. *(a) If* (I, ϕ) *is a mixing system with* $I \in \mathfrak{I}$ *then there exists* $K \in \mathbb{N}$ *such that it is the factor of any* $N - M$ *wedge with* $N, M \geq K$.

(b) For an $N - M$ *wedge with* M, N *relatively prime there exists* K *such that it is a factor of any loop of length* $L \geq K$.

(c) If (I, ϕ) *is an* $N - M$ *wedge with* M, N *relatively prime then* ϕ *is mixing.*

PROOF. (a) There exists K_1 so that $\phi^N = I \times I$ for any $N \geq K_1$. There exists a loop of length K_2 which maps onto ϕ. It is then easy to see that if $M, N \geq K_1 + K_2$ and $i_0 \in I$ then there are loops of length M and N each of which maps onto ϕ via maps which send an arbitrary point in each to i_0. We can put these together to map the $N - M$ wedge onto ϕ.

(b) If $L = aM + bN$ with $a, b \in \mathbb{N}$ then any loop of length L maps onto an $N - M$ wedge. So the result follows from Lemma 4.12.

(c) Choose K as in (b). Let $L \geq K + M + N$. If $i, j \in I$ then there is a path of length $q \leq M + N$ from i to j and there is a loop of length $L - q \geq K$ from i to i. Combine to get a path of length L from i to j. □

Corollary 4.14. *(a) If $\{(N_n, M_n)\}$ is a sequence of relatively prime pairs of positive integers with $N_n \to \infty, M_n \to \infty$ as $n \to \infty$ then the $N_n - M_n$ wedges generate $\Gamma(CM(X))$.*

(b) If K is a positive integer and $\{M_n\}$ is a sequence of positive integers relatively prime to K and $M_n \to \infty$ as $n \to \infty$ then the $K - M_n$ wedges generate $\Gamma(CM(X) \cap C_s(X; Per \supset \{K\}))$.

PROOF. (a) From Proposition 4.13 (c) it follows that if N, M are relatively prime then an $N - M$ wedge is mixing. From Theorem 3.1 and Proposition 4.13(a) it follows that $\Gamma(CM(X))$ is generated by the sequence of $N_n - M_n$ wedges.

(b) Assume (I, ϕ) is mixing with $|\phi^K| \neq \emptyset$. If $i \in |\phi^K|$ then there is a K loop which maps into ϕ which begins and ends at i. If ϕ is mixing then it is a factor of L loops for sufficiently large L. Hence, it is a factor of a $K - M_n$ wedge if n is sufficiently large. Again Theorem 3.1 implies that any $K - M$ wedge with K, M relatively prime represented by some mixing homeomorphism f with $K \in Per(f)$. □

Recall that $f \in C_s(X)$ is weak mixing when $f \times f$ is topologically transitive on $X \times X$. A factor of a weak mixing system is weak mixing. On the other hand, a nontrivial loop is not a weak mixing system. It follows that a weak mixing system does not have any nontrivial loop factors and so is chain transitive.

Theorem 4.15. *$CM(X)$ is a conjugacy transitive subset of $C_s(X)$. There exists $f \in H(X)$ with (X, f) minimal, which is a conjugacy transitive element of $CM(X)$. If f is a conjugacy transitive element for $CM(X)$ then f does not admit any periodic points.*

The set of weak mixing, minimal homeomorphisms which are conjugacy transitive form a dense G_δ subset of $CM(X)$.

PROOF. Again it follows from Theorem 3.2 that $CM(X)$ is conjugacy transitive. In detail for each ϕ mixing, the theorem implies that $\Gamma^{-1}(\phi) \cap CM(X)$ contains a dense set of topologically mixing homeomorphisms. Hence, the G_δ set $\Gamma^{-1}(\phi) \cap WM(X)$ is dense in $CM(X)$. Intersecting over the countable set of mixing ϕ we obtain a G_δ set, dense in $CM(X)$, each member of which is a transitive element of $CM(X)$.

We will construct a conjugacy transitive element of $CM(X)$ which is minimal and so does not admit periodic points. Just as in Corollary 4.11 it follows that no transitive element admits periodic points.

Let $N_1 = 1 = M_1$ and let $A = \begin{pmatrix} a & b \\ c & d \end{pmatrix}$ be a matrix of positive integers with determinant ± 1. This implies that A^{-1} is an integer matrix. Inductively, let

$$(4.8) \qquad \begin{pmatrix} M_{n+1} \\ N_{n+1} \end{pmatrix} = \begin{pmatrix} a & b \\ c & d \end{pmatrix} \cdot \begin{pmatrix} M_n \\ N_n \end{pmatrix}.$$

Clearly, $min(M_{n+1}, N_{n+1}) \geq M_n + N_n > max(M_n, N_n)$ and so $M_n, N_n \to \infty$. Since A^{-1} is an integer matrix it follows that if d is a common divisor of M_{n+1} and N_{n+1} then it is a common divisor of M_n and N_n. So, by induction, (N_n, M_n) is a relatively prime pair for each n.

Let $p_{n+1,n}$ map each loop of the $N_{n+1} - M_{n+1}$ wedge onto the $N_n - M_n$ wedge with the common wedge point mapped to the wedge point. For example, the N_{n+1} loop maps a times around the N_n loop and b times around the M_n loop. By starting both loops around N_n and ending both around M_n it follows that, with the other choices arbitrary, $p_{n+1,n}$ realizes a $\pm$ determined lift. Hence, we obtain an invertible, pointed Shimomura sequence whose inverse limit (X, f) is a conjugacy transitive element of $CM(X)$.

We show that for every $x \in X$ $\{f^k(x) : k \in \mathbb{Z}\}$ is dense in X. Let $\alpha_n : (X, f) \to ([N_n + M_n - 1], \phi_{N_n - 0 - M_n})$ be the projection to the wedge at level n. Either $\alpha_{n+1}(x)$ or $\alpha_{n+1}(f(x))$ is not at the wedge point. Then $\alpha_{n+1}(f^k(x))$ for $k \in [-(N_{n+1} + M_{n+1} - 1), N_{n+1} + M_{n+1} - 1]$ maps at least onto one of the loops. Then $\alpha_n = p_{n+1,n} \circ \alpha_{n+1}$ maps onto both loops. It follows that (X, f) is a minimal system.

By Proposition 1.6 the set $MM(X)$ of minimal maps on X is a G_δ subset of $C_s(X)$. Such a map is chain transitive and so is chain mixing exactly when it does not factor over a nontrivial loop. Hence, $H(X) \cap MM(X) \cap CM(X)$ is the G_δ set of such homeomorphisms. Since it is conjugacy invariant and contains f as constructed above, it is dense in $CM(X)$. Since $WM(X)$ is a dense G_δ subset of $CM(X)$, the intersection $H(X) \cap MM(X) \cap WM(X)$ is dense in $CM(X)$. The conjugacy transitive elements $Trans(CM(X))$ is also a G_δ, dense since it is nonempty. Thus, $H(X) \cap MM(X) \cap WM(X) \cap Trans(CM(X))$ is a dense G_δ subset of $CM(X)$. □

Define $HM(X; 1!)$ to be the set of homeomorphisms f on X which admit a unique fixed point and if $x \in X$ is not fixed by f then the $\pm$ orbit $\{f^i(x) : i \in \mathbb{Z}\}$ is dense in X. Since X is perfect, it follows that such a homeomorphism is topologically transitive (see the Remark after Proposition 1.6). Since it has a fixed point it is chain mixing.

Theorem 4.16. *There exists $f \in HM(X; 1!)$ which is a transitive element for $CM(X; 1)$. The set $HM(X; 1!) \cap WM(X)$ is a dense, G_δ subset of $CM(X; 1)$.*

PROOF. As in Theorem 4.15, the set $WM(X; 1)$ is a dense G_δ subset of $CM(X; 1)$ by Theorem 3.2.

The $1 - M$ wedge, or *pointed loop* of length M, is given by $([1, M], \phi_{1-M})$ with

$$\phi_{1-M} = \{(i, i+1) : 1 \leq i < M\} \cup \{(1, 1), (M, 1)\}.$$

Let $\{N_n\}$ be a sequence in $\mathbb{N}$ with $N_1 = 1$ and $N_{n+1} \geq 2N_n + 2$ for all $n \in \mathbb{N}$. We define $p_{n+1,n} : ([1, N_{n+1}], \phi_{1-N_{n+1}}) \twoheadrightarrow ([1, N_n], \phi_{1-N_n})$ so that

- $p_{n+1,n}(i) = 1$ for $i = 1, 2, N_{n+1} - 1, N_{n+1}$.
- $p_{n+1,n}^{-1}(j)$ has at least two elements for all $j \in [1, N_n]$.

That is, the N_{n+1} loop at level $n+1$ is wrapped by $p_{n+1,n}$ at least twice around the N_n loop at level n. It follows from these two conditions that each $p_{n+1,n}$ realizes a $\pm$ directional lift. Hence, $\{p_{n+1,n} : ([1, N_{n+1}], \phi_{1-N_{n+1}}) \twoheadrightarrow ([1, N_n], \phi_{1-N_n})\}$ is an invertible, pointed Shimomura sequence. Let (X, f) be the limit with $\alpha_n : (X, f) \twoheadrightarrow ([1, N_n], \phi_{1-N_n})$ the projection to the nth coordinate.

Let $e \in X$ be the point with $\alpha_n(e) = 1$ for all n. Clearly, e is a fixed point for f. If $x \neq e$ then $\alpha_n(x) \neq 1$ for some n and so $\alpha_m(x) \neq 1$ for all $m \geq n$. It follows that α_m maps $\{f^i(x) : |i| \leq N_m$ onto $[1, N_m]$ for all $m \geq n$. This implies that the $\pm$ orbit of x is dense in X. That is, $f \in H(X; 1!) \subset CM(X; 1)$.

Since $\phi_{1-N_n} \in \Gamma(f)$ for all n, it follows from Corollary 4.14 (b) that $\Gamma(f) = \Gamma(CM(X; 1))$ and so f is a conjugacy transitive element of $CM(X; 1)$.

By Proposition 1.6 the set $H(X; 1!)$ is a G_δ subset of $CM(X; 1)$. Since it contains f and is conjugacy invariant it is dense in $CM(X; 1)$. Hence, $H(X; 1) \cap WM(X)$ is dense in $CM(X; 1)$ □

We did not bother considering $Trans(CM(X; 1))$ because, as we will now see, $CM(X; 1)$ is conjugacy minimal, i.e. every element of $CM(X; 1)$ is conjugacy transitive for $CM(X; 1)$.

We call a map f *periodic* when for some $n \in \mathbb{N}$, $|f^n| = X$, or, equivalently, $f^n = 1_X$.

Theorem 4.17. *(Shimomura) If $f \in C_s(X)$ is not periodic then the closure of the conjugacy class of f contains $CM(X; 1)$.*

PROOF. By Theorem 3.14 it suffices to show that $\Gamma(CM(X; 1)) \subset \Gamma(f)$. By Corollary 2.9 we can replace f, if necessary, by its natural lift $\tilde{f}$ to a homeomorphism. Notice that if $f^n(x) \neq x$ and $\tilde{x}$ is a lift of x then $\tilde{f}^n(\tilde{x}) \neq \tilde{x}$. In particular, if f is not periodic then $\tilde{f}$ is not.

So we assume $f \in H(X)$ is not periodic. We show that for every $M \in \mathbb{N}$, there is a $1 - M$ wedge in $\Gamma(f)$. By Corollary 4.14 (b) this will imply $\Gamma(CM(X; 1)) \subset \Gamma(f)$.

Since f is not periodic, there exists a point $x \in X$ such that $x, f(x), ..., f^{M+1}(x)$ are distinct points and so there is a clopen set U with $x \in U$ such that $\{f^i(U) : i = 0, ..., M + 1\}$ are pairwise disjoint. Let $\alpha : X \to I = \{1, ..., M\}$ with $\alpha^{-1}(i) = f^{i-1}(U)$ for $i = 2, ..., M$ and $\alpha^{-1}(1) = X \setminus \bigcup_{i=1}^{M-1} f^i(U)$. Clearly, f^α is the $1 - M$ wedge. □

Remark: Using a more delicate proof, Shimomura shows, in [**16**], that if f is not periodic and $g \in CM(X)$ then g is in the orbit closure of f iff $Per(f) \subset Per(g)$.

Corollary 4.18. *In $C_s(X)$ the sets $\{1_X\}$ and $CM(X; 1)$ are the only conjugacy minimal subsets.*

PROOF. If $f \in C_s(X)$ is chain transitive then it is not periodic. In particular, the conjugacy class of every element of $CM(X;1)$ is dense in $CM(X;1)$. That is, $CM(X;1)$ is a conjugacy minimal set.

The identity map commutes with every homeomorphism on X and so $\{1_X\}$ is a fixed point for the $H(X)$ action. It is therefore a conjugacy minimal set.

Clearly, $\Gamma(1_X) = \{(I, 1_I) : I \in \mathcal{I}\}$. It follows that 1_X is in the orbit closure of f iff for every positive integer N there exists a decomposition of X of cardinality N with each member an f invariant set.

Now suppose that that f is periodic with $f^n = 1_X$ for some $n \in \mathbb{N}$. Let $d_f(x,y) = max_{i=0}^{n-1} d(f^i(x), f^i(y))$. This replaces d with the topologically equivalent ultrametric d_f and the latter is f invariant, i.e. f is an isometry. Choose $\epsilon > 0$ small enough that the decomposition $\{V_\epsilon(x) : x \in X\}$ contains at least nN elements. Then, because f is an isometry, $\{\bigcup_{i=0}^{n-1} V_\epsilon(f^i(x)) : x \in X\}$ is a decomposition containing at least N elements each of which is f invariant. Hence, if f is periodic, then 1_X is in the closure of its conjugacy class.

Thus, every closed, conjugacy invariant set contains either $\{1_X\}$ or $CM(X;1)$. It follows that these are the only two conjugacy minimal sets. □

Notice that in Theorems 4.15 and 4.16 We did not describe any residual transitive elements. I conjecture that they do not exist. In the construction of the former result there is such a wide range of choices that it is hard to imagine a construction can yield a Shimomura sequence which satisfies the factoring property. The attempt in the latter case leads to an interesting semigroup.

Let $w = w(e, L)$ denote a finite word in the letters e, L of length N, denoted $|w|$, so that $w \in \{e, L\}^N$. Let k_w (and K_w) denote the number of letters e (resp. the number of letters L) in the word w. Define the affine function ℓ_w by $\ell_w(x) = k_w + K_w x$. Let $\mathcal{S}$ denote the set of words with $K_w > 0$. We define composition in $\mathcal{S}$ by $w = w_1 * w_2$ with $w(e, L) = w_2(e, w_1(e, L))$. That is, substitute for every occurrence of L in w_2 the word w_1. If $m = \ell_w(n)$ then w determines a unique map $p_{(w;m,n)} : ([1,m], \phi_{1-m}) \twoheadrightarrow ([1,n], \phi_{1-n})$ which starts with a map of 1 to 1. As one moves along the word, each e indicates a map of a digit to 1 and a move to the next digit in the domain, and each L indicates a map of n digits in order to $[1, n]$ followed by a move to the next digit, except that after the final letter of w is a stop instead of a move to the next digit. Conversely, it is clear that if $p : ([1,m], \phi_{1-m}) \twoheadrightarrow ([1,n], \phi_{1-n})$ for any $m \geq n$, then there is a unique word w so that $p = p_{(w;m,n)}$ which implies $m = \ell_w(n)$. It is easy to see that

$$
\begin{aligned}
\ell_{w_1 * w_2} &= \ell_{w_2} \circ \ell_{w_1}, \\
p_{(w_1 * w_2; m_2, n)} &= p_{(w_1; m_1, n)} \circ p_{(w_2; m_2, m_1)}, \\
\text{with} \quad m_1 &= \ell_{w_1}(n), \quad m_2 = \ell_{w_2}(m_1).
\end{aligned} \tag{4.9}
$$

For example, the word L is the identity in $\mathcal{S}$ with $\ell_L(x) = x$ and $p_{(L;n,n)}$ the identity on $([1,n], \phi_{1-n})$.

Let $\mathcal{S}'$ be the subsemigroup consisting of words w which begin and end with e and with $K_w \geq 2$. Let $\{w_n\}$ be a sequence of not necessarily distinct elements of $\mathcal{S}'$. Let $N_1 = 1$ and inductively define $N_{n+1} = \ell_{w_n}(N_n)$. Let $p_{n+1,n} = p_{(w_n; N_{n+1}, N_n)}$. This defines an invertible, pointed Shimomura sequence. The condition that the words begin and end with e is needed to get a $\pm$ directional lift. If the limit is (X,f) then $f \in H(X;1!)$.

The factoring property for the Shimomura sequence associated with $\{w_n\}$ is equivalent to following factorization property in the semigroup:

For every $n \in \mathbb{N}$ and $w \in \mathcal{S}$ there exists $m \geq n$ and $\tilde{w} \in \mathcal{S}$ such that

$$w * \tilde{w} \quad = \quad w_n * w_{n+1} * \cdots * w_m. \tag{4.10}$$

However, this is impossible for any sequence in $\mathcal{S}'$. Consider the finite list of positive integers which occur as the length of a run of L's in $w_a \in \mathcal{S}'$. Because w_a begins and ends with e it follows that for any $w_b \in \mathcal{S}$ these are exactly the length of runs in $w_a * w_b$. In particular, given n, if we choose $w \in \mathcal{S}'$ so that some length of runs occurs in w but not in w_n then the factorization is not possible.

Nonetheless, the semigroup is of interest in studying $H(X;1!)$.

Theorem 4.19. *With $N_1 = 1$ construct the pointed, invertible Shimomura sequence associated with the sequence $\{w_n = w : n \in \mathbb{N}\}$ for $w \in \mathcal{S}'$. Let (X,f) denote the limit so that $f \in H(X;1!)$.*

(a) If $w = eLeLLe$ or $w = eLLLe$ then (X,f) is topologically mixing.

(b) If $w = eLLe$, the simplest word in $\mathcal{S}'$, or $w = eLeLe$ then (X,f) is not weak mixing.

Proof. Let w^n be the nth power of the element w in the semigroup. Let $\alpha_n : X \to [1, N_n]$ be the projection map from the inverse limit.

(a): For $w = eLeLLe$ we have $\ell_w(x) = 3 + 3x$ so that $N_{n+1} = 3 + 3N_n$ which implies $N_n = \frac{1}{2}[5 \cdot 3^{n-1} - 3]$.

Claim: In w^n there occur between successive L's runs of e's of every length from 0 up to $2n-1$.

Proof of Claim: This is true by inspection for $n = 1$ since w contains runs of length 1 and 0 between successive L's. Since $w^{n+1} = w * w^n$ we replace the L's in w^n by w's to get w^{n+1}. This replaces each run of length k by a run of length $k+2$. Thus, we obtain runs of e's of length $2, \dots, 2n+1 = 2(n+1)-1$. Within each w occur runs of length 0 and 1. So the Claim follows by induction.

To show that (X,f) is topologically mixing, it suffices to show that for every $n \geq 2$ and every $i \in [1, N_n]$ there exists $K \in \mathbb{N}$ so that the hitting time set $N(U_i, U_i)$ contains every integer greater than K where $U_i = \alpha^{-1}(i)$. Since $\alpha^{-1}(1)$ contains the fixed point the result is clear for $i = 1$.

Let $K = N_n$ and let $t = N_n + r$ with $0 \leq r \leq 2k-1$ for some $k \in \mathbb{N}$. Consider the map $p_{n+k,n}$ In w^k there is a run of e's of length r between two successive L's. If j is the location of the i position in the first L of the pair within N_{n+k} then there exists x with $\alpha_{n+k}(x) = j$. Then $\alpha_n(x) = i$ and $\alpha_n(f^t(x)) = i$. In detail, iterating f moves x to position 1 in $[1, N_n]$ where it then remains for for r steps and then it is moved back to position i around the other end of the loop. Thus, $t \in N(U_i, U_i)$. Hence, (X,f) is topologically mixing.

For $w = eLLLe$, $N_{n+1} = 2 + 3N_n$ and so $N_n = 2 \cdot 3^{n-1} - 1$ with N_n odd for all n. For $n > 1$, as in the Claim above, w^n consists of blocks LLL separated by runs of e's of even length from 2 to $2(n-1)$. If we have in w^k two blocks of LLL separated by a run of e's of length $2r$, then at level $n+k$ we choose x at position i in the third L of the first block of the pair. We have $f^{t_1}(x)$ and $f^{t_2}(x)$ are at position

i in the first and second L's of the second block of the pair when $t_1 = N_n + 2r$ and $t_2 = 2N_n + 2r$. Since N_n is odd we can choose $K = 2N_n + 2$ and proceed as above. We leave the details to the reader.

(b): Since $\ell_w(x) = 2+2x$ we see that $N_{n+1} = 2+2N_n$ and so $N_n = 3\cdot 2^{n-1} - 2$. In particular, every $N_n$is even for $n \geq 2$. In w^n there occur between successive L's runs of e's only of even length from 0 to $2n-2$.

For any level $n \geq 2$ let $2 < i < N_n$. Assume that $t_1 \in N(U_i, U_i)$ and $t_2 \in N(U_i, U_{i-1})$. Let $k > max(t_1, t_2)$. If $x_1, x_2 \in U_i$ with $f^{t_1}(x_1) \in U_i$ and $f^{t_2}(x_2) \in U_{i-1}$ then at level $n+k$, neither x_1 nor x_2 lies in the portion of $[1, N_{n+k}]$ labeled by the last L in w^k because this is followed by k copies of e. This means that the pairs $x_1, f^{t_1}(x_1)$ and $x_2, f^{t_2+1}(x_2)$ all lie in positions i of L's in w^k. But the length N_n to which each L maps via $p_{n+k,n}$ is even and the number of e's between them are even. This implies that t_1 and t_2+1 are even and so t_2 is odd. It follows that $N(U_i, U_i) \cap N(U_i, U_{i-1}) = \emptyset$ and so (X, f) is not weak mixing.

For $w = eLeLe$, $N_{n+1} = 3 + 2N_n$ and so $N_n = 2^{n+1} - 3$ with N_n odd for all n. The runs of e's in w^n between two successive L's have odd length from 1 to $2n-1$. Since each N_n is odd, the number of steps is even between the same location not equal to 1 in different L's. Proceed as above. Details to the reader. □

Example 5 - Non Residual Factor

We conclude by observing that a factor of a homeomorphism of residual type need not be of residual type.

We showed in Proposition 4.5 that $(\tilde{X}, \tilde{f})$ which extends $T \times 1_C$ to the two-point compactification of $\mathbb{Z} \times C$ is of residual type. If we let (X, f) be the extension to the one-point compactification then f is a chain transitive homeomorphism with a fixed point and so its conjugacy class is dense in the minimal set $CM(X; 1)$. The system (X, f) is obviously a factor of $(\tilde{X}, \tilde{f})$. On the other hand, it is not topologically transitive and so its conjugacy class is disjoint from the G_δ set of topologically transitive maps with fixed points. By Theorem 4.16 the latter set is dense in $CM(X; 1)$. By the Baire category theorem dense G_δ subsets meet. Hence, the dense conjugacy class of f cannot be G_δ. That is, f is not of residual type.

References

[1] Ethan Akin, *The general topology of dynamical systems*, Graduate Studies in Mathematics, vol. 1, American Mathematical Society, Providence, RI, 1993. MR1219737

[2] Ethan Akin, *Good measures on Cantor space*, Trans. Amer. Math. Soc. **357** (2005), no. 7, 2681–2722 (electronic), DOI 10.1090/S0002-9947-04-03524-X. MR2139523

[3] Ethan Akin, Mike Hurley, and Judy A. Kennedy, *Dynamics of topologically generic homeomorphisms*, Mem. Amer. Math. Soc. **164** (2003), no. 783, viii+130, DOI 10.1090/memo/0783. MR1980335

[4] Ethan Akin, Eli Glasner, and Benjamin Weiss, *Generically there is but one self homeomorphism of the Cantor set*, Trans. Amer. Math. Soc. **360** (2008), no. 7, 3613–3630, DOI 10.1090/S0002-9947-08-04450-4. MR2386239

[5] Ethan Akin and Jeffrey D. Carlson, *Conceptions of topological transitivity*, Topology Appl. **159** (2012), no. 12, 2815–2830, DOI 10.1016/j.topol.2012.04.016. MR2942654

[6] Tatiane Cardoso Batista, Juliano dos Santos Gonschorowski, and Fabio Armando Tal, *Density of the set of symbolic dynamics with all ergodic measures supported on periodic orbits*, Fund. Math. **231** (2015), no. 1, 93–99, DOI 10.4064/fm231-1-6. MR3361237

[7] Nilson C. Bernardes Jr. and Udayan B. Darji, *Graph theoretic structure of maps of the Cantor space*, Adv. Math. **231** (2012), no. 3-4, 1655–1680, DOI 10.1016/j.aim.2012.05.024. MR2964619

[8] S. Bezuglyi, A. H. Dooley and J. Kwiatkoski, *Topologies on the group of homeomorphisms of the Cantor set*, (2004) ArXiv 0410507.

[9] Eli Glasner and Benjamin Weiss, *The topological Rohlin property and topological entropy*, Amer. J. Math. **123** (2001), no. 6, 1055–1070. MR1867311

[10] Eli Glasner and Benjamin Weiss, *Topological groups with Rokhlin properties*, Colloq. Math. **110** (2008), no. 1, 51–80, DOI 10.4064/cm110-1-2. MR2353899

[11] Michael Hochman, *Genericity in topological dynamics*, Ergodic Theory Dynam. Systems **28** (2008), no. 1, 125–165, DOI 10.1017/S0143385707000521. MR2380305

[12] Alexander S. Kechris and Christian Rosendal, *Turbulence, amalgamation, and generic automorphisms of homogeneous structures*, Proc. Lond. Math. Soc. (3) **94** (2007), no. 2, 302–350, DOI 10.1112/plms/pdl007. MR2308230

[13] Aleksandra Kwiatkowska, *The group of homeomorphisms of the Cantor set has ample generics*, Bull. Lond. Math. Soc. **44** (2012), no. 6, 1132–1146, DOI 10.1112/blms/bds039. MR3007646

[14] David Richeson and Jim Wiseman, *Chain recurrence rates and topological entropy*, Topology Appl. **156** (2008), no. 2, 251–261, DOI 10.1016/j.topol.2008.07.005. MR2475112

[15] Takashi Shimomura, *A topological dynamical system on the Cantor set approximates its factors and its natural extension*, Topology Appl. **159** (2012), no. 14, 3137–3142, DOI 10.1016/j.topol.2012.06.001. MR2948272

[16] Takashi Shimomura, *Aperiodic homeomorphisms approximate chain mixing endomorphisms on the Cantor set*, Tsukuba J. Math. **36** (2012), no. 2, 173–183. MR3058237

[17] Takashi Shimomura, *Special homeomorphisms and approximation for Cantor systems*, Topology Appl. **161** (2014), 178–195, DOI 10.1016/j.topol.2013.10.018. MR3132360

MATHEMATICS DEPARTMENT, THE CITY COLLEGE, 137 STREET AND CONVENT AVENUE, NEW YORK CITY, NEW YORK 10031

E-mail address: ethanakin@earthlink.net

Contemporary Mathematics
Volume **678**, 2016
http://dx.doi.org/10.1090/conm/678/13640

The mathematical work of John C. Oxtoby

Steve Alpern, Joseph Auslander, and Cesar E. Silva

ABSTRACT. We describe the mathematical work of John C. Oxtoby.

Figure 1. John C. Oxtoby at his desk in 1975. Photo by Bern Schwartz. Courtesy of Margaret Oxtoby.

1. Introduction

John Corning Oxtoby, whose worked spanned the years 1937–1989, made fundamental contributions to ergodic theory, measure theory and point set topology. Perhaps he is best known for the Oxtoby–Ulam theorem [**14**], which showed the prevalence of ergodicity for measure preserving homeomorphisms of compact manifolds.

2010 *Mathematics Subject Classification.* Primary 37A05, 37B05; Secondary 01A70.

This is a survey of the mathematical work of John Oxtoby. We cannot do justice to his whole work in the article and we concentrate on his best known work and the work that is most familiar to the authors.

2. Biographical Note

John C. Oxtoby was born September 14, 1910 in Saginaw, Michigan. He grew up in San Anselmo, California, where his father was President of San Francisco Theological Seminary. He studied physics and mathematics at the University of California, Berkeley, where he earned BA and MA degrees and went to Harvard for graduate school in mathematics. He was appointed to the Harvard Society of Fellows as a Junior Fellow, 1936–1939.

In 1927 he was struck with polio, affecting all four limbs, but learned to walk again over a period of a year. When asked in 1939 by a prospective employer if he could teach in the classroom, he replied "As for the mechanics of teaching, I have never found that lecturing involved any undue physical strain." He climbed up the three floors of stairs to his office at Bryn Mawr for most of his career, with the aid of leg braces. In 1945 he married Jean Shaffer and they had three children: Thomas, David, and Margaret.

Oxtoby was Class of 1897 Professor of Mathematics at Bryn Mawr College, where he chaired the department for 28 years and taught from 1939 to 1979. He also was a visiting professor at Yale University in 1960–1961. He was active in the American Mathematical Society, where he served on the Council (1952–1955) and was editor of its *Bulletin* (1956–1960); he wrote over 400 reviews for *Mathematical Reviews*, and was a much sought-after referee of mathematical papers. He was the Hedrick Lecturer of the Mathematical Association of America in 1956. In 1978, he received the Lindback Award for distinguished teaching. In 1980, he was awarded an honorary Doctor of Science degree by Widener University. John C. Oxtoby passed away January 2, 1991.

3. Category

A form of the Baire Category Theorem says that, in a complete metric space, the countable intersection of dense open sets is dense. Since a dense set cannot be empty, this result can be used to establish the existence of certain objects, particularly those that are defined by a countable number of conditions. One of the first such uses of the Baire Category Theorem was to show the existence of a continuous nowhere differentiable function. However, such functions were already known to exist by explicit construction.

Perhaps the first use of the Baire Category Theorem to establish the existence of an object whose existence was uncertain at the time, was Oxtoby's 1937 paper [**17**]. A continuous transformation T on a bounded closed region Ω of r-dimensional Euclidean space, $r \geq 2$, is called transitive (or topologically transitive) if the image of any neighborhood σ overlaps any other neighborhood σ' under some iterate of T. Besicovitch had earlier in the same year published a construction of a transitive homeomorphism of the plane [**2**], but for general bounded regions the existence of transitive homeomorphisms was unknown (Kerékjártó had a year earlier listed this as an important unsolved problem). Oxtoby's idea was to consider an apparently harder problem, the existence of a transitive measure-preserving homeomorphism, and to look at the space $M(\Omega)$ of such homeomorphisms on Ω, using a complete metric based on the uniform metrics for homeomorphisms and their inverses. This approach solved the first part of Problem 115 in the *Scottish Book* [**9**].

The main construction shows how to modify a homeomorphism $T \in M(\Omega)$ for which no iterate of a fixed neighborhood σ overlaps another given one σ', so that overlap occurs. Such neighborhoods can be reduced to a countable class by considering only rational spheres (rational coordinates of centers and rational radii), so this would show that such T are nowhere dense. This is accomplished by the following construction, in Oxtoby's words, [**17**]:

> Choose any point x_0 in σ and join any image $T^{k_0}x_0$ to σ' by a thin tube contained in the interior of Ω. Choose x_1 at distance nearly ε along this tube, and let it be so chosen so that an image $T^{k_1}(x_1)$ lies very near x_1. This can always be done, since points having this recurrence property are dense [this had been shown earlier by Birkhoff]. Choose a similar point x_2 nearly ε further along toward $\sigma'$and continue until a point x_p, in σ' is reached. Consider the finite sequence $x_0, T(x_0), \ldots, T^{k_0}(x_0), T(x_1), \ldots, T^{k_1}(x_1), \ldots, T(x_p), \ldots, T^{k_{p-1}}(x_{p-1}), x_p$. Join each $T^{k_i}(x_i)$ to $x_{i+1}$by a connected neighborhood of diameter less than ε, the neighborhoods being taken disjoint and each containing no other points of the sequence. ... Now within each neighborhood define a transformation S so as to take $T^{k_i}(x_i)$ into x_{i+1} while leaving everything outside the neighborhoods fixed. ...This S moves no point by more than ε, and is such that repeated application of ST takes x_0 into x_p. By choosing ε small, the transformation ST can be made to lie arbitrarily close to T, which completes the proof.
>
> It follows at once that transitive measure-preserving transformations of Ω exist, and the proof indicates how they can be defined step by step.

A later approach to obtaining the approximation ST is based on a 1971 paper of Peter Lax [**8**]. For simplicity, assume that Ω is the r dimensional unit cube, and $d_1, \ldots, d_n$ is a dyadic decomposition into closed subcubes sufficiently fine so that $d_1 \subset \sigma$ and $d_p \subset \sigma'$. Lax observed that the image of any set consisting of m dyadic cubes has volume m/n and therefore intersects at least m dyadic cubes. By the well-known Marriage Theorem of M. Hall, there is a permutation τ of the index set $N = \{1, \ldots, n\}$ such that $T(d_i)$ overlaps the set $d_{\tau(i)}$ for every i in N. Steve Alpern, in his dissertation supervised by Lax and Oxtoby, showed that it is easy to approximate τ by a cyclic permutation τ' in the sense that $d_{\tau(i)}$ is close to $d_{\tau'(i)}$ for all i in N. Let c_i denote the center of the cube d_i. Since $c_{\tau'(i)}$ is in the cube $d_{\tau'(i)}$ which is close to the cube $d_{\tau(i)}$ which overlaps $T(d_i)$, it follows that $c_{\tau'(i)}$ is close to $T(c_i)$. So we can easily define $S \in M(\Omega)$ so that S is uniformly small and maps each point $T(c_i)$ into the point $c_{\tau'(i)}$. Hence ST is close to T and cyclically permutes the dyadic centers c_i, $i \in N$. Clearly an iterate of the neighborhood σ under ST overlaps σ'. This argument is presented in more detail in the second edition of Oxtoby's classic book *Measure and Category* [**31**]. It is not significantly simpler than the original method.

After adopting Baire Category methods in this manner to show the existence and genericity of transitive homeomorphisms in $M(\Omega)$, Oxtoby worked with Stan Ulam to obtain the same result for ergodic homeomorphisms in $M(\Omega)$. This appeared in their 1941 article [**15**]. To obtain this result, it was required to approximate a given $T \in M(\Omega)$, or specifically $M(I^r)$, by a homeomorphism ST which had an orbit which not only entered both σ and σ' but moreover entered each cube d_i of a fine dyadic decomposition of I^r roughly the same number of times. This was accomplished by an elegant application of the Birkhoff Ergodic Theorem. Lax's idea can also be used, the same way as above, to produce the required approximation.

Figure 2. Oxtoby and Ulam as Junior Fellows at Harvard. Courtesy of Margaret Oxtoby.

Oxtoby and Ulam's theorem on the genericity of ergodicity among measure-preserving homeomorphisms was a landmark result in ergodic theory, showing that the so called "quasi ergodic hypothesis" was not unrealistic. This result was clearly ahead of its time, as it was not until 1944 that Halmos [**5**] was able to show that ergodicity was generic among measure-preserving automorphisms of a Lebesgue space. Later that same year Halmos established genericity for the smaller class of weak mixing automorphisms. In his ergodicity paper Halmos refers to the Oxtoby–Ulam paper and notes that his (Halmos) result was for a wider class of transformations and with a different topology (weak topology for automorphisms) than Oxtoby–Ulam (uniform convergence for homeomorphisms). Katok and Stepin [**7**] extended the Oxtoby–Ulam result for measure-preserving manifold homeomorphisms to the weak mixing case in 1970, thereby obtaining the measure-preserving homeomorphism version of Halmos's weak mixing genericity for automorphisms. That all four of these results were related by a more general theorem was established by Alpern in 1976 when he showed the following: any measure theoretic property generic for measure-preserving automorphisms (in the weak topology) is generic for measure-preserving homeomorphisms (with the topology of uniform convergence).

This Oxtoby–Ulam theory (i.e., generic for automorphisms is generic for measure-preserving homeomorphisms) has been extended to σ−compact manifolds. In 1979, Prasad [**35**] extended the result to R^n. Subsequent work of Alpern and Prasad between 1987 and 1990 showed that for general σ−compact manifolds, the dynamics induced by the homeomorphism at the ends of the manifold were critical in determining generic behavior for homeomorphisms. So, for example, ergodic homeomorphisms are always generic in the subspace of end-preserving homeomorphisms in $M(\Omega)$. Much of the Oxtoby–Ulam theory for both compact and non-compact manifolds was presented in 2000 in the book of Alpern and Prasad, *Typical Dynamics of Volume Preserving Homeomorphisms* [**1**].

In 1941, George D. Birkhoff wrote Oxtoby a note saying "I am writing to you and Ulam to express my warm congratulations on your recent paper in the *Annals of Mathematics*. It seems to me to be a splendid piece of work." Regarding his work with Oxtoby on this paper, Ulam described it in his autobiography [**36**] as "...one of the more important results that I had a part in."

4. Homeomorphic Measures

Let X and Y be topological spaces and μ and ν be measures defined on the Borel sets of X and Y, respectively. The measures μ and ν are said to be homeomorphic if there exits a homeomorphism h from X to Y that takes μ to ν (i.e., $\nu = \mu \circ h^{-1}$). In their *Annals* paper on genericity of ergodic homeomorphisms on manifolds [**15**] Oxtoby and Ulam also proved an interesting result on homeomorphic measures on the n-cube. As mentioned in [**15**, p. 886]: "The basic result ... was originally proposed by one of us in 1936 ... and a proof was obtained at that time by J. von Neumann, but was not published. The present proof, based on somewhat different considerations, was worked out subsequently. The result is here published for the first time." They proved that a Borel probability measure on the n-dimensional unit cube is homeomorphic to n-dimensional Lebesgue measure if and only if it is nonatomic, positive on nonempty open sets, and gives zero measure to the boundary. This solves Problem 170 in the *Scottish Book* [**9**]. Later, Oxtoby extended this to the metric space of irrationals in the unit interval in [**27**]. Also, with his doctoral student V. S. Prasad, they obtained the analog for homeomorphic measures on the Hilbert cube (in this case the third condition of giving zero measure to the boundary is not needed) [**33**]. Homeomorphic Bernoulli measures on the Cantor set were considered by his doctoral student Navarro-Bermúdez [**10**], and they returned to study them in [**11**]. More recently, homeomorphic Bernoulli measures have been characterized by Yingst in [**38**].

5. Ergodic Sets

The paper *Ergodic Sets* [**20**] relates topological dynamics and ergodic theory. It is based upon Oxtoby's invited address to a meeting of the American Mathematical Society in 1951. It contains material which is in the tool kit of every dynamicist.

The first several sections are an exposition of the theory of Kryloff and Bogoliouboff. (This was the first appearance in English of this theory.)

The setup is a self homeomorphism of a compact metric space. A point is called quasi-regular if the ergodic averages exist for all continuous functions, and so defines an invariant measure on the space. The set of quasi-regular points has invariant measure one as does the set of regular points, those for which the measure is ergodic and positive on open sets. Uniquely ergodic systems are defined and characterized.

Later sections consider generalizations to non-compact systems, and the decomposition of an arbitrary system as a direct sum of ergodic systems. There is also an elegant example of a homeomorphism that is minimal and not uniquely ergodic; this was later shown to have exactly two ergodic measures by Williams in [**37**].

The work of some of the giants of that era–Maharam, Ambrose, Halmos, Kakutani, Morse, Gottschalk, Hedlund, von Neumann–are referenced in this paper.

6. Banach–Mazur Game

In 1935, Mazur formulated the following two-person infinite game. The first player is dealt a subset A of a closed interval E, and the second player is dealt its complement $E \setminus A$. Then the first player chooses a closed interval I_1 in E and the second player chooses a closed interval $I_2 \subset I_1$. The players take turns choosing a sequence of intervals $I_1 \supset I_2 \supset I_3 \supset I_4 \supset \cdots$. The first player wins the game if the set $\bigcap_i I_i$ intersects A, otherwise the second player wins. Mazur conjectured that the first player has a winning strategy if and only if the set A is of first category, and the second player has a winning strategy if and only if for some subinterval I of E, $(E \setminus A) \cap I$ is of first category. Banach solved Mazur's conjecture but did not publish his proof. The first proof was published by Oxtoby [**22**], where he also generalized the game to arbitrary topological spaces. It follows that if A is the intersection of E with a Bernstein set, then neither A nor $E \setminus A$ contain an uncountable $\mathcal{G}_\delta$ set, so for all intervals I in E neither $A \cap I$ nor $(E \setminus A) \cap I$ is of first category. This means that the game is not determined. There is a beautiful exposition in Oxtoby's book [**31**]. This is listed as Problem 43 in the *Scottish Book* [**9**], where there is a discussion of current developments, in particular in the second edition.

7. Other work and *Measure and Category*

In [**23**], using the Continuum Hypothesis, Oxtoby constructs a Baire space (i.e., a space where the countable intersection of dense open sets is dense) such that its Cartesian square is not Baire. He also introduces the notion of pseudocomplete spaces and proves that a pseudocomplete space is Baire and products of pseudocomplete spaces are pseudocomplete. This has led to the definition of Oxtoby spaces [**34**]. Other work in point set topology appeared in [**24**], [**29**].

In [**6**], Kakutani and Oxtoby construct an extension of Lebesgue measure. Let Ω be the unit interval, $\mathcal{A}$ the σ-algebra of Lebesgue measurable sets in Ω and λ

be Lebesgue measure. It is known that the corresponding L^2 space of $(\Omega, \mathcal{A}, \lambda)$ is separable. The authors construct an extension $\bar{\mathcal{A}}$ of $\mathcal{A}$ and an extension $\bar{\lambda}$ of λ so that $(\Omega, \bar{\mathcal{A}}, \bar{\lambda})$ is non-separable and every measure-preserving automorphism of $(\Omega, \mathcal{A}, \lambda)$ extends to a measure-preserving automorphism of $(\Omega, \bar{\mathcal{A}}, \bar{\lambda})$ (in particular, the extension is translation invariant).

In [**4**], Erdős and Oxtoby answer a question of Maharam and A. Stone by showing that there is a partition of the plane where each set in the partition intersects in positive measure each positive measure rectangle of the plane, and each set in this partition retains the positive measure intersection property under arbitrary rotations. Further, they show that there cannot be a partition where each set in the partition retains this property under every area-preserving homeomorphism of the plane. They also characterize when this property holds for the product of two σ-finite measure spaces.

In the summer of 1968, his son David brought home several books on gas-liquid chromatography and showed them to his father, who recognized that the theory could be presented in a simpler and more elegant fashion based on the theory of stochastic processes. This led to his paper in *The Journal of Chemical Physics* [**26**], which John Oxtoby described as his paper with the most reprint requests.

His expository or partly expository work included [**12**], [**30**], [**25**], [**3**], [**32**]. He had two other joint works with Ulam, [**13**], [**14**]. He wrote a paper extending the nonsingular ergodic theorem of Hurewicz, [**19**]. His early work on Borel classes included [**16**]. We have not discussed his work on invariant measures in groups [**18**], flows on the torus [**21**], and a Luzin type theorem [**28**].

In 1957, Oxtoby was invited to be a Phillips Lecturer at Haverford College. This was the start for his book *Measure and Category* [**31**], that would be published by Springer Verlag in 1971 in both English and German, and that was translated into other languages. In it Oxtoby explores the use of the Baire Category Theorem as a method to prove existence, and the analogy between the notions of measure and (Baire) category. It is a model of elegance and exposition and has inspired generations of mathematicians.

Acknowledgement

The authors are indebted to Dan Mauldin for sharing with us the PowerPoint notes of his talk at the Oxtoby Centennial Conference, and to V.S. Prasad and the referee for a careful reading of the paper. For further information about that conference the reader may refer to web.williams.edu/Mathematics/csilva/Oxtoby_Centennial_Conf.htm.

We would like to thank John Oxtoby's daughter, Margaret, for sharing with us the photographs that we use and the originals of the letters from which we quote, and her copy of Ty Cunningham's note read at Bryn Mawr's faculty meeting on the occasion of John Oxtoby's passing.

References

[1] Steve Alpern and V. S. Prasad, *Typical dynamics of volume preserving homeomorphisms*, Cambridge Tracts in Mathematics, vol. 139, Cambridge University Press, Cambridge, 2000. MR1826331

[2] A. S. Besicovitch. A problem on topological transformations of the plane. *Fund. Math.*, 28:61–65, 1937.

[3] J. R. Choksi, S. J. Eigen, J. C. Oxtoby, and V. S. Prasad, *The work of Dorothy Maharam on measure theory, ergodic theory and category algebras*, Measure and measurable dynamics (Rochester, NY, 1987), Contemp. Math., vol. 94, Amer. Math. Soc., Providence, RI, 1989, pp. 57–71, DOI 10.1090/conm/094/1012978. MR1012978

[4] Paul Erdős and John C. Oxtoby, *Partitions of the plane into sets having positive measure in every non-null measurable product set*, Trans. Amer. Math. Soc. **79** (1955), 91–102. MR0072928

[5] Paul R. Halmos, *In general a measure preserving transformation is mixing*, Ann. of Math. (2) **45** (1944), 786–792. MR0011173

[6] Shizuo Kakutani and John C. Oxtoby, *Construction of a non-separable invariant extension of the Lebesgue measure space*, Ann. of Math. (2) **52** (1950), 580–590. MR0037335

[7] A. B. Katok and A. M. Stepin, *Metric properties of homeomorphisms that preserve measure* (Russian), Uspehi Mat. Nauk **25** (1970), no. 2 (152), 193–220. MR0260974

[8] Peter D. Lax, *Approximation of meausre preserving transformations*, Comm. Pure Appl. Math. **24** (1971), 133–135. MR0272983

[9] *The Scottish Book*, Birkhäuser, Boston, Mass., 1981. Mathematics from the Scottish Café; Including selected papers presented at the Scottish Book Conference held at North Texas State University, Denton, Tex., May 1979; Edited by R. Daniel Mauldin. MR666400

[10] Francisco J. Navarro-Bermúdez, *Topologically equivalent measures in the Cantor space*, Proc. Amer. Math. Soc. **77** (1979), no. 2, 229–236, DOI 10.2307/2042644. MR542090

[11] Francisco J. Navarro-Bermúdez and John C. Oxtoby, *Four topologically equivalent measures in the Cantor space*, Proc. Amer. Math. Soc. **104** (1988), no. 3, 859–860, DOI 10.2307/2046806. MR939966

[12] J. C. Oxtoby, *Horizontal chord theorems*, Amer. Math. Monthly **79** (1972), 468–475. MR0299735

[13] J. C. Oxtoby and S. M. Ulam. On the equivalence of any set of first category to a set of measure zero. *Fundamenta Mathematicae*, 31(1):201–206, 1938.

[14] J. C. Oxtoby and S. M. Ulam, *On the existence of a measure invariant under a transformation*, Ann. of Math. (2) **40** (1939), 560–566. MR0000097

[15] J. C. Oxtoby and S. M. Ulam, *Measure-preserving homeomorphisms and metrical transitivity*, Ann. of Math. (2) **42** (1941), 874–920. MR0005803

[16] J. C. Oxtoby, *The category and Borel class of certain subsets of $\mathcal{L}_p$*, Bull. Amer. Math. Soc. **43** (1937), no. 4, 245–248, DOI 10.1090/S0002-9904-1937-06525-6. MR1563516

[17] John C. Oxtoby. Note on transitive transformations. *Proc Natl Acad Sci U S A*, 23:443–446, 1937.

[18] John C. Oxtoby, *Invariant measures in groups which are not locally compact*, Trans. Amer. Math. Soc. **60** (1946), 215–237. MR0018188

[19] John C. Oxtoby. On the ergodic theorem of Hurewicz. *Ann. of Math. (2)*, 49:872–884, 1948.

[20] John C. Oxtoby, *Ergodic sets*, Bull. Amer. Math. Soc. **58** (1952), 116–136. MR0047262

[21] John C. Oxtoby, *Stepanoff flows on the torus*, Proc. Amer. Math. Soc. **4** (1953), 982–987. MR0060812

[22] John C. Oxtoby, *The Banach-Mazur game and Banach category theorem*, Contributions to the theory of games, vol. 3, Annals of Mathematics Studies, no. 39, Princeton University Press, Princeton, N. J., 1957, pp. 159–163. MR0093741

[23] John C. Oxtoby, *Cartesian products of Baire spaces*, Fund. Math. **49** (1960/1961), 157–166. MR0140638

[24] John C. Oxtoby, *Spaces that admit a category measure*, J. Reine Angew. Math. **205** (1960/1961), 156–170. MR0140637

[25] John C. Oxtoby, *On two theorems of Parthasarathy and Kakutani concerning the shift transformation*, Ergodic Theory (Proc. Internat. Sympos., Tulane Univ., New Orleans, La., 1961), Academic Press, New York, 1963, pp. 203–215. MR0160875

[26] John C. Oxtoby, *Stochastic models for chromatography*, J. Chem. Phys. **51** (1969), 3886–3890. MR0418797

[27] John C. Oxtoby, *Homeomorphic measures in metric spaces*, Proc. Amer. Math. Soc. **24** (1970), 419–423. MR0260961

[28] John C. Oxtoby, *Approximation by measure-preserving homeomorphisms*, Recent advances in topological dynamics (Proc. Conf. Topological Dynamics, Yale Univ., New Haven, Conn.,

1972; in honor of Gustav Arnold Hedlund), Springer, Berlin, 1973, pp. 206–217. Lecture Notes in Math., Vol. 318. MR0404585

[29] John C. Oxtoby, *The kernel operation on subsets of a T_1-space*, Fund. Math. **90** (1975/76), no. 3, 275–284. MR0431090

[30] John C. Oxtoby, *Diameters of arcs and the gerrymandering problem*, Amer. Math. Monthly **84** (1977), no. 3, 155–162. MR0433333

[31] John C. Oxtoby, *Measure and category*, 2nd ed., Graduate Texts in Mathematics, vol. 2, Springer-Verlag, New York-Berlin, 1980. A survey of the analogies between topological and measure spaces. MR584443

[32] John C. Oxtoby, *Biographical note*, Measure and measurable dynamics (Rochester, NY, 1987), Contemp. Math., vol. 94, Amer. Math. Soc., Providence, RI, 1989, pp. xvii–xviii. MR1012971

[33] John C. Oxtoby and Vidhu S. Prasad, *Homeomorphic measures in the Hilbert cube*, Pacific J. Math. **77** (1978), no. 2, 483–497. MR510936

[34] Roberto Pichardo-Mendoza, *Notes on Oxtoby spaces and pseudocompleteness*, Topology Proc. **46** (2015), 117–134. MR3224170

[35] V. S. Prasad, *Ergodic measure preserving homeomorphisms of* $\mathbf{R}^n$, Indiana Univ. Math. J. **28** (1979), no. 5, 859–867, DOI 10.1512/iumj.1979.28.28060. MR551151

[36] S. M. Ulam, *Adventures of a mathematician*, Charles Scribner's Sons, New York, 1976. MR0485098

[37] Susan Williams, *Toeplitz minimal flows which are not uniquely ergodic*, Z. Wahrsch. Verw. Gebiete **67** (1984), no. 1, 95–107, DOI 10.1007/BF00534085. MR756807

[38] Andrew Q. Yingst, *A characterization of homeomorphic Bernoulli trial measures*, Trans. Amer. Math. Soc. **360** (2008), no. 2, 1103–1131 (electronic), DOI 10.1090/S0002-9947-07-04431-5. MR2346485

Operational Research & Management Sciences Group, Warwick University
E-mail address: Steve.Alpern@wbs.ac.uk

Department of Mathematics, University of Maryland
E-mail address: jna@math.umd.edu

Department of Mathematics and Statistics, Williams College
E-mail address: csilva@williams.edu

Contemporary Mathematics
Volume **678**, 2016
http://dx.doi.org/10.1090/conm/678/13641

The Čech cohomology and the spectrum for 1-dimensional tiling systems

Tetyana I. Andress and E. Arthur Robinson, Jr.

Abstract. This paper discusses the relation between three groups: the dynamical cohomology $H(T)$ of a substitution shift dynamical system T, the first integer Čech cohomology group $\check{H}^1(Y)$ of the suspension space Y over T, and the continuous point-spectrum $\mathbb{E}(F)$ of the suspension flow F^t. In particular, the eigenfunctions of F^t embed as a subgroup of $\check{H}^1(Y)$. We also study a real-valued functional W on $\check{H}^1(Y)$, called the winding number, which assigns each eigenfunction its eigenvalue. In the case that W is injective, we call F^t *cohomological ergodic*, and if the image of W is equal to $\mathbb{E}(F)$, we say F^t has *cohomological pure point spectrum*.

1. Introduction

In this paper we consider 1-dimensional tiling flows F^t corresponding to 1-dimensional tile substitutions S. We also consider substitution shift dynamical systems T corresponding to primitive aperiodic substitutions σ. Our purpose is to explore the relation between two well known topological invariants for these dynamical systems: their point-spectrum, as studied, for example in [**Ho-86, Ra-90, FMN-96, So-97, AI-01, CSg-01, SiSo-02, HS-03, FiHR-03, Sg-03, CS-03, R-04, BK-06, BBK-06**], and their cohomology, studied, for example in [**AP-98, BD-01, FoHK-02, BD-02, CS-03, CS-06, S-08, BD-08, BKeS-12**]. The purpose of this paper is to survey what is known about this topic, as well as to add some new observations and results.

We proceed as follows. We let T be a strictly ergodic homeomorphism of a Cantor space X. For a continuous, positive, real-valued function g on X, we let F^t be the corresponding suspension flow, and let Y be the 1-dimensional suspension-space on which F^t acts. We study three abelian groups associated to the dynamical systems T and F^t. The first is the dynamical cohomology $H(T)$ of T, which is defined as the integer-valued continuous functions modulo the coboundaries. The second is the 1st integer Čech cohomology $\check{H}^1(Y)$ of the suspension space Y over T. And the third is the group $E(F)$ of continuous eigenfunctions for F^t, with pointwise multiplication. For each of these groups, there is a natural real-valued functional, and in each case we study its image in $\mathbb{R}$. The functional L on $H(T)$ is the integral,

2010 *Mathematics Subject Classification.* Primary 37B50; Secondary 37A30.

Key words and phrases. Substitution shift, tiling flow, spectrum, cohomology.

This research was partially supported by a grant from the Simons Foundation (award number 244739) to E. Arthur Robinson, Jr.

and its image, called the "measure group", is denoted $\mathbb{H}(X) \subseteq \mathbb{R}$. The functional W on $\check{H}^1(Y)$ is called the "Schwartzman winding number" (see Section 3.3). Its image turns out to be the same subgroup $\mathbb{H}(X) \subseteq \mathbb{R}$. The functional on the eigenfunctions $E(F)$ is the map V that assigns the eigenvalue $V(f)$ to to an eigenfunction f. For a flow F^t the set of eigenvalues $\mathbb{E}(F)$ is a subgroup of $\mathbb{R}$.

In the first part of the paper, we investigate the situation in which $\ker(L)$ (equivalently $\ker(W)$) is trivial. We call this "cohomological ergodicity". In this case, the subgroup $\mathbb{H}(T) \subseteq \mathbb{R}$ is isomorphic to $H(T)$ and to $\check{H}^1(Y)$. We show in Theorem 2.7 that cohomological ergodicity holds for T when the substitution σ is primitive, aperiodic, irreducible, and has a "common prefix". Examples T that fail to satisfy cohomological ergodicity are also discussed.

The second part of the paper is based on the observation from [**Sc-57**] that the eigenfunctions $E(F)$ of the flow F^t embed into $\check{H}^1(Y)$. Schwartzman's paper [**Sc-57**], published in 1957, was arguably the first serious application of algebraic topology to dynamical systems theory. We ask: how much larger is $\check{H}^1(Y)$ than $E(F)$? In the case that $\check{H}^1(Y)$ is entirely accounted for by eigenfunctions $E(F)$, we say F^t has *cohomological pure point spectrum.* In Theorem 4.2, we show that a primitive aperiodic Pisot substitution with a common prefix has cohomological pure point spectrum. We also give examples of cohomologically ergodic tile substitutions for which the inclusion of $E(F) \subseteq \check{H}^1(Y)$ is strict.

Although much of this paper is a survey, it is also an exposition of some previously unpublished results. Most of these come from the first author's 2007 Ph.D. dissertation [**An-07**]. Those parts of the theory needed to support the new results are worked out in more detail, whereas the other parts are relegated to the "Notes" throughout, where references are provided.

The second author wishes to thank Anne Siegel for some helpful suggestions. Both authors thank the referee for many useful comments that vastly improved this paper.

2. Discrete Dynamical Systems

2.1. Cantor Dynamical Systems. A *measured Cantor dynamical system* (X, T, μ) is a homeomorphism of T a Cantor space X (a space that is homeomorphic to the Cantor set), together with a T-invariant Borel probability measure μ on X. We always assume here that T is *strictly ergodic*, meaning T is *minimal* (all orbits are dense) and *uniquely ergodic* (μ is the unique T-invariant Borel probability measure).

Let $C(X, \mathbb{Z})$ denote the set of continuous integer-valued functions on X (non-trivial since X is a Cantor set). The *coboundaries* $B(T)$ are the functions of the form $n(x) = p(Tx) - p(x)$, where $p \in C(X, \mathbb{Z})$. We define the *dynamical cohomology group* of T by $H(T) := C(X, \mathbb{Z})/B(T)$. We define $L : C(X, \mathbb{Z}) \to \mathbb{R}$ by $L(n) = \int_X n \, d\mu$. Any $n(x) \in I(T) := \ker(L)$ is called an *infinitesimal.*

Central to this discussion will be the *measure group*, which we denote by $\mathbb{H}(T) := L(H(T))$, namely the image of L. Note that $\mathbb{H}(T)$ is the additive subgroup of $\mathbb{R}$ generated by the measures $\mu(E)$ of the clopen sets (or equivalently, measures of cylinder sets) $E \subseteq X$. Since μ is T-invariant, $B(T) \subseteq I(T)$, which implies $L : H(T) \to \mathbb{H}(T)$ is well defined and surjective.

DEFINITION 2.1. *If L is injective, we say T is* cohomologically ergodic[1] .

Clearly cohomological ergodicity is equivalent to $B(T) = I(T)$, and implies $\mathbb{H}(T)$ is isomorphic to $H(T)$. Later, we will give examples of strictly ergodic T that are, and examples that are not, cohomologically ergodic. Cohomological ergodicity is an instance of the general phenomenon called "stability" in [**KR-01**]. In particular, it means that $B(T)$ is closed in $C(X,\mathbb{Z})$.

NOTE 1. The dynamical cohomology group $H(T)$ is also sometimes called $K_0(T)$ because it is isomorphic to $K_0(C(X) \ltimes_T \mathbb{Z})$, the 0th K-group of the crossed product C^*-algebra for T (see e.g., [**AP-98**]). The unique ergodicity of T implies that there is a unique real-valued "trace" τ on this C^*-algebra, and the image of $K_0(C(X) \ltimes_T \mathbb{Z})$ under τ is called the *gap labeling group* because it is related to gaps in the spectrum of a corresponding discrete Schrödinger operator (see [**Be-92**]). The 1-dimensional *gap labeling theorem* [**Be-92**] says the image of τ equals $\mathbb{H}(T)$. The higher-dimensional version of this (for $\mathbb{Z}^d$ and $\mathbb{R}^d$ actions), was originally called Bellisard's *gap labeling conjecture*, but is now a theorem: [**BO-03, KaP-03, BeBG-06**].

Let (X, T, μ) be a strictly ergodic Cantor dynamical system. Define the *orbit* of $x \in X$ by $O_T(x) = \{T^n x : n \in \mathbb{Z}\}$. Another homeomorphism R on X is said to be *orbit equivalent* to T if $O_T(x) = O_R(x)$ for all $x \in X$. In this case there is a *cocycle* $n : X \to \mathbb{Z}$ such that $Rx = T^{n(x)}x$. Note that the cocycle $n(x)$ is not necessarily continuous. The *full group* of T, denoted $[T]$ is the set of all R that are orbit equivalent to T. It is easy to see that if $R \in [T]$ then (X, R, μ) is a strictly ergodic Cantor dynamical system. If $R \in [T]$, and the cocycle $n(x)$ is continuous, we say R is in the *topological full group* of T, and write $R \in [[T]]$. Equivalently, R and T are *topologically orbit equivalent.*

Two clopen sets E_1 and E_2 are called *T-equivalent* if $E_2 = RE_1$ for some $R \in [[T]]$. This implies $\mu(E_1) = \mu(E_2)$. Following [**BzK-00**], we say T is *saturated* if whenever $\mu(E_1) = \mu(E_2)$ for E_1, E_2 clopen, there exists $R \in [[T]]$ so that $E_2 = UE_1$.

LEMMA 2.2 ([**BzK-00**]). *A strictly ergodic Cantor dynamical system (X, T, μ) is cohomologically ergodic (i.e., satisfies $B(T) = I(T)$) if and only if it is saturated.*

NOTE 2. A strictly ergodic Cantor dynamical system T need not be cohomologically ergodic. It is shown in [**BzK-00**] that the Chacon transformation T, realized as a strictly ergodic shift on $X \subseteq \{0,1\}^{\mathbb{Z}}$ is not cohomologically ergodic. Similarly, argumets in [**BKeS-12**] show that the Morse substitution dynamical system T is not cohomologically ergodic. Examples of T that are cohomologically ergodic are discussed in Sectiom 5.

NOTE 3. For T a minimal homeomorphism of a Cantor set, but without the assumption of unique ergodicity, extend the previous definition of $I(T)$ be the set of integer functions with integral zero for all T-invariant measures. Define $G(T) = C(X, \mathbb{T})/I(T)$. Let $G(T)^+ \subseteq G(T)$ (and $H(T)^+ \subseteq H(T)$) be the semigroups generated by the positive functions. Let $[1]$ denote the class of 1. Then $(G(T), G(T)^+, [1])$ (and $(H(T), H(T)^+, [1])$) are called the *ordered dimension group* (and the *ordered cohomology group*) of T. It is shown in [**GPS-95**] that, up to order

[1]The reason calling this "ergodicity" will be discussed after the definition of the flow version, Definition 3.4.

isomorphism, these ordered groups are complete invariants for *strong orbit equivalence* (and for *orbit equivalence*). In particular, strong orbit equivalence (which lies between orbit equivalence and topological orbit equivalence) allows $n(x)$ to have at most one discontinuity.

NOTE 4. It is shown in [**Bo-83**] (see also [**BoT-98**]) that if $Rx = T^{n(x)}x$ for continuous $n(x)$ (topological orbit equivalence) then $R = V^{-1}TV$ or $R = V^{-1}T^{-1}V$ for some homeomorphism V. The latter is called *flip-conjugacy.*

2.2. Shifts. Let $\mathcal{A} = \{0, 1, \ldots, d-1\}$, $d > 1$. Let $\mathcal{A}^{\mathbb{Z}}$ be the 2-sided full-shift on $\mathcal{A}$, provided with the product topology, and let T be the *shift* (homeomorphism), defined $(Tx)_i = x_{i-1}$. A *shift space* $X \subseteq \mathcal{A}^{\mathbb{Z}}$ is a closed T-invariant subset of $\mathcal{A}^{\mathbb{Z}}$, and we restrict T to X. We always assume X is uncountable, which implies X is a Cantor space.

Let $\mathcal{A}^* = \cup_{n\geq 0}\mathcal{A}^n$ denote the set of all finite words in $\mathcal{A}$. The *language* of X is the subset $\mathcal{L} \subseteq \mathcal{A}^*$ of words that appear in some $x \in X$. Each $w = w_0 w_1 \ldots w_{n-1} \in \mathcal{L}$ defines the *cylinder set* $\lfloor w \rfloor = \{x \in X : x_j = w_j, 0 \leq j < n\}$. Cylinder sets are clopen and provide a basis for the topology on X.

NOTE 5. Any T-invariant Borel probability measure μ on a subshift X is completely determined by its values $\mu(E)$ on cylinder sets $E = \lfloor w \rfloor$, $w \in \mathcal{L}$. If T is also uniquely ergodic, then any $w \in \mathcal{L}$ occurs in every $x \in X$ with a common positive frequency, which we denote by $\ell(w) > 0$. These frequencies satisfy $\ell(w) = \mu(\lfloor w \rfloor)$ (see [**Qu-87**]). It follows that $\mathbb{H}(T)$ is the subgroup of $\mathbb{R}$ generated by the frequencies of words in $\mathcal{L}$. For this reason, $\mathbb{H}(T)$ is sometimes called the *frequency module* (see e.g., [**Be-92**]).

2.3. Substitutions. A *substitution* is a mapping $\sigma : \mathcal{A} \to \mathcal{A}^* := \cup_{n\geq 1}\mathcal{A}^n$. The *incidence matrix* Q for σ is the matrix with entries $q_{a,b}$ equal to the number of times a occurs in $\sigma(b)$. We assume $Q^n > 0$ for some $n \geq 1$, and call Q (and σ) *primitive.* For primitive Q, the *Perron-Frobenius Theorem* (see [**LM-95**]), says Q has a simple positive eigenvalue $\lambda > 1$, called the *Perron-Frobenius* eigenvalue, which satisfies $|\theta| < \lambda$ for all other eigenvalues. We normalize the corresponding *Perrron-Frobenius* eigenvectors: (left) $Q^t\mathbf{m} = \lambda\mathbf{m}$ and (right) $Q\mathbf{h} = \lambda\mathbf{h}$, so that $\mathbf{m} \cdot \mathbf{1} = 1$ and $\mathbf{h} \cdot \mathbf{m} = 1$.

Let $\mathcal{L}$ be the language generated by the set of all sub-words of $\{\sigma^n(a) : a \in \mathcal{A}, n \geq 0\}$. There is a unique shift space $X \subseteq \mathcal{A}^{\mathbb{Z}}$ whose language is $\mathcal{L}$ (see [**Qu-87**] or [**Fg-02**] for details, and and various alternative definitions of X). We let T be the shift map restricted to X. We say a primitive substitution σ is (shift) *aperiodic* if T has no periodic points. This implies X is uncountable, and thus a Cantor space. Since σ is a primitive, aperiodic substitution it follows that T is *strictly ergodic* (see [**Qu-87**]). Also, every eigenfunction f for T (see Section 3.5) can be chosen to be continuous [**Ho-86**]. We will refer to this property by saying T is *homogeneous* (see [**R-04**]). We call the strictly ergodic Cantor dynamical system (X, T, μ) the *substitution shift dynamical system* corresponding to σ.

Here are a few additional properties that σ may or may not have:

(1) σ is *irreducible* if the characteristic polynomial $q(z)$ of Q is irreducible. This implies λ is irrational.
(2) σ is *unimodular* if $\det(Q) = \pm 1$.

(3) σ is a *Pisot* substitution if the Perron-Frobenius eigenvalue λ is a Pisot number. A Pisot number is a real algebraic integer $\lambda > 1$ whose conjugates satisfy $|\theta| < 1$ (see e.g., [**AI-01, BD-02, HS-03**]).

(4) σ has a *common prefix* $p \in \mathcal{A}$ (or *common suffix* $s \in \mathcal{A}$) if for each $a \in \mathcal{A}$, there exists $u_a \in \mathcal{A}^*$ so that $\sigma(a) = pu_a$ (or $\sigma(a) = v_a s$).

NOTE 6. A substitution σ is called *proper* (see [**DHS-99**]) if it has both a common prefix and a common suffix. For any primitive aperiodic substitution σ with its dynamical system T, there is a proper substitution σ' with its dynamical system T' topologically conjugate to T. However, σ' generally has a much larger alphabet that σ, and is usually not irreducible, even if σ is (see [**DHS-99**]).

For $\mathbf{v} \in \mathbb{R}^d$, we write $\mathbb{Z}[\mathbf{v}] := \mathbf{v} \cdot \mathbb{Z}^d$, and note that $\mathbb{Z}[\mathbf{v}] \cong \mathbb{Z}^e$ for some $0 \leq e \leq d$. For $\lambda \in \mathbb{R}$, we write $\mathbb{Z}[\lambda]$ for the additive abelian group generated by $\{\lambda^k : k = 0, 1, 2, \dots\}$.

LEMMA 2.3. *Let Q be a primitive matrix with Perron-Frobenius eigenvalue λ and normalized left Perron-Frobenius eigenvector $\mathbf{m}$. Then $\lambda^{-k}\mathbb{Z}[\mathbf{m}] \subseteq \lambda^{-(k-1)}\mathbb{Z}[\mathbf{m}]$. If Q is irreducible, $\mathbb{Z}[\mathbf{m}] \cong \mathbb{Z}[\lambda] \cong \mathbb{Z}^d$. Moreover, Q is unimodular if and only if λ is a unit, if and only if $\lambda^{-k}\mathbb{Z}[\mathbf{m}] = \mathbb{Z}[\mathbf{m}]$ for all k.*

PROOF. For $\lambda^{-k}\mathbf{m} \cdot \mathbf{n} \in \lambda^{-k}\mathbb{Z}[\mathbf{m}]$, we have $\lambda^{-k}\mathbf{m} \cdot \mathbf{n} = \lambda^{-(k+1)}Q^t\mathbf{m} \cdot \mathbf{n} = \lambda^{-(k+1)}\mathbf{m} \cdot Q\mathbf{n} \in \mathbb{Z}[\lambda^{-(k+1)}\mathbf{m}]$. An elementary calculation shows that Q irreducible implies the entries of $\mathbf{m}$ are rationally independent. In the unimodular case, Q is invertible and $\lambda^{-k}\mathbb{Z}[\mathbf{m}] = \mathbb{Z}[\mathbf{m}]$ for all k. If Q is not unimodular, the inclusions $\lambda^{-k}\mathbb{Z}[\mathbf{m}] \subseteq \lambda^{-(k+1)}\mathbb{Z}[\mathbf{m}]$ are proper. □

We define the *Perron-Frobenius group* of a primitive matrix Q by

$$\mathrm{PF}_Q := \mathbb{Z}[\lambda^{-1}]\,\mathbb{Z}[\mathbf{m}] = \bigcup_{k=0}^{\infty} \lambda^{-k}\mathbb{Z}[\mathbf{m}]. \tag{2.1}$$

It follows from Lemma 2.3 that PF_Q is finitely generated if and only if Q is unimodular, and in this case $\mathrm{PF}_Q = \mathbb{Z}[\mathbf{m}] \cong \mathbb{Z}^d$.

NOTE 7. The Perron-Frobenius group is closely related to a version of the *dimension group*[2] Δ_Q that is studied in [**LM-95, Ki-98**]. Assume that Q is primitive and irreducible. This implies Q is nonsingular, and we have

$$\Delta_Q := \bigcup_{k=0}^{\infty} Q^{-k}\mathbb{Z}^d \subseteq \mathbb{Q}^d,$$

(this is simpler than the definition in [**LM-95, Ki-98**], which does not assume Q is nonsingular). In the unimodular case $\Delta_Q = \mathbb{Z}^d$. An easy calculation shows that the Perron Frobenius group (2.1) satisfies $\mathrm{PF}_Q = \mathbf{m} \cdot \Delta_Q$.

2.4. Kakutani-Rohlin partitions. A *semi-partition* $\mathcal{P} = \{P_0, \dots, P_{n-1}\}$ on a Cantor space X is a collection of pairwise disjoint clopen sets in X. A *partition* is a semi-partition such that $\cup_{P\in\mathcal{P}} P = X$. A partition $\mathcal{Q}$ *refines* partition $\mathcal{P}$, denoted $\mathcal{Q} \geq \mathcal{P}$, if for each $Q \in \mathcal{Q}$ there is a $P \in \mathcal{P}$ with $Q \subseteq P$. A sequence $\mathcal{P}_k$ of partitions is *refining* if $\mathcal{P}_{k+1} \geq \mathcal{P}_k$ for all k. A semi-partition of the form $\mathcal{P} = \{B, TB, T^2B, \dots, T^{h-1}B\}$, for B clopen, is called the *height h Rohlin tower*

[2] A dimension group usually also comes together with an order structure (like in Note 3), but we that ignore that structure here.

with *base* B for a Cantor dynamical system T. A *Kakutani-Rohlin partition* is a partition $\mathcal{P}$ that is a finite union of disjoint Rohlin towers. A sequence of partitions $\mathcal{P}_k$ on X is said to be *generating* if any clopen set E is the union of elements of $\mathcal{P}_k$ for k sufficiently large. Equivalently, $\mathcal{P}_k$ is generating if and only if for each $n \geq 1$, there is a $K \in \mathbb{N}$, so that for $k \geq K$, the function $x \mapsto x_{[-n,n]} : X \to \mathcal{A}^{2n+1}$ is constant on each $P \in \mathcal{P}^k$.

Now let T be primitive aperiodic substitution shift dynamical system, for substitution σ. For each $a \in \mathcal{A}$, $B_a^{(k)} := \sigma^k(\lfloor a \rfloor)$, is the base of a height $h_a := |\sigma^k(a)|$ Kakutani-Rohlin tower (see [**DHS-99**]) and together these towers form a Kakutani-Rohlin partition we denote $\mathcal{P}_k$. These partitions satisfy $\mathcal{P}_{k+1} \geq \mathcal{P}_k$, and if σ is also proper, then $\mathcal{P}_k$ is a generating sequence (see [**DHS-99**]). The following proposition generalizes these results.

PROPOSITION 2.4. *Let σ be a primitive, aperiodic substitution on $\mathcal{A}$, and $r_k \in \mathbb{Z}$ be a sequence. Fix k. Then for each $a \in \mathcal{A}$, $B_a^{(k)} := T^{r_k}\sigma^k(\lfloor a \rfloor)$ is the base of a Rohlin tower of height $\ell_a := |\sigma^k(a)|$. These towers are disjoint and their union, denoted $\mathcal{P}_k$, is a Kakutani-Rohlin partition. In the case that σ has a common prefix, the sequence r_k can be chosen so that the sequence $\mathcal{P}_k$ of partitions is generating.*

PROOF. It is easy to see (by [**DHS-99**]) that $\mathcal{P}_k$ is a Kakutani-Rohlin partition. Now suppose σ has a common prefix (the common suffix case is similar). Then for $a \in \mathcal{A}$, $\sigma(a) = pu_a$, and for any $ab \in \mathcal{L}$ (including, possibly aa), $\sigma(ab) = pu_a pu_b$. Thus for $k > 1$,

$$\sigma^k(ab) = \sigma^{k-1}(p)\sigma^{k-1}(u_a)\sigma^{k-1}(p)\sigma^{k-1}(u_b). \tag{2.2}$$

Let $e_k = |\sigma^{k-1}(p)|$ and let $d_k = \lfloor e_k/2 \rfloor$ (the integer part of $e_k/2$), and let $c_{a,k} = |\sigma^{k-1}(u_a)|$.

Fix $n \geq 0$ and take k large enough that $e_k \geq n$. This is possible because σ is primitive. Since $\sigma^k(a) = \sigma^{k-1}(p)\sigma^{k-1}(u_a)$, it follows that $\ell_{a,k} = |\sigma^k(a)| = |\sigma^{k-1}(p)| + |\sigma^{k-1}(u_a)| = e_k + c_{a,k}$, which implies $[0, 2e_k + c_{a,k}] = [0, e_k + \ell_{a,k}]$. Thus by (2.2), any $x \in \lfloor \sigma^k(a) \rfloor$ satisfies

$$x_{[0,e_k+\ell_{a,k}]} = \sigma^{k-1}(p)\sigma^{k-1}(u_a)\sigma^{k-1}(p) \tag{2.3}$$

Now $\sigma^k(\lfloor a \rfloor) \subseteq \lfloor \sigma^k(a) \rfloor$, so (2.3) holds for $x \in \sigma^k(\lfloor a \rfloor)$.

Let $Q \in \mathcal{P}_k$. Then there exists $a \in \mathcal{A}$ and $0 \leq j < \ell_{a,k}$ so that $Q = T^{j+d_k}\sigma^k(\lfloor a \rfloor)$. Any $x \in Q$ satisfies $T^{-j+d_k}x \in \sigma^k(\lfloor a \rfloor)$, so by (2.3),

$$x_{[-j-d_k,\ell_{a,k}-j]} = (T^{-j-d_k}x)_{[0,d_k+\ell_{a,k}]} = \sigma^{k-1}(p)\sigma^{k-1}(u_a)\sigma^{k-1}(p). \tag{2.4}$$

But $n < e_k \leq d_k/2 < d_k + j$, and since $j \leq \ell_{a,k}$, $n < e_k \leq d_k/2 < \ell$. Thus $[-n, n] \subseteq [-j - d_k, d_k + c_{a,k} - j]$, so $x_{[-n,n]}$ is constant on Q. □

Let σ be primitive aperiodic substitution, $r_k \in \mathbb{Z}$, and $\mathcal{P}_k$ be the sequence of Kakutain-Rohlin partitions from Proposition 2.4. Define the *tower base vectors* $\mathbf{m}^{(k)} = (m_0^{(k)}, m_1^{(k)}, \dots, m_{d-1}^{(k)})$, by $m_a^{(k)} = \mu(B_a^{(k)}) = \mu(\sigma^k(\lfloor a \rfloor))$.

LEMMA 2.5. *The tower base vectors satisfy $\mathbf{m}^{(k)} = \lambda^{-k}\mathbf{m}$, and these vectors generate* PF_Q.

PROOF. First note that $\mathbf{m}^{(0)} = \mathbf{m}$, and also, $Q^t\mathbf{m}^{(k)} = \mathbf{m}^{(k-1)}$ for all k. By induction, $\mathbf{m}^{(k)} = Q^t\mathbf{m}^{(k-1)} = \lambda^{k-1}Q^t\mathbf{m} = \lambda^k\mathbf{m}$. Then

$$\bigcup_{k=1}^{\infty} \mathbb{Z}[\mathbf{m}^{(k)}] = \bigcup_{k=1}^{\infty} \lambda^{-k}\mathbb{Z}[\mathbf{m}] = \mathrm{PF}_Q.$$

□

The next theorem characterizes the measure group $\mathbb{H}(T)$ in many cases. Examples of measure groups are computed in Section 5.1.

THEOREM 2.6. *Let T be the substitution shift for a primitive, aperiodic, irreducible substitution σ with a common prefix. Let Q be its incidence matrix, and $\mathbf{m}$ its left normalized Perron-Frobenius eigenvector. Then $\mathbb{H}(T) = \mathrm{PF}_Q = \mathbb{Z}[\lambda^{-1}]\,\mathbb{Z}[\mathbf{m}]$, with $\mathbb{H}(T)$ finitely generated if and only if Q is unimodular, in which case $\mathbb{H}(T) = \mathbb{Z}[\mathbf{m}] \cong \mathbb{Z}^d$.*

PROOF. Let $\mathcal{P}_k$ be the sequence of Kakutani-Rohlin towers from Proposition 2.4. It follows from Proposition 2.4 that for any clopen set E, there exists k, so that E is a union of levels of $\mathcal{P}_k$. Thus by Lemma 2.5, $\mu(E) \in \lambda^{-k}\mathbb{Z}[\mathbf{m}]$. □

NOTE 8. For σ a primitive, aperiodic, but without the common prefix assumption, the following characterization of $\mathbb{H}(T)$ was obtained in [**Be-92**]. Let X_2 be the 2-block shift over X (see [**Qu-87**], [**LM-95**]): $X_2 \subseteq \mathcal{A}_2^{\mathbb{Z}}$ where $\mathcal{A}_2 = \mathcal{L} \cap \mathcal{A}^2$ is the set of 2-blocks in $\mathcal{L}$. Then X_2 is the substitution shift for a primitive aperiodic substitution σ_2, with incidence matrix Q_2 having the same Perron-Frobenius eigenvalue as Q (see [**Qu-87**]). Let $\mathbf{m}^{[2]}$ be the left Perron-Frobenius eigenvector, normalized $\mathbf{m}^{[2]} \cdot \mathbf{1} = 1$. Then $\mathbb{H}(T)$ is the subgroup of $\mathbb{R}$ generated by $\{\lambda^{-k}(\mathbf{n}\cdot\mathbf{m} + \mathbf{k}\cdot\mathbf{m}^{[2]}) : k-1 \in \mathbb{N}, \mathbf{n} \in \mathbb{Z}^d, \mathbf{k} \in \mathbb{Z}^{d_2}\}$. In [**Be-92**] this is called the $\mathbb{Z}[\lambda^{-1}]$*-module generated by* $\mathbf{m}$ *and* $\mathbf{m}^{[2]}$.

The next result gives a class of substitution dynamical systems that are cohomologically ergodic.

THEOREM 2.7. *If σ is a primitive, aperiodic, irreducible substitution with a common prefix, then the corresponding substitution shift dynamical system T is cohomologically ergodic. In particular, $\mathbb{H}(T) \cong H(T)$.*

PROOF. Let E and F be clopen sets such that $\mu(E) = \mu(F)$. By Proposition 2.4, there exists k so that E and F are both unions of the levels of $\in \mathcal{P}_k$. The entries of $\mathbf{m}^{(k)}$ give the measures of the bases of $\mathcal{P}_k$, and all the levels in each tower have the same measure as the base. By Lemma 2.3, $\mathbf{m}^{(k)}$ has rationally independent entries, since by Lemma 2.5, $\mathbf{m}^{(k)} = \lambda^{-k}\mathbf{m}$. Thus E and F must consist of the same number of levels from each tower in $\mathcal{P}_k$. Clearly we can define a homeomorphism $U \in [[T]]$ that matches up components of E with components of F within each tower and that is constant on the rest of X. It follows that T is saturated. Lemma 2.2 implies T is cohomologically ergodic. □

3. Flows and tiling systems

3.1. Suspension flows. Let T be a strictly ergodic Cantor dynamical system on X, and let $g(x)$ be a continuous positive real-valued function on X such that $\int_X g\,d\mu = 1$. Define $\widetilde{T} : X \times \mathbb{R} \to X \times \mathbb{R}$ by $\widetilde{T}(x,s) = (Tx, s + g(x))$, and a flow $\widetilde{F}^t$

on $X \times \mathbb{R}$ by $\widetilde{F}^t(x,s) = (x, s+t)$. Define $Y = X \times \mathbb{R}/\sim_{\widetilde{T}}$, where $(x,s) \sim_{\widetilde{T}} (x',s')$ if $(x,s) = \widetilde{T}^n(x',s')$ for some n. Then Y is a 1-dimensional compact metric space, called the *suspension space* of T by g. We define continuous flow F^t on Y by $F^t = \widetilde{F}^t/\sim_{\widetilde{T}}$, called the *suspension flow* (or *flow under the function* g). The strict ergodicity of T implies strict ergodicity for F^t. In particular, the unique F^t invariant probability measure ν on Y is defined by $\nu(E \times [a,b]) = (b-a)\mu(E)$, where $E \subseteq X$ and $0 \le a < b < \min_{x \in E} g(x)$.

3.2. The Bruschlinsky group. In the suspension space Y, every point y has a neighborhood that is a product of an interval and a Cantor set. Such spaces are sometimes called *laminations* or *matchbox manifolds*. Such spaces have a well defined 1-dimensional integer Čech cohomology $\check{H}^1(Y)$ (see e.g., [**S-08**] for the definition in this context).

Let $C(Y, \mathbb{T})$ be the group of continuous unimodular complex functions on Y (where $\mathbb{T} = \{z \in \mathbb{C} : |z| = 1\}$). Define a "sum" as pointwise multiplication: $(f_0 + f_1)(y) = f_0(y) f_1(y)$ and $(-f)(y) = 1/f(y)$. Let $R(Y) = \{\exp(2\pi i p(y)) : p : Y \to \mathbb{R} \text{ continuous}\}$. It is easy to see that f_1 is homotopic to f_2, denoted $f_1 \sim f_2$, if and only if $f_0 - f_1 \in R(Y)$. Define the *Bruschlinsky group* by $Br(Y) := C(Y, \mathbb{T})/R(Y)$. We will need the following old result.

THEOREM 3.1 (Bruschlinsky's Theorem [**Br-34**]). *The group $Br(Y)$ is isomorphic to $\check{H}^1(Y)$.*

This theorem is often quoted in the dynamics and tiling literature (see e.g., [**Sc-57**, **PT-82**, **BoH-96**, **AP-98**, **Yi-10**]). The standard reference seems to be [**Hu-59**] (pp. 47-52 and Exercise C, p. 59), but this is only slightly more modern than [**Br-34**]. In Note 6 (below), we try to offer some insight into the theorem in the case that Y is a substitution tiling space. Following [**PT-82**]), from now on, we will sometimes write $\check{H}^1(Y)$ to really mean $Br(Y)$.

For $n(x) \in C(X, \mathbb{Z})$ define

$$f_n(x,r) = \exp(2\pi i\, r n(x)/g(x)) \in C(Y, \mathbb{T}). \tag{3.1}$$

LEMMA 3.2. *For any $f \in C(Y, \mathbb{T})$ there is $n \in C(X, \mathbb{Z})$ so that $f \sim f_n$. Moreover, $n \in B(T)$ if and only if $f_n \in R(Y)$. It follows that $H(T) \cong \check{H}^1(Y)$.*

PROOF. This is proved in the case $g(x) = 1$ in [**PT-82**]. The general case follows from the observation that the suspension flow F^t on the suspension space Y for an arbitrary continuos return time $g(x)$, and the unit suspension F_1^t on Y_1, (i.e., the suspension for the return time $g_1(x) = 1$), are topologically conjugate via the homeomorphism $S(x,r) = (x, r/g(x))$ between Y and Y_1. □

3.3. Winding numbers. Now suppose (Y, F, ν) is a uniquely ergodic flow on a compact metric space. We say that $f \in C(Y, \mathbb{T})$ is *continuously differentiable* at $y \in Y$ if

$$f'(y) := \lim_{t \to 0} \frac{1}{t}(f(F^t y) - f(y)) \in C(Y, \mathbb{T}). \tag{3.2}$$

We define the *winding number* of a continuously differentiable f by

$$W(f) = \frac{1}{2\pi i} \int_Y \frac{f'(y)}{f(y)}\, d\nu(y). \tag{3.3}$$

The winding number was defined by Schwartzman [**Sc-57**], who proved the following.

LEMMA 3.3. (Schwartzman, [**Sc-57**]) *Every homotopy class in $C(Y,\mathbb{T})$ contains a continuously differentiable function, and $f_1 \sim f_2$ implies $W(f_1) = W(f_2)$. Moreover, $W(f_1 + f_2) = W(f_1) + W(f_2)$. Thus $W : \check{H}^1(Y) \to \mathbb{R}$ is a well defined homomorphism (i.e., a real valued functional).*

DEFINITION 3.4. *We say F^t is cohomologically ergodic if W is injective.*

NOTE 9. We think of the functions f with $W(f) = 0$ as being "cohomologically invariant". Then cohomological ergodicity says that $W(f) = 0$ implies $f = 0$.

PROPOSITION 3.5. *The image of the winding number functional satisfies $W(\check{H}^1(Y)) = \mathbb{H}(T)$ in $\mathbb{R}$, and thus does not depend on the return time $g(x)$. Moreover, the suspension flow F^t is cohomologically ergodic if and only if T is cohomologically ergodic.*

PROOF. For $n \in C(X,\mathbb{Z})$ let f_n be defined by (3.1). Then the derivative satisfies $f_n'(x,t) := 2\pi i(n(x)/g(x))\, f_n(x,t)$ satisfies $f_n'/f_n = 2\pi i(n(x)/g(x))$. Thus $W(f_n) = \int_Y (n(x)/g(x))\, d\nu = \int_X n\, d\mu$. □

3.4. Eigenfunctions and eigenvalues. We say $\omega \in \mathbb{R}$ is an *eigenvalue* for F^t corresponding to an *eigenfunction* $f \in L^2(Y,\nu)$ if

$$f(F^t y) = \exp(2\pi i\omega t)\, f(y), \tag{3.4}$$

for all $t \in \mathbb{R}$ and ν a.e., $y \in Y$. We denote the set of all eigenvalues by $\mathbb{E}(F)$. We assume F^t is *homogeneous* (in the sense of [**R-04**]). This means there is a continuous eigenfunction f for each eigenvalue $\omega \in \mathbb{E}(F)$. Since F^t is strictly ergodic, each eigenvalue ω is simple, each eigenfunction f is $\mathbb{T}$-valued, and $\mathbb{E}(F)$ is a countable subgroup of $\mathbb{R}$. Recall that F^t is *weakly mixing* if $\mathbb{E}(F) = \{1\}$, and has *pure point spectrum* if $E(F)$ generates $L^2(Y,\nu)$.

Let $E(F)$ denote the group of all (continuous) eigenfunctions modulo constant multiples. To make it an additive group we define $(f_1 + f_2)(y) = f_1(y)f_2(y)$. For $f \in E(F)$ define $V(f)$ to be the corresponding eigenvalue in $\mathbb{R}$. Then V is a homomorphism, and $\mathbb{E}(F)$ is the image of $E(F)$ under V. We call V the *eigenvalue functional.*

LEMMA 3.6 (Schwartzman [**Sc-57**]). *Continuous eigenfunctions are continuously differentiable, and $V(f) = W(f)$ for $f \in E(F)$.*

THEOREM 3.7. *Suppose F^t is a suspension of a strictly ergodic Cantor dynamical system T, for some positive continuous return time function g, (so that F^t is strictly ergodic). Assume F^t is homogeneous. Then $E(F) \subseteq Br(Y) = \check{H}^1(Y)$, and $V = W|_{E(F)}$. Thus the eigenvalue group $\mathbb{E}(F)$ is a subgroup of the measure group $\mathbb{H}(T)$.*

PROOF. We may assume eigenfunctions satisfy $f \in C(Y,\mathbb{T})$. Two eigenfunctions for the same eigenvalue are constant multiples, and thus homotopic. So there is a natural map $E(F) \to R(Y)$. Now by definition, any distinct $f_1, f_2 \in E(F)$ satisfy $V(f_1) \neq V(f_2)$. By Lemma 3.6, $W(f_1) \neq W(f_2)$, and by Lemma 3.3, $f_1 \not\sim f_2$. So $(f_1 - f_2) \notin R(Y)$. Thus the map $E(F) \to R(Y)$ is injective. □

Since $W(\check{H}^1(Y)) = \mathbb{H}(T)$, it follows that the image of W does not depend on the return time g. We will see in the Section 5, however, that $\mathbb{E}(F)$ generally does depend on g. We interpret the inclusion $\mathbb{E}(F) \subseteq \mathbb{H}(T)$ to mean that $\mathbb{H}(T)$ provides an "upper bound" on what eigenvalues $\mathbb{E}(F)$ a suspension can have. Recall that $\mathbb{H}(T)$ is the subgroup of $\mathbb{R}$ generated by the measures of clopen (or cylinder) sets, so these measures determine all the possible eigenvalues. But sometimes, not all these eigenvalues occur.

NOTE 10. It is shown in [**I-O-07**] that $E(F)$ embeds in $K_0(C(X) \ltimes_T \mathbb{Z})$ (which is $\cong \check{H}^1(Y)$). Similarly, in [**BKeS-12**] the "maximal equicontinuous" factor (G, K^t) of (Y, F^t) is considered. The maximal equicontinuous is the unique *Kronecker flow* K^t (i.e., a rotation action on a metric compact abelian group G) that satisfies $\mathbb{E}(K) = \mathbb{E}(F)$. They show $\check{H}^1(G) \cong \mathbb{E}(K)$ $(= \mathbb{E}(F))$, and that $\pi^* : \check{H}^1(G) \to \check{H}^1(Y)$ is injective, where π^* is the map induced by the factor map $\pi : Y \to G$.

DEFINITION 3.8. *We say a strictly ergodic, homogeneous flow F^t has* cohomological pure point spectrum *if it is cohomologically ergodic and $\check{H}^1(Y) = E(F)$, or equivalently it is cohomologically ergodic and $\mathbb{H}(T) = \mathbb{E}(F)$.*

In dynamical systems, one studies both topological and metric pure point spectrum. Topological pure point spectrum is equivalent to F^t being equicontinuous, or equivalently, topologically conjugate to a strictly ergodic Kronecker flow K^t. Such a flow always has cohomological pure point spectrum. But substitution tiling flows can never be equicontinuous because they always have proximal points. Metric pure point spectrum means F^t is metrically isomorphic to a strictly ergodic Kronecker flow K^t. Many examples of substitution tiling flows with metric pure point spectrum are known, including some discussed in Section 5. All of those examples cohomological pure point spectrum.

It is reasonable to conjecture that metric pure point spectrum is equivalent to cohomological pure point spectrum for a suspension of a strictly ergodic Cantor dynamical system T, but at this time we do not know either implication. We do not even know whether a strictly ergodic Cantor dynamical system T with metric pure point spectrum must be cohomologically ergodic.

3.5. The unit suspension and spectrum of T. In this section we reconsider the case of a strictly ergodic and homogeneous Cantor dynamical system (X, T, μ). An *eigenvalue* η for T, corresponding to an *eigenfunction* $f \in C(X, \mathbb{C})$, is an $\eta \in \mathbb{T}$ satisfying $f(Tx) = \eta f(x)$ for μ a.e. $x \in X$. We denote the set of all eigenvalues of T by $\mathbb{D}(T)$. Since T is strictly ergodic, each eigenvalue η is simple, each eigenfunction f is $\mathbb{T}$-valued, and $\mathbb{D}(T)$ is a countable subgroup of $\mathbb{T}$. We define $\mathbb{E}(T)$ to be the set of all $\omega \in \mathbb{R}$ so that $e^{2\pi i\omega} \in \mathbb{D}(T)$.

The *unit suspension* of T is the suspension flow F_1^t on Y_1 corresponding to the constant return time function $g(x) = 1$. The flow F_1^t is strictly ergodic and homogeneous. It is cohomologically ergodic if and only if T is cohomologically ergodic. It is well known that $\mathbb{E}(T) = \mathbb{E}(F_1)$. We say that T has *cohomological pure point spectrum* the unit suspension F_1^t has cohomological pure point spectrum.

4. Substitution tilings

Let σ be a primitive, aperiodic substitution on $\mathcal{A} = \{0, 1, \ldots, d-1\}$ and let (X, T, μ) be the corresponding strictly ergodic substitution shift dynamical system.

Let $\mathbf{m}$ and $\mathbf{h}$ be the normalized left and right Perron-Frobenius eigenvectors of the incidence matrix Q. Let $\mathbf{g} = (g_0, g_1, \ldots, g_{d-1})$ be any positive vector satisfying $\mathbf{g} \cdot \mathbf{m} = 1$. Define $g : X \to \mathbb{R}$ by $g(x) = \sum \mathbf{g}_{x_0}$ where $x = \ldots x_{-1}.x_0x_1x_2\ldots$, and let F^t on Y be the corresponding suspension flow. Then F^t is strictly ergodic and homogeneous (see [**CS-03**]). We call the pair $S = (\sigma, \mathbf{g})$ a *tile substitution*. We call Y a 1-dimensional *substitution tiling space*, and we call F^t a *substitution tiling flow*.

Tile substitutions have natural a geometric interpretation (see also [**CS-03**, **R-04**, **KSS-05**, **FR-08**]). Let $\mathcal{T} = \{I_0, I_1, \ldots, I_{d-1}\}$ be a set of labeled, half-closed intervals $I_a = [0, g_a)_a$, called *prototiles*. For any $u = u_0u_2\ldots u_{n-1} \in \mathcal{A}^*$, let $t_0 = 0$ and for $j \geq 1$, $t_j = g_{u_0} + \cdots + g_{u_{j-1}}$, and define a labeled *tiling*

$$I_u = [t_0, t_1)_{u_0}[t_1, t_2)_{u_1}\cdots[t_{n-1}, t_n)_{u_n}$$

of the interval $[0, t_n)$ by (translations of the tiles in) $\mathcal{T}$.

Let $\mathcal{T}^*$ denote the set of all finite tilings of intervals by $\mathcal{T}$. We can interpret the tile substitution $S = (\sigma, \mathbf{g})$ as a mapping $S : \mathcal{T} \to \mathcal{T}^*$ defined by $S(I_a) = I_{\sigma(a)}$.[3] For example, consider the *golden mean substitution* σ defined by $\sigma(0) = 01$, $\sigma(1) = 0$. Take $\mathbf{g} = \mathbf{h} = (\lambda + 1)^{-1}(1, \lambda)$, the normalized right Perron-Frobenius eigenvector for Perron-Frobenius eigenvalue $\lambda = (1 + \sqrt{5})/2$. Then the tile substitution S, realized as a map on prototiles, looks like this:

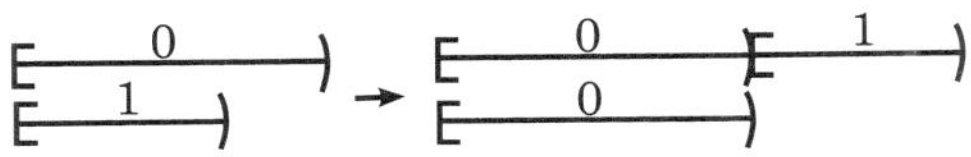

Let Z be the set of all tilings z of $\mathbb{R}$ by (translations of the tiles in) $\mathcal{T}$ satisfying the additional requirement that every finite subtiling of z is a translate of a subtiling of some $S^k(I_a)$, $a \in \mathcal{A}$, $k \geq 0$. We call Z the *substitution tiling space* corresponding to S. This space has a natural compact metric topology (see e.g., [**R-04**]), and there is a *tiling flow* E^t, which acts on Z by translation. One easily shows that (Z, E^t) is topologically conjugate to (Y, F^t), and from now on we identify them. A portion of a tiling z for the Fibonacci substitution S is shown below (where the dot indicates the position of $0 \in \mathbb{R}$):

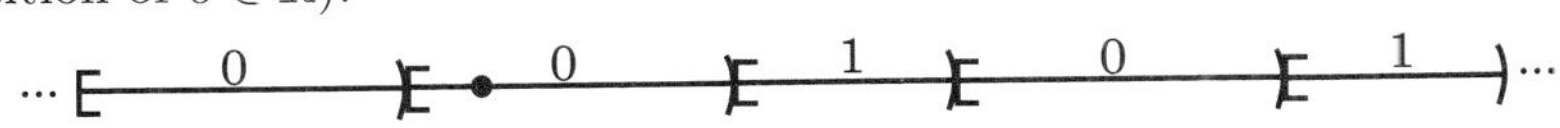

We say that S is *primitive, aperiodic, irreducible, unimodular, Pisot*, or has a *common prefix* according to σ. For a given σ there are two special tile substitutions corresponding to two special tile length vectors $\mathbf{g}$. We call $S_1 = (\sigma, \mathbf{1})$, $\mathbf{g} = \mathbf{1} = (1, 1, \ldots, 1)$, the *unit* tile substitution (all the tiles have length 1). The corresponding substitution tiling flow F_1^t is the unit suspension of T. We call $S_{\text{PF}} = (\sigma, \mathbf{h})$, where $\mathbf{g} = \mathbf{h}$ is the normalized right Perron-Frobenius eigenvector, a *self-similar* tile substitution. It is *self-similar* in the sense of ([**So-97**]), which means for each $a \in \mathcal{A}$, the length of $S_{\text{PF}}(I_a)$ is λ times the length of I_a, where λ is the Perron-Frobenius eigenvalue. We sometimes write F_1^t and F_{PF}^t for the corresponding substitution tiling flows.

THEOREM 4.1. *Let F^t be the substitution tiling flow corresponding to a primitive, aperiodic tile substitution S. Then F^t is strictly ergodic and homogeneous. If*

[3]The substitution S also evidently defines a mapping on Y which, as a result of σ being aperiodic, is a homeomorphism (see [**R-04**]).

S is also irreducible, and has a common prefix, then F^t is cohomologically ergodic and $\mathbb{E}(F) \subseteq \mathbb{H}(T)$.

PROOF. As previously noted, strict ergodicity and homogeneity are proved in [**CS-03**]. Cohomological ergodicity follows from Proposition 3.5, and the inclusion $\mathbb{E}(F) \subseteq \mathbb{H}(T)$ follows from Theorem 3.7. □

Note that if Q is the incidence matrix and $\mathbf{m}$ the normalized left Perron-Frobenius eigenvector, $\mathbb{H}(T) = \mathrm{PF}_Q = \mathbb{Z}[\lambda^{-1}]\,\mathbb{Z}[\mathbf{m}]$ which is $\cong \check{H}^1(Y)$ in the cohomologically ergodic case. If S is also unimodular, then $\mathrm{PF}_Q = \mathbb{Z}[\mathbf{m}] \cong \mathbb{Z}^d$. Now we turn to the point spectrum.

THEOREM 4.2. *Let F^t be the substitution tiling flow corresponding to a primitive, aperiodic, irreducible, unimodular, Pisot tile substitution $S = (\sigma, \mathbf{g})$ with a common prefix. Then F^t has cohomological pure point spectrum. In particular, F^t is cohomologically ergodic, and $\mathbb{E}(F) = \mathbb{H}(T) = \mathrm{PF}_Q \cong \mathbb{Z}[\mathbf{m}]$.*

COROLLARY 4.3. *If σ is a primitive, aperiodic, unimodular, Pisot substitution with a common prefix, then T has cohomological pure point spectrum. In particular, T is cohomologically ergodic and $\mathbb{D}(T) = \exp(2\pi i\,\mathrm{PF}_Q)$.*

Cohomological ergodicity and the inclusion $\mathbb{E}(F) \supseteq \mathbb{H}(T)$ follow from Theorem 4.1. The proof of $\subseteq$ takes up the remainder of this section. We prove it first in the self-similar case $\mathbf{g} = \mathbf{h}$, then in general.

For a word $u = u_0 u_1 \dots u_{n-1} \in \mathcal{L}$, the *population vector* is defined $\mathbf{p}_u = (p_0, p_1, \dots, p_{d-1}) \in \mathbb{Z}^d$, where $p_a = |\{j = 0, \dots, n-1 : u_j = a\}|$. We say $u \in \mathcal{L}$ is a *recurrence word* if $u = x_{[r,s]}$, $r \le s$, for some $x \in X$ such that $x_r = x_{s+1}$. For a real number t define $\{t\} = t - \lfloor t \rfloor$. We begin with a general lemma from [**CS-03**] (see also [**Ho-86**] and [**So-98**]).

LEMMA 4.4. *If S is a primitive, aperiodic tile substitution, then $\omega \in \mathbb{E}(F)$ if and only if for every recurrence word u,*

$$\lim_{n\to\infty} \{\omega\, \mathbf{g} \cdot Q^n \mathbf{p}_u\} = 0. \tag{4.1}$$

For a Perron-Frobenius substitution $S_{\mathbf{PF}}$ (corresponding to $\mathbf{g} = \mathbf{h}$, the Perron-Frobenius eigenvector), we have the following.

LEMMA 4.5. *Suppose $S_{\mathbf{PF}}$ is a primitive, aperiodic Perron-Frobenius tile substitution. If for each $a \in \mathcal{A}$, $\omega \in \mathbb{R}$ satisfies*

$$\lim_{n\to\infty} \{\omega\, h_a\, \lambda^n\} = 0, \tag{4.2}$$

where λ is the Perron-Frobenius eigenvalue, then $\omega \in \mathbb{E}(F)$.

PROOF. It suffices to prove (4.1) for $\mathbf{p}_a$. But $\omega\mathbf{h} \cdot Q^n \mathbf{p}_a = \omega (Q^t)^n \mathbf{h} \cdot \mathbf{p}_a = \omega\lambda^n \mathbf{h} \cdot \mathbf{p}_a = \omega\lambda^n h_a$, since $\mathbf{p}_a$ is the ath standard basis vector. □

LEMMA 4.6. *Suppose σ is primitive, irreducible, Pisot substitution. Then Q has 1-dimensional expanding subspace $L^u(Q)$ and a $d-1$ dimensional contracting subspace $L^s(Q)$, and the same is true for Q^t. Moreover, $L^u(Q) \perp L^s(Q^t)$ and $L^u(Q^t) \perp L^s(Q)$. If we define $P\mathbf{v} := \mathbf{v} - (\mathbf{h} \cdot \mathbf{v})\mathbf{m}$ for $\mathbf{v} \in \mathbb{R}^d$, then $P\mathbf{v} \in L^s(Q^t)$.*

PROOF. Since Q is irreducible, it has a left $\mathbf{m}_1, \mathbf{m}_2, \ldots, \mathbf{m}_d$ and right $\mathbf{h}_1, \mathbf{h}_2, \ldots, \mathbf{h}_d$ eigenbasises. Let $\mathbf{m}_1 = \mathbf{m}$, $\mathbf{h}_1 = \mathbf{h}$, which are positive real. But other eigenvectors $\mathbf{m}_j$ and $\mathbf{h}_j$, $j > 1$ may be complex. Then $\mathbf{h}_i \cdot \mathbf{m}_j = 0$, for $i \neq j$, because $\lambda_i \mathbf{h}_i \cdot \mathbf{m}_j = (Q^t \mathbf{h}_i) \cdot \mathbf{m}_j = \mathbf{h}_i \cdot (Q\mathbf{m}_j) = \lambda_j \mathbf{h}_i \cdot \mathbf{m}_j$ for $\lambda_j \neq \lambda_j$ (note that here: $\mathbf{v} \cdot \mathbf{w} := v_1\overline{w_1} + v_2\overline{w_2} + \cdots + v_d\overline{w_d}$). Define $L^u(Q)$ and $L^u(Q^t)$ to be the spans of $\mathbf{m}$ and $\mathbf{h}$. The complex eigenvalues, if they occur at all, occur in conjugate pairs. We replace such a pair with the new real pair consisting of the real and imaginary parts of the conjugate pair. Both of these lie in the complex span of the original pair. We then define $L^s(Q)$ and $L^s(Q^t)$ to be the real parts of the spans of $\mathbf{m}_2, \ldots, \mathbf{m}_d$ and $\mathbf{h}_2, \ldots, \mathbf{h}_d$. For the second assertion, $\mathbf{h} \cdot P\mathbf{v} = \mathbf{h} \cdot (\mathbf{v} - (\mathbf{h} \cdot \mathbf{v})\mathbf{m}) = \mathbf{h} \cdot \mathbf{v} - (\mathbf{h} \cdot \mathbf{v})(\mathbf{h} \cdot \mathbf{m}) = 0$ since $\mathbf{h} \cdot \mathbf{m} = 1$. □

The next result is one direction of Theorem 4.2 in the case of $\mathbf{g} = \mathbf{h}$.

PROPOSITION 4.7. *Suppose S_{PF} is a primitive, aperiodic, irreducible Pisot, Perron-Frobenius tile substitution. Then the corresponding substitution tiling flow F^t satisfies $\mathbb{H}(T) \subseteq \mathbb{E}(F)$.*

PROOF. For $\mathbf{v}, \mathbf{k} \in \mathbb{R}^n$, let $t = \mathbf{v} \cdot \mathbf{h}$, and $\alpha = \mathbf{k} \cdot \mathbf{m}$. Then

$$\begin{aligned} t\alpha\lambda^n &= \lambda^n(\mathbf{v} \cdot \mathbf{h})(\mathbf{k} \cdot \mathbf{m}) = \mathbf{k} \cdot (\mathbf{v} \cdot \mathbf{h})\lambda^n \mathbf{m} \\ &= \mathbf{k} \cdot (\mathbf{v} \cdot \mathbf{h})(Q^t)^n \mathbf{m} = Q^n \mathbf{k} \cdot (\mathbf{v} \cdot \mathbf{h})\mathbf{m} \\ &= Q^n \mathbf{k} \cdot (\mathbf{v} - P\mathbf{v}) = Q^n \mathbf{k} \cdot \mathbf{v} - \mathbf{k} \cdot (Q^t)^n P\mathbf{v}. \end{aligned}$$

If $\mathbf{k} = \mathbf{p}_a$ and $\mathbf{v} = \mathbf{p}_b$, then

$$\{m_b h_a \lambda^n\} = \{\mathbf{p}_a \cdot Q^n \mathbf{p}_b - \mathbf{p}_a \cdot (Q^t)^n P\mathbf{p}_b\} = \{-\mathbf{p}_a \cdot (Q^t)^n P\mathbf{p}_b\}, \tag{4.3}$$

since $Q^n \mathbf{p}_a \cdot \mathbf{p}_b \in \mathbb{Z}$, and thus $\{m_b h_a \lambda^n\} \to 0$ since $P\mathbf{p}_b \in L^s(Q)$. This shows $\mathbb{Z}[\mathbf{m}] \subseteq \mathbb{E}(F)$. □

NOTE 11. In [**BK-06**], S_{PF} is assumed to be unimodular, but without necessarily, the assumption of a common prefix. The main result there is that $\mathbb{Z}[\mathbf{m}] = \mathbb{E}(F)$. The inclusion $\mathbb{Z}[\mathbf{m}] \subseteq \mathbb{E}(F)$ is said to be well known, and attributed to [**BT-86**]. But a direct proof is also given, which constructs a Kronecker factor (G, K^t) of (Y, F^t) with $\mathbb{E}(K) = \mathbb{Z}[\mathbf{m}] \subseteq \mathbb{E}(F)$ (but note that [**BK-06**] uses a different normalization: $\mathbf{m} \cdot \mathbf{m} = 1$ and $\mathbf{h} \cdot \mathbf{m} = 1$). The opposite inclusion, $\mathbb{E}(F) \subseteq \mathbb{Z}[\mathbf{m}]$, is the main result of [**BK-06**]. For us, this follows from Theorem 4.1 (using the assumption that S has a common prefix).

The next result, which completes the proof of Theorem 4.2, shows that for a primitive, aperiodic, irreducible, Pisot tile substitutions, the Perron-Frobenius eigenvector $\mathbf{h}$ can be replaced by any $\mathbf{g}$ satisfying $\mathbf{g} \cdot \mathbf{m} = 1$. This result appears in [**CS-03**], but we include a proof for completeness.

PROPOSITION 4.8 ([**CS-03**]). *Suppose $S_{\mathrm{PF}} = (\sigma, \mathbf{h})$ is a primitive, aperiodic, irreducible, Pisot, Perron-Frobenious tile substitution. Let $S = (\sigma, \mathbf{g})$ be any other tile substitution based on the same discrete substitution σ. Then $\mathbb{E}(F) = \mathbb{E}(F_{\mathrm{PF}})$.*

PROOF. Lemma 4.4 implies $\{\omega\mathbf{h} \cdot Q^n \mathbf{p}_u\} \to 0$ for every $\omega \in \mathbb{E}(F_{\mathrm{PF}})$ and every return word u. Then

$$\begin{aligned} \{\omega\mathbf{h} \cdot Q^n \mathbf{p}_u\} &= \{\omega(\mathbf{g} - (\mathbf{g} - \mathbf{h})) \cdot Q^n \mathbf{p}_u\} \\ &= \{\omega\mathbf{g} \cdot Q^n \mathbf{p}_u - \omega(Q^t)^n(\mathbf{g} - \mathbf{h}) \cdot \mathbf{p}_u\}. \end{aligned}$$

Since $(\mathbf{g}-\mathbf{h})\cdot\mathbf{m} = \mathbf{g}\cdot\mathbf{m}-\mathbf{h}\cdot\mathbf{m} = 0$, it follows that $\mathbf{g}-\mathbf{h}\in L^s(Q^t)$ and $(Q^t)^n(\mathbf{g}-\mathbf{h})\to\mathbf{0}$. Thus $\{\omega\mathbf{g}\cdot Q^n\mathbf{p}_u\}\to 0$ and $\omega\in\mathbb{E}(F)$. □

5. Examples

5.1. Metallic and alloy substitutions. For $n\geq 1$ we define the nth *metallic* substitution $\sigma: 0\to 0^n1, 1\to 0$. Also for $m\geq 3$, we define the mth *alloy* substitution $\sigma: 0\to 0^{m-1}1^{m-2}, 1\to 01$.

PROPOSITION 5.1. *Each metallic or alloy substitution is primitive, aperiodic, irreducible, unimodular, Pisot, and has a common prefix. The corresponding substitution shift T and, for any $\mathbf{g}$, the substitution tiling flow F^t corresponding to $S=(\sigma,\mathbf{g})$, is cohomologically ergodic, with $H(T)\cong\mathbb{H}(T)=\mathbb{Z}[\mathbf{m}]\cong\check{H}^1(Y)$.*

PROOF. The characteristic polynomials of Q in the metallic case are $p_n(z)=z^2-nz-1$, and in the alloy case are $q_m(z)=z^2-mz+1$. □

The positive roots λ_n of $p_n(z)$ are sometimes called *metallic numbers*: $\lambda_1=(1/2)(1+\sqrt{5})$ is called the *golden mean*, $\lambda_2=1+\sqrt{2}$ is called the *silver mean*, etc. We call the roots λ'_m of $q_m(z)$ *alloy numbers* because the incidence matrices satisfy $Q'_{n+2}=Q_1Q_n$ (gold, together with a base metal). Any monic quadratic polynomial over $\mathbb{Z}$ with constant term ±1 is either $p_n(\pm z)$, $q_n(\pm z)$ or $r(z)=z^2\pm1$. Thus the metallic and alloy numbers make up all the quadratic units. All quadratic units are Pisot.

THEOREM 5.2. *Any metallic or alloy substitution tiling flow F^t has cohomological pure point spectrum: $\mathbb{E}(F)=\mathbb{H}(T)$. In the metallic case, $\mathbb{E}(F)=(\lambda+1)^{-1}\mathbb{Z}[\lambda]\cong\mathbb{Z}^2$ (with $\mathbb{E}(F)=\mathbb{Z}[\lambda]$ in the Golden mean case, where $\lambda+1$ is a unit). In the alloy case, $\mathbb{E}(F)=(\lambda-(m-2))^{-1}\mathbb{Z}[\lambda]\cong\mathbb{Z}^2$. For the susbtitution shifts T, $\mathbb{D}(T)=\exp(2\pi i\,\mathbb{E}(F))$, and every such T has* pure point spectrum.

PROOF. In the metallic case, $\mathbf{m}=(\lambda+1)^{-1}(1,\lambda)$, so $\mathbb{E}(F)=\mathbb{Z}[\mathbf{m}]=(\lambda+1)^{-1}\mathbb{Z}[\lambda]$. In the alloy case, $m\geq 3$, $\mathbf{m}=(\lambda-m+2)^{-1}(1,\lambda-m+1)$, and $\mathbb{Z}[\lambda-m+1]=\mathbb{Z}[\lambda]$. □

5.2. A unimodular non-Pisot substitution. A *non-Pisot* substitution is a primitive, aperiodic, irreducible substitution σ, such that λ is not a Pisot number. The non-Pisot substitution σ: $0\to 0313, 1\to 031313, 2\to 03223, 3\to 0323$, studied in [**FiHR-03**], is unimodular and has a common prefix. It is shown in [**FiHR-03**] that F^t_{PF} is weakly mixing, which means $\mathbb{E}(F_{\text{PF}})=\{0\}$. Clearly $\mathbb{E}(F_1)=\mathbb{Z}$. However, in both cases, $\check{H}^1(Y)\cong\mathbb{H}(T)=\mathbb{Z}[\mathbf{m}]\cong\mathbb{Z}^4$, so neither has cohomological pure point spectrum.

5.3. The completely non-Pisot case. We call σ *completely non-Pisot* if all the eigenvalues θ of Q satisfy $|\theta|\geq 1$. Primitive, aperiodic, completely non-Pisot substitutions σ, are studied in [**CS-03**] with the additional assumption that there is a *full recurrence word*: $u\in\mathcal{L}$ such that $\mathbf{p}_u, Q\mathbf{p}_u,\ldots,Q^{d-1}\mathbf{p}_u$ are linearly independent. An example is the substitution $0\to 0111, 1\to 0$ from [**FR-08**]. If $\mathbf{g}=(g_0,g_1,\ldots,d_{d-1})$ has rationally independent entries, then F^t is weakly mixing, [**CS-03**]. This provides many examples of substitutions without cohomological pure point spectrum, since $\mathbb{E}(F)=\{0\}$ is always a proper subgroup of $\mathbb{H}(T)$, (which, by Theorem 2.6, is not finitely generated). It is also shown in [**CS-03**] that if $g_a/g_b\in\mathbb{Q}\ \forall a,b\in\mathcal{A}$, for a primitive, aperiodic, completely non-Pisot tile substitution then $\mathbb{E}(F) \subseteq g_0\mathbb{Q}$. In the two-letter primitive, aperiodic, completely non-Pisot substitutions σ, it is shown in [**KSS-05**] that F^t is topologically mixing (which implies weak mixing), if and only if $g_0/g_1 \notin \mathbb{Q}$.

6. Appendix: Remarks on Bruschlinsky's theorem

Let Y be be a substitution tiling space for a primitive, aperiodic tile substitution S. Our goal in this section is to provide some insight why the Bruschlinsky group $Br(Y)$ is isomorphic to the the integer Čech cohomology $\check{H}^1(Y)$. As we will see, this follows from the fact that Y can be expressed as an inverse limit of an endomorphism $S : \Gamma \to \Gamma$ of a directed graph Γ. The integer Čech cohomology of a graph Γ is the same as its integer simplicial cohomology, and we will show below that the simplicial cohomology is isomorphic to $Br(\Gamma)$. The desired result, that $\check{H}^1(Y) \cong Br(Y)$, then follows from the fact that the Čech cohomology of an inverse limit is the direct limit of the Čech cohomologies.

The graph Γ mentioned above is the *Anderson-Putnam complex*. The edge set $\mathcal{E}$ of Γ is equal to the (directed) prototiles. The vertices $\mathcal{V}$ are defined by the condition that the front end of one tile meets the back end of another at a vertex $v \in \mathcal{V}$ if and only if that pair also meets somewhere in some tiling in Γ. The substitution S then acts as an endomorphism of this graph; a piecewise linear, one-dimensional dynamical system. For technical reasons one needs to first replace[4] the substitution with a "collared" 3-higher block substitution S, and take Γ to be the complex with the "collared" prototiles as its edges (see [**AP-98**] for details). Using this setup, the inverse limit characterization

$$Y = \lim_{\leftarrow}(\Gamma, S)$$

is obtained in [**AP-98**]. The corresponding direct limit

$$\check{H}^1(Y) = \lim_{\rightarrow}(\check{H}^1(\Gamma), S^*)$$

is a standard property of Čech cohomology.

Here is an example of S and Γ from [**AP-98**]. Start with the Fibonacci substitution σ: $0 \to 01$, $1 \to 0$, and let $\mathcal{A}_3 = \{a, b, c, d\}$ where $a = 0.01$, $b = 1.00$, $c = 1.01$ and $d = 0.10$. Then the collared substitution σ_3 is given by $a = 0.01 \to 01.010 = cd$, $b \to dc$, $c \to d$ and $d \to db$. Without loss of generality, we take $S = S_1$, the unit tile substitution (where all tiles have length 1). Then $\mathcal{E} = \{a, b, c, d\}$, viewed as (labeled) prototiles. Here $\mathcal{V}$ is a three element set, corresponding to pairs of adjacent prototiles, which we write as $\mathcal{V} = \{\alpha, \beta, \gamma\}$. This complex and the action of S on it are shown below.

---

[4]If the original substitution S "forces its border", see [**AP-98**], then S itself be used.

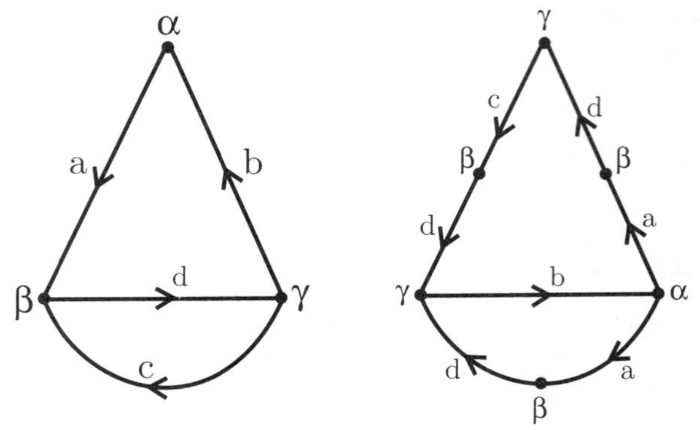

In general, the the sets of 0- and 1-simplicial cochains are, respectively, $\mathbb{Z}^{\mathcal{E}}$ and $\mathbb{Z}^{\mathcal{V}}$. So in the Fibonacci case they are $\mathbb{Z}^3$ and $\mathbb{Z}^4$. The coboundary operator is a $|\mathcal{V}| \times |\mathcal{E}|$ matrix δ with entries $d_{v,e} = \pm 1$, depending on whether v enters or leaves e, or $d_{v,e} = 0$ if v is not incident on e. Then $\check{H}^1(\Gamma) = H^1(\Gamma) := \mathbb{Z}^{\mathcal{V}}/\delta(\mathbb{Z}^{\mathcal{E}})$.

The coboundary matrix for the Fibonacci case is

$$\delta = \begin{bmatrix} -1 & 1 & 0 \\ 1 & 0 & -1 \\ 0 & 1 & -1 \\ 0 & -1 & 1 \end{bmatrix},$$

which has rank 2, and the columns are primitive vectors. Thus $\check{H}^1(\Gamma) = H^1(\Gamma) := \mathbb{Z}^4/\delta(\mathbb{Z}^3) \cong \mathbb{Z}^2$. This calculation concludes with

$$\check{H}^1(Y) = \varinjlim(\check{H}^1(\Gamma), S^*) \cong \mathbb{Z}^2$$

(see [**AP-98**]), which follows from the fact that the incidence matrix Q for the collared tile substitution S has rank 2, and its projection S^* to $\mathbb{Z}^4/\delta(\mathbb{Z}^3) \cong \mathbb{Z}^2$ is an abelian group isomorphism.

View each edge in Γ is a labeled unit interval. Thus if $y \in \Gamma$ is not at a vertex, it corresponds to $t = t(y) \in (0,1)$ and to a label $\ell(y) \in \mathcal{A}_3$, where the function $\ell(y)$ is constant on the interiors of the edges. Given $\mathbf{n} \in \mathbb{Z}^{\mathcal{E}}$ define $f_{\mathbf{n}} \in C(\Gamma, \mathbb{T})$ by $f_{\mathbf{c}}(y) = 1$ for $y \in \mathcal{V}$ and $f_{\mathbf{n}}(y) = \exp(2\pi i \mathbf{n}_{\ell(y)} t(y))$ for $y \in \Gamma \backslash \mathcal{V}$. Conversely, if $f \in C(\Gamma, \mathbb{T})$ it is clear that f is homotopic to some $f_1 \in C(\Gamma, \mathbb{T})$ with $f_1(v) = 1$ for each $v \in \mathcal{V}$. And then f_1 is homotopic to $f_{\mathbf{n}}$ for some $\mathbf{n} \in \mathbb{Z}^{\mathcal{E}}$. In particular, if $\mathbf{n} = (n_1, \ldots, n_{|\mathcal{E}|})$, then n_e is the degree of $f_1 : e = [0,1] \to \mathbb{T}$, viewed as a circle homeomorphism. And if $f \in R(\Gamma)$ then f is homotopic to $1 = f_{\mathbf{0}}$, and $\mathbf{0} \in \delta(\mathbb{Z}^{\mathcal{V}})$.

LEMMA 6.1. *If* $\mathbf{n}$ *is one of the columns of* δ*, then* $r_{\mathbf{n}}(y) = \mathbf{n}_{\ell(y)} t(y)$ *is a continuous real valued function, so* $f_{\mathbf{n}}(y) = \exp(2\pi i r_{\mathbf{n}}(y)) \in R(\Gamma)$.

To see this, suppose $\mathbf{n}$ is column v of δ. Then the function $r_{\mathbf{n}}$ defines a "height function", where the height of v is 0, and every other vertex has height 1. Going from v to one of its adjacent vertices, $r_{\mathbf{n}}$ increases linearly. Between any other two vertices, it is constant. It follows that $f_{\mathbf{n}} \in R(\Gamma)$.

COROLLARY 6.2. $Br(\Gamma) = \check{H}^1(\Gamma)$.

References

[An-07] Andress, Tetyana I, *The spectrum and the first Čech cohomology group of a one dimensional tiling dynamical system*, Ph.D. Dissertation, The George Washington University, (2007).

[AI-01] Pierre Arnoux and Shunji Ito, *Pisot substitutions and Rauzy fractals* (English, with English and French summaries), Bull. Belg. Math. Soc. Simon Stevin **8** (2001), no. 2, 181–207. Journées Montoises d'Informatique Théorique (Marne-la-Vallée, 2000). MR1838930

[AP-98] Jared E. Anderson and Ian F. Putnam, *Topological invariants for substitution tilings and their associated C^*-algebras*, Ergodic Theory Dynam. Systems **18** (1998), no. 3, 509–537, DOI 10.1017/S0143385798100457. MR1631708

[BBK-06] Veronica Baker, Marcy Barge, and Jaroslaw Kwapisz, *Geometric realization and coincidence for reducible non-unimodular Pisot tiling spaces with an application to β-shifts*, Ann. Inst. Fourier (Grenoble) **56** (2006), no. 7, 2213–2248. Numération, pavages, substitutions. MR2290779

[BD-01] Marcy Barge and Beverly Diamond, *A complete invariant for the topology of one-dimensional substitution tiling spaces*, Ergodic Theory Dynam. Systems **21** (2001), no. 5, 1333–1358, DOI 10.1017/S0143385701001638. MR1855835

[BD-02] Marcy Barge and Beverly Diamond, *Coincidence for substitutions of Pisot type* (English, with English and French summaries), Bull. Soc. Math. France **130** (2002), no. 4, 619–626. MR1947456

[BD-08] Marcy Barge and Beverly Diamond, *Cohomology in one-dimensional substitution tiling spaces*, Proc. Amer. Math. Soc. **136** (2008), no. 6, 2183–2191, DOI 10.1090/S0002-9939-08-09225-3. MR2383524

[BKeS-12] Marcy Barge, Johannes Kellendonk, and Scott Schmieding, *Maximal equicontinuous factors and cohomology for tiling spaces*, Fund. Math. **218** (2012), no. 3, 243–268, DOI 10.4064/fm218-3-3. MR2982777

[BK-06] Marcy Barge and Jaroslaw Kwapisz, *Elements of the theory of unimodular Pisot substitutions with an application to β-shifts*, Algebraic and topological dynamics, Contemp. Math., vol. 385, Amer. Math. Soc., Providence, RI, 2005, pp. 89–99, DOI 10.1090/conm/385/07192. MR2180231

[Be-92] Jean Bellissard, *Gap labelling theorems for Schrödinger operators*, From number theory to physics (Les Houches, 1989), Springer, Berlin, 1992, pp. 538–630. MR1221111

[BeBG-06] Jean Bellissard, Riccardo Benedetti, and Jean-Marc Gambaudo, *Spaces of tilings, finite telescopic approximations and gap-labeling*, Comm. Math. Phys. **261** (2006), no. 1, 1–41, DOI 10.1007/s00220-005-1445-z. MR2193205

[BO-03] Moulay-Tahar Benameur and Hervé Oyono-Oyono, *Gap-labelling for quasi-crystals (proving a conjecture by J. Bellissard)*, Operator algebras and mathematical physics (Constanţa, 2001), Theta, Bucharest, 2003, pp. 11–22. MR2018220

[BT-86] E. Bombieri and J. E. Taylor, *Which distributions of matter diffract? An initial investigation*, J. Physique **47** (1986), no. 7, Suppl. Colloq. C3, C3-19–C3-28. International workshop on aperiodic crystals (Les Houches, 1986). MR866320

[Bo-83] McBlaine Michael Boyle, *Topological orbit equivalence and factor maps in symbolic dynamics*, ProQuest LLC, Ann Arbor, MI, 1983. Thesis (Ph.D.)–University of Washington. MR2632783

[BoH-96] Mike Boyle and David Handelman, *Orbit equivalence, flow equivalence and ordered cohomology*, Israel J. Math. **95** (1996), 169–210, DOI 10.1007/BF02761039. MR1418293

[BoT-98] Mike Boyle and Jun Tomiyama, *Bounded topological orbit equivalence and C^*-algebras*, J. Math. Soc. Japan **50** (1998), no. 2, 317–329, DOI 10.2969/jmsj/05020317. MR1613140

[Br-34] N. Bruschlinsky, *Stetige Abbildungen und Bettische Gruppen der Dimensionszahlen 1 und 3* (German), Math. Ann. **109** (1934), no. 1, 525–537, DOI 10.1007/BF01449153. MR1512908

[BzK-00] Sergey Bezuglyi and Jan Kwiatkowski, *The topological full group of a Cantor minimal system is dense in the full group*, Topol. Methods Nonlinear Anal. **16** (2000), no. 2, 371–397. MR1820514

[CSg-01] Vincent Canterini and Anne Siegel, *Geometric representation of substitutions of Pisot type*, Trans. Amer. Math. Soc. **353** (2001), no. 12, 5121–5144, DOI 10.1090/S0002-9947-01-02797-0. MR1852097

[CS-03] Alex Clark and Lorenzo Sadun, *When size matters: subshifts and their related tiling spaces*, Ergodic Theory Dynam. Systems **23** (2003), no. 4, 1043–1057, DOI 10.1017/S0143385702001633. MR1997967

[CS-06] Alex Clark and Lorenzo Sadun, *When shape matters: deformations of tiling spaces*, Ergodic Theory Dynam. Systems **26** (2006), no. 1, 69–86, DOI 10.1017/S0143385705000623. MR2201938

[DHS-99] F. Durand, B. Host, and C. Skau, *Substitutional dynamical systems, Bratteli diagrams and dimension groups*, Ergodic Theory Dynam. Systems **19** (1999), no. 4, 953–993, DOI 10.1017/S0143385799133947. MR1709427

[FMN-96] Sébastien Ferenczi, Christian Mauduit, and Arnaldo Nogueira, *Substitution dynamical systems: algebraic characterization of eigenvalues*, Ann. Sci. École Norm. Sup. (4) **29** (1996), no. 4, 519–533. MR1386224

[Fg-02] N. Pytheas Fogg, *Substitutions in dynamics, arithmetics and combinatorics*, Lecture Notes in Mathematics, vol. 1794, Springer-Verlag, Berlin, 2002. Edited by V. Berthé, S. Ferenczi, C. Mauduit and A. Siegel. MR1970385

[FoHK-02] A. H. Forrest, J. R. Hunton, and J. Kellendonk, *Cohomology of canonical projection tilings*, Comm. Math. Phys. **226** (2002), no. 2, 289–322, DOI 10.1007/s002200200594. MR1892456

[FiHR-03] Thomas L. Fitzkee, Kevin G. Hockett, and E. Arthur Robinson Jr., *A weakly mixing tiling dynamical system with a smooth model*, Theoret. Comput. Sci. **303** (2003), no. 2-3, 447–462, DOI 10.1016/S0304-3975(02)00501-7. MR1990776

[FR-08] Natalie Priebe Frank and E. Arthur Robinson Jr., *Generalized β-expansions, substitution tilings, and local finiteness*, Trans. Amer. Math. Soc. **360** (2008), no. 3, 1163–1177 (electronic), DOI 10.1090/S0002-9947-07-04527-8. MR2357692

[GPS-95] Thierry Giordano, Ian F. Putnam, and Christian F. Skau, *Topological orbit equivalence and C^*-crossed products*, J. Reine Angew. Math. **469** (1995), 51–111. MR1363826

[Hu-59] Sze-tsen Hu, *Homotopy theory*, Pure and Applied Mathematics, Vol. VIII, Academic Press, New York-London, 1959. MR0106454

[Ho-86] B. Host, *Valeurs propres des systèmes dynamiques définis par des substitutions de longueur variable* (French), Ergodic Theory Dynam. Systems **6** (1986), no. 4, 529–540, DOI 10.1017/S0143385700003679. MR873430

[HS-03] Michael Hollander and Boris Solomyak, *Two-symbol Pisot substitutions have pure discrete spectrum*, Ergodic Theory Dynam. Systems **23** (2003), no. 2, 533–540, DOI 10.1017/S0143385702001384. MR1972237

[I-O-07] Benjamín A. Itzá-Ortiz, *Eigenvalues, K-theory and minimal flows*, Canad. J. Math. **59** (2007), no. 3, 596–613, DOI 10.4153/CJM-2007-025-5. MR2319160

[KaP-03] Jerome Kaminker and Ian Putnam, *A proof of the gap labeling conjecture*, Michigan Math. J. **51** (2003), no. 3, 537–546. MR2021006

[KR-01] Anatole Katok, *Cocycles, cohomology and combinatorial constructions in ergodic theory*, Smooth ergodic theory and its applications (Seattle, WA, 1999), Proc. Sympos. Pure Math., vol. 69, Amer. Math. Soc., Providence, RI, 2001, pp. 107–173, DOI 10.1090/pspum/069/1858535. In collaboration with E. A. Robinson, Jr. MR1858535

[KeP-06] Johannes Kellendonk and Ian F. Putnam, *The Ruelle-Sullivan map for actions of $\mathbb{R}^n$*, Math. Ann. **334** (2006), no. 3, 693–711, DOI 10.1007/s00208-005-0728-1. MR2207880

[KSS-05] Richard Kenyon, Lorenzo Sadun, and Boris Solomyak, *Topological mixing for substitutions on two letters*, Ergodic Theory Dynam. Systems **25** (2005), no. 6, 1919–1934, DOI 10.1017/S0143385705000349. MR2183301

[Ki-98] Bruce P. Kitchens, *Symbolic dynamics*, Universitext, Springer-Verlag, Berlin, 1998. One-sided, two-sided and countable state Markov shifts. MR1484730

[LM-95] Douglas Lind and Brian Marcus, *An introduction to symbolic dynamics and coding*, Cambridge University Press, Cambridge, 1995. MR1369092

[PT-82] William Parry and Selim Tuncel, *Classification problems in ergodic theory*, London Mathematical Society Lecture Note Series, vol. 67, Cambridge University Press, Cambridge-New York, 1982. Statistics: Textbooks and Monographs, 41. MR666871

[Qu-87] Martine Queffélec, *Substitution dynamical systems—spectral analysis*, Lecture Notes in Mathematics, vol. 1294, Springer-Verlag, Berlin, 1987. MR924156

[Ra-90] G. Rauzy, *Sequences defined by iterated morphisms*, Sequences (Naples/Positano, 1988), Springer, New York, 1990, pp. 275–286. MR1040317

[R-04] E. Arthur Robinson Jr., *Symbolic dynamics and tilings of* $\mathbb{R}^d$, Symbolic dynamics and its applications, Proc. Sympos. Appl. Math., vol. 60, Amer. Math. Soc., Providence, RI, 2004, pp. 81–119, DOI 10.1090/psapm/060/2078847. MR2078847

[S-08] Lorenzo Sadun, *Topology of tiling spaces*, University Lecture Series, vol. 46, American Mathematical Society, Providence, RI, 2008. MR2446623

[Sc-57] Sol Schwartzman, *Asymptotic cycles*, Ann. of Math. (2) **66** (1957), 270–284. MR0088720

[Sg-03] Anne Siegel, *Représentation des systèmes dynamiques substitutifs non unimodulaires* (French, with English and French summaries), Ergodic Theory Dynam. Systems **23** (2003), no. 4, 1247–1273, DOI 10.1017/S0143385702001232. MR1997975

[So-97] Boris Solomyak, *Dynamics of self-similar tilings*, Ergodic Theory Dynam. Systems **17** (1997), no. 3, 695–738, DOI 10.1017/S0143385797084988. MR1452190

[So-98] B. Solomyak, *Nonperiodicity implies unique composition for self-similar translationally finite tilings*, Discrete Comput. Geom. **20** (1998), no. 2, 265–279, DOI 10.1007/PL00009386. MR1637896

[SiSo-02] V. F. Sirvent and B. Solomyak, *Pure discrete spectrum for one-dimensional substitution systems of Pisot type*, Canad. Math. Bull. **45** (2002), no. 4, 697–710, DOI 10.4153/CMB-2002-062-3. Dedicated to Robert V. Moody. MR1941235

[Yi-10] Inhyeop Yi, *Ordered group invariants for one-dimensional spaces*, Fund. Math. **170** (2001), no. 3, 267–286, DOI 10.4064/fm170-3-5. MR1880904

Department of Mathematics, George Washington University, Washington, DC 20052
E-mail address: andr@gwu.edu

Department of Mathematics, George Washington University, Washington, DC 20052
E-mail address: robinson@gwu.edu

Contemporary Mathematics
Volume **678**, 2016
http://dx.doi.org/10.1090/conm/678/13642

Markov diagrams for some non-Markovian systems

Kathleen Carroll and Karl Petersen

ABSTRACT. Markov diagrams provide a way to understand the structures of topological dynamical systems. We examine the construction of such diagrams for subshifts, including some which do not have any nontrivial Markovian part, in particular Sturmian systems and some substitution systems.

1. Introduction

F. Hofbauer [**12**] and J. Buzzi [**2**] defined Markov diagrams in order to study the structures and invariant measures of dynamical systems, especially those with a Markovian aspect, for example piecewise monotonic interval maps and other possibly nonunifomly expanding maps. Here we examine further the construction of these diagrams for subshifts, including some that are minimal and have zero entropy. Such subshifts may be considered to be highly non-Markovian, since they have some long-range order, indeed infinite memory. We hope that Markov diagrams will be useful also for understanding and classifying such systems, for example besides helping to identify measures of maximal entropy as in [**3, 12, 13**] also to determine complexity functions, estimate return times to cylinders, and so on.

In Sections 3 and 4 we provide a construction of Hofbauer-Buzzi Markov diagrams for Sturmian systems. In particular, in Theorem 4.13 we show that the Hofbauer-Buzzi Markov diagram of a Sturmian system can be constructed solely from its left special sequence. In Section 5 we discuss properties of Hofbauer-Buzzi Markov diagrams that hold for any subshift. We show that given a one-sided subshift X^+ there is a correspondence between those paths on the Hofbauer-Buzzi Markov diagram of X^+ that start with a vertex of length one and points in X^+ (Theorem 5.8). Corollary 5.9 relates the number of such paths to the complexity function of X^+. We prove that the eventually Markov part of the natural extension of any one-sided subshift is empty provided that the natural extension is an infinite minimal subshift (Proposition 5.13). In Section 6 we construct the Hofbauer-Buzzi Markov diagram for the Morse minimal subshift by showing that the vertices are precisely those blocks in the language of the subshift that are of the form 0 or 1 followed by a block that can be extended to the left in two ways. Theorem 7.1 applies the Hofbauer-Buzzi Markov diagrams of Sturmian subshifts to identify their lexicographically maximal elements.

2010 *Mathematics Subject Classification.* Primary 37B10.

2. Background

We recall some of the basic terminology and notation from topological and symbolic dynamics; for more details, see for example [**17**] or [**21**]. A *topological dynamical system* is a pair (X, T), where X is a compact Hausdorff space (usually metric) and $T : X \to X$ is a continuous mapping. We focus on topological dynamical systems which are *shift dynamical systems.* Let $\mathcal{A}$ be a finite set, called an *alphabet*, whose elements are called *symbols*. For us often $\mathcal{A} = \{0, 1, \ldots, n-1\}$, in fact often $\mathcal{A} = \{0, 1\}$. A *sequence* is a one-sided infinite string of symbols (a function $\mathbb{N} \to \mathcal{A}$) and a *bisequence* is an infinite string of symbols that extends in two directions (a function $\mathbb{Z} \to \mathcal{A}$). We will use the word "sequence" to apply also to bisequences, depending on the context to clarify the meaning. The *full n-shift* is $\Sigma_n = \{0, 1, \ldots, n-1\}^{\mathbb{Z}}$, the collection of all bisequences of symbols from $\mathcal{A} = \{0, 1, \ldots, n-1\}$. The *one-sided full n-shift* is $\Sigma_n^+ = \{0, 1, \ldots, n-1\}^{\mathbb{N}}$. We also define the *shift transformation* $\sigma : \Sigma(\mathcal{A}) \to \Sigma(\mathcal{A})$ and $\Sigma^+(\mathcal{A}) \to \Sigma^+(\mathcal{A})$ by $(\sigma x)_i = x_{i+1}$ for all i. The pair (Σ_n, σ) is called the *n-shift dynamical system.* We give $\mathcal{A}$ the discrete topology and $\Sigma(\mathcal{A})$ and $\Sigma^+(\mathcal{A})$ the product topology. The topologies on $\Sigma(\mathcal{A})$ and $\Sigma^+(\mathcal{A})$ are compatible with the metric $d(x, y) = 1/2^n$, where $n = \inf\{|k| \mid x_k \neq y_k\}$. A *subshift* is a pair (X, σ) (or (X^+, σ)), where $X \subset \Sigma_n$ (or $X^+ \subset \Sigma_n^+$) is a nonempty, closed, shift-invariant set. A finite string of letters from $\mathcal{A}$ is called a *block* and the length of a block B is denoted $|B|$. Furthermore, a block of length n is an n-block. Given a subshift (X, σ) of a full shift, $\mathcal{L}_n(X)$ denotes the set of all n-blocks that occur in points in X. The *language* of X is the collection $\mathcal{L}(X) = \bigcup_{n=0}^{\infty} \mathcal{L}_n(X)$. A *shift of finite type* is a subshift consisting of all sequences none of whose subblocks are in some finite collection of blocks. A topological dynamical system is *minimal* if every orbit is dense. The orbit closure of a sequence is minimal if and only if the sequence is *syndetically recurrent*: every block that appears in the sequence appears with bounded gaps see [**21**, p. 136]. The *complexity function* of a sequence u, denoted p_u, maps each natural number n to the number of blocks of length n that appear in u. If X is a subshift, then $p_X(n)$ is the number of blocks of length n that appear in $\mathcal{L}(X)$.

The construction of Hofbauer-Buzzi Markov diagrams involves the use of *follower sets.* There are several ways to define follower sets. The (*block to block*) *follower set* of a block $w \in \mathcal{L}(X)$ is $F_X(w) = \{v \in \mathcal{L}(X) | wv \in \mathcal{L}(X)\}$. Another approach defines the *future* F_X of a left-infinite sequence λ in X to be the collection of all right-infinite sequences ρ such that $\lambda\rho \in X$. This is a *ray to ray* follower set. It is also possible to define *block to ray* or *ray to block* follower sets. The definition of follower set (3.2) used in constructing Hofbauer-Buzzi diagrams is slightly different from both of these.

Follower sets have been particularly useful in examining sofic systems. A *sofic shift* is a shift space that is a factor of a shift of finite type [**25**]. Alternatively, a sofic shift consists of all sequences that are labels of infinite walks on a finite graph with labeled edges (see [**17**]). Fischer [**8**] and Krieger [**15**] used follower sets to construct covers for sofic shifts. A *presentation* of a sofic shift X is a finite labeled graph $\mathcal{G}$ for which $X_{\mathcal{G}} = X$. A presentation is *right-resolving* if for each vertex I of G the edges starting at I carry different labels. A *minimal right-resolving presentation* of a sofic shift X is a right-resolving presentation of X having the fewest vertices among all right-resolving presentations of X. Fischer proved that any two minimal right-resolving presentations are isomorphic as labeled graphs; the minimal right-resolving presentation of a sofic shift X is called the *Fischer cover* [**8**, **17**].

Given an irreducible (topologically transitive) sofic shift X over a finite alphabet $\mathcal{A}$, the Fischer cover can be constructed using the follower sets defined above. Let $\mathcal{C}_X$ be the collection of all (block to block) follower sets in X. We write $\mathcal{C}_X = \{F_X(w) | w \in \mathcal{L}(X)\}$ a finite set since X is sofic. Now construct a labeled graph $\mathcal{G} = (G, L)$ as follows. The vertices of G are the elements in $\mathcal{C}_X$. Let $c = F_X(w)$ be an element in $\mathcal{C}_X$ and $a \in \mathcal{A}$. If $wa \in \mathcal{L}(X)$, let $c' = F_X(wa) \in \mathcal{C}_X$ and draw an edge labeled a from c to c'. If $wa \notin \mathcal{L}(X)$, do nothing. Continuing this process for all elements in $\mathcal{C}_X$ yields a labeled graph $\mathcal{G}$ called the *follower set graph*. The Fischer cover of X is the labeled subgraph of the follower set graph formed by using only the follower sets of intrinsically synchronizing blocks. Here a block w in $\mathcal{L}(X)$ is *intrinsically synchronizing* if whenever $uw, wv \in \mathcal{L}(X)$ then $uwv \in \mathcal{L}(X)$ [**17**].

The Krieger cover is constructed using the *futures*, as defined above, of left-infinite sequences in X. We define the *future cover* as follows. Let $\mathcal{G}$ be the labeled graph whose vertices are the futures of left-infinite sequences. For a in $\mathcal{A}$, if λ and λa are left-infinite sequences in X, then there is an edge labeled a from $F_X(\lambda)$ to $F_X(\lambda a)$. The graph $\mathcal{G}$ is the *future* or Krieger cover of the subshift X [**15**, **17**]. The Krieger cover can be constructed for any subshift X, but it usually leads to non-irreducible and often uncountable graphs. Nevertheless, the Krieger cover is canonically associated to the subshift X. This is proved for the sofic case in [**16**] and in general in [**6**].

3. Hofbauer-Buzzi Markov diagrams

Franz Hofbauer [**12**] constructed Markov diagrams to determine measures of maximal entropy for piecewise monotone functions on the interval. In 1997, Buzzi extended the construction of the Hofbauer Markov diagram to arbitrary smooth interval maps [**1**], and in 2010 to any subshift [**2**]. The Hofbauer-Buzzi Markov diagram is a slight variation of Hofbauer's original Markov diagram. We will refer to such diagrams as *HB diagrams*. In order to describe the construction, we introduce the following definitions from [**2**].

Let $\mathcal{A}$ be a finite alphabet and $X^+ \subset \mathcal{A}^{\mathbb{N}}$ a one-sided subshift. Furthermore, let $X^{+-} \subset \mathcal{A}^{\mathbb{Z}}$ be its natural extension

$$X^{+-} = \{x \in \mathcal{A}^{\mathbb{Z}} | \text{ for all } p \in \mathbb{Z} \;\; x_p x_{p+1} ... \in X^+\},$$

with the action of the shift σ defined by $(\sigma x)_n = x_{n+1}$ for all $x \in X, n \in \mathbb{Z}$.

DEFINITION 3.1. Let π_{X^+} denote the continuous shift commuting projection from X^{+-} to X^+ defined by

$$\pi_{X^+}(x) = x_0 x_1 x_2 ...,$$

where $x = ...x_{-1}.x_0 x_1 x_2 ....$

DEFINITION 3.2. The *follower set* of a block $w = a_{-n} a_{-n+1} ... a_0$ in $\mathcal{L}(X^{+-})$ is

$$\text{fol}(a_{-n} a_{-n+1} ... a_0) = \{b_0 b_1 ... \in X^+ | \text{ there exists } b \in X^{+-} \text{ with } b_{-n} ... b_0 = a_{-n} ... a_0\}.$$

In other words, the follower set of a block $a_{-n} a_{-n+1} ... a_0$ consists of all one-sided rays $b_0 b_1 \dots$ which have an extension to a two-sided sequence $b = \dots b_{-n} \dots b_0 b_1 \dots$ in the natural extension X^{+-} of X^+ such that $b_{-n} \dots b_0 = a_{-n} \dots a_0$.

DEFINITION 3.3. A *significant block* of X^{+-} is $a_{-n}a_{-n+1}...a_0$ such that

$$\text{fol}(a_{-n}a_{-n+1}...a_0) \subsetneq \text{fol}(a_{-n+1}a_{-n+2}...a_0).$$

DEFINITION 3.4. The *significant form* of $a_{-n}a_{-n+1}...a_0$ in X^{+-} is

$$\text{sig}(a_{-n}...a_0) = a_{-k}...a_0,$$

where $k \leq n$ is maximal such that $a_{-k}...a_0$ is significant.

It is apparent that these definitions are tailored for one-sided subshifts. However, we can easily extend such definitions to an arbitrary two-sided subshift $X \subset \Sigma_n$ by letting X^+ denote the set of right rays that appear in points in X. Then X is equal to the natural extension X^{+-} of X^+.

We define the *HB diagram* $\mathcal{D}$ of a one or two-sided subshift X with natural extension X^{+-} to be the oriented graph whose vertices are the significant blocks of X^{+-} and whose arrows are defined by

$$a_{-n}...a_0 \to b_{-m}...b_0$$

if and only if $a_{-n}...a_0b_0 \in \mathcal{L}(X^{+-})$ and

$$b_{-m}...b_0 = \text{sig}(a_{-n}...a_0b_0).$$

(In Hofbauer's construction of Markov diagrams the vertices are the follower sets, not the significant blocks [**2**, **12**].)

Let $\mathcal{D}$ be the HB diagram of any one or two-sided subshift X. The following definitions from [**2**] relate $\mathcal{D}$ to X^{+-}.

DEFINITION 3.5. Given an HB diagram $\mathcal{D}$ of a subshift X with vertex set $V_{\mathcal{D}}$ (which may be infinite), the corresponding *Markov shift* is the set of all bi-infinite paths that occur on $\mathcal{D}$,

$$X(\mathcal{D}) = \{\alpha \in V_{\mathcal{D}}^{\mathbb{Z}} \mid \text{for all } p \in \mathbb{Z} \ \ \alpha_p \to \alpha_{p+1} \text{ on } \mathcal{D}\} \subset (V_{\mathcal{D}})^{\mathbb{Z}},$$

together with the shift map σ.

Note that the alphabet $V_{\mathcal{D}}$ may be infinite, and the HB diagram of an arbitrary subshift may not have paths that continue infinitely in two directions.

We relate $X(\mathcal{D})$ to X^{+-} as follows.

DEFINITION 3.6. Let $\hat{\pi}$ denote the natural continuous projection defined by

$$\hat{\pi} : \alpha \in X(\mathcal{D}) \mapsto a \in X^{+-}$$

with a_n the last symbol of the block α_n for all $n \in \mathbb{Z}$.

DEFINITION 3.7. Let $X(\mathcal{D})_v^+$ denote the space of one-sided infinite paths starting at vertex v on $\mathcal{D}$, and let $X(\mathcal{D})^+$ denote the space of one-sided infinite paths starting at a vertex v of length 1 on $\mathcal{D}$.

DEFINITION 3.8. Let $\hat{\pi}^+$ denote the projection defined by

$$\hat{\pi}^+ : \alpha \in X(\mathcal{D})^+ \mapsto a \in X^+$$

with a_n the last symbol of the block α_n for all $n \in \mathbb{N}$.

In case we want to project a finite path $\alpha_0 \to \alpha_1 \to \cdots \to \alpha_n$ on $\mathcal{D}$ to a block in $\mathcal{L}(X^+)$, we write $\hat{\pi}(\alpha_0...\alpha_n) = a_0...a_n$, where a_i is the last letter of α_i.

We state a few preliminary results that apply to any subshift.

LEMMA 3.9. *If $a_1...a_n$ is a signficant block of a subshift X, then $a_1...a_{n-1}$ is also a significant block of X.*

PROOF. If $a_1 \dots a_{n-1}$ is not significant, then $\text{fol}(a_1 \dots a_{n-1}) = \text{fol}(a_2 \dots a_{n-1})$. We want to show that $\text{fol}(a_1 \dots a_n) = \text{fol}(a_2 \dots a_n)$, so that $a_1 \dots a_n$ is not significant. Let $b_n b_{n+1} \dots \in \text{fol}(a_2 \dots a_n)$, so that there is $b \in X^{+-}$ with $b_2 \dots b_n = a_2 \dots a_n$. Then $b_{n-1} b_n b_{n+1} \dots \in \text{fol}(a_2 \dots a_{n-1}) = \text{fol}(a_1 \dots a_{n-1})$, so there is $c \in X^{+-}$ with $c_{n-1} c_n c_{n+1} \dots = b_{n-1} b_n b_{n+1} \dots$ and $c_1 \dots c_{n-1} = a_1 \dots a_{n-1}$. Therefore $c_n c_{n+1} \dots = b_n b_{n+1} \dots \in \text{fol}(a_1 \dots a_n)$. □

The following Proposition is an immediate consequence of Lemma 3.9.

PROPOSITION 3.10. *If there are infinitely many significant blocks of a subshift X, then for all $n \in \mathbb{N}$ there exists a significant block of X of length n.*

4. Markov diagrams for one-sided Sturmian systems

4.1. Basic properties of Sturmian sequences. We recall the definition and basic properties of Sturmian sequences; see [**9**, Ch. 6] for details. A one or two-sided sequence u with values in a finite alphabet is called *Sturmian* if it has complexity function (defined above) $p_u(n) = n + 1$ for all n. If u is Sturmian, then $p_u(1) = 2$. This implies that Sturmian sequences are over a two-letter alphabet, so we fix the alphabet $\mathcal{A} = \{0, 1\}$. Given a one-sided Sturmian sequence u, we let X_u^+ be the closure of $\{\sigma^n u | n \in \mathbb{N}\}$. Then (X_u^+, σ) is the *Sturmian system* associated with u.

EXAMPLE 4.1 The Fibonacci substitution is defined by:

$$\begin{aligned} \phi : 0 &\mapsto 01 \\ 1 &\mapsto 0. \end{aligned}$$

The fixed point of the Fibonacci substitution, f=0100101001001010010100100101..., is a Sturmian sequence, and (X_f^+, σ) is the Sturmian system associated with f (see [**18**]).

An infinite sequence u is *periodic* (respectively *eventually periodic*) if there exists a positive integer M such that for every n, $u_n = u_{n+M}$ (respectively there exists $m \in \mathbb{N}$ such that for all $|n| \geq m$, $u_n = u_{n+M}$). (Note that there is some possibility for variation in this definition). A set S of blocks is *balanced* if for any pair of blocks u, v of the same length in S, $||u|_1 - |v|_1| \leq 1$, where $|u|_1$ is the number of occurrences of 1 in u and $|v|_1$ is the number of occurrences of 1 in v. It follows that if a sequence u is balanced and not eventually periodic then it is Sturmian. This is a result of the fact that if u is aperiodic, then $p_u(n) \geq n + 1$ for all n, and if u is balanced then $p_u(n) \leq n + 1$ for all n. In fact, it can be proved that a sequence u is balanced and aperiodic if and only if it is Sturmian [**18**]. Furthermore, any shift of a Sturmian sequence is also Sturmian.

Sturmian sequences also have a natural association to lines with irrational slope. To see this, we introduce the following definitions. Let α and β be real numbers with $0 \leq \alpha, \beta \leq 1$. We define two infinite sequences $x_{\alpha,\beta}$ and $x'_{\alpha,\beta}$ by

$$\begin{aligned} (x_{\alpha,\beta})_n &= \lfloor \alpha(n+1) + \beta \rfloor - \lfloor \alpha n + \beta \rfloor \\ (x'_{\alpha,\beta})_n &= \lceil \alpha(n+1) + \beta \rceil - \lceil \alpha n + \beta \rceil \end{aligned}$$

for all $n \geq 0$. The sequence $x_{\alpha,\beta}$ is the *lower mechanical sequence* and $x'_{\alpha,\beta}$ is the *upper mechanical sequence* with slope α and intercept β. The use of the words slope and intercept in the above definitions stems from the following graphical interpretation. The points with integer coordinates that sit just below the line $y = \alpha x + \beta$ are $F_n = (n, \lfloor \alpha n + \beta \rfloor)$. The straight line segment connecting two consecutive points F_n and F_{n+1} is horizontal if $x_{\alpha,\beta} = 0$ and diagonal if $x_{\alpha,\beta} = 1$. The lower mechanical sequence is a coding of the line $y = \alpha x + \beta$ by assigning to each line segment connecting F_n and F_{n+1} a 0 if the segment is horizontal and a 1 if the segment is diagonal. Similarly, the points with integer coordinates that sit just above this line are $F'_n = (n, \lceil \alpha n + \beta \rceil)$. Again, we can code the line $y = \alpha x + \beta$ by assigning to each line segment connecting F'_n and F'_{n+1} a 0 if the segment is horizontal and a 1 if the segment is diagonal. This coding yields the upper mechanical sequence [**18**].

A mechanical sequence is *rational* if the line $y = \alpha x + \beta$ has rational slope and *irrational* if $y = \alpha x + \beta$ has irrational slope. In [**18**] it is proved that a sequence u is Sturmian if and only if u is irrational mechanical. In the following example we construct a lower mechanical sequence with irrational slope, thus producing a Sturmian sequence.

EXAMPLE 4.2. Let $\alpha = 1/\tau^2$, where $\tau = (1+\sqrt{5})/2$ is the golden mean, and $\beta = 0$. The lower mechanical sequence $x_{\alpha,\beta}$ is constructed as follows:

$$\begin{aligned}
(x_{\alpha,\beta})_0 &= \lfloor 1/\tau^2 \rfloor = 0 \\
(x_{\alpha,\beta})_1 &= \lfloor 2/\tau^2 \rfloor - \lfloor 1/\tau^2 \rfloor = 0 \\
(x_{\alpha,\beta})_2 &= \lfloor 3/\tau^2 \rfloor - \lfloor 2/\tau^2 \rfloor = 1 \\
(x_{\alpha,\beta})_3 &= \lfloor 4/\tau^2 \rfloor - \lfloor 3/\tau^2 \rfloor = 0 \\
(x_{\alpha,\beta})_4 &= \lfloor 5/\tau^2 \rfloor - \lfloor 4/\tau^2 \rfloor = 0 \\
(x_{\alpha,\beta})_5 &= \lfloor 6/\tau^2 \rfloor - \lfloor 5/\tau^2 \rfloor = 1 \\
&\vdots
\end{aligned}$$

Further calculation shows that $x_{\alpha,\beta} = 0010010100\ldots = 0f$, and $x'_{\alpha,\beta} = 1010010100\ldots = 1f$ (recall (Example 4.1) that f denotes the fixed point of the Fibonacci substitution), hence the fixed point f is a shift of the lower and upper mechanical sequences with slope $1/\tau^2$ and intercept 0.

We now consider the language of a Sturmian sequence u. It is easy to show that while Sturmian sequences are aperiodic, they are syndetically recurrent [**9**]. As a result, any block in $\mathcal{L}_n(u)$ appears past the initial position and can thus be extended on the left. Since there are $n+1$ blocks of length n, it must be that exactly one of them can be extended to the left in two ways. In a Sturmian sequence u, the unique block of length n that can be extended to the left in two different ways is called a *left special block*, and is denoted $L_n(u)$. The sequence $l(u)$ which has the $L_n(u)$'s as prefixes is called the *left special sequence* or *characteristic word* of X_u^+ [**9, 18**]. Similarly, in a Sturmian sequence u, the unique block of length n that can be extended to the right in two different ways is called a *right special block*, and is denoted $R_n(u)$. The block $R_n(u)$ is precisely the reverse of $L_n(u)$ [**9**].

4.2. The left special sequence. Since every Sturmian sequence u is irrational mechanical, there is a line with irrational slope α associated to u. This α can be used to determine the left special sequence of X_u^+.

Let $(d_1, d_2, ..., d_n, ...)$ be a sequence of integers with $d_1 \geq 0$ and $d_n > 0$ for $n > 1$. We associate a sequence $(s_n)_{n \geq -1}$ of blocks to this sequence by

$$s_{-1} = 1, \qquad s_0 = 0, \qquad s_n = s_{n-1}^{d_n} s_{n-2}.$$

The sequence $(s_n)_{n \geq -1}$ is a *standard sequence*, and $(d_1, d_2, ..., d_n, ..)$ is its *directive sequence*. We can then determine the left special sequence of X_u^+ with the following proposition stated in [**18**].

PROPOSITION 4.3. *Let $\alpha = [0, 1+d_1, d_2, ...]$ be the continued fraction expansion of an irrational α with $0 < \alpha < 1$, and let (s_n) be the standard sequence associated to $(d_1, d_2, ...)$. Then every s_n, $n \geq 1$, is a prefix of the left special sequence l of X_u^+ and*

$$l = \lim_{n \to \infty} s_n.$$

This is illustrated in the following two examples.

EXAMPLE 4.4. Let $\alpha = 1/\tau^2$, where $\tau = (1 + \sqrt{5})/2$ is the golden mean. The continued fraction expansion of $1/\tau^2$ is $[0, 2, \overline{1}]$ where $\overline{1} = 1^{\infty} = 111 \ldots$ By the above proposition $d_1 = 1, d_2 = 1, d_3 = 1, d_4 = 1, ...$ The standard sequence associated to $(d_1, d_2, ...)$ is constructed as follows:

$$\begin{aligned}
s_1 =& s_0^{d_1} s_{-1} = 01 \\
s_2 =& s_1^{d_2} s_0 = 010 \\
s_3 =& s_2^{d_3} s_1 = 01001 \\
s_4 =& s_3^{d_4} s_2 = 01001010 \\
&\vdots
\end{aligned}$$

Continuing this process, the left special sequence of X_u^+, where u is a coding of a line with slope $1/\tau^2$, is

$$l = 010010100100101001... = f.$$

It follows that the left special sequence of X_f^+ is f.

EXAMPLE 4.5. Let $\alpha = \pi/4$. The continued fraction expansion of $\pi/4$ is

$$[0, 1, 3, 1, 1, 1, 15, 2, 72, ...].$$

By Proposition 4.3 $d_1 = 0, d_2 = 3, d_3 = 1, d_4 = 1, ...$ Then,

$$\begin{aligned}
s_1 =& s_0^{d_1} s_{-1} = 1 \\
s_2 =& s_1^{d_2} s_0 = 1110 \\
s_3 =& s_2^{d_3} s_1 = 11101 \\
s_4 =& s_3^{d_4} s_2 = 111011110 \\
&\vdots
\end{aligned}$$

Continuing this process, the left special sequence of X_u^+, where u is a coding of a line with slope $\pi/4$, is

$$l = 111011110111101111011110....$$

4.3. Significant blocks of a one-sided Sturmian system. In order to construct the HB diagram of a Sturmian system, it is necessary to identify the significant blocks of the system. We first note that if X is any subshift of Σ_d, then $0, 1, ..., d-1$, and d are significant blocks provided $0, 1, ..., d-1$, and d are all in $\mathcal{L}(X^{+-})$. Hence, 0 and 1 are significant blocks of any Sturmian system. Let (X_u^+, σ) be a Sturmian system with $l = l_1 l_2 l_3 ...$ the left special sequence of u. In the next two propositions we prove that given $n \geq 1$, there are exactly two significant blocks of $\tilde{X}_u$ with length n.

The first proposition applies to any subshift of Σ_2.

PROPOSITION 4.6. *Let $X \subset \Sigma_2$ be a subshift. Suppose $a_{-n+1}a_{-n+2}...a_{-1}a_0$ is a block of length n in $\mathcal{L}(X^{+-})$. If $a_{-n+1}a_{-n+2}...a_{-1}a_0$ is significant, then $0a_{-n+2}...a_{-1}a_0$ and $1a_{-n+2}...a_{-1}a_0$ are in $\mathcal{L}(X^{+-})$.*

PROOF. Assume that the block $a_{-n+1}a_{-n+2}...a_{-2}a_{-1}a_0$ is significant, so that there is a ray $b_0 b_1 \cdots \in \text{fol}(a_{-n+1} \dots a_0) \subsetneq \text{fol}(a_{-n+2} \dots a_0)$. Then there is $b \in X^{+-}$ such that $b_{-n+2} \dots b_0 = a_{-n+2} \dots a_0$, and we must have $b_{-n+1} \neq a_{-n+1}$. □

Before we state the second proposition, recall that since u is recurrent, every block is extendable to the left. Hence, $\mathcal{L}(u) = \mathcal{L}(X_u^+) = \mathcal{L}(X_u^{+-})$. We abbreviate $L_k = L_k(u)$, the left special block of length k in u.

PROPOSITION 4.7. *Let u be a Sturmian sequence, with (X_u^+, σ) its associated Sturmian system. Then for each $n \geq 2$ there are exactly two significant blocks of $\tilde{X}_u$ of length n. The two significant blocks of length n are $0L_{n-1} = 0l_1...l_{n-1}$ and $1L_{n-1} = 1l_1...l_{n-1}$.*

PROOF. By Proposition 4.6 we know that if the block $a_{-n+1}a_{-n+3}...a_{-2}a_{-1}a_0$ is significant then $0a_{-n+2}...a_{-2}a_{-1}a_0$ and $1a_{-n+2}...a_{-2}a_{-1}a_0$ are in $\mathcal{L}(X_u^{+-})$. Thus the only possible significant blocks of length n are those blocks $a_{-n+1}...a_{-2}a_{-1}a_0$ such that $a_{-n+2}...a_{-2}a_{-1}a_0$ can be extended to the left in two ways. That is, the possible significant blocks are $a_{-n+1}a_{-n+2}...a_{-2}a_{-1}a_0$ with

$$a_{-n+2}...a_{-2}a_{-1}a_0 = L_{n-1}.$$

Let $L_{n-1} = a_{-n+2}...a_0$ and $\nu \in \mathcal{L}(u)$. Since there is exactly one right special block of each length $n \in \mathbb{N}$, $0L_{n-1}\nu$ and $1L_{n-1}\nu$ cannot both be right special blocks. We first prove that $1L_{n-1}$ is significant by considering the following two cases.

Case 1: There exists $\nu \in \mathcal{L}(u)$ such that $0L_{n-1}\nu$ is right special. Then $1L_{n-1}\nu$ is not right special. This implies that $1L_{n-1}\nu 1$ is not in $\mathcal{L}(u)$, since

$$||0L_{n-1}\nu 0|_1 - |1L_{n-1}\nu 1|_1| = 2,$$

which is not permitted as u is balanced. Thus there exists a ray $a_0\nu 1...$ in $\text{fol}(L_{n-1})$ that is not in $\text{fol}(1L_{n-1})$. Hence, $1L_{n-1}$ is significant.

Case 2: There does not exist $\nu \in \mathcal{L}(u)$ such that $0L_{n-1}\nu$ is right special. This implies that there exists exactly one ray $b_1 b_2 ...$ that can follow $0L_{n-1}$. We claim that because u is Sturmian such a case cannot occur.

Let $u = c_0c_1c_2c_3....$ Since $0L_{n-1} \in \mathcal{L}(u)$, we know that $0L_{n-1}$ appears in u infinitely many times. Suppose $0L_{n-1}$ appears for the first time starting at position c_{m+1}. Letting $0L_{n-1} = a_{-n+2}...a_0$, we have

$$u = c_0c_1c_2c_3...c_m0L_{n-1}b_1b_2... = c_0c_1c_2c_3...c_m0a_{-n+2}...a_0b_1b_2...$$

Furthermore, there exists $r \in \mathbb{N}$ such that $0L_{n-1}$ appears again starting at b_{r+1}. As $0L_{n-1}$ can be followed only by $b_1b_2...$, this implies that

$$u = c_0c_1c_2c_3...c_m0a_{-n+2}...a_0b_1b_2...b_r0a_{-n+2}...a_0b_1b_2...b_r0a_{-n+2}...a_0b_1b_2...b_r...$$

Letting $B = 0a_{-n+2}...a_0b_1b_2...b_r$, we have that

$$u = c_0c_1c_2c_3...c_mBBBB....$$

Thus, u is eventually periodic. This, however, is a contradiction as Sturmian sequences are not eventually periodic. Hence $1L_{n-1}$ is a significant block of X_u^{+-}.

By the same argument, it can be shown that $0L_{n-1}$ is also a significant block of X_u^{+-}. □

4.4. Construction of the diagram. Recall that the HB diagram of a one-sided Sturmian system is defined to be an oriented graph whose vertices are the significant blocks of $\tilde{X}_u$ and whose arrows are defined by

$$a_{-n}...a_0 \to b_{-m}...b_0$$

if and only if $a_{-n}...a_0b_0 \in \mathcal{L}(\tilde{X}_u)$ and

$$b_{-m}...b_0 = \operatorname{sig}(a_{-n}...a_0b_0).$$

Having determined the significant blocks of $\tilde{X}_u$, it remains only to determine the arrows. This will give us a complete description of the HB diagram of an arbitrary one-sided Sturmian system.

Let $l = l_1l_2l_3...$ be the left special sequence of X_u^+. We first consider the arrows leaving the significant blocks of length 1.

LEMMA 4.8. *If* $l_1 = 0$, *then* $0 \to 1$, $0 \to 00$ *and* $1 \to 10$. *If* $l_1 = 1$, *then* $1 \to 0$, $1 \to 11$ *and* $0 \to 01$.

PROOF. Suppose $l_1 = 0$. By definition

$$0 \to b_{-m}...b_0$$

if and only if $0b_0 \in \mathcal{L}(\tilde{X}_u)$ and

$$b_{-m}...b_0 = \operatorname{sig}(0b_0).$$

As $b_0 \in \{0,1\}$, we consider sig(00) and sig(01). Proposition 4.7 implies that the significant blocks of length two are 00 and 10, since $l_1 = 0$. Thus sig(00) = 00 and sig(01) = 1. Hence, $0 \to 1$ and $0 \to 00$. Additionally, consider $\operatorname{sig}(1b_0)$. Since 0 is the unique right special block of length one, the balance property implies that $11 \notin \mathcal{L}(\tilde{X}_u)$. Thus there is exactly one arrow leaving the block 1, $1 \to \operatorname{sig}(10)$, where sig(10) = 10 by Proposition 4.7.

Similarly, if $l_1 = 1$ we consider sig(10) and sig(11). In this case, Proposition 4.7 implies that the significant blocks of length two are 10 and 11. Thus sig(10) = 0 and sig(11) = 11 and $1 \to 0$ and $1 \to 11$. Furthermore, the only arrow leaving 0 is given by $0 \to \operatorname{sig}(01)$, where sig(01) = 01 by Proposition 4.7.

□

Now consider an arbitrary significant block $xl_1l_2...l_{n-1}$, where x is either 0 or 1. Again,

$$xl_1l_2...l_{n-1} \to b_{-m}...b_0$$

if and only if $xl_1l_2...l_{n-1}b_0 \in \mathcal{L}(\tilde{X}_u)$ and

$$b_{-m}...b_0 = \text{sig}(xl_1l_2...l_{n-1}b_0).$$

We consider what may follow $xl_1l_2...l_{n-1}$. There can be at most two arrows out of $xl_1l_2...l_{n-1}$, as $b_0 \in \{0,1\}$. It is always the case that $xl_1l_2...l_{n-1}l_n \in \mathcal{L}(\tilde{X}_u)$. Letting $b_0 = l_n$, we get

$$xl_1l_2...l_{n-1} \to b_{-m}...b_{-1}l_n$$

if and only if

$$b_{-m}...b_{-1}l_n = \text{sig}(xl_1l_2...l_{n-1}l_n).$$

However, $xl_1l_2...l_{n-1}l_n$ is significant; thus it must be that $b_{-m}...b_0 = xl_1l_2...l_{n-1}l_n$. Hence, we are guaranteed the arrow

$$xl_1l_2...l_{n-1} \to xl_1l_2...l_n.$$

This is stated below.

LEMMA 4.9. *Let $xL_{n-1} = xl_1l_2...l_{n-1}$, $n > 1$, $x \in \{0,1\}$, be a significant block of $\tilde{X}_u$. Then*

$$xl_1l_2...l_{n-1} \to xl_1l_2...l_n.$$

It follows from Lemma 4.9 that the left special sequence of X_u^+ is seen in the diagram by reading off the last symbol in the paths $0l_1 \to 0l_1l_2 \to 0l_1l_2l_3 \to \cdots$ and $1l_1 \to 1l_1l_2 \to 1l_1l_2l_3 \to \cdots$.

Now suppose $xl_1l_2...l_{n-1}y \in \mathcal{L}(\tilde{X}_u)$ and $y \neq l_n$. This occurs if and only if $xl_1l_2...l_{n-1}$ is a right special block. Since $R_n(u)$ is the reverse of $L_n(u) = l_1l_2...l_n$, we get the following lemma.

LEMMA 4.10. *Let $xl_1l_2...l_{n-1}$ be a significant block of $\tilde{X}_u$. In the HB diagram of X_u^+, two arrows leave $xl_1l_2...l_{n-1}$ if and only if $xl_1l_2...l_{n-1}$ is a right special block, equivalently if and only if $xl_1l_2...l_{n-1} = l_nl_{n-1}...l_2l_1$.*

Suppose $xl_1l_2...l_{n-1}$ is a right special significant block, where $n > 1$. Let $wl_1...l_{m-1}$, $1 \leq m < n$ be the *previous* right special significant block of $\tilde{X}_u$. In other words, there is no right special significant block of length greater than m and less than n. By definition $xl_1l_2...l_{n-1} = l_nl_{n-1}...l_2l_1$ and $wl_1...l_{m-1} = l_ml_{m-1}...l_2l_1$. We claim the following.

LEMMA 4.11. *Let $xl_1l_2...l_{n-1}$ and $wl_1...l_{m-1}$ be consecutive right special significant blocks as described and suppose $y \neq l_n$. Then*

$$\text{sig}(xl_1l_2...l_{n-1}y) = \text{sig}(wl_1...l_{m-1}y).$$

PROOF. Since $y \neq l_n$ it follows that $\text{sig}(xl_1l_2...l_{n-1}y) \neq xl_1l_2...l_{n-1}y$. Suppose to the contrary that

$$\text{sig}(xl_1l_2...l_{n-1}y) = \text{sig}(l_nl_{n-1}...l_2l_1y) = l_{m+i}...l_ml_{m-1}...l_2l_1y,$$

for some $i \geq 1$ with $m + i < n$.

Then $l_{m+i}...l_m l_{m-1}...l_2 l_1 y = z l_1 ... l_{m+i}$ for some $z \in \{0,1\}$, since $l_{m+i}...l_m l_{m-1} ...l_2 l_1 y$ is significant. However, this implies that

$$l_{m+i}...l_m l_{m-1}...l_2 l_1 = z l_1 ... l_{m+i-1}$$

is right special. This is a contradiction, since $l_{m+i}...l_m l_{m-1}...l_2 l_1$ is a right special block of length $m+i$, with $m < m+i < n$, and $wl_1...l_{m-1} = l_m l_{m-1}...l_2 l_1$ is the previous right special significant block. Hence,

$$sig(xl_1 l_2 ... l_{n-1} y) = sig(wl_1 ... l_{m-1} y).$$

□

We use the following lemma to determine the remaining arrows.

LEMMA 4.12. *Let $xl_1l_2...l_{n-1}$ and $wl_1l_2...l_{m-1}$, $1 \leq m < n$ be consecutive right special significant blocks as described and suppose $y \neq l_n$. If $x \neq w$, then*

$$xl_1l_2...l_{n-1} \to wl_1l_2...l_{m-1}l_m.$$

If $x = w$ then,

$$xl_1l_2...l_{n-1} \to \mathrm{sig}(wl_1l_2...l_{m-1}y).$$

PROOF. By Lemma 4.11,

$$\mathrm{sig}(xl_1l_2...l_{n-1}y) = \mathrm{sig}(wl_1...l_{m-1}y) = \mathrm{sig}(l_m l_{m-1}...l_2 l_1 l_m).$$

By Proposition 4.7

$$\mathrm{sig}(wl_1...l_{m-1}l_m) = wl_1...l_{m-1}l_m.$$

This gives us the arrow

$$xl_1l_2...l_{n-1} \to wl_1...l_{m-1}l_m.$$

Suppose $x \neq w$. We know that $xl_1l_2...l_{n-1} = l_n l_{n-1}...l_2 l_1$ and $wl_1...l_{m-1} = l_m l_{m-1}...l_2 l_1$, so $x \neq w$ implies that $l_n \neq l_m$. Additionally, $y \neq l_n$ implies $y = l_m$.

Now suppose $x = w$. Then $l_n = l_m$, and thus $y \neq l_m$. Therefore

$$xl_1l_2...l_{n-1} \to \mathrm{sig}(wl_1...l_{m-1}y).$$

That is, there is an arrow leaving $xl_1l_2...l_{n-1}$ that points to the same significant block as one of the arrows leaving wL_{m-1}. □

We summarize the construction of the HB diagram of an arbitrary one-sided Sturmian system in the following theorem.

THEOREM 4.13. *Let X_u^+ be a one-sided Sturmian system, with $l = l_1l_2l_3...$ the left special sequence of u, and $L_n = l_1 \dots l_n$ for each $n \geq 1$. The HB diagram of X_u^+ is the directed graph with vertices $0, 1, 0L_n$, and $1L_n$, $n \geq 1$, and whose arrows are defined by*

(a) $0 \to 1$, $0 \to 00$, *and* $1 \to 10$ *if* $l_1 = 0$, *and* $1 \to 0$, $1 \to 11$, *and* $0 \to 01$ *if* $l_1 = 1$,

(b) $0L_n \to 0L_{n+1}$, $1L_n \to 1L_{n+1}$,

(c) *If xL_n and wL_m, $n \geq m$, are consecutive right special blocks*

  (i) $xL_n \to wL_{m+1}$ *if* $x \neq w$

  (ii) $xL_n \to \mathrm{sig}(wL_m y)$, $y \neq l_{m+1}$, *if* $x = w$.

We describe the construction of the HB diagrams of two Sturmian systems. Recall from Example 4.1 that the Fibonacci Sturmian system is (X_f^+, σ), where $f = 0100101001001010010100....$ In Example 4.4 it is shown that the left special sequence of X_f^+ is f. By Proposition 4.7, the significant blocks of $\tilde{X}_f$ are

$$0, 1, 00, 10, 001, 101, 0010, 1010, 00100, 10100, ...$$

Furthermore, the first few right special significant blocks are $0, 10$, 0010, and 1010010. Following Theorem 4.13, we construct a portion of the HB diagram of X_f^+, as depicted in Figure 1.

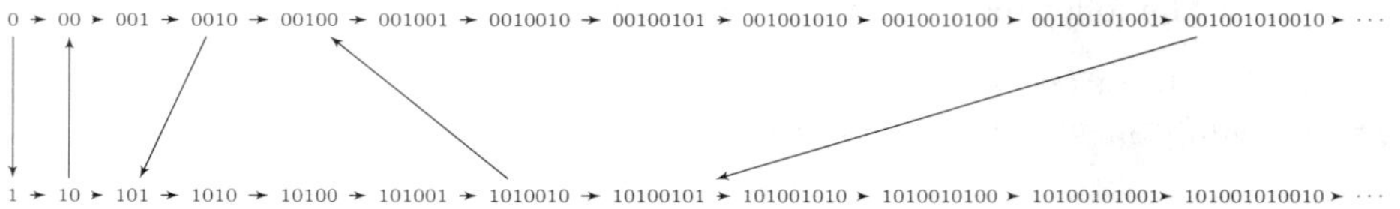

FIGURE 1. The HB diagram of X_f.

Next consider the sequence u, where u is the upper or lower mechanical sequence with slope $\alpha = \pi/4$ and intercept β, $\beta \leq 1$. Let (X_u^+, σ) be the Sturmian system associated with the sequence u. Earlier we found that the left special sequence of X_u^+ is

$$l = 1110111101110111011110....$$

Applying Proposition 4.7, the significant blocks of $\tilde{X}_u$ are

$$0, 1, 01, 11, 011, 111, 0111, 1111, 01110, 11110, ...,$$

and the first few right special significant blocks are $1, 11, 111$, 0111 and 11110111. Following Theorem 4.13, we begin construction of the HB diagram of X_u^+, as depicted in Figure 2.

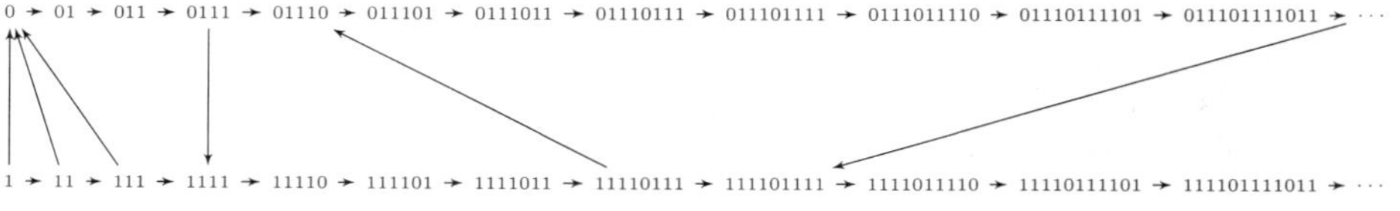

FIGURE 2. The HB diagram of X_u^+ where u is the upper or lower mechanical sequence with slope $\alpha = \pi/4$.

Unlike the situation with Markovian systems, in these diagrams there are no closed paths (since there are no periodic sequences in the subshifts).

5. General properties of HB diagrams

We consider next what HB diagrams can tell us about their associated systems. We first consider the properties of the HB diagram that hold for any subshift.

Let X^+ be a one-sided subshift with natural extension X^{+-} as previously defined.

LEMMA 5.1. *Suppose $c_kc_{k-1}...c_1c_0$ is a block in $\mathcal{L}(X^{+-})$. Then*

$$\mathrm{sig}(\mathrm{sig}(c_kc_{k-1}...c_1)c_0) = \mathrm{sig}(c_kc_{k-1}...c_1c_0).$$

PROOF. Let $|\mathrm{sig}(c_kc_{k-1}...c_1)|$ denote the length of $\mathrm{sig}(c_kc_{k-1}...c_1)$. Then

$$|\mathrm{sig}(\mathrm{sig}(c_kc_{k-1}...c_1)c_0)| \leq |\mathrm{sig}(c_kc_{k-1}...c_1c_0)|,$$

since $|\mathrm{sig}(c_kc_{k-1}...c_1)| \leq |c_kc_{k-1}...c_1|$.

Suppose on the contrary that

$$|\mathrm{sig}(\mathrm{sig}(c_kc_{k-1}...c_1)c_0)| < |\mathrm{sig}(c_kc_{k-1}...c_1c_0)|.$$

Furthermore, suppose $\mathrm{sig}(c_kc_{k-1}...c_1) = c_j...c_1$, where $1 \leq j \leq k$.

Then

$$\mathrm{sig}(\mathrm{sig}(c_kc_{k-1}...c_1)c_0) = \mathrm{sig}(c_j...c_1c_0) = c_m...c_1c_0$$

with $0 \leq m \leq j$, and

$$\mathrm{sig}(c_kc_{k-1}...c_1c_0) = c_r...c_1c_0$$

with $m < r \leq k$.

Since $c_r...c_1c_0$ is significant, we know that $\mathrm{fol}(c_r...c_1c_0) \subsetneq \mathrm{fol}(c_{r-1}...c_1c_0)$. Thus there exists a one-sided ray $b_0b_1b_2...$ in $\mathrm{fol}(c_{r-1}...c_1c_0)$ that is not in $\mathrm{fol}(c_r...c_1c_0)$. Furthermore, $b_0b_1b_2...$ is such that there exists a two-sided ray b in the natural extension of X^+, with $b_{-r+1}...b_0 = c_{r-1}...c_0$. However, this implies that $b_{-1}b_0b_1...$ is a one-sided ray in $\mathrm{fol}(c_{r-1}...c_1)$ that is not in $\mathrm{fol}(c_r...c_1)$. Hence, $c_r...c_1$ is significant.

It follows that $r \leq j$. If not $\mathrm{sig}(c_kc_{k-1}...c_1) = c_r...c_1$. However, we have assumed $r > m$. This contradicts $\mathrm{sig}(c_j...c_1c_0) = c_m...c_1c_0$, since $c_r...c_1c_0$ is a longer significant block than $c_m...c_1c_0$ that is also a suffix of $c_j...c_1c_0$. Thus,

$$|\mathrm{sig}(\mathrm{sig}(c_kc_{k-1}...c_1)c_0)| = |\mathrm{sig}(c_kc_{k-1}...c_1c_0)|,$$

and so

$$\mathrm{sig}(\mathrm{sig}(c_kc_{k-1}...c_1)c_0) = \mathrm{sig}(c_kc_{k-1}...c_1c_0).$$

□

The following corollary, which is stated as an exercise in [**2**], follows from Lemma 5.1 by induction.

COROLLARY 5.2. *Let $\alpha_0 \to \alpha_1 \to \cdots \to \alpha_n$ be a finite path on a HB diagram $\mathcal{D}$. Suppose $\alpha_0 = b_{-k}...b_0$ and let a_i be the last letter of α_i. Then for all $n \in \mathbb{N}$*

$$\alpha_n = \mathrm{sig}(b_{-k}...b_0a_1a_2...a_n).$$

Shifting our attention to the paths on $\mathcal{D}$, we define the *length* of a path to be the number of vertices in the path. From Corollary 5.2 we get the following result.

THEOREM 5.3. *Let X be a subshift with HB diagram $\mathcal{D}$. If $\alpha_0 \to \alpha_1 \to \cdots \to \alpha_{n-1}$ and $\beta_0 \to \beta_1 \to \cdots \beta_{n-1}$ are two paths of length n on $\mathcal{D}$ with α_0 and β_0 blocks of length 1, then $\hat{\pi}(\alpha_0\alpha_1...\alpha_{n-1}) = \hat{\pi}(\beta_0\beta_1...\beta_{n-1})$ if and only if $\alpha_i = \beta_i$ for all $0 \leq i \leq n-1$.*

PROOF. Let a_i and b_i be the last letters of α_i and β_i respectively.

Suppose $\hat{\pi}(\alpha_0\alpha_1...\alpha_{n-1}) = \hat{\pi}(\beta_0\beta_1...\beta_{n-1})$. By Corollary 5.2,

$$\alpha_i = \text{sig}(\alpha_0 a_1...a_i) = \text{sig}(a_0 a_1...a_i)$$

and

$$\beta_i = \text{sig}(\beta_0 b_1...b_i) = \text{sig}(b_0 b_1...b_i).$$

Furthermore, $a_0a_1...a_i = \hat{\pi}(\alpha_0\alpha_1...\alpha_i)$ and $b_0b_1...b_i = \hat{\pi}(\beta_0\beta_1...\beta_i)$. As $\hat{\pi}(\alpha_0\alpha_1...\alpha_i)$ and $\hat{\pi}(\beta_0\beta_1...\beta_i)$ are prefixes of $\hat{\pi}(\alpha_0\alpha_1...\alpha_{n-1})$ and $\hat{\pi}(\beta_0\beta_1...\beta_{n-1})$ respectively, it follows that

$$\hat{\pi}(\alpha_0\alpha_1...\alpha_i) = \hat{\pi}(\beta_0\beta_1...\beta_i).$$

Hence,

$$\alpha_i = \beta_i = \text{sig}(\hat{\pi}(\alpha_0\alpha_1...\alpha_i)) = \text{sig}(\hat{\pi}(\beta_0\beta_1...\beta_i)).$$

Now suppose $\alpha_i = \beta_i$ for all $0 \leq i \leq n-1$. Then $a_i = b_i$ for all $0 \leq i \leq n-1$. Thus $a_0...a_{n-1} = b_0...b_{n-1}$. By definition, $\hat{\pi}(\alpha_0\alpha_1...\alpha_{n-1}) = a_0...a_{n-1}$ and $\hat{\pi}(\beta_0\beta_1...\beta_{n-1}) = b_0...b_{n-1}$. Hence $\hat{\pi}(\alpha_0\alpha_1...\alpha_{n-1}) = \hat{\pi}(\beta_0\beta_1...\beta_{n-1})$. □

It follows from Theorem 5.3 that on the HB diagram $\mathcal{D}$ of a subshift X each distinct path of length n starting at a block of length one projects to a distinct block of length n in $\mathcal{L}(X)$.

Now suppose B is a block in $\mathcal{L}(X)$. We ask, does there exist a path on $\mathcal{D}$ starting with a block of length one that projects to the block B? In general, this is not the case.

EXAMPLE 5.4. Let $u = 1\bar{0}$ and consider the system (X^+, σ) where X^+ is the orbit closure of $\{\sigma^n u | n \in \mathbb{N}\}$. We claim that $1 \notin \mathcal{L}(X^{+-})$. For 1 to be in $\mathcal{L}(X^{+-})$, 1 must appear in a two-sided sequence b such that $b_p b_{p+1}... \in X^+$ for all $p \in \mathbb{Z}$. Suppose

$$b = ...b_{-n-2}1b_{-n}...b_{-2}b_{-1}b_0b_1b_1....$$

Then b_{-n-2} cannot equal 0 since $01 \notin \mathcal{L}(X^+)$, and b_{-n-2} cannot equal 1 since $11 \notin \mathcal{L}(X^+)$. Thus, 1 does not appear in any two-sided sequence b with the property that $b_p b_{p+1}... \in X^+$ for all $p \in \mathbb{Z}$. Hence, $1 \notin \mathcal{L}(X^{+-})$. This implies that 1 is not a vertex in the HB diagram $\mathcal{D}$ for X^+. As a result, there is no path on $\mathcal{D}$ starting with a block of length one that projects to any block in $\mathcal{L}(X^+)$ that begins with a 1.

In this example, we see that the relationship between $\mathcal{L}(X)$ and $\mathcal{L}(X^{+-})$ depends on the paths that appear on $\mathcal{D}$. If X is a two-sided subshift, obviously $X^{+-} = X$ and $\mathcal{L}(X) = \mathcal{L}(X^{+-})$. In contrast, if X^+ is a one-sided subshift it is not as easy to determine whether $\mathcal{L}(X^+) = \mathcal{L}(X^{+-})$. Recall that

$$X^{+-} = \{x \in \mathcal{A}^{\mathbb{Z}} | \text{ for all } p \in \mathbb{Z}, \ x_p x_{p+1}... \in X^+\}.$$

Hence, $\mathcal{L}(X^{+-}) \subset \mathcal{L}(X^+)$. We provide a construction of X^{+-} and thus a necessary and sufficient condition on X^+ such that $\mathcal{L}(X^+) = \mathcal{L}(X^{+-})$.

Let $a^{(n)} = a_0^{(n)} a_1^{(n)} a_2^{(n)} a_3^{(n)}...$ be points in X^+. We construct a sequence $(x_n(a^{(n)}))$ of two-sided sequences as follows. Let $b^{(n)} = 0^\infty.a^{(n)}$ and set $x_n(a^{(n)}) = \sigma^n b^{(n)}$.

PROPOSITION 5.5. *Let X^+ and $(x_n(a^{(n)}))$ be as described above. Then X^{+-} is the set of limit points of all $(x_n(a^{(n)}))$, $a^{(n)} \in X^+$ for all $n \geq 0$.*

PROOF. By compactness, the sequence $(x_n(a^{(n)}))$ has a limit point x. That is, there exists a subsequence $(x_{n_k}(a^{(n_k)}))$ of $(x_n(a^{(n)}))$ that converges to x. We claim that any such limit point is in X^{+-}. Suppose on the contrary that $x \notin X^{+-}$. Then there exists $p \in \mathbb{Z}$ such that $x_p x_{p+1} x_{p+2}... \notin X^+$. This, however, is impossible since the initial blocks of any right ray in x can be found as the initial blocks of a point in X^+ by construction. Hence, $x_p x_{p+1} x_{p+2}...$ is in the closure of X^+ and thus is in X^+.

Conversely, let $b = ...b_{-3}b_{-2}b_{-1}.b_0b_1b_2b_3...$ be an arbitrary bisequence in X^{+-}. We show that b is a limit point of a subsequence of $(x_n(a^{(n)}))$, for some $a^{(n)} \in X^+$. Let
$a^{(n)} = \pi_{X^+}(\sigma^{-n}b)$, where π_{X^+} is as defined earlier. That is

$$\begin{aligned} a^{(1)} &= \pi_{X^+}(\sigma^{-1}b) = b_{-1}b_0b_1... \\ a^{(2)} &= \pi_{X^+}(\sigma^{-2}b) = b_{-2}b_{-1}b_0... \\ &\vdots \\ a^{(n)} &= \pi_{X^+}(\sigma^{-n}b) = b_{-n}b_{-n+1}b_{-n+2}.... \end{aligned}$$

It follows that,

$$b = \lim_{n\to\infty} x_n(a^{(n)}).$$

Thus any point in X^{+-} is a limit point of a subsequence of $(x_n(a^{(n)}))$. □

COROLLARY 5.6. *$\mathcal{L}(X^+) = \mathcal{L}(X^{+-})$ if and only if for every block B in $\mathcal{L}(X^+)$ and for all $n \geq 0$ there exists $a^{(n)} \in X^+$ such that B appears in $a^{(n)}$ starting at position n.*

PROOF. Suppose that $B \in \mathcal{L}(X^+)$ and for all $n \geq 0$ there exists $a^{(n)} = a_0^{(n)}a_1^{(n)}a_2^{(n)}... \in X^+$ such that B appears in $a^{(n)}$ starting at position n. Construct the sequence $(x_n(a^{(n)}))$ as defined previously. For all $n \geq 0$,

$$x_n(a^{(n)}) = 0^\infty a_0^{(n)}...a_{n-1}^{(n)}.Ba_{|B|+n}^{(n)}....$$

Let x be any limit point of the sequence $(x_n(a^{(n)}))$. Then $x \in X^{+-}$ by Proposition 5.5 and $x = ...x_{-1}x_0.Bx_{|B|}....$ That is, B appears in x starting at position 0. Hence $B \in \mathcal{L}(X^{+-})$.

Now assume $\mathcal{L}(X^+) = \mathcal{L}(X^{+-})$ and $B \in \mathcal{L}(X^+)$ and $n \geq 0$ are given. Since $B \in \mathcal{L}(X^{+-})$, B appears in some $x \in X^{+-}$. Futhermore, there exists $m \in \mathbb{Z}$ such that $\sigma^m x$ has B appearing in position n. By definition, the ray $\pi_{X^+}(\sigma^m x)$ is in X^+. Setting $a^{(n)} = \pi_{X^+}(\sigma^m x)$, we have the desired result. □

The following Corollary is an immediate consequence of Corollary 5.6.

COROLLARY 5.7. *$\mathcal{L}(X^+) = \mathcal{L}(X^{+-})$ if and only if every block in $\mathcal{L}(X^+)$ is left extendable. In particular, if X^+ is minimal, then $\mathcal{L}(X^+) = \mathcal{L}(X^{+-})$.*

We now focus our attention on subshifts X such that $\mathcal{L}(X) = \mathcal{L}(X^{+-})$.

THEOREM 5.8. *Let X be a one or two-sided subshift with $\mathcal{L}(X) = \mathcal{L}(X^{+-})$. Let
$w = w_0w_1...w_n$ be a block in $\mathcal{L}(X)$. Then there exists a unique path $\alpha_0 \to \alpha_1 \to \cdots \to \alpha_n$ in the HB diagram of X with $\alpha_0 = w_0$ and $\hat{\pi}(\alpha_0\alpha_1...\alpha_n) = w$.*

PROOF. Let $w = w_0 w_1 ... w_n$ be a block in $\mathcal{L}(X)$. Since $w_0 ... w_i$ appears in w for $0 \leq i \leq n$, it follows that $w_0 ... w_i \in \mathcal{L}(X) = \mathcal{L}(X^{+-})$ for all $0 \leq i \leq n$. Set $\alpha_i = \text{sig}(w_0 w_1 ... w_i)$. By definition $\text{sig}(w_0 w_1 ... w_i)$ is a significant block in X^{+-} that ends with the letter w_i. It follows that if $\alpha_0 \to \alpha_1 \to \cdots \to \alpha_n$ is a path on $\mathcal{D}$ then $\hat{\pi}(\alpha_0 \alpha_1 ... \alpha_n) = w_0 w_1 ... w_n$.

It remains to show that for $0 \leq i \leq n-1$ there exist arrows from $\alpha_i \to \alpha_{i+1}$ in the HB diagram of X. By definition $\alpha_i \to \alpha_{i+1}$ if and only if $\alpha_{i+1} = \text{sig}(\alpha_i w_{i+1})$. Since $\alpha_i = \text{sig}(w_0 w_1 ... w_i)$, it suffices to show that $\alpha_{i+1} = \text{sig}(\text{sig}(w_0 w_1 ... w_i) w_{i+1})$. Lemma 5.2 implies that

$$\text{sig}(\text{sig}(w_0 w_1 ... w_i) w_{i+1}) = \text{sig}(w_0 w_1 ... w_i w_{i+1}),$$

where

$$\text{sig}(w_0 w_1 ... w_i w_{i+1}) = \alpha_{i+1}.$$

Thus, the desired path exists. Furthermore, this path is unique by Theorem 5.3. □

Here we have shown that if B is a block in $\mathcal{L}(X)$ then there exists a path on $\mathcal{D}$ starting with a block of length one that projects to the block B, provided $\mathcal{L}(X) = \mathcal{L}(X^{+-})$ (this condition is necessary—see Example 5.4). It immediately follows that if X is a subshift with $\mathcal{L}(X) = \mathcal{L}(X^{+-})$, then for any point x in the one-sided subshift X^+ there exists a unique path α in the HB diagram of X starting with a block of length one such that $\hat{\pi}^+(\alpha) = x$. Furthermore, from Theorems 5.3 and 5.8 we get the following corollary.

COROLLARY 5.9. *Let X be a one or two-sided subshift with $\mathcal{L}(X) = \mathcal{L}(X^{+-})$ and let p_X be the complexity function of X. Then the number of distinct paths of length n that occur on $\mathcal{D}$, the HB diagram of X, that begin with a block of length one is equal to $p_X(n)$.*

We provide an alternate statement and proof of Corollary 5.9 that is specific to Sturmian systems.

THEOREM 5.10. *Let (X_u^+, σ) be a Sturmian system. In the HB diagram of X_u^+, for $n \geq 1$ there are $p_u(n) = n+1$ paths of length n starting from either the vertex labeled 0 or the vertex labeled 1.*

PROOF. We denote the number of paths of length n by P_n. Let $n = 1$. As any path can begin with 0 or 1, $P_1 = 2$. Next let $n = 2$. If 0 is right special, the paths of length two are $0 \to 00, 0 \to 1$, and $1 \to 10$. If 1 is right special, the paths of length two are $1 \to 11, 1 \to 0$, and $0 \to 01$. In either case, $P_2 = 3$.

We proceed by induction. Fix $n \geq 2$ and assume $P_{n-1} = p_u(n-1)$. Then $P_{n-1} = n$. We wish to show that $P_n = n+1$. Consider the n distinct paths of length $n-1$. Because each of these paths can be continued, there are at least n paths of length n. Suppose there are $n+2$ paths of length n. From Theorem 5.3, each distinct path of length n starting with either 0 or 1 yields a distinct block of length n by reading off the last symbol of every vertex encountered. Then, that there are $n+2$ paths of length n implies that there are $n+2$ distinct blocks of length n in $\mathcal{L}(X_u^+)$. This contradicts $p_u(n) = n+1$, so $n \leq P_n < n+2$.

To prove that $P_n = n+1$, we show that exactly one of the paths of length $n-1$ can be continued in two ways. Consider the n paths of length $n-1$ with initial vertex 0 or 1. Each of these paths projects to a distinct block of length $n-1$, hence there is a path corresponding to every block in $\mathcal{L}_n(u)$. It follows that exactly one of these blocks is right special. Call this block $w = l_{n-1}l_{n-2}...l_1$, where $l = l_1l_2l_3...$ is the left special sequence of u. Let $\alpha_0 \to \cdots \to \alpha_{n-2}$ be the path that projects to w. That is $w = \hat{\pi}(\alpha_0...\alpha_{n-2}) = l_{n-1}l_{n-2}...l_1$. By Corollary 5.2, $\alpha_{n-2} = \mathrm{sig}(l_{n-1}l_{n-2}...l_0)$. However, $\mathrm{sig}(l_{n-1}l_{n-2}...l_0) = l_m l_{m-1}...l_0$, $0 \leq m \leq n-1$. Thus α_{n-2} is a right special significant block. This implies that the path $\alpha_0 \to \cdots \to \alpha_{n-2}$ can be continued in two ways. Thus, there are exactly $n+1$ paths of length n, as desired. □

In this proof, we not only show that there are $p_u(n)$ paths of length n with initial vertex 0 or 1, but we identify the path of length n that extends in two ways.

It follows from Corollary 5.9 that given a subshift X with the property that $\mathcal{L}(X) = \mathcal{L}(X^{+-})$, we can recover the complexity function for X by counting paths in the HB diagram of X.

Let X be any subshift and $\mathcal{D}$ its HB diagram. Recall that $X(\mathcal{D})$ is the set of all bi-infinite paths that occur on $\mathcal{D}$.

DEFINITION 5.11. A sequence $a \in X^{+-}$ is *eventually Markov* at time $p \in \mathbb{Z}$ if there exists $N = N(x,p)$ such that for all $n \geq N$

$$\mathrm{fol}(a_{p-n}...a_p) = \mathrm{fol}(a_{p-N}...a_p).$$

The *eventually Markov part* $X_M^{+-} \subset X^{+-}$ is the set of $a \in X^{+-}$ which are eventually Markov at all times $p \in \mathbb{Z}$.

The following theorem, due to Hofbauer and Buzzi, shows that $\hat{\pi} : X(\mathcal{D}) \to X_M^{+-}$ is an isomorphism [**1**, **2**, **12**].

THEOREM 5.12. *The natural projection $\hat{\pi}$ from the Hofbauer shift $X(\mathcal{D})$ to the subshift X^{+-} defined by*

$$\hat{\pi} : \alpha \in X(\mathcal{D}) \mapsto a \in X^{+-}$$

with a_n the last symbol of the block α_n for all $n \in \mathbb{Z}$ is well defined and is a Borel isomorphism from $X(\mathcal{D})$ to X_M^{+-}.

Hence, one could say that $X(\mathcal{D})$ is "partially isomorphic" to X^{+-}.

It is apparent that the HB diagram of a Sturmian system X_u^+ does not contain any bi-infinite paths, thus $X_u^+(\mathcal{D})$ is the empty set. This may seem alarming, but it turns out that the eventually Markov part of X_u^{+-} is also empty. In fact, we show that if the natural extension of a subshift is infinite and minimal, then the eventually Markov part of the natural extension is empty. Thus, it will follow that if X^+ is minimal the isomorphism in Theorem 5.12 is between two copies of the empty set. Nevertheless, Theorem 5.8 gives an isomorphism between $X(\mathcal{D})^+$ (paths in the HB diagram that start with blocks of length one, see Definition 3.7) and X^+.

PROPOSITION 5.13. *If X^+ is a subshift such that X^{+-} is infinite and minimal, then the eventually Markov part of X^{+-} is empty.*

PROOF. Suppose on the contrary that there exists $x \in X^{+-}$ that is eventually Markov at time $p \in \mathbb{Z}$. Then there exists $N = N(x,p)$ such that for all $n \geq N$,

$$\mathrm{fol}(x_{p-n}...x_p) = \mathrm{fol}(x_{p-N}...x_p).$$

Let $B = x_{p-N}...x_p$. Since x is left recurrent there exists $n > 2N+1$ such that $x_{p-n}...x_{p-n+N} = B$. Let $A = x_{p-n+N+1}...x_{p-N-1}$.

By the definition of X^{+-}, $x_{p-n}x_{p-n+1}... = BABx_{p+1}x_{p+2}... \in X^+$. Thus there exists a ray r_1 in fol(B) that has $x_{p-n+N}AB = x_pAB$ as a prefix. Since

$$\text{fol}(B) = \text{fol}(x_{p-N}...x_p) = \text{fol}(x_{p-n}...x_p) = \text{fol}(BAB),$$

it follows that $r_1 \in \text{fol}(BAB)$. This implies that there exists an $a^{(1)} \in X^+$ with prefix $BABAB$.

Since $a^{(1)} \in X^+$, there exists a ray r_2 in fol(B) that has a_pABAB as a prefix. Then fol$(B) =$ fol(BAB) implies that $r_2 \in \text{fol}(BAB)$. Hence there exists $a^{(2)} \in X^+$ with prefix $BABABAB$. Continuing in this manner, we construct a sequence $(a^{(n)}) \subset X^+$ with

$$\lim_{n\to\infty} (a^{(n)}) = BABABABABA....$$

Let $b^{(n)} = 0^\infty.a^{(n)}$ and $x_n(a^{(n)}) = \sigma^n b^{(n)}$ as in Proposition 5.5. Then any limit point y of $(x_n(a^{(n)}))$ is a periodic bisequence in X^{+-}. This is a contradiction, since X^{+-} does not contain any periodic points. Thus x is not eventually Markov at any time $p \in \mathbb{Z}$ and X_M^{+-}, the eventually Markov part of X^{+-}, is empty. □

As previously discussed, a Sturmian sequence u is syndetically recurrent and is not periodic. Since X_u^+ is the orbit closure of the almost periodic sequence u it follows that X_u^+ is minimal. Furthermore, since u is not periodic, X_u^+ is infinite. However, a priori, we don't know that X_u^{+-} is minimal.

Proposition 5.14. *If X^+ is minimal, then the natural extension X^{+-} of X^+ is minimal. In fact, for any $x \in X^{+-}$ both the forward orbit $\mathcal{O}^+(x) = \{\sigma^n x | n \geq 0\}$ and the backward orbit $\mathcal{O}^-(x) = \{\sigma^{-n}x | n \geq 0\}$ are dense in X^{+-}.*

Proof. Since X^+ is minimal, every block $B \in \mathcal{L}(X^+)$ appears with bounded gap in each $a \in X^+$. By Corollary 5.7, $\mathcal{L}(X^+) = \mathcal{L}(X^{+-})$. Therefore each block $B \in \mathcal{L}(X^{+-})$ appears in each long-enough block in $\mathcal{L}(X^{+-})$. Hence for all $x \in X^{+-}$, the block B appears with bounded gap to the left and the right in x. Thus $\mathcal{O}^+(x)$ and $\mathcal{O}^-(x)$ are dense in X^{+-}. □

Remark 5.15. Since any infinite minimal subshift contains no periodic points, it follows from Proposition 5.14 that if X^+ is both minimal and infinite, then X^{+-} is minimal and infinite.

6. The Morse minimal subshift

To describe the construction of the HB diagram of one particular substitution system, the Morse minimal subshift, we have to recall some well-known properties of the Prouhet-Thue-Morse sequence,

$$\omega = .\omega_0\omega_1\omega_2... = .0110100110010110....$$

For the sake of simplicity, we shall refer to this sequence as the Morse sequence. The one-sided subshift associated with ω is the *Morse minimal subshift*. It is defined by the pair (X_ω^+, σ), where X_ω^+ is the closure of $\{\sigma^n\omega | n \in \mathbb{N}\}$.

This sequence has many interesting properties. Axel Thue, concerned with constructing bi-infinite sequences on two symbols with controlled repetitions, constructed the two-sided Morse sequence

$$M = ...0110100110010110.0110100110010110...,$$

which he defined as having the property that the sequence contains no blocks of the form BBb where B is a block and b is the first letter of B [**5, 7**]. Thue's results were published in 1912. In 1917, Marston Morse, not knowing of Thue's results, constructed the Morse sequence in his dissertation. In [**19**] Morse and Hedlund proved that every element in the Morse minimal set, the closure of $\{\sigma^n(M)\ n \in \mathbb{Z}\}$, has the no BBb property. It was later shown by Gottshalk and Hedlund that the elements of the Morse minimal set are the only bi-infinite sequences with the no BBb property [**11**]. While the Gottshalk and Hedlund result does not carry over to the one-sided Morse sequence [**7**], it is still the case that the one-sided Morse sequence ω has the no BBb property.

The Morse sequence is also generated by iterating a substitution. Following Chapter 5 of [**9**], we recall how this is done and how the construction allows us to deduce important properties of the sequence. Let ζ be the substitution map defined by $\zeta(0) = 01$ and $\zeta(1) = 10$. The Morse sequence is the infinite sequence which begins with $\zeta^n(0)$ for every $n \in \mathbb{N}$. It follows from this construction that the Morse sequence is syndetically recurrent and neither periodic nor eventually periodic.

6.1. Recognizability of the Morse substitution. Since the Morse sequence arises from a substitution map, it is natural to consider how to "decompose" or "desubstitute" a block that occurs in $\mathcal{L}(X_\omega^+)$. The notion of recognizability deals with this problem [**23**].

DEFINITION 6.1. A substitution γ over the alphabet $\mathcal{A}$ is *primitive* if there exists $k \in \mathbb{N}$ such that for all $a, b \in \mathcal{A}$ the letter a occurs in $\gamma^k(b)$.

In the context of recognizability we consider only primitive substitutions. Note that the Morse substitution ζ is primitive since 0 and 1 both appear in $\zeta(0)$ and $\zeta(1)$.

Let $u = u_0u_1...$ be any fixed point of an aribitrary primitive substitution γ.

DEFINITION 6.2. For every $k \geq 1$, $E_k = \{0\} \cup \{|\gamma^k(u_0u_1...u_{p-1})| \mid p > 0\}$ is the set of *cutting bars of order k*.

DEFINITION 6.3. The substitution γ is said to be *recognizable* if there exists an integer $K > 0$ such that

$$n \in E_1 \text{ and } u_nu_{n+1}...u_{n+K} = u_mu_{m+1}...u_{m+K} \text{ implies } m \in E_1.$$

The smallest integer K satisfying this is the *recognizability index* of γ.

In other words, a substitution is recognizable if it is possible to determine if u_m is the first letter of a substituted block by examining the K terms that follow it. The Morse substitution ζ is recognizable with recognizability index 3. This means that it is possible to determine if 0 (or 1) is the first letter of $\zeta(0)$ (or $\zeta(1)$) by examining the three letters which follow it.

Note that this definition of recognizability does not satisfactorily guarantee desubstitution in the general setting. Even very simple primitive, aperiodic substitutions may fail to have the recognizability property. For example, the substitution γ on the alphabet $\{0, 1\}$ defined by $\gamma(0) = 010$ and $\gamma(1) = 10$ fails to be recognizable. Brigitte Mossé introduces another notion of recognizability, bilateral recognizability, in [**20**].

DEFINITION 6.4. A substitution γ is said to be *bilaterally recognizable* if there exists an integer $L > 0$ such that

$$n \in E_1 \text{ and } u_{n-L}...u_{n+L} = u_{m-L}...u_{m+L} \text{ implies } m \in E_1.$$

One advantage of Mossé's definition is that every primitive aperiodic substitution is bilaterally recognizable. Furthermore, if u is a fixed point of a primitive aperiodic substitution γ and $X_u^+ = \text{cl}\{\sigma^k u | k \in \mathbb{N}\}$ (cl denotes closure), then any block in $\mathcal{L}(X_u^+)$ can be "desubstituted" up to some prefix and some suffix at the ends of the block [**9**, **20**, **23**]. Since the Morse substitution is recognizable, we do not rely on bilateral recognizability. However, the consequences of bilateral recognizability could be useful in extending the results for the Morse minimal subshift to general substitution systems.

We now consider the decomposition of blocks appearing in the fixed point of a substitution. Let $b = u_i ... u_{i+|w|-1}$ be a block appearing in u. Since $\gamma(u) = u$ there exists an index j, a length l, a suffix S of $\gamma(u_j)$ and a prefix P of $\gamma(u_{j+l+1})$ such that

$$b = S\gamma(u_j + 1)...\gamma(u_{j+l})P.$$

DEFINITION 6.5. Let b be as above. The *1-cutting at the index i of b* is

$$S \dagger \gamma(u_{j+1}) \dagger ... \dagger \gamma(u_{j+l}) \dagger P,$$

and we say that b comes from the block $u_j ... u_{j+l+1}$. The block $u_j ... u_{j+l+1}$ is the *ancestor block* of b [**9**].

Note that S and P are not necessarily proper suffixes and prefixes, respectively. Furthermore, the 1-cutting yields a string on an enlarged alphabet. For the Morse sequence this alphabet is $\{0, 1, \dagger\}$.

To illustrate this, consider the block $\omega_4 ... \omega_9 = 1001100$ appearing in the Morse sequence. Let $S = 10 = \zeta(\omega_2)$, and $P = 0$, the one letter prefix of $\zeta(\omega_5)$. Then

$$1001100 = 10 \dagger 01 \dagger 10 \dagger 0 = 10 \dagger \zeta(\omega_3) \dagger \zeta(\omega_4) \dagger 0,$$

and has $\omega_2 ... \omega_5 = 1010$ as an ancestor block.

In this example, it is apparent that the 1-cutting of the block 1001100 partitions the block into a concatenation of the subblocks 10 and 01 with daggers in between. We define the *1-blocks* of the Morse sequence to be the blocks 01 and 10. By partitioning a block into its 1-blocks, it is possible to determine its ancestor block. The following lemma, found in [**9**], is a result of the recognizability of the Morse sequence.

LEMMA 6.6. *In the Morse sequence, every block of length at least five has a unique 1-cutting, or decomposition into 1-blocks, possibly beginning with the last letter of a 1-block and possibly ending with the first letter of a 1-block.*

REMARK 6.7. If a block has a unique 1-cutting, then the block has a unique ancestor block. The only blocks of length less than five appearing in the Morse sequence which do not have a unique partition into 1-blocks are 010, 101, 0101, and 1010, each of which has two possible ancestor blocks. Furthermore, a block has a unique 1-cutting if and only if that block has either 00 or 11 as a subblock.

We denote the *dual* of a letter $a \in \{0, 1\}$ by $\overline{a}$. If $a = 0$ then $\overline{a} = 1$ and vice versa. Note that each 1-block consists of a pair of dual letters.

LEMMA 6.8. *Let $a_{-n}a_{-n+1}a_{-n+2}...a_0$ be a block of length n in $\mathcal{L}(X_\omega^+)$ that has a unique 1-cutting. If the unique 1-cutting has a dagger immediately to the right of a_{-n+1}, then $a_{-n} = \overline{a}_{-n+1}$. That is, a_{-n} is uniquely determined by a_{-n+1}.*

PROOF. Since $a_{-n+1}a_{-n+2}...a_0$ has a unique decomposition into 1-blocks, $a_{-n+1}a_{-n+2}...a_0$ has 00 or 11 as a subblock. Without loss of generality, suppose 00 is a subblock of $a_{-n+1}a_{-n+2}...a_0$. Since 00 is not a 1-block, it follows that in the 1-cutting of $a_{-n+1}a_{-n+2}...a_0$ there is a dagger in the middle of 00. Furthermore, since there is a dagger immediately to the right of a_{-n+1}, it follows that there exists k, $1 \leq k \leq n/2$, such that $a_{-n-1+2k}a_{-n+2k} = 00$.

Now consider $a_{-n}a_{-n+1}...a_0 \in \mathcal{L}(X_\omega^+)$. Since 00 is a subblock of $a_{-n}a_{-n+1}...a_0$, there exists a unique 1-cutting of $a_{-n}a_{-n+1}a_{-n+2}...a_0$. Additionally, there must be a dagger in between $00 = a_{-n-1+2k}a_{-n+2k}$ in the unique 1-cutting. Thus there is a dagger between $a_{-n-1+2i}a_{-n+2i}$ for all $1 \leq i \leq n/2$. Letting $i = 1$, this implies that there is a dagger between $a_{-n+1}a_{-n+2}$. Hence, $a_{-n}a_{-n+1}$ must be a 1-block and $a_{-n} = \overline{a_{-n+1}}$. □

6.2. Significant blocks of the Morse minimal subshift. Using the properties of the Morse sequence detailed above, we determine the significant blocks of the Morse minimal subshift. Note that it follows from the minimality of the Morse sequence that $\mathcal{L}(X_\omega^+) = \mathcal{L}(X_\omega^{+-})$.

PROPOSITION 6.9. *Let $a_{-n+1}a_{-n+2}...a_{-2}a_{-1}a_0$ be a block of length $n \in \mathbb{N}$, $n \geq 2$, in $\mathcal{L}(X_\omega^+)$. Then $a_{-n+1}a_{-n+2}...a_{-2}a_{-1}a_0$ is significant if and only if $0a_{-n+2}...a_{-2}a_{-1}a_0$ and $1a_{-n+2}...a_{-2}a_{-1}a_0$ are in $\mathcal{L}(X_\omega^+)$.*

PROOF. One direction is proved in Proposition 4.6.

We first prove the converse for 01,10, 010, 101, 0101, and 1010. By examining the sequence ω it is apparent that each of these blocks satisfies the hypothesis that $0a_{-n+2}...a_{-2}a_{-1}a_0$ and $1a_{-n+2}...a_{-2}a_{-1}a_0$ are in $\mathcal{L}(X_\omega^+)$. To prove that each block is significant we construct a ray for each block that is in $\text{fol}(a_{-n+2}...a_{-2}a_{-1}a_0)$, but not in $\text{fol}(a_{-n+1}a_{-n+2}...a_{-2}a_{-1}a_0)$.

Consider 010. Since the Morse sequence contains no blocks of the form BBb, $01010 \notin \mathcal{L}(X_\omega^+)$. However, $1010 \in \mathcal{L}(X_\omega^+)$. Thus the ray $\omega_3\omega_4\omega_5... = 0100110...$ is in fol(10) but not fol(010). Hence 010 is significant. Similarly, it can be shown that

$$\begin{aligned}
\omega_2\omega_3\omega_4... &= 101001... \in \text{fol}(1) \text{ but } 101001... \notin \text{fol}(01),\\
\omega_{10}\omega_{11}\omega_{12}... &= 010110... \in \text{fol}(0) \text{ but } 010110... \notin \text{fol}(10),\\
\omega_{11}\omega_{12}\omega_{13}... &= 101101... \in \text{fol}(01) \text{ but } 101101... \notin \text{fol}(101),\\
\omega_4\omega_5\omega_6... &= 100110... \in \text{fol}(101) \text{ but } 100110... \notin \text{fol}(0101),\\
\omega_{12}\omega_{13}\omega_{14}... &= 011010... \in \text{fol}(010) \text{ but } 011010... \notin \text{fol}(1010).
\end{aligned}$$

Therefore, 01, 10, 101, 0101, and 1010 are all significant.

We now prove the converse for the remaining blocks $a_{-n+1}a_{-n+2}...a_{-2}a_{-1}a_0$ satisfying the property that $0a_{-n+2}...a_{-2}a_{-1}a_0$ and $1a_{-n+2}...a_{-2}a_{-1}a_0$ are in $\mathcal{L}(X_\omega^+)$. To prove that each of these blocks is significant, we explicitly construct a ray that is in the follower set of $a_{-n+2}...a_{-2}a_{-1}a_0$ but not in the follower set of $a_{-n+1}a_{-n+2}...a_{-2}a_{-1}a_0$. This is done by repeatedly desubstituting $a_{-n+1}a_{-n+2}...a_{-2}a_{-1}a_0$ and choosing a ray based on the ancestor block of $a_{-n+1}a_{-n+2}...a_{-2}a_{-1}a_0$.

It follows from Lemma 6.6 and Remark 6.7 that each of the remaining blocks has a unique 1-block decomposition. If $n > 3$, then partition $a_{-n+1}a_{-n+2}...a_{-2}a_{-1}a_0$ into 1-blocks. Lemma 6.8 implies that if the unique partition of $a_{-n+1}a_{-n+2}...a_{-2}a_{-1}a_0$ has a dagger directly after a_{-n+2}, then a_{-n+1} is uniquely determined by a_{-n+2}, and hence $0a_{-n+2}...a_{-2}a_{-1}a_0$ and $1a_{-n+2}...a_{-2}a_{-1}a_0$ are not both in $\mathcal{L}(X_\omega)$. It follows that for $n > 3$ the partition of $a_{-n+1}a_{-n+2}...a_{-2}a_{-1}a_0$ is

$$a_{-n+1} \dagger a_{-n+2}a_{-n+3} \dagger \cdots \dagger a_{-2}a_{-1} \dagger a_0$$

if n is even, and

$$a_{-n+1} \dagger a_{-n+2}a_{-n+3} \dagger \cdots \dagger a_{-3}a_{-2} \dagger a_{-1}a_0$$

if n is odd.

Next we map each 1-block in the partition to its preimage under ζ, or ancestor block. Let $s_{1,i/2}$ denote the preimage of the 1-block $a_{-n+i}a_{-n+i+1}$, where i is even and $2 \le i \le n-1$. Since there is a dagger placed directly after a_{-n+1}, a_{-n+1} uniquely determines the letter that can precede it. Let a_{-n} be this letter and let $s_{1,0}$ denote the preimage of $a_{-n}a_{-n+1}$. Similarly, when n is even a_0 uniquely determines a_1. In this case let $s_{1,n/2}$ denote the preimage of a_0a_1. The resulting block is $s_{1,0}s_{1,1}...s_{1,(n-1)/2}$ if n is odd, and $s_{1,0}s_{1,1}...s_{1,n/2}$ if n is even. For ease of notation, denote $s_{1,0}s_{1,1}...s_{1,(n-1)/2}$ or $s_{1,0}s_{1,1}...s_{1,n/2}$ by S_1. Note that S_1 is the ancestor block of $a_{-n+1}a_{-n+2}...a_{-2}a_{-1}a_0$.

At this stage, consider the length of S_1. If $|S_1| \ge 4$ and S_1 has a unique 1-block partition, then map each 1-block to its preimage under ζ. That is, map S_1 to its ancestor block. If $|S_1| < 4$ or $S_1 = 0101$ or 1010, then do nothing. We continue this process so that in general, if $|S_j| \ge 4$ and S_j can be uniquely partitioned into 1-blocks, we map S_j to its ancestor block, and otherwise do nothing. During this process, if the 1-block decomposition of S_j has a dagger between $s_{j,0}$ and $s_{j,1}$, then there is only one letter that can precede $s_{j,0}$. Let $s_{j,-1}$ be this letter. Then $s_{j+1,0}$ is defined to be the preimage of the 1-block $s_{j,-1}s_{j,0}$.

EXAMPLE 6.10. We illustrate this process for the block $00110100 \in \mathcal{L}(X_\omega^+)$. The unique 1-block decomposition of 00110100 is

$$0 \dagger 01 \dagger 10 \dagger 10 \dagger 0.$$

Mapping each 1-block to its preimage under ζ, we get $S_1 = 10110$. Since $|10110| = 5$, we partition 10110 into 1-blocks as follows:

$$1 \dagger 01 \dagger 10.$$

Again, map each 1-block to its preimage under ζ to get $S_2 = 001$. As $|001| < 4$, the decomposition process is complete.

We claim that there exists an $m \in \mathbb{N}$ such that $S_m = s_{m,0}s_{m,1}s_{m,2}$, where $s_{m,1}s_{m,2} \in \{01, 10\}$; or $S_m = s_{m,0}s_{m,1}s_{m,2}s_{m,3}$, where $s_{m,1}s_{m,2}s_{m,3} \in \{010, 101\}$. Since the decomposition process is repeated until $3 \le |S_m| \le 4$ we need only show that

$$S_m \in \{001, 101, 010, 110, 0010, 1010, 0101, 1101\}.$$

Suppose on the contrary that S_m is not one of the above blocks. Since $S_m \in \mathcal{L}(X_\omega^+)$, it follows that

$$S_m \in \{100, 011, 1001, 0110\}.$$

Each of these blocks has a unique partitioning in which there is a dagger directly to the right of $s_{m,1}$. Hence $s_{m,0}$ is uniquely determined by $s_{m,1}$, and $s_{m,0} = \overline{s_{m,1}}$, by Lemma 6.8. We prove that if $s_{m,0}$ is uniquely determined by $s_{m,1}$ then there is only one possible value for a_{-n+1}.

First suppose that for all $1 \leq k < m$ the partition of S_k has a dagger between $s_{k,0}$ and $s_{k,1}$. That is,

$$s_{k,0} \dagger s_{k,1}s_{k,2} \dagger \cdots \dagger s_{k,j} \qquad \text{and} \qquad s_{k,0} \dagger s_{k,1}s_{k,2} \dagger \cdots \dagger s_{k,j}s_{k,j+1}$$

are the 1-cuttings for $1 \leq k < m$ when $|S_k|$ is even and odd respectively, but

$$s_{m,0}s_{m,1} \dagger s_{m,2} \qquad \text{and} \qquad s_{m,0}s_{m,1} \dagger s_{m,2}s_{m,3}$$

are the 1-cuttings for $s_{m,0}$ when S_m has length 3 and 4 respectively.

If this is the case, then $\zeta(s_{k+1,0}) = s_{k,-1}s_{k,0}$ for all $1 \leq k < m$. Then

$$\zeta^2(s_{k+1,0}) = \zeta(s_{k,-1})\zeta(s_{k,0}) = \zeta(s_{k,-1})s_{k-1,-1}s_{k-1,0};$$

and in general

$$\zeta^j(s_{k+1,0}) = \zeta^{j-1}(s_{k,-1})\zeta^{j-2}(s_{k-1,-1})...\zeta(s_{k-j+2,-1})s_{k-j+1,-1}s_{k-j+1,0}$$

for $j \leq k$. Hence

$$\begin{aligned}\zeta^m(s_{m,0}) &= \zeta^{m-1}(s_{m-1,-1})\zeta^{m-2}(s_{m-2,-1})...\zeta^2(s_{2,-1})\zeta(s_{1,-1})\zeta(s_1,0)\\ &= \zeta^{m-1}(s_{m-1,-1})\zeta^{m-2}(s_{m-2,-1})...\zeta^2(s_{2,-1})\zeta(s_{1,-1})\overline{a_{-n+1}}a_{-n+1}.\end{aligned}$$

Thus the last letter of $\zeta^m(s_{m,0})$ is a_{-n+1}. Since each desubstitution is unique, this implies that a_{-n+1} is uniquely determined by $s_{m,0}$.

Let $S_m = s_{m,0}...s_{m,q}$, where $q \in \{2,3\}$. By a similar argument to that used above, it can be shown that the block $a_{-n+2}a_{-n+3}..a_1a_0$ is a prefix of $\zeta^m(s_{m,1}...s_{m,q})$. That is,

$$\zeta^m(S_m) = \zeta^m(s_{m,1}...s_{m,q}) = a_{-n+2}a_{-n+3}...a_1a_0C,$$

where $C \in \mathcal{L}(X_\omega^+)$. Since each desubstitution of $a_{-n+2}a_{-n+3}...a_1a_0$ is unique, this implies that $s_{m,1}$ is uniquely determined by the block $a_{-n+2}a_{-n+3}...a_1a_0$.

Furthermore, since we have assumed that $s_{m,0}$ is uniquely determined by $s_{m,1}$, it follows that $\zeta^m(s_{m,0}) = \zeta^m(\overline{s_{m,1}})$. Hence, a_{-n+1}, the last letter of $\zeta^m(s_{m,0})$, is uniquely determined by $s_{m,1}$. Thus, there is only one possible value for a_{-n+1}. Therefore, $0a_{-n+2}a_{-n+3}...a_1a_0$ and $1a_{-n+2}a_{-n+3}...a_1a_0$ cannot both be in $\mathcal{L}(X_\omega^+)$, contradicting $a_{-n-1}a_{-n+2}a_{-n+3}...a_1a_0$ being a significant block.

Now suppose that there exists $1 \leq k < m$ such that the 1-block decomposition of S_k has a dagger after $s_{k,1}$. Then there exists an r, $1 \leq r \leq k$ such that for all $1 \leq j < r$ the partition of S_j has a dagger between $s_{j,0}$ and $s_{j,1}$, but the partition of S_r has a dagger after $s_{r,1}$. By the previous argument, it follows that a_{-n+1} is uniquely determined by $s_{r,0}$ and thus by $s_{r,1}$. Hence, $0a_{-n+2}a_{-n+3}..a_1a_0$ and $1a_{-n+2}a_{-n+3}..a_1a_0$ cannot both be in $\mathcal{L}(X_\omega^+)$. Therefore,

$$S_m \notin \{100, 011, 1001, 0110\}.$$

REMARK 6.11. Let $S_m = s_{m,0}...s_{m,q}$, where $q \in \{2,3\}$. By the above argument it must be the case that for all $1 \leq k < m$ the partition of S_k has a dagger between $s_{k,0}$ and $s_{k,1}$. This implies that a_{-n+1} is the last letter of $\zeta^m(s_{m,0})$ and $a_{-n+2}a_{-n+3}..a_1a_0$ is a prefix of $\zeta^m(s_{m,1}...s_{m,q})$, as discussed.

Having established the existence of an $m \in \mathbb{N}$ such that $S_m = s_{m,0}s_{m,1}s_{m,2}$, where $s_{m,1}s_{m,2} \in \{01, 10\}$, or $S_m = s_{m,0}s_{m,1}s_{m,2}s_{m,3}$, where $s_{m,1}s_{m,2}s_{m,3} \in \{010, 101\}$, define the ray ν as follows.
If $s_{m,0} = 0$ then,

$$\nu = \begin{cases} \omega_5\omega_6\omega_7... = 0011001..., & \text{if } s_{m,1}s_{m,2} = 01 \\ \omega_6\omega_7\omega_8... = 0110010..., & \text{if } s_{m,1}s_{m,2s_{m,3}} = 010 \\ \omega_4\omega_5\omega_6... = 1001100..., & \text{if } s_{m,1}s_{m,2} = 10 \\ \omega_5\omega_6\omega_7... = 0011001..., & \text{if } s_{m,1}s_{m,2s_{m,3}} = 101 \end{cases}$$

If $s_{m,0} = 1$ then,

$$\nu = \begin{cases} \omega_{13}\omega_{14}\omega_{15}... = 1101001..., & \text{if } s_{m,1}s_{m,2} = 10 \\ \omega_{14}\omega_{15}\omega_{16}... = 1010010..., & \text{if } s_{m,1}s_{m,2s_{m,3}} = 101 \\ \omega_{12}\omega_{13}\omega_{14}... = 0110100..., & \text{if } s_{m,1}s_{m,2} = 01 \\ \omega_{13}\omega_{14}\omega_{15}... = 1101001..., & \text{if } s_{m,1}s_{m,2s_m,3} = 010 \end{cases}$$

Note that the sequence ν is defined so that $s_{m,1}s_{m,2}\nu$ (or $s_{m,1}s_{m,2}s_{m,3}\nu$) is in X_ω^+, but $s_{m,0}s_{m,1}s_{m,2}\nu$ (or $s_{m,0}s_{m,1}s_{m,2}s_{m,3}\nu$) is not in X_ω^+. We provide an example of this.

EXAMPLE 6.12. Let $s_{m,0}s_{m,1}s_{m,2} = 001$. Then $s_{m,0}s_{m,1}s_{m,2}\nu = 0010011001....$ Consider the 1-block decomposition of the prefix block 00100. Since a dagger must be placed between consecutive zeros we get $0 \dagger 01 \dagger 0 \dagger 0$. This, however, is not an allowed 1-block decomposition. Thus 00100 is not in $\mathcal{L}(X_\omega^+)$ and $s_{m,0}s_{m,1}s_{m,2}\nu \notin X_\omega^+$.

Suppose $s_{m,1}s_{m,2} \in \{01, 10\}$. Let $\zeta^m(s_{m,1}s_{m,2}\nu) = d_1d_2d_3....$ Note that the first $n-1$ letters of $\zeta^m(s_{m,1}s_{m,2}\nu)$ form the block $a_{-n+2}a_{-n+3}...a_0$. We claim that $d_{n-1}d_nd_{n+1}...$ is in $\text{fol}(a_{-n+2}a_{-n+3}...a_0)$ but not in $\text{fol}(a_{-n+1}a_{-n+2}...a_0)$. Since $s_{m,1}s_{m,2}\nu \in X_\omega^+$, it follows that $\zeta^m(s_{m,1}s_{m,2}\nu) \in X_\omega^+$, as ω is fixed under the substitution ζ. Thus, $d_{n-1}d_nd_{n+1}...$ is in $\text{fol}(a_{-n+2}a_{-n+3}...a_0)$. It remains to show that $a_{-n+1}\zeta^m(s_{m,1}s_{m,2}\nu)$ is not in X_ω^+.

Let $\nu = \nu_1\nu_2\nu_3...$ and consider the block $a_{-n+1}\zeta^m(s_{m,1}s_{m,2}\nu_1\nu_2)$. Since the first $n-1$ terms of $\zeta^m(s_{m,1}s_{m,2}\nu_1\nu_2)$ form the block $a_{-n+2}a_{-n+3}...a_0$, the block $a_{-n+1}a_{-n+2}a_{-n+3}...a_0$ is a prefix of the block $a_{-n+1}\zeta^m(s_{m,1}s_{m,2}\nu_1\nu_2)$. Implementing the decomposition process m times, the resulting block is $s_{m,0}s_{m,1}s_{m,2}\nu_1\nu_2$. However, $s_{m,0}s_{m,1}s_{m,2}\nu_1\nu_2 \notin \mathcal{L}(X_\omega^+)$. As the ancestor block of any block in $\mathcal{L}(X_\omega^+)$ is in $\mathcal{L}(X_\omega^+)$ it follows that $a_{-n+1}\zeta^m(s_{m,1}s_{m,2}\nu_1\nu_2) \notin \mathcal{L}(X_\omega^+)$. Therefore, $a_{-n+1}\zeta^m(s_{m,1}s_{m,2}\nu)$ is not in X_ω^+. Therefore, $a_{-n+1}a_{-n+2}a_{-n+3}...a_0$ is significant.

Using similar arguments, it can be shown that $a_{-n+1}a_{-n+2}...a_0$ is significant in the case that

$$s_{m,1}s_{m,2}s_{m,3} \in \{010, 101\}.$$

□

REMARK 6.13. We relate this result to Proposition 4.7 from the Sturmian case. Since there is not a unique left special sequence as in the Sturmian case, let m_n be the number of blocks of length n in $\mathcal{L}(X_\omega^+)$ that can be extended to the left in two ways. Denote the set of such blocks by $\{L_n^i\}$, $1 \le i \le m_n$. Then for each $n \ge 2$ the significant blocks of length n of X_ω^+ are $\{0L_{n-1}^i, 1L_{n-1}^i\}$, $1 \le i \le m_{n-1}$.

6.3. HB diagram of the Morse minimal subshift. Since there is no left special sequence to direct us to the blocks in $\mathcal{L}(X_\omega^+)$ that can be extended to the left in two ways, the process of determining the significant blocks is more tedious. Nevertheless, it was possible above to use the no BBb property and desubstitution to identify the significant blocks.

After determining the significant blocks of the Morse minimal subshift, the next step is to determine the arrows. As in the Sturmian case, only those blocks that can be extended to the right in two ways will have two arrows out. However, for the Morse minimal subshift there is no easy technique for determining the significant blocks that satisfy this property. Hence the process of constructing the HB diagram of the Morse minimal subshift is not nearly as streamlined as for the Sturmians. We construct the HB diagram in the following way.

Begin by generating a list of signficant blocks. Start with the blocks 0 and 1. Since $00, 10, 01$, and 11 are all in $\mathcal{L}(X_\omega^+)$, it follows that 0 and 1 can both be extended to the left in two ways. Hence $00, 10, 01$, and 11 are all significant. Next consider $\mathcal{L}_2(X_\omega^+) = \{00, 01, 10, 11\}$, the blocks of length 2 in $\mathcal{L}(X_\omega^+)$. As the Morse minimal subshift has no blocks of the form BBb, 000 and 111 are not in $\mathcal{L}(X_\omega^+)$. Hence 00 and 11 cannot be extended to the left in two ways. However, 10 and 01 can be extended to the left in two ways. Thus $110, 010, 001$, and 101 are the significant blocks of length 3.

Now consider $\mathcal{L}_3(X_\omega^+) = \{001, 010, 011, 100, 101, 110\}$. By the no BBb property, it follows that 0001 and 1110 are not in $\mathcal{L}(X_\omega^+)$. The remaining blocks can all be extended to the left in two ways. Thus the significant blocks of length 4 are 0010, 1010, 0011, 1011, 0100, 1100, 0101, and 1101. Continuing in this manner, we are able to generate a list of significant blocks of X_ω^+;

0,1, 00, 10, 01, 11, 110, 010, 001, 101, 0010, 1010, 0011, 1011, 0100, 1100, 0101, 1101, 00110, 01001,10110, 11010,... .

To determine the arrows in the HB diagram of X_ω^+, we consider the right extensions of each significant block. We illustrate the process of determining the arrows by considering those arrows that start at a significant block of length 4. It is easily seen that $0011, 1011, 0100$, and 1100 can only be extended to the right in one way. Thus there is exactly one arrow out of each of these blocks, and these arrows are:

$$
\begin{aligned}
0011 &\to \mathrm{sig}(00110) = 00110 \\
1011 &\to \mathrm{sig}(10110) = 10110 \\
0100 &\to \mathrm{sig}(01001) = 01001 \\
1100 &\to \mathrm{sig}(11001) = 11001.
\end{aligned}
$$

Additionally, 1010 and 0101 can be extended to the right in only one way, as the blocks 10101 and 01010 are of the form BBb. The arrows out of these blocks are:

$$
\begin{aligned}
1010 &\to \mathrm{sig}(10100) = 0100 \\
0101 &\to \mathrm{sig}(01011) = 1011.
\end{aligned}
$$

Lastly, consider the blocks 1101 and 0010. Instead of using the no BBb property, we consider the 1-block decomposition of each block. This gives us $1\dagger10\dagger1$ and $0\dagger01\dagger0$. Since extending each block to the right must yield a legal 1-block decomposition, it follows that 1101 can be followed only by a 0, and 0010 can be followed only by a 1. Thus, the arrows out of these blocks are:

$$1101 \to \mathrm{sig}(11010) = 1010$$
$$0010 \to \mathrm{sig}(00101) = 0101.$$

Note that although there is exactly one arrow out of each significant block of length 4, this is not the case in general. For example, using the same process it can be shown that all four significant blocks of length 5 can be extended to the right in two ways.

Figure 3 depicts a portion of the HB diagram of the Morse minimal subshift that has been constructed using the process described above.

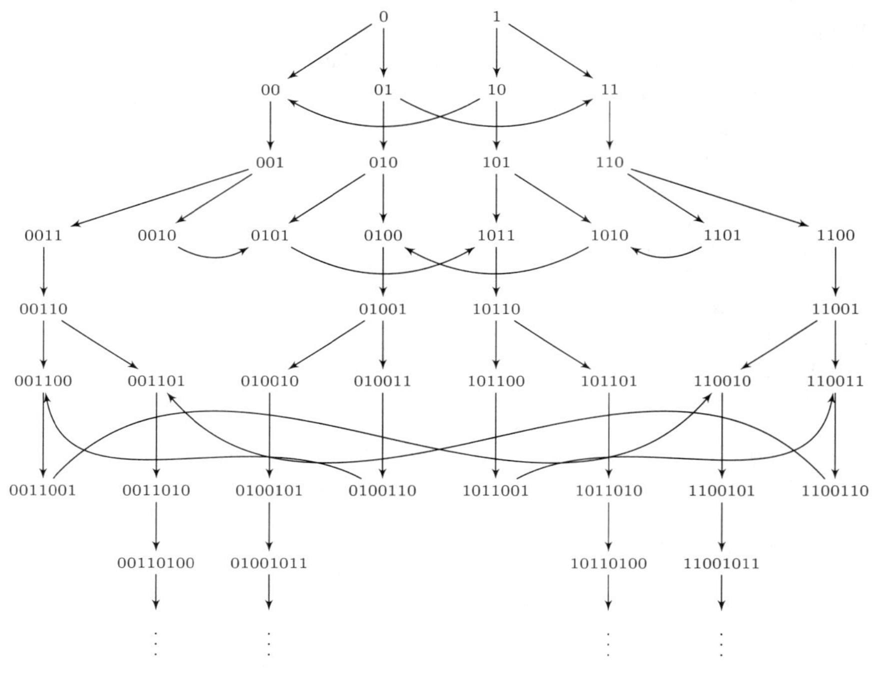

FIGURE 3. The HB diagram of the Morse minimal subshift.

7. Conclusion and Further Directions

We have described the construction of HB diagrams for some highly non-Markovian systems, that is, systems with long-range memory. These HB diagrams provide a way to visualize all the possibilities for extending any given block and present useful information about the languages of such systems and therefore about the structures of the systems themselves. Our examples of Sturmian systems and the Morse system are among the most-studied symbolic dynamical systems, so it is unlikely that these diagrams can lead to any new results about them or new proofs of known properties (indeed the construction of the diagrams used many known facts about the sequences and systems). For example, the complexity functions of subshifts can be read off the diagrams by counting paths that begin at significant blocks of length 1 (Corollary 5.9). The complexity function of the Morse minimal subshift is given by (see [**9**, Ch. 5]) $p_\omega(1) = 2$, $p_\omega(2) = 4$ and for $n \geq 3$ if $n = 2^r + q + 1$, $r \geq 0$, $0 < q \leq 2^r$, then

$$p_\omega(n) = \begin{cases} 6(2^{r-1}) + 4q & \text{if } 0 < q \leq 2^{r-1} \\ 8(2^{r-1}) + 2q & \text{if } 2^{r-1} < q \leq 2^r. \end{cases}$$

It is apparent in examining the portion of the HB diagram of the Morse minimal subshift shown in Figure 3 that the number of paths with initial vertex 0 or 1 is equal to $p_\omega(n)$ for $n \leq 8$. But to use the diagram to prove the formula for all lengths would probably require the same properties used to prove the formula directly (in addition to the many properties used to establish the diagram), so here the diagrams seem not to provide any advantage.

K. Johnson [**14**] (see also [**10, 22**]) observed that every one-sided subshift (X, σ) with alphabet $\{0, 1, \ldots, d-1\}$ has a unique lexicographically maximal element $\omega_{\max}(X, \sigma)$, which, since it lexicographically dominates all its shifts, is the expansion of 1 base β for a unique $\beta \in (1, \infty)$ and thus determines a smallest beta-shift that contains (X, σ). In particular, there is a one-to-one correspondence between minimal subshifts on alphabets $\{0, 1, \ldots, d-1\}$ and real $\beta \in (1, \infty)$ for which the expansions of 1 base β are uniformly recurrent (syndetic, also sometimes called almost periodic). HB diagrams provide an efficient way to identify the lexicographically maximal element in a subshift: start at the vertex $d-1$, and whenever two arrows leave a significant block follow the one with the larger label (i.e., the one whose target vertex has largest final entry). Johnson gave algorithms for constructing the maximal elements in constant-length and some variable-length substitution systems, showing in particular that the maximal element in the Fibonacci system is $1f$, where $f = 010\ldots$ is the fixed point of the substitution, and the maximal element in the Morse system is $\sigma\omega$, the shift of the Morse sequence ω. Applying our method to the part of the HB diagram that we have constructed for the Morse system corroborates Johnson's result, but proving that the maximal path in the diagram determines the Morse sequence would probably involve arguments similar to those of [**14**]. However, for Sturmian systems (including the Fibonacci) we can use the diagrams to identify the maximal elements.

THEOREM 7.1. *Let (X_u^+, σ) be a Sturmian subshift with left special sequence $l = l_1 l_2 \ldots$. Then the lexicographically maximal sequence in X_u^+ is $\omega_{\max}(X_u^+, \sigma) = 1l$.*

PROOF. Let L_0 denote the empty block and for each $n = 0, 1, 2, \ldots$ let $L_n = l_1 l_2 \ldots l_n$ denote the left special block of length n in u. By Proposition 4.7, the significant blocks (vertices of the diagram) are the blocks $0L_{n-1}$ and $1L_{n-1}, n = 1, 2, \ldots$. So our path following edges with maximal labels begins at the block $1L_0$. By Lemma 4.9, we always have arrows $xL_{k-1} \to xL_k, x \in \{0, 1\}$, so our maximally labeled path starting at $1L_0$ will be forced to proceed, as long as there is only one exiting arrow from each of its vertices, along vertices $1L_k$ until it reaches a vertex $1L_{n-1}$ which has two exiting arrows. By Lemma 4.10, then $1L_{n-1}$ is right special, and $1L_{n-1} = 1l_1 \ldots l_{n-1} = l_n l_{n-1} \ldots l_1$, so that $l_n = 1$ and the edge labeled 1 (i.e., to the target vertex with final entry 1) is to $1L_n$. Continuing in this way always to follow the arrow with the largest label whenever there is a choice then produces the maximal element $\omega_{\max}(X_u^+, \sigma) = 1l$. □

A similar argument shows that the lexicographically *minimal* element in a Sturmian subshift with left special sequence l is $\omega_{\min}(X_u^+, \sigma) = 0l$.

The diagrams might be put to further use in various other ways. How can invariant measures be represented on the diagrams? Can we detect unique ergodicity or minimality from these diagrams? In Section 6 we were able to construct the HB diagram of the Morse minimal subshift because the recognizability property of the Morse substitution allowed us to say precisely which blocks are significant. Can this result be generalized to any recognizable, or bilaterally recognizable, substitution? It is known that beta-shifts, as well as their factors, have unique measures of maximal entropy [**4**, **12**, **24**]. The HB diagram of a beta-shift turns out to be just a relabeling of the well-known beta-shift graph (see, e.g., [**14**, **22**]). Is there a simple way to transform the HB diagram of a subshift to produce the HB diagram of one of its factors? A relation between the two diagrams could help to understand factor maps, and in particular to identify measures of maximal entropy or maximal relative entropy.

Acknowledgement

This paper is based on the UNC-Chapel Hill master's project of the first author, written under the direction of the second author. We thank the referee for several corrections and simplifications.

References

[1] Jérôme Buzzi, *Intrinsic ergodicity of smooth interval maps*, Israel J. Math. **100** (1997), 125–161, DOI 10.1007/BF02773637. MR1469107

[2] Jérôme Buzzi, *Entropy theory on the interval* (English, with English and French summaries), École de Théorie Ergodique, Sémin. Congr., vol. 20, Soc. Math. France, Paris, 2010, pp. 39–82. MR2856509

[3] Jérôme Buzzi and Pascal Hubert, *Piecewise monotone maps without periodic points: rigidity, measures and complexity*, Ergodic Theory Dynam. Systems **24** (2004), no. 2, 383–405, DOI 10.1017/S0143385703000488. MR2054049

[4] Vaughn Climenhaga and Daniel J. Thompson, *Intrinsic ergodicity beyond specification: β-shifts, S-gap shifts, and their factors*, Israel J. Math. **192** (2012), no. 2, 785–817, DOI 10.1007/s11856-012-0052-x. MR3009742

[5] Ethan M. Coven, Michael Keane, and Michelle Lemasurier, *A characterization of the Morse minimal set up to topological conjugacy*, Ergodic Theory Dynam. Systems **28** (2008), no. 5, 1443–1451, DOI 10.1017/S0143385707000831. MR2449536

[6] Doris Fiebig and Ulf-Rainer Fiebig, *Covers for coded systems*, Symbolic dynamics and its applications (New Haven, CT, 1991), Contemp. Math., vol. 135, Amer. Math. Soc., Providence, RI, 1992, pp. 139–179, DOI 10.1090/conm/135/1185086. MR1185086

[7] Earl D. Fife, *Binary sequences which contain no BBb*, Trans. Amer. Math. Soc. **261** (1980), no. 1, 115–136, DOI 10.2307/1998321. MR576867

[8] Roland Fischer, *Sofic systems and graphs*, Monatsh. Math. **80** (1975), no. 3, 179–186. MR0407235

[9] N. Pytheas Fogg, *Substitutions in dynamics, arithmetics and combinatorics*, Lecture Notes in Mathematics, vol. 1794, Springer-Verlag, Berlin, 2002. Edited by V. Berthé, S. Ferenczi, C. Mauduit and A. Siegel. MR1970385

[10] Shaobo Gan, *Sturmian sequences and the lexicographic world*, Proc. Amer. Math. Soc. **129** (2001), no. 5, 1445–1451 (electronic), DOI 10.1090/S0002-9939-00-05950-5. MR1814171

[11] W. H. Gottschalk and G. A. Hedlund, *A characterization of the Morse minimal set*, Proc. Amer. Math. Soc. **15** (1964), 70–74. MR0158386

[12] Franz Hofbauer, *On intrinsic ergodicity of piecewise monotonic transformations with positive entropy*, Israel J. Math. **34** (1979), no. 3, 213–237 (1980), DOI 10.1007/BF02760884. MR570882

[13] Franz Hofbauer and Peter Raith, *Topologically transitive subsets of piecewise monotonic maps, which contain no periodic points*, Monatsh. Math. **107** (1989), no. 3, 217–239, DOI 10.1007/BF01300345. MR1008681

[14] Kimberly Christian Johnson, *Beta-shift dynamical systems and their associated languages*, ProQuest LLC, Ann Arbor, MI, 1999. Thesis (Ph.D.)–The University of North Carolina at Chapel Hill. MR2699401

[15] Wolfgang Krieger, *On sofic systems. I*, Israel J. Math. **48** (1984), no. 4, 305–330, DOI 10.1007/BF02760631. MR776312

[16] Wolfgang Krieger, *On sofic systems. II*, Israel J. Math. **60** (1987), no. 2, 167–176, DOI 10.1007/BF02790789. MR931874

[17] Douglas Lind and Brian Marcus, *An introduction to symbolic dynamics and coding*, Cambridge University Press, Cambridge, 1995. MR1369092

[18] M. Lothaire, *Algebraic Combinatorics on Words*, Encyclopedia of Mathematics and its Applications, vol. 90, Cambridge University Press, Cambridge, 2002, A collective work by Jean Berstel, Dominique Perrin, Patrice Seebold, Julien Cassaigne, Aldo De Luca, Steffano Varricchio, Alain Lascoux, Bernard Leclerc, Jean-Yves Thibon, Veronique Bruyere, Christiane Frougny, Filippo Mignosi, Antonio Restivo, Christophe Reutenauer, Dominique Foata, Guo-Niu Han, Jacques Desarmenien, Volker Diekert, Tero Harju, Juhani Karhumaki and Wojciech Plandowski, With a preface by Berstel and Perrin. MR1905123 (2003i:68115)

[19] Marston Morse and Gustav A. Hedlund, *Unending chess, symbolic dynamics and a problem in semigroups*, Duke Math. J. **11** (1944), 1–7. MR0009788

[20] Brigitte Mossé, *Puissances de mots et reconnaissabilité des points fixes d'une substitution* (French), Theoret. Comput. Sci. **99** (1992), no. 2, 327–334, DOI 10.1016/0304-3975(92)90357-L. MR1168468

[21] Karl Petersen, *Ergodic theory*, Cambridge Studies in Advanced Mathematics, vol. 2, Cambridge University Press, Cambridge, 1989. Corrected reprint of the 1983 original. MR1073173

[22] Karl Petersen, *Information compression and retention in dynamical processes*, Dynamics and randomness (Santiago, 2000), Nonlinear Phenom. Complex Systems, vol. 7, Kluwer Acad. Publ., Dordrecht, 2002, pp. 147–217, DOI 10.1007/978-94-010-0345-2_6. MR1975578

[23] Martine Queffélec, *Substitution dynamical systems—spectral analysis*, 2nd ed., Lecture Notes in Mathematics, vol. 1294, Springer-Verlag, Berlin, 2010. MR2590264

[24] Peter Walters, *Equilibrium states for β-transformations and related transformations*, Math. Z. **159** (1978), no. 1, 65–88. MR0466492

[25] Benjamin Weiss, *Subshifts of finite type and sofic systems*, Monatsh. Math. **77** (1973), 462–474. MR0340556

Department of Mathematics, CB 3250 Phillips Hall, University of North Carolina, Chapel Hill, North Carolina 27599

Current address: Department of Prosthetics and Orthotics, University of Hartford, 200 Bloomfield Avenue, West Hartford, Connecticut 06117

E-mail address: `kacarroll@hartford.edu`

Department of Mathematics, CB 3250 Phillips Hall, University of North Carolina, Chapel Hill, North Carolina 27599

E-mail address: `petersen@math.unc.edu`

Contemporary Mathematics
Volume **678**, 2016
http://dx.doi.org/10.1090/conm/678/13643

Contractive spaces and relatively contractive maps

Darren Creutz

Abstract. We present an exposition of contractive spaces and of relatively contractive maps. Contractive spaces are the natural opposite of measure-preserving actions and relatively contractive maps the natural opposite of relatively measure-preserving maps. These concepts play a central role in the work of the author and J. Peterson on the rigidity of actions of semisimple groups and their lattices and have also appeared in recent work of various other authors. We present detailed definitions and explore the relationship of these phenomena with other aspects of the ergodic theory of group actions, proving along the way several new results, with an eye towards explaining how contractiveness is intimately connected with rigidity phenomena.

1. Introduction

Contractive spaces were introduced by Jaworski [**Jaw94**] originally under the name strongly approximately transitive (SAT) actions and are the natural opposite of measure-preserving actions: if G is a group acting on a probability space (X, ν) then the action is measure-preserving when for every measurable set B and every $g \in G$ it holds that $\nu(gB) = \nu(B)$; the action is contractive when for every measurable set B with $\nu(B) > 0$ it holds that $\sup_g \nu(gB) = 1$.

In the case when the acting group G is amenable, if G acts continuously (or merely Borel) on a compact metric space then a simple compactness argument (in weak*) shows the existence of probability measures preserved by the action. In contrast, if G is nonamenable then there always exists spaces on which there is no preserved measure. However, there is a natural way to associate to G a contractive space, the Poisson boundary, that is intimately connected with its action on any compact metric space.

The central idea in the (amenability half of) the rigidity theorem for actions of semisimple groups and their lattices developed in [**SZ94**] and completed in [**CP14b**] is that if (X, ν) is any probability space on which G acts measurably and (B, β) is any Poisson boundary of G then $(B \times X, \beta \times \nu) \to (X, \nu)$ is a relatively contractive map (defined below). A crucial feature of relatively contractive maps is the uniqueness theorem (Theorem 4.13) and the various factor theorems (Theorems 5.1 and 5.2) that follow from it. In this sense, contractiveness (and its relativized version) are powerful ideas in the theory of actions of nonamenable groups.

2010 *Mathematics Subject Classification.* Primary 37A40; Secondary 28D99.

In the context of stationary actions, making use of the Poisson boundary and the fact that the product space is a relatively contractive extension, Furstenberg and Glasner developed the beginnings of a structure theory for non-measure-preserving actions of groups [**FG10**] (though they did not have the machinery of relatively contractive maps and relied entirely on the contractive nature of the Poisson boundary in their proof).

Our purpose here is to explore the concepts of contractive spaces and relatively contractive maps and their relation to other concepts in the ergodic theory of group actions, notably to relatively measure-preserving maps and to joinings. Many of the results presented appear (implicitly or explicitly) in [**CP14b**] and [**Cre13**] but we include some new results. In particular, we present an example (based on an observation of Glasner and Weiss [**GW15**]) of a contractive stationary space that is not a Poisson boundary of the acting group, the first such example known to the author. The existence of such a space demonstrates the usefulness of studying contractive spaces and relatively contractive maps in their own right, and not merely as a part of the study of boundary actions.

We conclude the paper with a discussion of some of the main applications of relatively contractive maps to rigidity phenomena of lattices. In particular, we give an outline of how the notions discussed here play a key role in the proofs of results such as:

THEOREM (Theorems 6.1 and 6.2). *Let G be a semisimple group with trivial center and no compact factors with at least one factor being a connected (real) Lie group with property (T).*

Let $\Gamma < G$ be an irreducible lattice (meaning that the projection of Γ is dense in every proper normal subgroup of G). Then:

(i) *every measure-preserving action $\Gamma \curvearrowright (X, \nu)$ on a nonatomic probability space is essentially free; and*

(ii) *if $\pi : \Gamma \to \mathcal{U}(M)$ is a representation into the unitary group of a finite factor M such that $\pi(\Gamma)'' = M$ then either M is finite-dimensional or π extends to an isomorphism of the group von Neumann algebra $L\Gamma \simeq M$.*

2. Group Actions on Probability Spaces

Throughout the paper G will denote a locally compact second countable topological group and Γ will be reserved for countable discrete groups. Often $\Gamma < G$ will be a lattice.

2.1. G-Spaces and G-Maps.

DEFINITION 2.1. A **G-space** is a standard Borel probability space (X, ν) that is equipped with an action of G such that ν is quasi-invariant under the action (the class of null sets is preserved by the action). This will be written $G \curvearrowright (X, \nu)$.

DEFINITION 2.2. Let $G \curvearrowright (X, \nu)$ be a G-space. The **translate of ν by $g \in G$** is the probability measure $g\nu$ defined by $g\nu(E) = \nu(g^{-1}E)$ for all measurable sets E.

DEFINITION 2.3. Let (X, ν) and (Y, η) be G-spaces. A measurable map $\pi : X \to Y$ such that $\pi_*\nu = \eta$ is a **G-map** when π is G-equivariant: $\pi(gx) = g\pi(x)$ for all $g \in G$ and almost every $x \in X$ (here $\pi_*\nu$ is the **pushforward measure** defined by, for E a measurable subset of Y, $\pi_*\nu(E) = \nu(\pi^{-1}(E))$).

DEFINITION 2.4. Let $\pi : (X,\nu) \to (Y,\eta)$ be a G-map of G-spaces. The **disintegration** of ν over η is the almost everywhere unique map $D_\pi : Y \to P(X)$ such that the support of $D_\pi(y)$ is contained in $\pi^{-1}(y)$ and that $\int_Y D_\pi(y)\, d\eta(y) = \nu$.

2.2. Point Realizations.

DEFINITION 2.5. Let $G \curvearrowright (X,\nu)$ be a G-space. A **point realization** (also called a **compact model**) of the action is a compact metric space X_0 equipped with a continuous G-action $G \curvearrowright X_0$ and a Borel probability measure (with full support) ν_0 such that the action $G \curvearrowright (X_0,\nu_0)$ is measurably isomorphic to $G \curvearrowright (X,\nu)$ (meaning there is a measurable isomorphism defined almost everywhere).

DEFINITION 2.6. Let $\pi : (X,\nu) \to (Y,\eta)$ be a G-map of G-spaces. A **point realization** of π is a continuous map $\pi_0 : X_0 \to Y_0$ of compact metric spaces such that $G \curvearrowright (X_0,\nu_0)$ and $G \curvearrowright (Y_0,\eta_0)$ are point realizations of the G-spaces and such that $\pi_0 : (X_0,\nu_0) \to (Y_0,\eta_0)$ is measurably isomorphic to $\pi : (X,\nu) \to (Y,\eta)$ (the obvious diagram of maps commutes in the category of measurable G-maps).

The following result, due to Mackey [**Mac62**], states that in the case of locally compact second countable groups (the class we restrict ourselves to), measurable actions can always be realized as actions on points:

THEOREM 2.7 *When G is locally compact second countable, there exists a point realization of every G-map of G-spaces.*

The proof is somewhat technical and the reader is referred to [**CP14b**] or [**Cre13**] for a detailed proof.

2.3. Measure-Preserving Spaces and Relatively Measure-Preserving Maps.

DEFINITION 2.8. Let (X,ν) be a G-space. Then (X,ν) is **measure-preserving** when $g\nu = \nu$ for all $g \in G$.

DEFINITION 2.9. Let G be a locally compact second countable group and $\pi : (X,\nu) \to (Y,\eta)$ a G-map of G-spaces. Then π is **relatively measure-preserving** when the disintegration of ν over η via π, $D_\pi : Y \to P(X)$, is G-equivariant: $D_\pi(gy) = gD_\pi(y)$ for all $g \in G$ and almost every $y \in Y$.

2.4. Joinings. Joinings between G-spaces are one of the main tools in the ergodic theory of group actions. The reader is referred to [**Gla03**] for a detailed introduction to the theory of joinings and how they can be used to define a notion of orthogonality (or disjointness) of actions of groups on probability spaces. In the context of measure-preserving actions, joinings have proven their usefulness in a variety of ways, e.g. [**Fur67**], [**dJR87**], [**Gla03**], [**dlR11**].

DEFINITION 2.10. Let (X,ν) and (Y,η) be G-spaces. Let $\alpha \in P(X \times Y)$ such that $(\mathrm{pr}_X)_*\alpha = \nu$, $(\mathrm{pr}_Y)_*\alpha = \eta$ and α is quasi-invariant under the diagonal G-action. The space $(X \times Y, \alpha)$ with the diagonal G-action is called a **joining** of (X,ν) and (Y,η).

DEFINITION 2.11. A joining α of G-spaces is **G-invariant** when α is G-measure-preserving under the diagonal action.

DEFINITION 2.12. Let (X,ν) and (Y,η) be G-spaces. The space $(X \times Y, \nu \times \eta)$ with the diagonal G-action is the **independent joining** of (X,ν) and (Y,η).

Proposition 2.4.1. *Let $\alpha \in P(X \times Y)$ be a joining of the G-spaces (X, ν) and (Y, η). Consider the projection $p : X \times Y \to Y$. The disintegration of α over η via p is of the form $D_p(y) = \alpha_y \times \delta_y$ for some $\alpha_y \in P(X)$ almost surely.*

Definition 2.13 ([**Gla03**, Definition 6.9]). Let (X, ν), (Y, η) and (Z, ζ) be G-spaces and let α be a joining of (X, ν) and (Y, η) and β be a joining of (Y, ν) and (Z, ζ). Let $\alpha_y \in P(X)$ and $\beta_y \in P(Z)$ be the projections of the disintegrations of α and β over η. The measure $\rho \in P(X \times Z)$ by

$$\rho = \int_Y \alpha_y \times \beta_y \, d\eta(y)$$

is the **composition** of α and β.

Proposition 2.4.2 ([**Gla03**, Proposition 6.10]). *The composition of two joinings is a joining. If two joinings are G-invariant then so is their composition.*

2.5. Relative Joinings.

Definition 2.14. Let (X, ν) and (Y, η) be G-spaces with a common G-quotient (Z, ζ), that is a diagram of G-maps and G-spaces as follows:

$$\begin{array}{ccc} & & (X,\nu) \\ & & \big\downarrow{\scriptstyle \pi} \\ (Y,\eta) & \xrightarrow{\varphi} & (Z,\zeta) \end{array}$$

Treat $X \times Y$ as a G-space with the diagonal action. A G-quasi-invariant Borel probability measure $\rho \in P(X \times Y)$ is a **relative joining** of (X, ν) and (Y, η) over (Z, ζ) when the following diagram of G-maps commutes:

$$\begin{array}{ccc} X \times Y, \rho & \xrightarrow{p_X} & (X,\nu) \\ \big\downarrow{\scriptstyle p_Y} & & \big\downarrow{\scriptstyle \pi} \\ (Y,\eta) & \xrightarrow{\varphi} & (Z,\zeta) \end{array}$$

where p_X and p_Y are the natural projections from $X \times Y$ to X and Y, respectively.

In general, the product $\nu \times \eta$ is not a relative joining of (X, ν) and (Y, η) over (Z, ζ) unless (Z, ζ) is trivial since we require that $\pi \circ p_X = \varphi \circ p_Y$ almost everywhere. However, there is a notion of independent joining in the relative case:

Definition 2.15. Let (X, ν) and (Y, η) be G-spaces with common G-quotient (Z, ζ). Let $\pi : (X, \nu) \to (Z, \zeta)$ and $\varphi : (Y, \eta) \to (Z, \zeta)$ be the quotient maps. The probability measure $\rho \in P(X \times Y)$ given by

$$\rho = \int_Z D_\pi(z) \times D_\varphi(z) \, d\zeta(z)$$

is the **independent relative joining** of (X, ν) and (Y, η) over (Z, ζ).

Of course, the independent relative joining is a relative joining. We also note that the independent joining $\nu \times \eta$ is the independent relative joining over the trivial system.

PROPOSITION 2.5.1. *Let $\pi : (X, \nu) \to (Y, \eta)$ be a G-map of G-spaces. Then the independent relative joining of (X, ν) and (Y, η) over (Y, η) is G-isomorphic to (X, ν).*

PROOF. The independent relative joining is $(X \times Y, \alpha)$ where

$$\alpha = \int_Y D_\pi(y) \times \delta_y \, d\eta(y).$$

Let $p : X \times Y \to X$ be the projection to X. Let $\alpha_x \in P(X \times Y)$ by $\alpha_x = \delta_x \times \delta_{\pi(x)}$. Then

$$\begin{aligned}\int_X \alpha_x \, d\nu(x) &= \int_Y \int_X \delta_x \times \delta_{\pi(x)} \, dD_\pi(y)(x) \, d\eta(y) \\ &= \int_Y \int_X \delta_x \times \delta_y \, dD_\pi(y)(x) \, d\eta(y) \\ &= \int_Y D_\pi(y) \times \delta_y \, d\eta(y) = \alpha\end{aligned}$$

and α_x is supported on $p^{-1}(x) = \{x\} \times Y$. Therefore $D_p(x) = \alpha_x$ by uniqueness of disintegration. Since α_x is a point mass, then p is an isomorphism so $(X \times Y, \alpha)$ is isomorphic to (X, ν). □

2.6. Lattices and Induced Actions. A countable subgroup $\Gamma < G$ in a locally compact second countable group is a **lattice** when it is discrete in the G-topology and there exists a finite-Haar-measure fundamental domain for G/Γ. The main examples of lattices are the arithmetic points of algebraic groups, for example $\mathbb{SL}_n(\mathbb{Z}) < \mathbb{SL}_n(\mathbb{R})$.

We recall now the construction of the induced action from a lattice to the ambient group, see, e.g., [**Zim84**]. Let $\Gamma < G$ be a lattice in a locally compact second countable group and let (X, ν) be a Γ-space. Take a fundamental domain F for G/Γ such that $e \in F$. Let $m \in P(F)$ be the Haar measure of G restricted to F and normalized to be a probability measure on F. Define the cocycle $\alpha : G \times F \to \Gamma$ by $\alpha(g, f) = \gamma$ such that $gf\gamma \in F$ and observe that such a γ is unique so this is well-defined. Note that $\alpha(gh, f) = \alpha(h, f)\alpha(g, hf\alpha(h, f))$ meaning α is indeed a cocycle. We also remark that $\alpha(f, e) = e$ for $f \in F$ and that $\alpha(\gamma, e) = \gamma^{-1}$ for $\gamma \in \Gamma$. Consider now the action of G on $F \times X$ given by

$$g \cdot (f, x) = (gf\alpha(g, f), \alpha(g, f)^{-1}x)$$

and observe that the measure $m \times \nu$ is quasi-invariant under this action. So $(F \times X, m \times \nu)$ is a G-space.

Also consider the Γ-action on $(G \times X, \text{Haar} \times \nu)$ given by

$$\gamma \cdot (g, x) = (g\gamma^{-1}, \gamma x)$$

and observe that this is quasi-invariant as well. Since the Γ-action on G/Γ is proper the space of Γ-orbits of $G \times X$ under that action is well-defined and we denote it by $G \times_\Gamma X$ and write elements as $[g, x]$. Define a G-action on $G \times_\Gamma X$ by $h \cdot [g, x] = [hg, x]$.

Define the map $\tau : F \times X \to G \times_\Gamma X$ by $\tau(f, x) = [f, x]$ and the map $\rho : G \times_\Gamma X \to F \times X$ by $\rho([g, x]) = (g\alpha(g, e), \alpha(g, e)^{-1}x)$. Observe that ρ is well-defined since $\alpha(g\gamma, e) = \gamma^{-1}\alpha(g, e)$.

Clearly, $\tau(\rho([g,x])) = [g,x]$ and $\rho(\tau(f,x)) = (f,x)$ so these maps invert one another. Moreover,

$$\tau(g \cdot (f,x)) = [gf\alpha(g,f), \alpha(g,f)^{-1}x] = [gf,x] = g \cdot [f,x] = g \cdot \tau(f,x)$$

and similarly, $\rho(h \cdot [g,x]) = h \cdot \rho([g,x])$ so τ and ρ are inverse G-isomorphisms of $(F \times X, m \times \nu)$ and $(G \times_\Gamma X, \alpha)$ where $\alpha = \tau_*(m \times \nu)$.

These isomorphisms show that the construction defined is independent of the fundamental domain chosen and we define the **induced action to G of $\Gamma \curvearrowright (X,\nu)$** to be the G-space $(G \times_\Gamma X, \alpha)$.

More generally, one can induce a Γ-map of Γ-spaces. Let $\pi : (X,\nu) \to (Y,\eta)$ be a Γ-map of Γ-spaces. Fix a fundamental domain F for G/Γ and $m \in P(F)$ as above. Define the map $\Phi : (F \times X, m \times \nu) \to (F \times Y, m \times \eta)$ by $\Phi(f,x) = (f, \pi(x))$. Then

$$\Phi(g \cdot (f,x)) = (gf\alpha(g,f), \pi(\alpha(g,f)^{-1}x)) = (gf\alpha(g,f), \alpha(g,f)^{-1}\pi(x)) = g \cdot \Phi(f,x)$$

so Φ is a G-map of G-spaces. Let $\Pi : (G \times_\Gamma X, \alpha) \to (G \times_\Gamma Y, \beta)$ be the image of Φ over the canonical isomorphisms defined above for the induced actions. The G-map Π between the induced G-spaces is referred to as the **induced G-map from the Γ-map π**.

3. Contractive Spaces

The study of contractive spaces was initiated by Jaworski [**Jaw94**], [**Jaw94**] as part of his proof of the Choquet-Deny Theorem for nilpotent groups. The key observation in his work is that Poisson boundaries enjoy a property he termed strong approximate transitivity (SAT), a name based on the fact that it can be viewed as stronger version of the approximate transitivity property of Connes and Woods.

Later work indicated that contractive is a better term for this property as it is both more descriptive and removes the somewhat misleading appearance of a connection to the AT property (the reader is referred to [**CS14**], [**CP14b**] and [**Cre13**]). Contractiveness is the natural opposite of measure-preserving and is orthogonal to measure-preserving in a variety of ways (presented in the sequel).

3.1. The Definition of Contractive Actions.

DEFINITION 3.1 (Jaworski [**Jaw94**]). An action $G \curvearrowright (X,\nu)$ is a **contractive** when for every measurable set B with $\nu(B) > 0$ it holds that $\sup_g \nu(gB) = 1$.

This is easily seen to be equivalent to the statement that given any positive measure set B there exists a sequence $\{g_n\}$ of group elements such that $\nu(g_n B) \to 1$ which can be thought of saying that under $\{g_n^{-1}\}$ the whole space A contracts to the set B.

PROPOSITION 3.1.1. *If $G \curvearrowright (X,\nu)$ is both contractive and measure-preserving, it is the trivial (one-point) system.*

PROOF. Let B be any positive measure set. For any $\epsilon > 0$ there exists $g \in G$ such that $\nu(gB) > 1 - \epsilon$ since the action is contractive. But $\nu(gB) = \nu(B)$ since the action is measure-preserving. Therefore $\nu(B) > 1 - \epsilon$ for all ϵ. We conclude that $\nu(B) = 1$ for any positive measure set meaning that (X,ν) is (measurably) a single point. □

PROPOSITION 3.1.2 (Jaworski [**Jaw94**]). *If $G \curvearrowright (X, \nu)$ is contractive then it is ergodic.*

PROOF. Let B be a positive measure G-invariant set. Since the action is contractive, for any $\epsilon > 0$ there exists $g \in G$ such that $\nu(gB) > 1 - \epsilon$. As B is G-invariant, $gB = B$ and so $\nu(B) > 1 - \epsilon$. We then conclude that $\nu(B) = 1$ meaning the action is ergodic. □

3.2. Properties of Contractive Actions. We now state various equivalent characterizations of contractiveness and properties of such spaces. These facts are all special cases of results presented in the following section on relatively contractive maps and so we omit proofs here in favor of presenting the more general proofs in the sequel.

THEOREM 3.2 (Jaworski [**Jaw94**]). *$G \curvearrowright (X, \nu)$ is contractive if and only if the map $L^\infty(X, \nu) \to L^\infty(G, \mathrm{Haar})$ given by $f \mapsto \widehat{f}$ where*

$$\widehat{f}(g) = g\nu(f) = \int f(gx)\, d\nu(x)$$

is an isometry.

Note that for measure-preserving systems, the map $f \mapsto \widehat{f}$ has image precisely equal to the constants and that this is an equivalent characterization of measure-preserving.

THEOREM 3.3 (Creutz-Shalom [**CS14**]). *Contractiveness is a property of the measure class: if $G \curvearrowright (X, \nu)$ is contractive and η is a probability measure on X in the same measure class as ν then $G \curvearrowright (X, \eta)$ is contractive. In fact, the same sequence of elements in G contracts both measures: if $\nu(g_n B) \to 1$ then $\eta(g_n B) \to 1$.*

Viewing contractive actions at the level of point realizations plays a key role in the proof of the uniqueness theorem for relatively contractive maps. To this end, we present a definition and result due to Furstenberg and Glasner on point realizations of contractive actions:

DEFINITION 3.4 (Furstenberg-Glasner [**FG10**]). Let $G \curvearrowright X$ be a continuous action of a group G on a compact metric space X and let ν be a Borel probability measure on X with full support. The action $G \curvearrowright (X, \nu)$ is **contractible** when for every point $x \in X$ there exists a sequence $\{g_n\}$ in G such that $g_n\nu \to \delta_x$ in weak* (here δ_x is the point mass at x).

THEOREM 3.5 (Furstenberg-Glasner [**FG10**]). *A G-space $G \curvearrowright (X, \nu)$ is contractive if and only if every point realization of the action is contractible.*

The generalization of the above statement due to the author and J. Peterson [**CP14b**] is the main technical result in the proof of the uniqueness theorem and factor theorems for relatively contractive maps. This result also justifies the intuitive idea that contractive spaces have the property that "the group contracts the measure to every possible point mass".

3.3. Examples of Contractive Actions. Poisson boundaries are the main examples of contractive actions. Introduced by Furstenberg [**Fur63**] as a means for performing harmonic analysis on Lie groups, Poisson boundaries have led to a variety of deep results in the rigidity theory of lattices in Lie groups, including the celebrated Normal Subgroup Theorem of Margulis [**Mar79**]. As our purpose here is to focus on contractive spaces, we include minimal details about Poisson boundaries and refer the reader to the work of Bader and Shalom [**BS06**] (Section 2) for an excellent overview of the construction of the Poisson boundary and its various properties.

DEFINITION 3.6 (Furstenberg [**Fur63**]). Let Γ be a countable discrete group and μ a probability measure on Γ (an element of $\ell^1\Gamma$ that is nonnegative and has norm one) such that the support of μ generates Γ and such that μ is symmetric: $\mu(\gamma) = \mu(\gamma^{-1})$ for all $\gamma \in \Gamma$. Consider the random walk on Γ with law μ: the space $(\Gamma^{\mathbb{N}}, \mu^{\mathbb{N}})$ where Γ acts by multiplication on the left of the first element. The map $T : \Gamma^{\mathbb{N}} \to \Gamma^{\mathbb{N}}$ by $T(\omega_1, \omega_2, \ldots) = (\omega_1\omega_2, \omega_3, \ldots)$ (multiplying the first two elements) commutes with the Γ-action. The space of T-ergodic components (equivalently, the tail σ-algebra of the space under the filtration by coordinates) is the **Poisson boundary** for Γ with the measure μ.

The above definition can be extended in the obvious way to locally compact groups provided the measure μ on the group G is taken to be nonsingular with respect to the Haar measure (it is enough for some convolution power to be nonsingular).

THEOREM 3.7 (Jaworksi [**Jaw94**]). *Let (B, β) be any Poisson boundary for G (any meaning for any probability measure on G). Then $G \curvearrowright (B, \beta)$ is contractive.*

Since contractiveness is a property of the measure class, if (B, β) is a Poisson boundary of G and ν is a probability measure in the same class as β then $G \curvearrowright (X, \nu)$ is also contractive.

The Poisson boundary enjoys an additional property, namely that of stationarity: if $G \curvearrowright (X, \nu)$ is a G-space and μ is a probability measure on G then the convolution $\mu * \nu$ is the probability measure on X defined by $\mu * \nu(B) = \int_G g\nu(B)\, d\mu(g)$; a G-space is μ-stationary when $\mu * \nu = \nu$. The fact that Poisson boundaries are stationary follows from the fact that $T_*\mu^{\mathbb{N}} = \mu * \mu^{\mathbb{N}}$. Stationary spaces have received much attention since the same argument showing that amenable groups always have invariant measures can be used to show that if G is any group acting on a compact metric space X then there always exists a stationary measure on X.

These facts lead to the natural question of whether there exist contractive spaces admitting a stationary measure in the measure class (for some probability measure on the group with full support) that are not quotients of the Poisson boundary (and also to the question of whether there exist contractive actions admitting no stationary measures at all). To the best of our knowledge, this question has been unanswered previously and we now present an example of such a space.

THEOREM 3.8. *Let $G = \mathbb{R} \rtimes \mathbb{R}^+$ be the "$ax + b$ group" consisting of all maps $\mathbb{R} \to \mathbb{R}$ of the form $x \mapsto ax + b$ for constants $a, b \in \mathbb{R}$. Then the natural action of G on $\mathbb{R}$ (equipped with an probability measure in the class of the Lebesgue measure on $\mathbb{R}$) is contractive but is not a quotient of the Poisson boundary of G for any symmetric measure μ on G that is nonsingular with respect to the Haar measure.*

PROOF. That the action is contractive is a consequence of the work of Jaworski [**Jaw94**]. Glasner and Weiss ([**GW15**] Example 3.5) observed that the action is not doubly ergodic in the sense of Kaimanovich [**Kai03**]. However, Kaimanovich showed that the action of a group on any quotient of any Poisson boundary (with respect to a stationary admissible measure) is doubly ergodic. We therefore conclude that this action is contractive but is not a boundary action. □

4. Relatively Contractive Maps

Relatively contractive maps were introduced in [**CP14b**] as a generalization of contractive spaces to quotient maps. We present here the definition of such maps and state (either with proof or with reference to where proofs can be found) the various properties that make them useful. The main application of these results are the factor theorems appearing in the following section which in turn are the central result needed in the rigidity theorems presented in the final section of the paper.

4.1. Conjugates of Disintegration Measures. The principal notion in formulating the idea of relatively contractive maps is to "conjugate" the disintegration measures. For a G-map of G-spaces $\pi : (X, \nu) \to (Y, \eta)$, the disintegration of ν over η can be summarized as saying that for almost every $y \in Y$ there is a unique measure $D_\pi(y) \in P(X)$ such that $D_\pi(y)$ is supported on the fiber over y and $\int_Y D_\pi(y)\, d\eta(y) = \nu$.

For $g \in G$ and $y \in Y$, we have that $D_\pi(gy)$ is supported on the fiber over gy, that is, on $\pi^{-1}(gy) = g\pi^{-1}(y)$, and that for any Borel $B \subseteq X$, we have that $gD_\pi(y)(B) = D_\pi(y)(g^{-1}B)$ meaning that $gD_\pi(y)$ is supported on $g\pi^{-1}(y)$. Therefore we can formulate the following:

DEFINITION 4.1. Let $\pi : (X, \nu) \to (Y, \rho)$ be a G-map of G-spaces. The **conjugated disintegration measure** over π at a point $y \in Y$ by the group element $g \in G$ is

$$D_\pi^{(g)}(y) = g^{-1} D_\pi(gy).$$

The preceding discussion shows that $D_\pi^{(g)}(y)$ is supported on $g^{-1}g\pi^{-1}(y) = \pi^{-1}(y)$. Hence:

PROPOSITION 4.1.1. *Let $\pi : (X, \nu) \to (Y, \eta)$ be a G-map of G-spaces and fix $y \in Y$. The conjugated disintegration measures*

$$\mathcal{D}_y = \{g^{-1} D_\pi(gy) : g \in G\}$$

are all supported on $\pi^{-1}(y)$.

Another approach to the conjugates of disintegration measures is to observe that:

PROPOSITION 4.1.2. *Let $\pi : (X, \nu) \to (Y, \eta)$ be a G-map of G-spaces. For any $g \in G$ then $\pi : (X, g^{-1}\nu) \to (Y, g^{-1}\eta)$ is also a G-map of G-spaces. Let $D_\pi : Y \to P(X)$ be the disintegration of ν over η. Then $D_\pi^{(g)}$ is the disintegration of $g^{-1}\nu$ over $g^{-1}\eta$.*

PROOF. To see that π maps $(X, g^{-1}\nu)$ to $(Y, g^{-1}\eta)$ follows from π being G-equivariant.

We have already seen that $g^{-1}D_\pi(gy)$ is supported on $\pi^{-1}(y)$ so to prove the proposition it remains only to show that $\int g^{-1}D_\pi(gy)\, dg^{-1}\eta(y) = g^{-1}\nu$. This is clear as

$$\int_Y g^{-1}D_\pi(gy)\;dg^{-1}\eta(y) = g^{-1}\int_Y D_\pi(gg^{-1}y)\;d\eta(y)$$
$$= g^{-1}\int_Y D_\pi(y)\;d\eta(y) = g^{-1}\nu$$

since D_π disintegrates ν over η. □

A basic fact we will need in what follows is that the conjugated disintegration measures are mutually absolutely continuous to one another (over a fixed point y of course, as y varies they have disjoint supports):

PROPOSITION 4.1.3. *Let $\pi : (X,\nu) \to (Y,\eta)$ be a G-map of G-spaces. For almost every y the set*

$$\mathcal{D}_y = \{g^{-1}D_\pi(gy) : g \in G\}$$

is a collection of mutually absolutely continuous probability measures supported on $\pi^{-1}\{y\}$.

PROOF. We will prove this in the case when G is countable, the reader is referred to [**CP14b**] for a proof in the locally compact case. For $g \in G$ write

$$A_g = \{y \in Y : D_\pi(y) \text{ and } g^{-1}D_\pi(gy) \text{ are not in the same measure class}\}.$$

Then A_g is a Borel set for each $g \in G$ since $D_\pi : Y \to P(X)$ is a Borel map and the equivalence relation on $P(X)$ given by $\alpha \sim \beta$ if and only if α and β is in the same measure class is Borel.

Since $g^{-1}D_\pi(gy)$ is the disintegration of $g^{-1}\nu$ over $g^{-1}\eta$ and $g^{-1}\nu$ is in the same measure class as ν, Lemma 4.1.4 (following the proof) gives that $\eta(A_g) = 0$ for each $g \in G$. Therefore $\eta(\bigcup_{g\in G} A_g) = 0$ since G is countable, proving the theorem. □

LEMMA 4.1.4. *Let (X,ν) be a probability space and $\pi : (X,\nu) \to (Y,\pi_*\nu)$ a measurable map to a probability space. Let α be a probability measure in the same measure class as ν. Let $D(y)$ denote the disintegration of ν over $\pi_*\nu$ via π and let $D'(y)$ denote the disintegration of α over $\pi_*\alpha$ via π. Then for almost every $y \in Y$, $D(y)$ and $D'(y)$ are in the same measure class.*

PROOF. Since α and ν are in the same measure class, the Radon-Nikodym derivative $\frac{d\alpha}{d\nu}$ exists and is in $L^1(X,\nu)$. Let f_y be the restriction of $\frac{d\alpha}{d\nu}$ to $\pi^{-1}(y)$. Then by the uniqueness of the disintegration, $D'(y) = f_y D(y)$ almost surely. Therefore $D'(y)$ is absolutely continuous with respect to $D(y)$ almost surely. By a symmetric argument, $D(y)$ is absolutely continuous with respect to $D'(y)$ so they are in the same measure class. □

4.2. The Definition of Relatively Contractive Maps.

DEFINITION 4.2. Let $\pi : (X,\nu) \to (Y,\eta)$ be a G-map of G-spaces. We say π is **relatively contractive** when for almost every $y \in Y$ and any measurable $B \subseteq X$ with $D_\pi(y)(B) > 0$ it holds that $\sup_{g\in G} D_\pi^{(g)}(y)(B) = 1$.

Just as measure-preserving actions are precisely those actions which are relatively measure-preserving extensions of a point (an easy exercise from the definition of relative measure-preserving), relatively contractive maps generalize contractive spaces:

PROPOSITION 4.2.1. *A G-space (X, ν) is contractive if and only if it is a relatively contractive extension of a point.*

PROOF. In the case where $(Y, \eta) = 0$ is the trivial one point system, the disintegration measure is always ν and so being a relatively contractive extension reduces to the definition of contractive: $g^{-1} D_\pi(g \cdot 0) = g^{-1}\nu$ for all $g \in G$ since $g \cdot 0 = 0$ and therefore $\sup_g D_\pi^{(g)}(0)(B) = 1$ implies $\sup_g g^{-1} D_\pi(0)(B) = 1$ so $\sup_g g^{-1}\nu(B) = 1$ for all measurable B with $\nu(B) > 0$. □

We remark that if $\pi : (X, \nu) \to (Y, \eta)$ is a relatively measure-preserving G-map of G-spaces then the conjugated disintegration measures have the property that $D_\pi^{(g)}(y) = D_\pi(y)$ for all $g \in G$. From this, it is easy to obtain the first indication that relatively contractive maps are orthogonal to relatively measure-preserving maps:

PROPOSITION 4.2.2. *Let $\pi : (X, \nu) \to (Y, \eta)$ be a G-map of G-spaces that is both relatively measure-preserving and relatively contractive. Then π is an isomorphism.*

PROOF. Let B be a measurable subset of X. Since π is relatively contractive, for almost every y such that $D_\pi(y)(B) > 0$ there exists $\{g_n\}$ in G such that $D_\pi^{(g_n)}(y)(B) \to 1$. Since π is relatively measure-preserving, $D_\pi^{(g_n)})(y)(B) = D_\pi(y)(B)$. Therefore $D_\pi(y)(B) = 1$ for almost every y such that $D_\pi(y)(B) > 0$. Then

$$\nu(B) = \int_Y D_\pi(y)(B)\, d\eta(y) = \int_Y \mathbb{1}_{\pi(B)}(y)\, d\eta(y) = \eta(\pi(B)).$$

As this holds for every measurable set B, π is an isomorphism. □

4.3. The Algebraic Characterization. Generalizing Jaworksi [**Jaw94**], we may characterize relatively contractive maps algebraically:

PROPOSITION 4.3.1. *Let $\pi : (X, \nu) \to (Y, \rho)$ be a G-map of G-spaces. Then π is relatively contractive if and only if the map $f \mapsto D_\pi^{(g)}(y)(f)$ is an isometry between $L^\infty(X, D_\pi(y))$ and $L^\infty(G, \mathrm{Haar})$ for almost every $y \in Y$ (here $D_\pi^{(g)}(y)(f)$ is a function of g).*

PROOF. Assume π is relatively contractive. Take y in the measure one set where the disintegration measures are relatively contractive. Let $f \in L^\infty(X, D_\pi(y))$ with $\|f\| = 1$. Fix $\epsilon > 0$ and let B be a measurable set such that $f(x) > 1 - \epsilon$ for $x \in B$ (replacing f with $-f$ if necessary) and such that $D_\pi(y)(B) > 0$. As π is relatively contractive, there exists $g \in G$ such that $D_\pi^{(g)}(y)(B) > 1 - \epsilon$. Then $D_\pi^{(g)}(f) > 1 - 2\epsilon$. As ϵ was arbitrary, this shows that the map is an isometry on the norm one functions hence is an isometry as claimed.

Conversely, assume the map is an isometry for almost every y. For such a y, let $B \subseteq \pi^{-1}(y)$ with $D_\pi(y)(B) > 0$ and then $1 = \|\mathbb{1}_B\|_\infty = \sup_g D_\pi^{(g)}(y)(B)$ so π relatively contractive. □

Note that π is relatively measure-preserving if and only if the map that would be isometric for relatively contractive, $f \mapsto D_\pi^{(g)}(y)(f)$, is simply the map $f \mapsto D_\pi(y)(f)$ which is the projection to the "constants" on each fiber.

We remark that in effect there is a zero-one law for relatively contractive extensions. Namely, if $\pi : (X,\nu) \to (Y,\eta)$ is a G-map of ergodic G-spaces then the set of y such that $D_\pi^{(g)}(y)$ induces an isometry $L^\infty(X, D_\pi(y)) \to L^\infty(G, \mathrm{Haar})$ has either measure zero or measure one. This follows from the fact that the set of such y must be G-invariant and hence follows by ergodicity: if $D_\pi^{(g)}(y)$ induces an isometry then for any $h \in G$ and $f \in L^\infty(X,\nu)$

$$\sup_{g\in G} \left|D_\pi^{(g)}(hy)(f)\right| = \sup_{g\in G} \left|D_\pi^{(gh)}(y)(h\cdot f)\right| = \sup_{g\in G} \left|D_\pi^{(g)}(y)(h\cdot f)\right| = \|h\cdot f\| = \|f\|.$$

4.4. Relatively Contractible Spaces. Generalizing the idea of contractible spaces as point realizations of contractive actions, in [**CP14b**] the notion of relatively contractible (point) spaces was introduced and used to characterize relatively contractive maps.

DEFINITION 4.3. Let $\pi : (X,\nu) \to (Y,\eta)$ be a G-map of G-spaces. A point realization $\pi_0 : (X_0,\nu_0) \to (Y_0,\eta_0)$ for this map is **relatively contractible** when for η_0-almost every $y \in Y_0$ and every x in the support of $D_\pi(y_0)$ there exists a sequence $g_n \in G$ such that $D_{\pi_0}^{(g_n)}(y) \to \delta_x$ in weak*.

This definition gives rise to the intuitive view of relatively contractive maps as those where almost every fiber can be contracted to any point mass under the group action (though it must be kept in mind that the sequence which contracts one fiber to a point mass may not contract the other fibers, this is especially critical when G is uncountable).

THEOREM 4.4. *Let $\pi : (X,\nu) \to (Y,\eta)$ be a G-map of G-spaces. Then π is relatively contractive if and only if every continuous compact model of π is relatively contractible.*

PROOF. Assume that π is relatively contractive. By Proposition 4.3.1, there is a measure one set of y such that $f \mapsto D_\pi^{(g)}(y)(f)$ is an isometry from $L^\infty(X, D_\pi(y))$ and $L^\infty(G,\mathrm{Haar})$. Fix y in that set and let $x \in X$ be in the support of $D_\pi(y)$. Choose $f_n \in C(X)$ such that $0 \le f_n \le 1$, $\|f_n\| = 1$ and $f_n \to \mathbb{1}_{\{x\}}$ pointwise (possible since $C(X)$ separates points) and such that $f_{n+1} \le f_n$. Since π is relatively contractive (and x is in the support of $D_\pi(y)$), $\sup_g D_\pi^{(g)}(y)(f_n) = 1$ for each n. Choose $g_n \in G$ such that

$$1 - \frac{1}{n} < D_\pi^{(g_n)}(y)(f_n)$$

and observe then that, since $f_{n+1} \le f_n$,

$$1 - \frac{1}{n+1} < D_\pi^{(g_{n+1})}(y)(f_{n+1}) \le D_\pi^{(g_{n+1})}(y)(f_n)$$

and therefore $\lim_{m\to\infty} D_\pi^{(g_m)}(f_n) = 1$ for each fixed n.

Now $P(X)$ is compact so there exists a limit point $\zeta \in P(X)$ such that $\zeta = \lim_j D_\pi^{(g_{n_j})}(y)$ along some subsequence. Now $\zeta(f_n) = 1$ for each n by the above and $f_n \to \mathbb{1}_{\{x\}}$ is pointwise decreasing so by bounded convergence $\zeta(\{x\}) = \lim \zeta(f_n) = 1$. This means that for almost every y, the conclusion holds for all $x \in \pi^{-1}(y)$.

For the converse, first consider any continuous compact model such that for almost every $y \in Y$ and every x in the support of $D_\pi(y)$, there exists a sequence $\{g_n\}$ such that $D_\pi^{(g_n)}(y) \to \delta_x$. Let $f \in C(X)$. Then the supremum of f on supp $D_\pi(y)$ is attained at some $x \in \pi^{-1}(y)$ since supp $D_\pi(y)$ is a closed, hence compact, set. Take g_n such that $g_n^{-1}D_\pi(g_n y) \to \delta_x$. Then $g_n^{-1}D_\pi(g_n y)(f) \to f(x) = \|f\|_{L^\infty(D_\pi(y))}$. Hence for $f \in C(X)$ the map is an isometry.

Now assume that for every continuous compact model for π and for almost every y and every $x \in \operatorname{supp} D_\pi(y)$ there is a sequence $g_n \in G$ such that $g_n^{-1}D_\pi(g_n y) \to \delta_x$.

Suppose that π is not relatively contractive. Then there exists a measurable set $A \subseteq X$ with $\nu(A) > 0$ and $1 > \delta > 0$ such that

$$B = \{y \in Y : D_\pi(y)(A) > 0 \text{ and } \sup_g D_\pi^{(g)}(y)(A) \leq 1 - \delta\} > 0$$

has $\eta(B) > 0$.

Fix $\epsilon > 0$. Let $\psi_n \in C_c(G)$ be an approximate identity (ψ_n are nonnegative continuous functions with $\int \psi_n dm = 1$ where m is a Haar measure on G such that the compact supports of the ψ_n are a decreasing sequence and $\cap_n \operatorname{supp} \psi_n = \{e\}$; the reader is referred to [**FG10**] Corollary 8.7). Define $f_n = \mathbb{1}_A * \psi_n = \int_G \mathbb{1}_A(hx)\psi_n(h)\, dm(h)$. Then the f_n are G-continuous functions by [**FG10**] Lemma 8.6.

By Proposition 4.4.1 (below),

$$\lim_n \|\mathbb{1}_A * \psi_n\|_{L^\infty(X, D_\pi(y))} = 1$$

for all $y \in B$.

There then exists a set $B_1 \subseteq B$ with $\eta(B_1) > \eta(B) - \epsilon$ and $N \in \mathbb{N}$ such that for all $y \in B_1$ and all $n \geq N$, $\|\mathbb{1}_A * \psi_n\|_{L^\infty(X, D_\pi(y))} > 1 - \epsilon$. Let V be a compact set neighborhood of the identity in G such that $|\eta(B_1 \cap h^{-1}B_1) - \eta(B_1)| < \epsilon$ for all $h \in V$ (possible as the G-action is continuous on the algebra of measurable sets). Choose $n \geq N$ such that the support of $\psi = \psi_n$ is contained in V.

Set $f = \mathbb{1}_A * \psi$. Since f is G-continuous there exists a continuous compact model on which $f \in C(X)$ by [**FG10**] Theorem 8.5. Hence, for almost every $y \in Y$,

$$\sup_g D_\pi^{(g)}(y)(f) = \|f\|_{L^\infty(X, D_\pi(y))}.$$

Removing a null set from B_1, then for all $y \in B_1$ there exists $g_y \in G$ such that

$$D_\pi^{(g_y)}(y)(f) > \|f\|_{L^\infty(X, D_\pi(y))} - \epsilon > 1 - 2\epsilon.$$

Observe that

$$\begin{aligned}
(1 - 2\epsilon)\eta(B_1) &\leq \int_{B_1} D_\pi^{(g_y)}(f)\, d\eta(y) \\
&= \int_{B_1}\int_X f(g_y^{-1}x)\, dD_\pi(g_y y)(x)\, d\eta(y) \\
&= \int_{B_1}\int_X\int_G \mathbb{1}_A(hg_y^{-1}x)\psi(h)\, dm(h)\, dD_\pi(g_y y)\, d\eta(y) \\
&= \int_G\int_{B_1} D_\pi(g_y y)(g_y h^{-1}A)\, d\eta(y)\psi(h)\, dm(h) \\
&= \int_G\int_{hB_1} D_\pi(g_y h^{-1}y)(g_y h^{-1}A)\, dh\eta(y)\psi(h)\, dm(h)
\end{aligned}$$

$$\begin{aligned}
&= \int_G \int_{hB_1} D_\pi^{(g_y h^{-1})}(y)(A)\, dh\eta(y)\psi(h)\, dm(h) \\
&\leq \int_G \int_{hB_1} \sup_g D_\pi^{(g)}(y)(A)\, dh\eta(y)\psi(h)\, dm(h) \\
&= \int_G \Big(\int_{hB_1 \setminus B_1} \sup_g D_\pi^{(g)}(y)(A)\, dh\eta(y) \\
&\qquad + \int_{hB_1 \cap B_1} \sup_g D_\pi^{(g)}(y)(A)\, dh\eta(y) \Big) \psi(h)\, dm(h) \\
&\leq \int_G \big(h\eta(hB_1 \setminus B_1) + (1-\delta) h\eta(hB_1 \cap B_1) \big) \psi(h)\, dm(h) \\
&= \int_G \big(h\eta(hB_1) - \delta h\eta(hB_1 \cap B_1) \big) \psi(h)\, dm(h) \\
&= \eta(B_1) - \delta \int_G \eta(B_1 \cap h^{-1}B_1)\psi(h)\, dm(h).
\end{aligned}$$

Now the support of ψ is contained in V and $|\eta(B_1 \cap h^{-1}B_1) - \eta(B_1)| < \epsilon$ for all $h \in V$. Therefore

$$-2\epsilon\eta(B_1) \leq -\delta \int_G (\eta(B_1) - \epsilon)\psi(h)\, dm(h) = -\delta\eta(B_1) + \delta\epsilon.$$

Hence

$$\delta\eta(B_1) \leq \epsilon(2\eta(B_1) + \delta).$$

Then

$$\delta\eta(B) \leq \delta(\eta(B_1) + \epsilon) \leq 2\epsilon(\eta(B_1) + \delta) \leq 2\epsilon(\eta(B) + \delta).$$

Since δ is fixed and this holds for all $\epsilon > 0$, $\eta(B) = 0$ contradicting that π is not relatively contractive. $\square$

We remark briefly on how to construct the approximate identity used in the proof. Let $F : [0, \infty) \to [0, 1]$ be a continuous monotone decreasing function such that $F(0) = 1$ and $F(t) = 0$ for $t \geq 1$. Such functions are easily constructed. For each $n \in \mathbb{N}$ define $F_n(t) = F(nt)$. Then $F_n : [0, \infty) \to [0, 1]$ is continuous and $F_n(0) = 1$ and $F_n(t) = 0$ for $t \geq \frac{1}{n}$. Therefore $F_n(t) \to 0$ for $t > 0$ and $F_n(0) \to 1$. Also, since F is decreasing, $F_{n+1}(t) = F((n+1)t) \leq F(nt) = F_n(t)$. Now returning to our compact metric space (X, d), fix $x_0 \in X$ and define $f_n(x) = F_n(d(x, x_0))$. Then $f_n \in C(X)$ since $d(\cdot, x_0) \in C(X)$ and $F_n \in C([0, \infty))$. Clearly $f_{n+1}(x) \leq f_n(x)$ and $f_n(x) \to 0$ for $x \neq x_0$ and $f_n(x_0) \to 1$. This sequence $\{f_n\}$ is then the approximate identity used in the proof.

The following fact was used in the above proof and a detailed proof can be found in [**CP14b**] so we opt to omit it here.

PROPOSITION 4.4.1. *Let $\pi : (X, \nu) \to (Y, \eta)$ be a G-map of G-spaces. Let $\psi_n \in C_c(G)$ be an approximate identity (the ψ_n are nonnegative continuous functions with decreasing compact supports V_n such that $\cap V_n = \{e\}$ and $\int \psi_n dm = 1$ for m a Haar measure on G). Then for any measurable set $A \subseteq X$ and almost every $y \in Y$ such that $D_\pi(y)(A) > 0$,*

$$\lim_n \|\mathbb{1}_A * \psi_n\|_{L^\infty(X, D_\pi(y))} = 1.$$

4.5. Relatively Contractive Maps and Dense Subgroups. In general, the map $g \mapsto D_\pi^{(g)}(y)$ is not continuous (however, it can be shown to be continuous almost everywhere for almost every y) which can be seen by considering an induced action from a lattice to a locally compact second countable group (for such an action, there cannot exist point realizations making the map continuous, a fact left to the reader). This fact accounts for the difficulty in the proof of the following statement.

THEOREM 4.5. *Let $\pi : (X,\nu) \to (Y,\eta)$ be a relatively contractive G-map of G-spaces. Let G_0 be a countable dense subgroup of G. Then π is a relatively contractive G_0-map.*

PROOF. Suppose that π is not G_0-relatively contractive. By the proof of Theorem 4.4, there then exists a continuous compact model for $\pi : X \to Y$, a positive measure set $A \subseteq Y$, a nonnegative continuous function $f \in C(X)$ and $\delta > 0$ such that for all $y \in A$,

$$\sup_{g_0 \in G_0} D_\pi^{(g_0)}(y)(f) \leq \|f\|_{L^\infty(X,D_\pi(y))} - \delta.$$

Let $\epsilon > 0$ such that $\eta(A) > \epsilon$. Since π is G-relatively contractive, there is a conull Borel set Y_{00} such that for every $y \in Y_{00}$, $\sup_g D_\pi^{(g)}(y)(f) = \|f\|_{L^\infty(X,D_\pi(y))}$.

Consider the set

$$E = \{(g,y) \in G \times Y_{00} : D_\pi^{(g)}(y)(f) \geq \|f\|_{L^\infty(X,D_\pi(y))} - \epsilon\}.$$

Since D_π is a Borel map, this is a Borel set. By the von Neumann Selection Theorem [**vN49**] there then exists a conull Borel set $Y_0 \subseteq Y_{00}$ such that the map $\mathrm{pr}_Y : E \to Y_{00}$ admits a Borel section on Y_0. Choose a Borel section $g_y \in G$ for $y \in Y_0$ such that $D_\pi^{(g_y)}(y) \geq \|f\|_{L^\infty(X,D_\pi(y))} - \epsilon$.

Consider the Borel function $Y \to P(X)$ given by $y \mapsto D_\pi(g_y y)$. By Lusin's Theorem, there exists a measurable set $D \subseteq Y$ with $\eta(D) > 1 - \epsilon$ and a continuous map $F : Y \to P(X)$ such that $F(y) = D_\pi(g_y y)$ for $y \in D$.

For $y \in Y_0$, choose $\{g_n\}$ in G_0 such that $g_n \to g_y$. Then $\|g_y \cdot f - g_n \cdot f\|_\infty \to 0$ since G acts continuously on $C(X)$ and $F(g_y^{-1} g_n y) \to F(y)$ in weak* hence $F(g_y^{-1} g_n y)(g_y \cdot f) \to F(y)(g_y \cdot y)$. Therefore

$$\begin{aligned}
&\left|F(y)(g_y \cdot f) - F(g_y^{-1} g_n y)(g_n \cdot f)\right| \\
&\quad\leq \left|F(y)(g_y \cdot f) - F(g_y^{-1} g_n y)(g_y \cdot f)\right| + \left|F(g_y^{-1} g_n y)(g_y \cdot f) - F(g_y^{-1} g_n y)(g_n \cdot f)\right| \\
&\quad\leq \left|F(y)(g_y \cdot f) - F(g_y^{-1} g_n y)(g_y \cdot f)\right| + \int_X \left|f(g_y^{-1} x) - f(g_n^{-1} x)\right| dF(g_y^{-1} g_n y)(x) \\
&\quad\leq \left|F(y)(g_y \cdot f) - F(g_y^{-1} g_n y)(g_y \cdot f)\right| + \|g_y \cdot f - g_n \cdot f\|_\infty \to 0.
\end{aligned}$$

Observe that for $y \in D$,

$$F(y)(g_y \cdot f) = D_\pi(g_y y)(g_y \cdot f) = D_\pi^{(g_y)}(y)(f) > \|f\|_{L^\infty(X,D_\pi(y))} - \epsilon.$$

Consider the set $D'_n = \{y \in A : g_y^{-1} g_n y \in D\}$. Then for $y \in D'_n$,

$$F(g_y^{-1} g_n y)(g_n \cdot f) = D_\pi(g_y g_y^{-1} g_n y)(g_n \cdot f) = D_\pi^{(g_n)}(y)(f) \leq \|f\|_{L^\infty(X,D_\pi(y))} - \delta.$$

Consider the sets $E_n = D \cap D'_n$. Since $g_n \to g_y$, $\eta(E_n) \to \eta(D \cap A) > 0$. For $y \in E_n$,

$$\left|F(y)(g_y \cdot f) - F(g_y^{-1} g_n y)(g_n \cdot f)\right| \geq \delta - \epsilon.$$

But $\left|F(y)(g_y\cdot f)-F(g_y^{-1}g_ny)(g_n\cdot f)\right|\to 0$ as $n\to\infty$ for every y. This contradiction means that π is relatively contractive for G_0. □

4.6. Examples of Relatively Contractive Maps. Let (X,ν) and (Y,η) be contractive G-spaces. In general it need not hold that $(X\times Y,\nu\times\eta)$ is contractive (with the diagonal G-action), however:

EXAMPLE 4.6. *Let (X,ν) be a contractive G-space and (Y,η) be a G-space. The map $\mathrm{pr}_Y:(X\times Y,\nu\times\eta)\to(Y,\eta)$ is relatively contractive ($X\times Y$ has the diagonal G-action).*

PROOF. The disintegration measures $D_\pi(y)$ are supported on $X\times\delta_y$ and have the form $D_\pi(y)=\nu\times\delta_y$. Clearly

$$D_\pi^{(g)}(y)=g^{-1}(\nu\times\delta_{gy})=g^{-1}\nu\times\delta_y$$

and since (X,ν) is contractive then π is relatively contractive. □

More generally, the following holds:

EXAMPLE 4.7. *Let $\pi:(X,\nu)\to(Y,\eta)$ be a relatively contractive G-map of G-spaces. Let (Z,ζ) be a G-space. The map $\pi\times\mathrm{id}:(X\times Z,\nu\times\zeta)\to(Y\times Z,\eta\times\zeta)$ is relatively contractive (where $X\times Z$ and $Y\times Z$ have the diagonal G-action).*

PROOF. Since the disintegration of the identity is point masses, for almost every $(y,z)\in Y\times X$, it holds that $D_{\pi\times\mathrm{id}}^{(g)}(y,z)=D_\pi^{(g)}(y)\times\delta_z$. Then π being relatively contractive implies $\pi\times\mathrm{id}$ is relatively contractive. □

Inducing contractive actions also gives rise to relatively contractive maps:

THEOREM 4.8. *Let $\Gamma<G$ be a lattice in a locally compact second countable group. Let (X,ν) be a contractive Γ-space and $p:G\times_\Gamma X\to G/\Gamma$ be the G-map that is the natural projection from the induced G-space over (X,ν) to G/Γ. Then p is a relatively contractive G-map.*

PROOF. Treat $G\times_\Gamma X$ as $(F\times X,m\times\nu)$ for F a fundamental domain for G/Γ with cocycle $\alpha:G\times F\to\Gamma$. Consider $p:F\times X\to F$ the projection. The disintegration $D_p(f)$ of $m\times\nu$ over m is of the form $D_p(f)=\delta_f\times\nu$. For $g\in G$,

$$D_p^{(g)}(f)=g^{-1}D_p(gf\alpha(g,f))=g^{-1}(\delta_{gf\alpha(g,f)}\times\nu)=\delta_f\times\alpha(g,f)\nu.$$

Fix $(f_0,x_0)\in F\times X$ and choose $\gamma_n\in\Gamma$ such that $\gamma_n\nu\to\delta_{x_0}$. Set $g_n=\gamma_n^{-1}f_0^{-1}$. Then $\alpha(g_n,f_0)=\gamma_n$ so $D_p^{(g_n)}(f_0)=\delta_{f_0}\times\gamma_n\nu\to\delta_{f_0}\times\delta_{x_0}$ meaning p is relatively contractive. □

4.7. Factorization of Contractive Maps. We now present the first step in the uniqueness theorem for relatively contractive maps which describes the possible intermediate quotient maps.

THEOREM 4.9. *Let $\pi:(X,\nu)\to(Y,\eta)$ and $\varphi:(Y,\eta)\to(Z,\rho)$ be G-maps of G-spaces. If $\varphi\circ\pi$ is relatively contractive then both φ and π are relatively contractive.*

PROOF. We use Theorem 4.4 and take a continuous compact model for π to do so. First observe, for all $g\in G$ and almost every z, that $\pi_*D_{\varphi\circ\pi}^{(g)}(z)=D_\varphi^{(g)}(z)$.

For such z where also $\overline{\text{conv}}\,\{D^{(g)}_{\varphi\circ\pi}(z)\} = P((\varphi\circ\pi)^{-1}(z))$ and every x in the support of $D_{\varphi\circ\pi}(z)$, there is $g_n \in G$ such that $D^{(g_n)}_{\varphi\circ\pi}(z) \to \delta_x$. Therefore

$$D^{(g_n)}_{\varphi}(z) = \pi_* D^{(g_n)}_{\varphi\circ\pi}(z) \to \pi_*\delta_x = \delta_{\pi(x)}$$

and so for every y in the support of $D_\varphi(z)$, the point mass δ_y is a limit point of $D^{(g)}_\varphi(z)$. Hence φ is relatively contractive.

Suppose that π is not relatively contractive. Then, by the proof of Theorem 4.4, there exists a continuous compact model for $\pi : X \to Y$ such that $f \mapsto |D^{(g)}_\pi(y)(f)|$ is not an isometry from $C(X)$ to $L^\infty(G)$ for a positive measure set of $y \in Y$.

Observe that if the map is an isometry on a countable dense set $C_0 \subseteq C(X)$ then for any $f \in C(X)$ there exists $f_n \in C_0$ with $f_n \to f$ in sup norm, hence

$$|D^{(g)}_\pi(y)(f)| = |D^{(g)}_\pi(y)(f - f_n) + D^{(g)}_\pi(f_n)| \geq |D^{(g)}_\pi(y)(f_n)| - \|f - f_n\|_\infty.$$

For $\epsilon > 0$, choose n such that $\|f - f_n\|_\infty < \epsilon$. Choose g such that $D^{(g)}_\pi(y)(f_n)| > \|f_n\| - \epsilon$. Then

$$|D^{(g)}_\pi(y)(f)| > \|f_n\| - \epsilon - \epsilon > \|f\| - 3\epsilon$$

and so the map is an isometry for f as well.

Therefore, there is a positive measure set of y such that $f \mapsto |D^{(g)}_\pi(y)(f)|$ is not an isometry on C_0. Hence, since C_0 is countable, there is some $f \in C_0$ and a positive measure set of y such that $\sup_g |D^{(g)}_\pi(y)(f)| < \|f\|_{L^\infty(X,D_\pi(y))}$. So there is some $\delta > 0$ and a measurable set $A \subseteq Y$ with $\eta(A) > 0$ such that $\sup_g |D^{(g)}_\pi(y)(f)| < \|f\|_{L^\infty(X,D_\pi(y))} - \delta$ for all $y \in A$. We may assume (by taking a subset) that A is closed. Since η is a Borel measure, it is regular, hence we may assume A is closed (by taking a subset).

Now there exists a positive measure set $B \subseteq Z$ on which $D_\varphi(z)(A) > 0$ for $z \in B$. For $z \in B$ such that z is in the measure one set on which $\varphi \circ \pi$ contracts to point masses,

$$\begin{aligned}
D^{(g)}_{\varphi\circ\pi}(z)(f) &= \int_{\varphi^{-1}(z)} D^{(g)}_\pi(y)(f)\; dD^{(g)}_\varphi(z)(y) \\
&= \int_{\varphi^{-1}(z)\cap A} D^{(g)}_\pi(y)(f)\; dD^{(g)}_\varphi(z)(y) + \int_{\varphi^{-1}(z)\setminus A} D^{(g)}_\pi(y)(f)\; dD^{(g)}_\varphi(z)(y) \\
&\leq \int_{\varphi^{-1}(z)\cap A} \|f\|_{L^\infty(X,D_\pi(y))} - \delta\; dD^{(g)}_\varphi(z)(y) \\
&\qquad + \int_{\varphi^{-1}(z)\setminus A} \|f\|_{L^\infty(X,D_\pi(y))}\; dD^{(g)}_\varphi(z)(y) \\
&\leq \|f\|_{L^\infty(X,D_{\varphi\circ\pi}(z))} - \delta D^{(g)}_\varphi(z)(A).
\end{aligned}$$

Now for any x in the support of $D_{\varphi\circ\pi}(z)$, there exists g_n such that $D^{(g_n)}_{\varphi\circ\pi}(z) \to \delta_x$. Hence also $D^{(g_n)}_\varphi(z) \to \delta_{\pi(x)}$. Choose $x \in \pi^{-1}(A) \cap (\text{supp } D_{\varphi\circ\pi}(z))$ such that $f(x) = \|f\|_{L^\infty(X,D_{\varphi\circ\pi}(z))}$ (possible since $\pi^{-1}(A) \cap (\text{supp } D_{\varphi\circ\pi}(z))$ is closed, hence compact, and f is continuous). Then

$$\begin{aligned}
f(x) = \lim_n D^{(g_n)}_{\varphi\circ\pi}(z)(f) &\leq \lim_n \|f\|_{L^\infty(X,D_{\varphi\circ\pi}(z))} - \delta D^{(g_n)}_\varphi(z)(A) \\
&= \|f\|_{L^\infty(X,D_{\varphi\circ\pi}(z))} - \delta\delta_{\pi(x)}(A) = \|f\|_{L^\infty(X,D_{\varphi\circ\pi}(z))} - \delta
\end{aligned}$$

is a contradiction. Hence π is relatively contractive. $\square$

The above statement is the analogue of one direction of the similar well-known fact about relative measure-preserving:

THEOREM 4.10. *Let $\pi : (X,\nu) \to (Y,\eta)$ and $\varphi : (Y,\eta) \to (Z,\zeta)$ be G-maps of G-spaces such that $\varphi \circ \pi : (X,\nu) \to (Z,\zeta)$ is relatively measure-preserving. Then π and φ are both relatively measure-preserving. Conversely, if π and φ are relatively measure-preserving then so is $\varphi \circ \pi$.*

COROLLARY 4.11. *Any G-factor of a contractive G-space is a contractive G-space. Any G-factor of a measure-preserving G-space is a measure-preserving G-space.*

PROOF. Let (X,ν) be a contractive G-space and $\pi : (X,\nu) \to (Y,\eta)$ be a G-map of G-spaces. Take $\varphi : (Y,\eta) \to 0$ to be the G-map to the trivial one-point space. Then $\varphi \circ \pi : (X,\nu) \to 0$ is relatively contractive since (X,ν) is contractive and therefore φ is relatively contractive since its composition with π is and so (Y,η) is contractive. The same argument applied to relative measure-preserving maps shows the second statement. □

COROLLARY 4.12. *Let (X,ν) be a G-space such that $\pi : (X,\nu) \to (Y,\eta)$ is a relatively contractive G-map of G-spaces and $\varphi : (X,\nu) \to (Z,\zeta)$ is a relatively measure-preserving G-map of G-spaces. Then $\pi \times \varphi : (X,\nu) \to (Y \times Z, (\pi \times \varphi)_*\nu)$ by $(\pi \times \varphi)(x) = (\pi(x), \varphi(x))$ is a G-isomorphism.*

PROOF. Consider the G-map $\mathrm{pr}_Y \circ (\pi \times \varphi) = \pi$. Since π is relatively contractive then both the projection map to Y and $\pi \times \varphi$ are relatively contractive (Theorem 4.9). Likewise $\mathrm{pr}_Z \circ (\pi \times \varphi) = \varphi$ is relatively measure-preserving so the projection to Z and $\pi \times \varphi$ are relatively measure-preserving. By Proposition 4.2.2 then $\pi \times \varphi$ is an isomorphism. □

4.8. Uniqueness of Relatively Contractive Maps. We are now in a position to present the uniqueness theorem for relatively contractive maps. The proof is somewhat technical and can be found in [**CP14b**], we opt to omit it in the interest of brevity.

THEOREM 4.13. *Let (X,ν) be a contractive G-space and (Y,η) be a measure-preserving G-space. Let $\psi : (X \times Y, \nu \times \eta) \to (Y,\eta)$ be the natural projection map (treating $(X \times Y, \nu \times \eta)$ as G-space with the diagonal action). Let $\pi : (X \times Y, \nu \times \eta) \to (Z,\alpha)$ be a G-map of G-spaces and let $\pi' : (X \times Y, \nu \times \eta) \to (Z,\alpha')$ be a G-map of G-spaces such that α' is in the same measure class as α. Let $\varphi : (Z,\alpha) \to (Y,\eta)$ and $\varphi' : (Z,\alpha') \to (Y,\eta)$ be G-maps such that $\varphi \circ \pi = \psi$ and $\varphi' \circ \pi' = \psi$. That is, we consider the following commutative diagram of G-maps and G-spaces:*

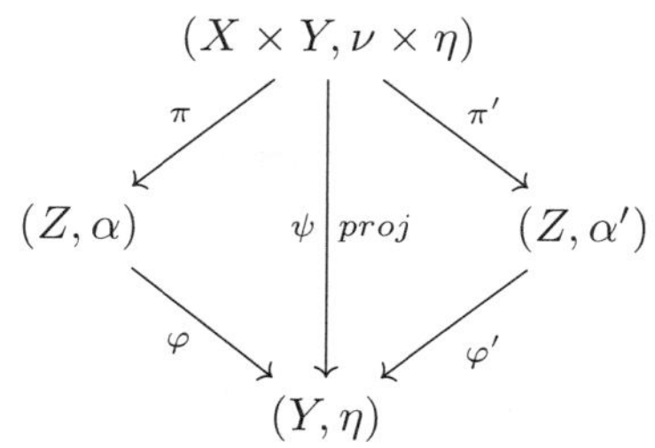

Assume that the disintegrations $D_\varphi(y)$ of α over η via φ and the disintegrations $D_{\varphi'}(y)$ of α' over η via φ' have the property that $D_\varphi(y)$ and $D_{\varphi'}(y)$ are in the same measure class almost surely. Then $\pi = \pi'$ almost everywhere, $\varphi = \varphi'$ almost everywhere and $\alpha = \alpha'$.

COROLLARY 4.14 (Creutz-Shalom [**CS14**]). *Let (X, ν) be a contractive G-space and let $\pi : (X, \nu) \to (Z, \alpha)$ and $\pi' : (X, \nu) \to (Z, \alpha')$ be G-maps of G-spaces such that α and α' are in the same measure class. Then $\pi = \pi'$ almost surely and $\alpha = \alpha'$.*

PROOF. Consider the composition of maps $\varphi \circ \pi : (X, \nu) \to 0$ where $\varphi : (Z, \eta) \to 0$ is the map to the trivial system. Since (X, ν) is contractive, the preceding theorem gives the result. □

4.9. Joinings With Contractive Spaces.

THEOREM 4.15. *Let (X, ν) be a contractive G-space and let (Y, η) be a G-space. Then there is at most one joining $(X \times Y, \alpha)$ of (X, ν) and (Y, η) such that the projection to X is relatively measure-preserving.*

PROOF. Let $f \in L^\infty(Y, \eta)$ and define

$$F(x) = D_{\mathrm{pr}_X}(x)(f \circ \mathrm{pr}_Y).$$

Taking compact models for X and Y such that π is continuous makes clear that F is a bounded Borel function on X. Then for any $g \in G$ we have that, using that pr_X is relatively measure-preserving,

$$\begin{aligned} g\nu(F) &= \int_X F(gx)\, d\nu(x) \\ &= \int_X \int_{X \times Y} f(\mathrm{pr}_Y(z, y))\, dD_{\mathrm{pr}_X}(gx)(z, y)\, d\nu(x) \\ &= \int_X \int_{X \times Y} f(g\mathrm{pr}_Y(z, y))\, dD_{\mathrm{pr}_X}(x)(z, y)\, d\nu(x) \\ &= \int_{X \times Y} f(g\mathrm{pr}_Y(z, y))\, d\alpha(z, y) \\ &= \int_Y f(gy)\, d(\mathrm{pr}_Y)_*\alpha(y) = \int_Y f(gy)\, d\eta(y) = g\eta(f). \end{aligned}$$

Suppose now that $(X \times Y, \alpha_1)$ and $(X \times Y, \alpha_2)$ are both joinings such that pr_X is relatively measure-preserving. Fix $f \in L^\infty(Y, \eta)$ and let $F_1(x) = D^1_{\mathrm{pr}_X}(x)(f \circ \mathrm{pr}_Y)$ and $F_2(x) = D^2_{\mathrm{pr}_X}(x)(f \circ \mathrm{pr}_Y)$ where $D^j_{\mathrm{pr}_X}$ is the disintegration of α_j over ν. Set $F(x) = F_1(x) - F_2(x)$. Then F is a bounded Borel function on X and by the above we have that $g\nu(F) = g\nu(F_1) - g\nu(F_2) = g\eta(f) - g\eta(f) = 0$ for all $g \in G$. Since (X, ν) is contractive we also know that $\|F\|_{L^\infty(X,\nu)} = \sup_g |g\nu(F)| = 0$. Therefore $F(x) = 0$ almost surely and so $F_1(x) = F_2(x)$ almost surely. As this holds for all $f \in L^\infty(Y, \eta)$ we conclude that

$$(\mathrm{pr}_Y)_* D^1_{\mathrm{pr}_X}(x) = (\mathrm{pr}_Y)_* D^2_{\mathrm{pr}_X}(x)$$

for almost every $x \in X$. The conclusion now follows since the measures have the same disintegration. □

COROLLARY 4.16. *Let (X, ν) be a contractive G-space and (Y, η) be a measure-preserving G-space. Then the independent joining is the only joining of (X, ν) and (Y, η) such that pr_X is relatively measure-preserving.*

PROOF. Observe that the independent joining $(X \times Y, \nu \times \eta)$ is a joining and that $D_{\mathrm{pr}_X}(x) = \delta_x \times \eta$. Since (Y, η) is measure-preserving,

$$D_{\mathrm{pr}_X}(gx) = \delta_{gx} \times \eta = (g\delta_x) \times \eta = (g\delta_x) \times (g\eta) = g(\delta_x \times \eta) = gD_{\mathrm{pr}_X}(x)$$

so pr_X is relatively measure-preserving. By the previous theorem then the independent joining is the unique such joining. □

COROLLARY 4.17. *Let (X, ν) be a G-space such that $\pi : (X, \nu) \to (Y, \eta)$ is a relatively measure-preserving G-map of G-spaces and $\varphi : (X, \nu) \to (Z, \zeta)$ is a relatively contractive G-map of G-spaces where (Y, η) is a contractive G-space and (Z, ζ) is a measure-preserving G-space. Then (X, ν) is G-isomorphic to $(Y \times Z, \eta \times \zeta)$.*

PROOF. By Corollary 4.12 the map $\pi \times \varphi$ is a G-isomorphism of (X, ν) with $(Y \times Z, (\pi \times \varphi)_*\nu)$. Now $(\mathrm{pr}_Y)_*(\pi \times \varphi)_*\nu = \pi_*\nu = \eta$ and likewise $(\mathrm{pr}_Z)_*(\pi \times \varphi)_*\nu = \zeta$ so $(\pi \times \varphi)_*\nu$ is a joining of (Y, η) and (Z, ζ). Since π is relatively measure-preserving and $\pi = \mathrm{pr}_X \circ (\pi \times \varphi)$ we have that pr_X is relatively measure-preserving. The previous corollary then says that it is the independent joining. □

COROLLARY 4.18. *Let (X, ν) be a contractive G-space and $\pi : (X, \nu) \to (Y, \eta)$ a G-map of G-spaces. Then the only joining of (X, ν) and (Y, η) such that the projection to X is relatively measure-preserving is the joining $(X \times Y, \tilde{\pi}_*\nu)$ where $\tilde{\pi}(x) = (x, \pi(x))$.*

PROOF. Let $D(x)$ be the disintegration of $\tilde{\pi}_*\nu$ over ν. Then $D(x)$ is supported on $\{x\} \times Y \cap \operatorname{supp} \tilde{\pi}_*\nu = \{(x, \pi(x))\}$. Therefore $D(x) = \delta_{(x,\pi(x))}$. So $D(gx) = \delta_{(gx,\pi(gx))} = \delta_{g(x,\pi(x))} = g\delta_{(x,\pi(x))} = gD(x)$. By the previous theorem this is then the unique joining with projection to X being relatively measure-preserving. □

More generally:

THEOREM 4.19. *Let (X, ν) be a contractive G-space and $\pi : (X, \nu) \to (Y, \eta)$ a G-map of G-spaces. Let $\zeta \in P(X \times Y)$ be a joining of (X, ν) and (Y, η') for some η' absolutely continuous with respect to η such that the projection to X of ζ to ν is relatively measure-preserving. Then $\zeta = \tilde{\pi}_*\nu$ where $\tilde{\pi}(x) = (x, \pi(x))$ and in particular, $\eta' = \eta$.*

PROOF. Let D be the disintegration of ζ over ν. Then $D(x) = \delta_x \times \zeta_x$ for some $\zeta_x \in P(Y)$ for almost every x. Note that $D(gx) = gD(x)$ for $g \in G$ since the projection is relatively measure-preserving and therefore $\zeta_{gx} = g\zeta_x$ for all $g \in G$. Let $f \in C(Y)$. Define $F \in L^\infty(X, \nu)$ by

$$F(x) = f(\pi(x)) - \zeta_x(f).$$

Let $\epsilon > 0$ and take $x_0 \in X$ such that $|F(x_0)| > \|F\|_{L^\infty(X,\nu)} - \epsilon$. Since (X, ν) is contractive, there exists $g_n \in G$ such that $g_n\nu \to \delta_{x_0}$. Observe that, using that $\zeta_{gx} = g\zeta_x$,

$$\begin{aligned} g_n\nu(F) &= \int_X f(\pi(g_n x)) - \zeta_{g_n x}(f) \, d\nu(x) \\ &= \int_X f(g_n\pi(x)) - g_n\zeta_x(f) \, d\nu(x) \\ &= g_n\eta(f) - g_n\eta'(f) \end{aligned}$$

since $\int_X \zeta_x \, d\nu(x) = \eta'$.

Now η' is absolutely continuous with respect to η and $g_n\eta = \pi_* g_n \nu \to \pi_* \delta_{x_0} = \delta_{\pi(x_0)}$. Since (Y, η) is contractive, being a factor of a contractive space, by Corollary 4.14 (the proof of which goes through even when η' is only absolutely continuous with respect to and not necessarily in the same measure class as η), $g_n\eta' \to \delta_{\pi(x_0)}$ also. Therefore

$$g_n\nu(F) = g_n\eta(f) - g_n\eta'(f) \to f(\pi(x_0)) - f(\pi(x_0)) = 0$$

since $f \in C(Y)$. So we have that $\|F\| < \epsilon$. This holds for all $\epsilon > 0$ so $F(x) = 0$ almost surely. As this holds for all $f \in C(Y)$ we then have that $\zeta_x = \delta_{\pi(x)}$ almost surely. This means that $D(x) = \delta_x \times \delta_{\pi(x)} = \delta_{\tilde{\pi}(x)}$ almost surely so $\zeta = \tilde{\pi}_*\nu$ as claimed. Since $\text{proj}_Y \ \tilde{\pi}_*\nu = \pi_*\nu = \eta$, then $\eta' = \text{proj}_Y \ \zeta = \eta$. □

We also obtain a special case of a result of Furstenberg and Glasner. Proposition 3.1 in [**FG10**] states that there is a unique stationary joining between a G-boundary and an arbitrary G-space; we obtain another proof of this fact when the G-space is measure-preserving:

COROLLARY 4.20 (Furstenberg-Glasner [**FG10**]). *Let G be a group and $\mu \in P(G)$ a probability measure on G. Let (B, β) be the (G, μ)-boundary and (X, ν) a measure-preserving G-space. Then the only joining $(B \times X, \alpha)$ of (B, β) and (X, ν) such that $\mu * \alpha = \alpha$ is the independent joining.*

PROOF. Let $\pi : G^{\mathbb{N}} \to B$ be the boundary map (see [**BS06**] section 2), meaning that $\beta_\omega = \lim_n \omega_1 \cdots \omega_n \beta = \delta_{\pi(\omega)}$ $\mu^{\mathbb{N}}$-almost surely and $\pi_*\mu^{\mathbb{N}} = \beta$. Since α is μ-stationary, $\alpha = \int \alpha_\omega \, d\mu^{\mathbb{N}}(\omega)$. Now $(\text{proj}_B)_*\alpha_\omega = \beta_\omega = \delta_{\pi(\omega)}$ and $(\text{proj}_X)_*\alpha_\omega = \nu_\omega = \nu$ since (X, ν) is measure-preserving. Therefore $\alpha_\omega = \delta_{\pi(\omega)} \times \nu$ and since $\pi_*\mu^{\mathbb{N}} = \beta$ then the disintegration of α over β is $D(b) = \delta_b \times \nu$ which is G-equivariant. Hence the projection to B is relatively measure-preserving so the claim follows by the previous corollaries. □

4.10. Relatively Contractive Maps and Finite Index Subgroups. Relative contractiveness is not affected by passage to finite index subgroups. The proof of the following fact can be found in [**CP14b**] and is a relatively easy exercise for the reader.

THEOREM 4.21. *Let G be a locally compact second countable group and $H < G$ be a finite index subgroup. Let $\pi : (X, \nu) \to (Y, \eta)$ be a relatively contractive G-map of ergodic G-spaces. Then, restricting the actions to H makes π a relatively contractive H-map.*

4.11. Contractive Actions and Lattices. The following is a generalization of Proposition 3.7 in [**CS14**] (which shows the same result only for Poisson boundaries):

THEOREM 4.22. *Let G be a locally compact second countable group and $\Gamma < G$ a lattice. Let (X, ν) be a contractive (G, μ)-space (meaning that $\mu * \nu = \nu$) for some symmetric $\mu \in P(G)$ such that the support of μ generates G. Then the restriction of the G-action to Γ makes (X, ν) a contractive Γ-space.*

LEMMA 4.11.1. *Let G be a locally compact second countable group and let (X, ν) be a contractive (G, μ)-space for some $\mu \in P(G)$ such that the support of μ generates G. Let $A \subseteq X$ be a measurable set with $\nu(A) > 0$. Then for every $\epsilon > 0$,*

$$\mu^{\mathbb{N}}(\{(\omega_1, \omega_2, \ldots) \in G^{\mathbb{N}} : \lim_{n \to \infty} \nu(\omega_n^{-1} \cdots \omega_1^{-1} A) > 1 - \epsilon\}) > 0.$$

PROOF. Let $\varphi \in L^\infty(G)$ be defined by $\varphi(g) = \nu(g^{-1}A)$. Then φ is a μ-harmonic nonnegative bounded function on G since $\mu * \nu = \nu$. As $G \curvearrowright (X, \nu)$ is contractive, $\|\varphi\|_\infty = \|\mathbb{1}_A\|_\infty = 1$ (because the map $L^\infty(X, \nu) \to L^\infty(G)$ is an isometry). Define the function $f \in L^\infty(G^{\mathbb{N}})$ by

$$f(\omega_1, \omega_2, \ldots) = \lim_{n \to \infty} \varphi(\omega_1 \cdots \omega_n)$$

which exists $\mu^{\mathbb{N}}$-almost everywhere by the Martingale Convergence Theorem (in fact f descends to an L^∞-function on the Poisson boundary of (G, μ)). Then $f \geq 0$ and $\|f\|_\infty = \|\varphi\|_\infty = 1$ since the mapping between L^∞ of the Poisson boundary and the harmonic functions on G is an isometry. Let $\epsilon > 0$. Then

$$\mu^{\mathbb{N}}(\{(\omega_1, \omega_2, \ldots) \in G^{\mathbb{N}} : f(\omega_1, \ldots) > 1 - \epsilon\}) > 0$$

since $\|f\|_\infty = 1$. Since

$$f(\omega_1, \ldots) = \lim_{n \to \infty} \varphi(\omega_1 \cdots \omega_n) = \lim_{n \to \infty} \nu(\omega_n^{-1} \cdots \omega_1^{-1} A),$$

this completes the proof. □

PROOF OF THEOREM 4.22. Let m be the invariant (Haar) probability measure on G/Γ. Let K_0 be a bounded open subset of G. Set $K = K_0\Gamma \subseteq G/\Gamma$. Then $m(K) > 0$ since K_0 is open. By the Random Ergodic Theorem (due to Kifer [**Kif86**] in general and Kakutani [**Kak51**] in the measure-preserving case), for m-almost every $z \in G/\Gamma$ and $\mu^{\mathbb{N}}$-almost every $(\omega_1, \omega_2, \ldots)$ it holds that

$$\lim_{N \to \infty} \frac{1}{N} \sum_{n=1}^{N} \mathbb{1}_K(\omega_n \cdots \omega_1 z) = m(K) > 0.$$

Pick $z \in G/\Gamma$ such that the above holds $\mu^{\mathbb{N}}$-almost everywhere. Then $\omega_n \cdots \omega_1 z \in K$ infinitely often $\mu^{\mathbb{N}}$-almost surely and so, as μ is symmetric, $\omega_n^{-1} \cdots \omega_1^{-1} z \in K$ infinitely often $\mu^{\mathbb{N}}$-almost surely.

Let z_0 be a representative of z in G. Let $B \subseteq X$ be a measurable set with $\nu(B) > 0$. Set $A = z_0 B$. Fix $\epsilon > 0$. Then $\nu(A) > 0$ since ν is quasi-invariant and so, by Lemma 4.11.1,

$$\mu^{\mathbb{N}}(\{(\omega_1, \omega_2, \ldots) \in G^{\mathbb{N}} : \lim_{n \to \infty} \nu(\omega_n^{-1} \cdots \omega_1^{-1} A) > 1 - \epsilon\}) > 0.$$

As the intersection of a positive measure set with a measure one set is nonempty, there then exists $(\omega_1, \omega_2, \ldots)$ such that $\omega_n^{-1} \cdots \omega_1^{-1} z \in K$ infinitely often and $\lim_{n \to \infty} \nu(\omega_n^{-1} \cdots \omega_1^{-1} A) > 1 - \epsilon$. Hence there exists n such that $\omega_n^{-1} \cdots \omega_1^{-1} z \in K$ and $\nu(\omega_n^{-1} \cdots \omega_1^{-1} A) > 1 - 2\epsilon$.

Observe that $\omega_n^{-1} \cdots \omega_1^{-1} z_0 \in K_0\Gamma$ since $z_0 \in z\Gamma$ and $\omega_n^{-1} \cdots \omega_1^{-1} z \in K = K_0\Gamma$. Write $\omega_n^{-1} \cdots \omega_1^{-1} z_0 = k\gamma$ for some $k \in K_0$ and $\gamma \in \Gamma$. Then

$$1 - 2\epsilon < \nu(\omega_n^{-1} \cdots \omega_1^{-1} A) = \nu(\omega_n^{-1} \cdots \omega_1^{-1} z_0 B) = \nu(k\gamma B).$$

As this holds for all $\epsilon > 0$, there then exists sequences $\{k_n\}$ in K_0 and $\{\gamma_n\}$ in Γ such that $\nu(k_n \gamma_n B) \to 1$. As K is bounded, $\overline{K}$ is compact so there exists a subsequence $\{k_{n_j}\}$ such that $k_{n_j} \to k_\infty \in G$.

Set $C = X \setminus B$ and set $C_j = k_{n_j}\gamma_{n_j}C$. Then $\nu(C_j) \to 0$. Since $k_{n_j}^{-1} \to k_\infty^{-1}$ and $\nu(C_j) \to 0$, by the continuity of the G-action on $L^1(X,\nu)$, it follows that $\nu(k_{n_j}^{-1}C_j) \to 0$. Therefore $\nu(\gamma_{n_j}C) \to 0$ meaning $\nu(\gamma_{n_j}B) \to 1$. As B was an arbitrary measurable set of positive measure, this shows that the Γ-action on (X,ν) is contractive. □

4.12. Inducing Relatively Contractive Maps.

THEOREM 4.23. *Let $\Gamma < G$ be a lattice in a locally compact second countable group. Let $\pi : (X,\nu) \to (Y,\eta)$ be a Γ-map of Γ-spaces and let $\Pi : G\times_\Gamma X \to G\times_\Gamma Y$ be the induced G-map of G-spaces. Then π is a relatively contractive Γ-map if and only if Π is a relatively contractive G-map.*

PROOF. Assume first that Π is relatively contractive. Fix a fundamental domain (F,m) for G/Γ as in the induced action construction and let $\alpha : G \times F \to \Gamma$ be the associated cocycle for the G-action on $F \times X$. Let $\Phi : (F \times X, m \times \nu) \to (F \times Y, m \times \eta)$ by $\Phi = \mathrm{id} \times \pi$. Then Φ is isomorphic to Π over the canonical isomorphisms $G \times_\Gamma X \simeq F \times X$ and $G \times_\Gamma Y \simeq F \times Y$ so Φ is relatively contractive. Consider the disintegration map $D_\Phi : F \times Y \to P(F \times X)$. Observe that for $(f,y) \in F \times Y$

$$D_\Phi(f,y) = \delta_f \times D_\pi(y)$$

since $\Phi = \mathrm{id} \times \pi$ and all the spaces have the product measure. Now consider the conjugates of the disintegration map: for $g \in G$ and $(f,y) \in F \times Y$,

$$\begin{aligned} D_\Phi^{(g)}(f,y) &= g^{-1}D_\Phi(g(f,y)) = g^{-1}D_\Phi(gf\alpha(g,f),\alpha(g,f)^{-1}y) \\ &= g^{-1}(\delta_{gf\alpha(g,f)} \times D_\pi(\alpha(g,f)^{-1}y)) \\ &= \delta_f \times \alpha(g^{-1},gf\alpha(g,f))^{-1}D_\pi(\alpha(g,f)^{-1}y) \\ &= \delta_f \times D_\pi^{(\alpha(g,f)^{-1})}(y). \end{aligned}$$

Now take $r \in L^\infty(X,\nu)$ and define $q(f,x) = r(x)$. Then for $m \times \eta$-almost every (f,y)

$$\|q\|_{L^\infty(F\times X, D_\Phi(f,y))} = \|r\|_{L^\infty(X,D_\pi(y))}$$

and since Φ is relatively contractive, for $m \times \eta$-almost every (f,y) there exists $g_n \in G$ such that

$$D_\Phi^{(g_n)}(f,y)(q) \to \|q\|_{L^\infty(F\times X,D_\Phi(f,y))}.$$

Therefore

$$\delta_f \times D_\pi^{(\alpha(g_n,f)^{-1})}(y)(q) \to \|r\|_{L^\infty(X,D_\pi(y))}$$

and by construction of q then

$$D_\pi^{(\alpha(g_n,f)^{-1})}(y)(r) \to \|r\|_{L^\infty(X,D_\pi(y))}.$$

Hence for η-almost every y there exists a sequence $\gamma_n = \alpha(g_n,f)^{-1} \in \Gamma$ (outside of possibly a measure zero set, which f is chosen is irrelevant) such that

$$D_\pi^{(\gamma_n)}(r) \to \|r\|_{L^\infty(X,D_\pi(y))}$$

which means that π is relatively contractive.

Now assume that π is relatively contractive. Let $x \in X$ and $f \in F$ and set $y = \pi(x)$. As above.

$$D_\Phi^{(g)}(f,y) = \delta_f \times D_\pi^{(\alpha(g,f)^{-1})}(y).$$

Since π is relatively contractive, there exists $\{\gamma_n\}$ such that $D_{\pi}^{(\gamma_n)}(y) \to \delta_x$. Set $g_n = f\gamma_n f^{-1}$. Then $\alpha(g_n, f) = \gamma_n^{-1}$ and so

$$D_{\Phi}^{(g_n)}(f, y) = \delta_f \times D_{\pi}^{(\gamma_n)}(y) \to \delta_{(f,x)}$$

meaning that Π is relatively contractive. □

4.13. Relative Joinings Over Relatively Contractive Maps. Relatively contractive maps were introduced in [**CP14b**] and used to show that any joining between a contractive space and a measure-preserving space such that the projection to the contractive space is relatively measure-preserving is necessarily the independent joining. We generalize this fact to the case of relative joinings and obtain an analogous result.

THEOREM 4.24. *Let (X, ν) and (Y, η) be G-spaces with a common G-quotient (Z, ζ) such that $\varphi : (Y, \eta) \to (Z, \zeta)$ is relatively contractive and $\pi : (X, \nu) \to (Z, \zeta)$ is a G-map. Then there exists at most one relative joining of (X, ν) and (Y, η) over (Z, ζ) such that the projection to (Y, η) is relatively measure-preserving.*

PROOF. For convenience, write $W = X \times Y$. Let ρ be a relative joining of (X, ν) and (Y, η) over (Z, ζ) such that $\varphi : (Y, \eta) \to (Z, \zeta)$ is relatively contractive, $p_Y : (W, \rho) \to (Y, \eta)$ is relatively measure-preserving and $p_X : (W, \rho) \to (X, \nu)$ and $\pi : (X, \nu) \to (Z, \zeta)$ are G-maps such that $\pi \circ p_X = \varphi \circ p_Y$ almost everywhere. Denote by $\psi : (W, \rho) \to (Z, \zeta)$ the composition: $\psi = \pi \circ p_X = \varphi \circ p_Y$.

Let $z \in Z$ and let $f \in L^\infty(\pi^{-1}(z), D_\pi(z))$ be arbitrary. Then $f \circ p_X \in L^\infty(\psi^{-1}(z), D_\psi(z))$ since $D_\psi(z) = \int D_{p_X}(x)\, dD_\pi(z)(x)$. Define

$$F(y) = D_{p_Y}(y)(f \circ p_X)$$

and observe that $F \in L^\infty(\varphi^{-1}(z), D_\varphi(z))$.

For an arbitrary $g \in G$, using that p_Y is relatively measure-preserving,

$$\begin{aligned}
D_{\varphi}^{(g)}(z)(F) &= \int_{\varphi^{-1}(z)} F(y)\, dg^{-1}D_\varphi(gz) \\
&= \int_{\varphi^{-1}(gz)} F(g^{-1}y)\, dD_\varphi(gz) \\
&= \int_{\varphi^{-1}(gz)} \int_{p_Y^{-1}(g^{-1}y)} f(p_X(w))\, dD_{p_Y}(g^{-1}y)(w)\, dD_\varphi(gz)(y) \\
&= \int_{\varphi^{-1}(gz)} \int_{p_Y^{-1}(g^{-1}y)} f(p_X(w))\, dg^{-1}D_{p_Y}(y)(w)\, dD_\varphi(gz)(y) \\
&= \int_{\varphi^{-1}(gz)} \int_{p_Y^{-1}(y)} f(p_X(g^{-1}w))\, dD_{p_Y}(y)(w)\, dD_\varphi(gz)(y) \\
&= \int_{\varphi^{-1}(gz)} \int_{p_Y^{-1}(y)} f(g^{-1}p_X(w))\, dD_{p_Y}(y)(w)\, dD_\varphi(gz)(y)
\end{aligned}$$

Now $\int_{\varphi^{-1}(gz)} D_{p_Y}(y)\, dD_\varphi(gz)(y) = D_\psi(gz)$ and therefore

$$\begin{aligned}
D_\varphi^{(g)}(z)(F) &= \int_{\psi^{-1}(gz)} f(g^{-1}p_X(w))\, dD_\psi(gz)(w) \\
&= \int_{p_X(\psi^{-1}(gz))} f(g^{-1}x)\, d((p_X)_* D_\psi(gz))(x) \\
&= \int_{\pi^{-1}(gz)} f(g^{-1}x)\, dD_\pi(gz)(x) \\
&= D_\pi^{(g)}(z)(f).
\end{aligned}$$

Now let ρ_1 and ρ_2 both be relative joinings over (Z,ζ). Since φ is relatively contractive, there is a measure one set of $z \in Z$ such that for all $F \in L^\infty(\varphi^{-1}(z), D_\varphi(z))$, we have that $\sup_{g\in G} |D_\varphi^{(g)}(F)| = \|F\|$. Fix z in this measure one set.

Let $f \in L^\infty(\pi^{-1}(z), D_\pi(z))$ be arbitrary. Let $D_{p_Y}^j$ and D_ψ^j for $j = 1,2$ denote the disintegrations of ρ_1 and ρ_2 over η and ζ, respectively. Define, for $j = 1,2$,

$$F_j(y) = D_{p_Y}^j(f \circ p_X)$$

and set $F(y) = F_1(y) - F_2(y)$. As above, $F \in L^\infty(\varphi^{-1}(z), D_\varphi(z))$. Now, by the above, for any $g \in G$,

$$D_\varphi^{(g)}(z)(F_1) = D_\pi^{(g)}(z)(f) = D_\varphi^{(g)}(z)(F_2)$$

and therefore $D_\varphi^{(g)}(z)(F) = 0$.

Since z is in the measure one set where that map is an isometry, $\|F\| = \sup_g |D_\varphi^{(g)}(z)(F)| = 0$. Therefore $F = 0$ almost everywhere. As this holds for all $f \in L^\infty(\pi^{-1}(z), D_\pi(z))$, we conclude that $D_\varphi^1(y) = D_\varphi^2(y)$ for $D_\varphi(z)$-almost-every $y \in \varphi^{-1}(z)$.

Now let $f \in L^\infty(\psi^{-1}(z), D_\psi(z))$ be arbitrary and observe that

$$\begin{aligned}
D_\psi^j(z)(f) &= \int_{\psi^{-1}(z)} f(x,y)\, dD_\psi^j(z)(x,y) \\
&= \int_{\varphi^{-1}(z)} \int_{p_Y^{-1}(y)} f(x,y)\, dD_{p_Y}^j(y)(x)\, dD_\varphi(z)(y).
\end{aligned}$$

Since $D_{p_Y}^1(y) = D_{p_Y}^2(y)$ for $D_\varphi(z)$-almost every y,

$$D_\psi^1(z)(f) = D_\psi^2(z)(f).$$

This holds for all $f \in L^\infty(\psi^{-1}(z), D_\psi(z))$ and so $D_\psi^1(z) = D_\psi^2(z)$.

Since the above holds for all z in a measure one set,

$$\rho_1 = \int_Z D_\psi^1(z)\, d\zeta(z) = \int_Z D_\psi^2(z)\, d\zeta(z) = \rho_2.$$

□

COROLLARY 4.25. *Let (X,ν) and (Y,η) be G-spaces with a common G-quotient (Z,ζ) such that $\varphi : (Y,\eta) \to (Z,\zeta)$ is relatively contractive and $\pi : (X,\nu) \to (Z,\zeta)$ is relatively measure-preserving. Then the only relative joining of (X,ν) and (Y,η) over (Z,ζ) such that the projection to (Y,η) is relatively measure-preserving is the independent relative joining.*

PROOF. By the previous theorem, we need only show that the independent relative joining $\rho = \int D_\pi \times D_\varphi \, d\zeta$ is a relative joining such that the projection to (Y, η) is relatively measure-preserving. Let D_{p_Y} be the disintegration of ρ over η. Observe that $p_Y^{-1}(y) = \pi^{-1}(\varphi(y)) \times \{y\}$ and that the support of $D_\pi(\varphi(y)) \times \delta_y$ is the same. Now

$$\begin{aligned} \int_Y D_\pi(\varphi(y)) \times \delta_y \, d\eta(y) &= \int_Z \int_Y D_\pi(z) \times \delta_y \, dD_\varphi(z)(y) \, d\eta(y) \\ &= \int_Z D_\pi(z) \times D_\varphi(z) \, d\zeta(z) = \rho \end{aligned}$$

so by uniqueness, $D_{p_Y}(y) = D_\pi(\varphi(y)) \times \delta_y$ almost everywhere. Then, using that π is relatively measure-preserving,

$$D_{p_Y}(gy) = D_\pi(\varphi(gy)) \times \delta_{gy} = gD_\pi(\varphi(y)) \times g\delta_y = gD_{p_Y}(y)$$

so p_Y is relatively measure-preserving. By the previous theorem, ρ is then the unique relative joining. $\square$

COROLLARY 4.26. *Let G be a locally compact second countable group and let $(X, \nu), (Y, \eta), (Z, \zeta)$ and (W, ρ) be G-spaces such that the following diagram of G-maps commutes:*

$$\begin{array}{ccc} (W, \rho) & \xrightarrow{\psi} & (X, \nu) \\ {\scriptstyle\tau}\downarrow & & \downarrow{\scriptstyle\pi} \\ (Y, \eta) & \xrightarrow{\varphi} & (Z, \zeta) \end{array}$$

If τ and π are relatively measure-preserving and ψ and φ are relatively contractive then (W, ρ) is G-isomorphic to the independent relative joining of (X, ν) and (Y, η) over (Z, ζ).

PROOF. Consider the map $p : W \to X \times Y$ by $p(w) = (\psi(w), \tau(w))$. Then $p_*\rho$ is a relative joining of (X, ν) and (Y, η) over (Z, ζ). Let $p_X : X \times Y \to X$ and $p_Y : X \times Y \to Y$ be the natural projections and observe that the following diagram commutes:

$$\begin{array}{ccccc} (W, \rho) & \xrightarrow{p} & (X \times Y, p_*\rho) & \xrightarrow{p_X} & (X, \nu) \\ & & {\scriptstyle p_Y}\downarrow & & \downarrow{\scriptstyle\pi} \\ & & (Y, \eta) & \xrightarrow{\varphi} & (Z, \zeta) \end{array}$$

since $p_X \circ p = \psi$ and $p_Y \circ p = \tau$.

Now ψ is relatively contractive so p and p_X are relatively contractive and likewise τ being relatively measure-preserving implies p and p_Y are relatively measure-preserving. Therefore p is an isomorphism (Proposition 4.2.2). Since φ is relatively contractive and p_Y is relatively measure-preserving and π is relatively measure-preserving, the previous corollary says that $p_*\rho$ is the independent relative joining. $\square$

5. The Factor Theorems

The results in the previous section on the uniqueness of relatively contractive maps and their (lack of) joinings with measure-preserving systems lead to the so-called factor theorems that are the main tool in the rigidity theorems presented in the final section.

5.1. The Intermediate Contractive Factor Theorem.

The first factor theorem we present appears in [**CP14b**] and is a generalization of the factor theorem for contractive actions in [**CS14**].

THEOREM 5.1. *Let $\Gamma < G$ be a lattice in a locally compact second countable group and let Λ contain and commensurate Γ (commensurate meaning that $\Gamma \cap \lambda\Gamma\lambda^{-1}$ has finite index in Γ for each $\lambda \in \Lambda$) and be dense in G. Let (X, ν) be a contractive (G, μ)-space, meaning that $\mu * \nu = \nu$, (for some $\mu \in P(G)$ such that the support of μ generates G) and (Y, η) be a measure-preserving G-space. Let $\pi : (X \times Y, \nu \times \eta) \to (Y, \eta)$ be the natural projection map from the product space with the diagonal action. Let (Z, ζ) be a Λ-space such that there exist Γ-maps $\varphi : (X \times Y, \nu \times \eta) \to (Z, \zeta)$ and $\rho : (Z, \zeta) \to (Y, \eta)$ with $\rho \circ \varphi = \pi$. Then φ and ρ are Λ-maps and (Z, ζ) is Λ-isomorphic to a G-space and over this isomorphism the maps φ and ρ become G-maps.*

PROOF. Write $(W, \rho) = (X \times Y, \nu \times \eta)$. Fix $\lambda \in \Lambda$. Define the maps $\varphi_\lambda : W \to Z$ and $\rho_\lambda : Z \to Y$ by $\varphi_\lambda(w) = \lambda^{-1}\varphi(\lambda w)$ and $\rho_\lambda(z) = \lambda^{-1}\rho(\lambda z)$. Then $\rho_\lambda \circ \varphi_\lambda(w) = \lambda^{-1}\rho(\lambda\lambda^{-1}\varphi(\lambda w)) = \lambda^{-1}\rho(\varphi(\lambda w)) = \lambda^{-1}\pi(\lambda w) = \pi(w)$ since π is Λ-equivariant. Let $\Gamma_0 = \Gamma \cap \lambda^{-1}\Gamma\lambda$. Then for $\gamma_0 \in \Gamma_0$, write $\gamma_0 = \lambda^{-1}\gamma\lambda$ for some $\gamma \in \Gamma$ and we see that $\varphi_\lambda(\gamma_0 w) = \lambda^{-1}\varphi(\lambda\gamma_0 w) = \lambda^{-1}\varphi(\gamma\lambda w) = \lambda^{-1}\gamma\varphi(\lambda w) = \gamma_0\lambda^{-1}\varphi(\lambda w) = \gamma_0\varphi_\lambda(w)$ meaning that φ_λ is Γ_0-equivariant. Likewise ρ_λ is Γ_0-equivariant. Hence φ, φ_λ, ρ and ρ_λ are all Γ_0-equivariant.

Since (X, ν) is a contractive (G, μ)-space and Γ_0 is a lattice in G, by Theorem 4.22, (X, ν) is a contractive Γ_0-space. By Theorem 4.13 applied to Γ, we can conclude that $\varphi_\lambda = \varphi$ and that $\rho_\lambda = \rho$ provided we can show that the disintegration measures $D_\rho(y)$ and $D_{\rho_\lambda}(y)$ are in the same measure class for almost every y. Assuming this for the moment, we then conclude that φ is Λ-equivariant since $\varphi_\lambda = \varphi$ for each λ. The σ-algebra of pullbacks of measurable functions on (Z, ζ) form a Λ-invariant sub-σ-algebra of $L^\infty(W, \rho)$ which is therefore also G-invariant (because Λ is dense in G) and so (Z, ζ) has a point realization as a G-space [**Mac62**] and likewise φ and ρ as G-maps.

It remains only to show that the disintegration measures have the required property. First note that $D_\rho(y) = \varphi_* D_{\rho\circ\varphi}(y)$ by the uniqueness of the disintegration measure and likewise that $D_{\rho_\lambda}(y) = (\varphi_\lambda)_* D_{\rho_\lambda \circ \varphi_\lambda}(y) = \lambda^{-1}\varphi_*\lambda D_{\rho\circ\varphi}(y) = \lambda^{-1}\varphi_* D_{\rho\circ\varphi}^{(\lambda^{-1})}(\lambda y)$. Now $\rho \circ \varphi = \pi$ is a Λ-map so $D_{\rho\circ\varphi}^{(\lambda^{-1})}(\lambda y)$ is in the same measure class as $D_{\rho\circ\varphi}(\lambda y)$. Therefore $D_{\rho_\lambda}(y)$ is in the same measure class as $\lambda^{-1}\varphi_* D_{\rho\circ\varphi}(\lambda y) = \lambda^{-1}D_\rho(\lambda y)$. Now $\lambda^{-1}D_\rho(\lambda y)$ disintegrates $\lambda^{-1}\zeta$ over $\lambda^{-1}\eta$ via ρ and $\lambda^{-1}\zeta$ is in the same measure class as ζ since (Z, ζ) is a Λ-space. Therefore, by Lemma 4.1.4, $\lambda^{-1}D_\rho(\lambda y)$ and $D_\rho(y)$ are in the same measure class for almost every y. Hence $D_{\rho_\lambda}(y)$ and $D_\rho(y)$ are in the same measure class for almost every y as needed. $\square$

5.2. The Intermediate Contractive Factor Theorem for Products. The second factor theorem we present is a strengthening of the Bader-Shalom Intermediate Factor Theorem [**BS06**] that first appeared in [**Cre13**]:

THEOREM 5.2. *Let $G = G_1 \times G_2$ be a product of two locally compact second countable groups and let $\mu_j \in P(G_j)$ be admissible probability measures for $j = 1, 2$. Set $\mu = \mu_1 \times \mu_2$.*

Let (B, β) be the Poisson boundary for (G, μ) and let (X, ν) be a measure-preserving G-space. Let (W, ρ) be a G-space such that there exist G-maps $\pi : (B \times X, \beta \times \nu) \to (W, \rho)$ and $\varphi : (W, \rho) \to (X, \nu)$ with $\varphi \circ \pi$ being the natural projection to X.

Let (W_1, ρ_1) be the space of G_2-ergodic components of (W, ρ) and let (W_2, ρ_2) be the space of G_1-ergodic components. Likewise, let (X_1, ν_1) and (X_2, ν_2) be the ergodic components of (X, ν) for G_2 and G_1, respectively.

Then (W, ρ) is G-isomorphic to the independent relative joining of $(W_1, \rho_1) \times (W_2, \rho_2)$ and (X, ν) over $(X_1, \nu_1) \times (X_2, \nu_2)$.

We opt to omit the proof as it involves both the relatively contractive maps and their relationship to joinings and also results due to Bader and Shalom [**BS06**] on the nature of the ergodic decomposition of spaces on which products of groups acts (which fall outside our scope). The reader is referred to [**Cre13**] for a detailed proof.

6. Rigidity of Actions of Lattices

To conclude our exposition, we present now the main results of [**CP14b**] and [**Cre13**], all of which rely in crucial fashion on the factor theorems developed above. These rigidity results are the main application of contractive spaces and relatively contractive maps and were the motivation for the development of these concepts. While the proofs are beyond the scope of our exposition (and can be found in the respective papers), we stress that the key ingredient in the proofs is the uniqueness property of relatively contractive maps in the form of the factor theorems.

THEOREM 6.1 ([**CP14b**], [**Cre13**]). *Let G be a semisimple group with trivial center and no compact factors with at least one factor being a connected (real) Lie group with property (T). Let $\Gamma < G$ be an irreducible lattice (meaning that the projection of Γ is dense in every proper normal subgroup of G). Then every measure-preserving action $\Gamma \curvearrowright (X, \nu)$ on a nonatomic probability space is essentially free.*

The proof strategy is to show that any action which is not essentially free is weakly amenable (meaning that the stabilizer subgroups of almost every point are co-amenable in Γ) which, combined with the (partial) property (T) like behavior of Γ forces the action to be atomic. The key idea is to study the relatively contractive map $B \times X \to X$ where B is a Poisson boundary of G (the map is relatively contractive since Γ is a lattice and B is a stationary space).

The intermediate factor theorem guarantees that any Γ-space A appearing in a chain $B \times X \to A \to X$ combined with the results on joinings with contractive spaces has the property that A is isomorphic to $C \times X$ where C is a quotient of B. Various results of Zimmer (see [**Cre13**] and [**CP14b**] for details) state that if $\Gamma \curvearrowright (X, \nu)$ is not weakly amenable then there exist spaces A not isomorphic to X

but sharing the same stabilizer subgroups. If the stabilizers are not trivial (i.e. the action is not essentially free), then the factor theorem leads to the conclusion that nontrivial subgroups of Γ stabilize points in C. However, it is an easy consequence of the construction of the Poisson boundary that the action on C is always essentially free if C is nontrivial.

The crucial fact in the above strategy is that one obtains a large amount of structural information about such spaces A from the fact that the map is relatively contractive, in particular, enough information to rule out nontrivial intermediate spaces.

We remark that the above theorem implies the Margulis Normal Subgroup Theorem in a direct way: if $N \lhd \Gamma$ is nontrivial then the Bernouli action of Γ/N, treated as a Γ-space, has stabilizer subgroups precisely equal to N and so the theorem states that in such a case, N must be finite index (the Bernoulli shift must be atomic).

In closing, we also mention that the notion of relatively contractive maps has been extended to the noncommutative setting of operator algebras by the author and J. Peterson [**CP14a**], leading to a sweeping generalization of the normal subgroup theorem:

THEOREM 6.2 ([**CP14a**]). *Let G be a semisimple group with trivial center and no compact factors with at least one factor being a connected (real) Lie group with property (T) and let $\Gamma < G$ be an irreducible lattice. Let $\pi : \Gamma \to \mathcal{U}(M)$ be a representation into the unitary group of a finite factor M such that $\pi(\Gamma)'' = M$. Then either M is finite-dimensional or π extends to an isomorphism of the group von Neumann algebra $L\Gamma \simeq M$.*

This result gives a form of operator-algebraic superrigidity in the sense that such a lattice Γ cannot be "separated" from its group von Neumann algebra in the same way that the Margulis-Zimmer superrigidty theorem states that it cannot be separated from G: if $\varphi : \Gamma \to H$ is a homomorphism into an algebraic group with $\overline{\varphi(\Gamma)}$ noncompact then φ extends to an isomorphism of G.

The result on operator algebraic superrigidity should be contrasted with the case of amenable groups: if Γ and Λ are amenable countable groups then $L\Gamma$ is always isomorphic to $L\Lambda$. In this sense, lattices in semisimple groups are as far from amenable as possible and the superrigidity theorem is a major indication of this.

Acknowledgments

The author wishes to thank the referee for numerous helpful suggestions and corrections and, in particular, for noticing an important issue with the definition of relatively contractible spaces and providing a solution.

References

[BS06] Uri Bader and Yehuda Shalom, *Factor and normal subgroup theorems for lattices in products of groups*, Invent. Math. **163** (2006), no. 2, 415–454, DOI 10.1007/s00222-005-0469-5. MR2207022

[CP14a] Darren Creutz and Jesse Peterson, *Character rigidity for lattices and commensurators*, Preprint. arXiv:1311.4513, 2014.

[CP14b] ———, *Stabilizers of ergodic actions of lattices and commensurators*, Transactions of the American Mathematical Society (2014), (to appear).

[Cre13] Darren Creutz, *Stabilizers of actions of lattices in products of groups*, Journal of Ergodic Theory and Dynamical Systems (2013), (to appear).

[CS14] Darren Creutz and Yehuda Shalom, *A normal subgroup theorem for commensurators of lattices*, Groups Geom. Dyn. **8** (2014), no. 3, 789–810, DOI 10.4171/GGD/248. MR3267524

[dJR87] A. del Junco and D. Rudolph, *On ergodic actions whose self-joinings are graphs*, Ergodic Theory Dynam. Systems **7** (1987), no. 4, 531–557, DOI 10.1017/S0143385700004193. MR922364

[dlR11] Thierry de la Rue, *Joinings in ergodic theory*, Mathematics of complexity and dynamical systems. Vols. 1–3, Springer, New York, 2012, pp. 796–809, DOI 10.1007/978-1-4614-1806-1_49. MR3220709

[FG10] Hillel Furstenberg and Eli Glasner, *Stationary dynamical systems*, Dynamical numbers—interplay between dynamical systems and number theory, Contemp. Math., vol. 532, Amer. Math. Soc., Providence, RI, 2010, pp. 1–28, DOI 10.1090/conm/532/10481. MR2762131

[Fur63] Harry Furstenberg, *A Poisson formula for semi-simple Lie groups*, Ann. of Math. (2) **77** (1963), 335–386. MR0146298

[Fur67] Harry Furstenberg, *Disjointness in ergodic theory, minimal sets, and a problem in Diophantine approximation*, Math. Systems Theory **1** (1967), 1–49. MR0213508

[Gla03] Eli Glasner, *Ergodic theory via joinings*, Mathematical Surveys and Monographs, vol. 101, American Mathematical Society, Providence, RI, 2003. MR1958753

[GW15] Eli Glasner and Benjamin Weiss, *Weak mixing properties for non-singular actions*, Ergodic Theory and Dynamical Systems **FirstView** (2015), 1–15.

[Jaw94] Wojciech Jaworski, *Strongly approximately transitive group actions, the Choquet-Deny theorem, and polynomial growth*, Pacific J. Math. **165** (1994), no. 1, 115–129. MR1285567

[Kai03] V. A. Kaimanovich, *Double ergodicity of the Poisson boundary and applications to bounded cohomology*, Geom. Funct. Anal. **13** (2003), no. 4, 852–861, DOI 10.1007/s00039-003-0433-8. MR2006560

[Kak51] Shizuo Kakutani, *Random ergodic theorems and Markoff processes with a stable distribution*, Proceedings of the Second Berkeley Symposium on Mathematical Statistics and Probability, 1950, University of California Press, Berkeley and Los Angeles, 1951, pp. 247–261. MR0044773

[Kif86] Yuri Kifer, *Ergodic theory of random transformations*, Progress in Probability and Statistics, vol. 10, Birkhäuser Boston, Inc., Boston, MA, 1986. MR884892

[Mac62] George W. Mackey, *Point realizations of transformation groups*, Illinois J. Math. **6** (1962), 327–335. MR0143874

[Mar79] G. A. Margulis, *Finiteness of quotient groups of discrete subgroups* (Russian), Funktsional. Anal. i Prilozhen. **13** (1979), no. 3, 28–39. MR545365

[SZ94] Garrett Stuck and Robert J. Zimmer, *Stabilizers for ergodic actions of higher rank semisimple groups*, Ann. of Math. (2) **139** (1994), no. 3, 723–747, DOI 10.2307/2118577. MR1283875

[vN49] John von Neumann, *On rings of operators. Reduction theory*, Ann. of Math. (2) **50** (1949), 401–485. MR0029101

[Zim84] Robert J. Zimmer, *Ergodic theory and semisimple groups*, Monographs in Mathematics, vol. 81, Birkhäuser Verlag, Basel, 1984. MR776417

DEPARTMENT OF MATHEMATICS, VANDERBILT UNIVERSITY, NASHVILLE, TENNESSEE 37240
E-mail address: darren.creutz@vanderbilt.edu

Contemporary Mathematics
Volume **678**, 2016
http://dx.doi.org/10.1090/conm/678/13644

Mandelpinski structures in the parameter planes of rational maps

Robert L. Devaney

ABSTRACT. In this paper we give three examples of "Mandelpinski structures" that arise in the parameter planes for maps of the form $z^n + \lambda/z^d$. These structures include Mandelpinski necklaces, Mandelpinski spokes, and a Mandelpinski maze. We use the term "Mandelpinski" here since each of these objects consists of a variety of curves in the parameter plane that alternately pass through a large number of Mandelbrot sets and Sierpinski holes.

In this paper we give a survey of three different types of "Mandelpinski structures" that arise in the parameter planes for maps of the form

$$F_\lambda(z) = z^n + \frac{\lambda}{z^d}$$

where $n, d \geq 2$. Roughly speaking, a Mandelpinski structure is a collection of curves along which alternate a large number of Mandelbrot sets and Sierpinski holes. A Sierpinski hole is a disk in the parameter plane in which each parameter corresponds to a map whose Julia set is a Sierpinski curve, i.e., the Julia set is homeomorphic to the well known Sierpinski carpet fractal.

The types of curves that the Mandelbrot sets and Sierpinski holes alternate along can be very different. The first case we shall consider are Mandelpinski necklaces. When $n = d > 2$, these consist of infinitely many disjoint simple closed curves surrounding the origin in the λ-plane that contain more and more Mandelbrot sets and Sierpinski holes as the necklaces get smaller. See Section 2.

The second case involves Mandelpinski spokes (Section 4). In this case we assume $n \geq 2$ is even and $d \geq 3$ is odd. We first describe an infinite collection of Mandelbrot sets $\mathcal{M}^k$. Then we show that there are infinitely many disjoint arcs extending away from each $\mathcal{M}^k$ in different directions, and each arc now passes alternately through infinitely many Mandelbrot sets and Sierpinski holes. Attaching these sets to the arcs yields the Mandelpinski spokes.

In the third case we again assume that $n \geq 2$ is even and $d \geq 3$ is odd and show that there is a Mandelpinski maze in the parameter plane for this family. This result is described in Section 5. This maze consists of a sequence of planar graphs of increasing complexity. At each stage, each vertex on the graph corresponds to a different Mandelbrot set and the middle of each edge corresponds to a Sierpinski

2010 *Mathematics Subject Classification.* Primary 37F10; Secondary 37F45.

Robert L. Devaney was partially supported by Simons Foundation Grant #208780.

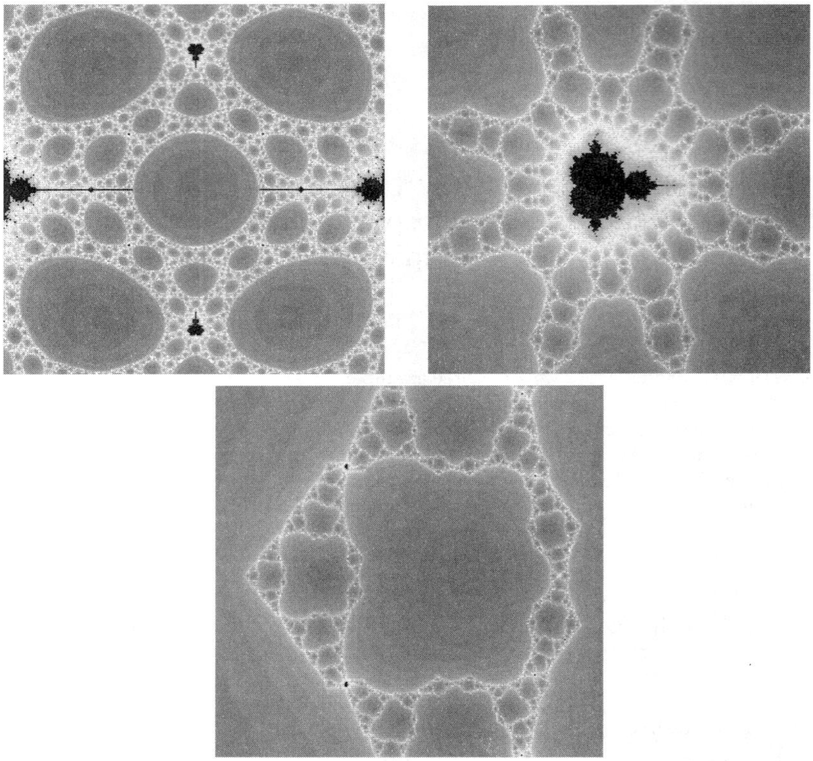

FIGURE 1. In this figure we display (a) some Mandelpinski necklaces for the family $z^3 + \lambda/z^3$; (b) the Mandelpinski spokes for the family $z^2 + \lambda/z^3$; and (c) a portion of the Mandelpinski maze for the family $z^2 + \lambda/z^3$. To see the Mandelbrot sets in each case usually requires many zooms into these images.

hole. To construct the next phase of the maze, at each vertex we "duplicate" the current graph centered at the given Mandelbrot set and then attach this additional graph to this Mandelbrot set to further extend the maze. See Figure 1 for a picture of each of these structures.

We will not provide complete proofs of all of these results in this paper. Rather, we will simply show how certain structures arise in the dynamical plane and then illustrate how ideas from complex analysis allow us to reproduce similar versions in the parameter plane. This follows what Adrien Douady often said: "In complex dynamics, we sow the seeds in the dynamical plane and reap the harvest in the parameter plane."

1. Preliminaries

In this paper we consider the family of rational maps given by

$$F_\lambda(z) = z^n + \frac{\lambda}{z^d}$$

where $n, d \geq 2$. When $|z|$ is large, we have that $|F_\lambda(z)| > |z|$, so the point at ∞ is an attracting fixed point in the Riemann sphere. We denote the immediate basin of attraction of ∞ by B_λ. There is also a pole at the origin for each of these maps, and so there is a neighborhood of the origin that is mapped into B_λ. If the preimage of B_λ surrounding the origin is disjoint from B_λ, we call this region the trap door and denote it by T_λ.

The Julia set of F_λ, $J(F_\lambda)$, has several equivalent definitions. $J(F_\lambda)$ is the set of all points at which the family of iterates of F_λ fails to be a normal family in the sense of Montel. Equivalently, $J(F_\lambda)$ is the closure of the set of repelling periodic points of F_λ, and it is also the boundary of the set of all points whose orbits tend to ∞ under iteration of F_λ, not just those in the boundary of B_λ. See [**18**].

One checks easily that there are $n + d$ critical points that are given by

$$c^\lambda = \left(\frac{d\lambda}{n}\right)^{\frac{1}{n+d}}$$

with the corresponding critical values given by

$$v^\lambda = \frac{(d+n)\lambda^{\frac{n}{n+d}}}{d^{\frac{d}{n+d}} n^{\frac{n}{n+d}}}.$$

Note that, when $n = d$, there are only two critical values given by $\pm 2\sqrt{\lambda}$. There are also $n + d$ prepoles given by

$$p^\lambda = (-\lambda)^{\frac{1}{n+d}}.$$

Let ω be an $(n+d)^{\text{th}}$ root of unity. Then we have $F_\lambda(\omega z) = \omega^n F_\lambda(z)$, and so it follows that the dynamical plane is symmetric under the rotation $z \mapsto \omega z$. In particular, all of the critical orbits have "similar" fates. If one critical orbit tends to ∞, then all must do so. If one critical orbit tends to an attracting cycle of some period, then all other critical orbits also tend to an attracting cycle, though these other cycles may have different periods. Nonetheless, the points on these attracting cycles are all symmetrically located with respect to the rotation by ω. As a consequence, each of B_λ, T_λ, and $J(F_\lambda)$ are symmetric under rotation by ω. Similarly, one checks easily that the parameter plane is symmetric under the rotation $\lambda \mapsto \nu\lambda$ where ν is an $(n-1)^{\text{st}}$ root of unity. The parameter plane is also symmetric under complex conjugation $\lambda \mapsto \overline{\lambda}$.

There is an Escape Trichotomy [**12**] for this family of maps. The first scenario in this trichotomy occurs when one and hence, by symmetry, all of the critical values lie in B_λ. In this case it is known that $J(F_\lambda)$ is a Cantor set [**12**]. The corresponding set of λ-values in the parameter plane is denoted by $\mathcal{C}$ and called the Cantor set locus. The second scenario is that the critical values all lie in T_λ (which we assume is disjoint from B_λ). In this case the Julia set is a Cantor set of simple closed curves surrounding the origin. This can only happen when $n, d \geq 2$ but not both equal to 2 [**17**]. We call the region $\mathcal{E}^1$ in parameter plane where this occurs the "McMullen domain"; it is known that $\mathcal{E}^1$ is an open disk surrounding the origin [**3**]. The third scenario is that the orbit of a critical point enters T_λ at iteration 2 or higher. Then, by the above symmetry, all such critical orbits do the same. In this case, it is known that the Julia set is a Sierpinski curve [**10**], i.e., a set that is homeomorphic to the well known Sierpinski carpet fractal. The regions in the parameter plane for which this happens are the open disks that we call Sierpinski holes [**8**], [**20**]. If the critical orbits do not escape to ∞, then it is known [**15**] that the Julia set is a connected set. Thus we call the set of parameters for which the critical orbits either do not escape or else enter the trap door at iteration 2 or higher the connectedness locus. This is the region between $\mathcal{C}$ and $\mathcal{E}^1$. See Figure 2.

Proving that a Julia set is a Sierpinski curve is often quite easy. By a theorem of Whyburn [**21**], any planar set that is compact, connected, locally connected, nowhere dense, and has the property that any pair of complementary domains are bounded by simple closed curves that are pairwise disjoint is homeomorphic to the Sierpinski carpet. In complex dynamics, verifying the first four of these properties is, in many cases, trivial. See [**4**], [**10**] for more details.

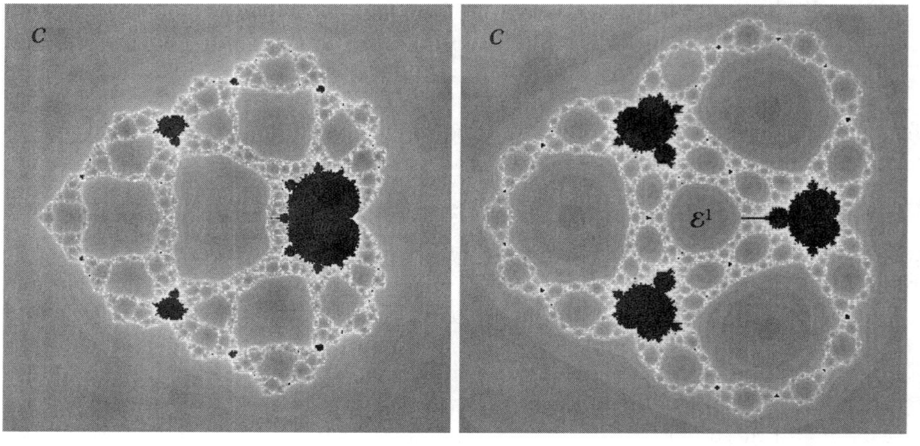

FIGURE 2. The parameter planes for the family $z^n + \lambda/z^d$ when $n = 2, d = 3$ and $n = 4, d = 3$. There is one principal Mandelbrot set in the first case and three symmetrically located such sets in the second. All of the red holes in these pictures (except the one surrounding the origin) are Sierpinski holes. $\mathcal{E}^1$ is too small to be seen in the first figure.

In [**2**] and [**9**] it has been shown that there are $n-1$ principal Mandelbrot sets in the parameter plane for these maps. These are symmetrically located by the rotation $z \mapsto \nu z$ around the origin and extend from the Cantor set locus down to the McMullen domain.

For more details about the dynamical properties of these maps and the structure of the parameter plane, see [**4**].

2. Mandelpinski Necklaces

In this section we restrict attention to the family

$$F_\lambda(z) = z^n + \frac{\lambda}{z^n}$$

where $n \geq 3$. In this case the critical points of F_λ are given by $\lambda^{1/2n}$, the two critical values by $\pm 2\sqrt{\lambda}$, and the prepoles by $(-\lambda)^{1/2n}$. There is an additional symmetry in the dynamical planes for these maps. Let H_λ be an involution given by $\lambda^{1/n}/z$.

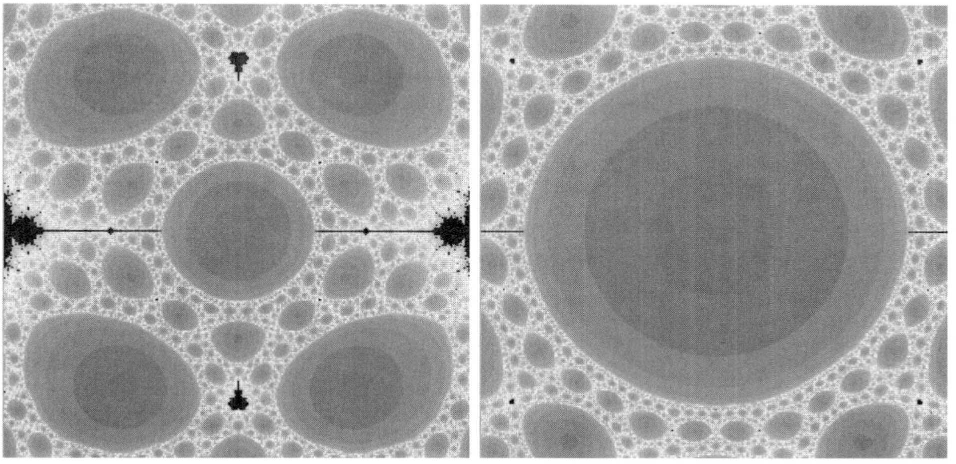

FIGURE 3. The rings around the McMullen domain in the family $z^3 + \lambda/z^3$. In each case, the McMullen domain is the central disk. The Sierpinski holes are visible in these rings, but most of the Mandelbrot sets are too small to be seen.

Then $F_\lambda(H_\lambda(z)) = F_\lambda(z)$, so the dynamical plane is symmetric under each of these involutions.

Recall that, if $|\lambda|$ is small, the Julia set of F_λ is a Cantor set of simple closed curves surrounding the origin and the set of parameters for which this holds is the McMullen domain $\mathcal{E}^1$.

We define the center of a Sierpinski hole to be the unique parameter for which the orbits of the critical points all land on ∞, and the escape time is the number of iterations that it takes for the critical orbits to enter the trap door. The center of a Mandelbrot set is the unique superstable parameter that lies in the main cardioid.

The following result was proved in [**13**].

Theorem. *For each $k \geq 0$ there exists a simple closed curve C_k which surrounds the McMullen domain and the C_k converge to the boundary of this domain as $k \to \infty$. Each C_k passes alternately through exactly $(n-2)n^k + 1$ centers of Sierpinski holes with escape time $k+2$ and centers of baby Mandelbrot sets, i.e., very small copies of the usual Mandelbrot set.*

The center of a Sierpinski hole is the unique parameter for which the orbits of the critical points all land on ∞, and the escape time is the number of iterations that it takes for the critical orbits to enter the trap door. The center of a Mandelbrot set is the unique superstable parameter that lies in the main cardioid.

We call each C_k a *Mandelpinski necklace.* As a remark, this result also holds when $n = 2$ [**1**], but then the formula $(n-2)n^k + 1$ shows that each C_k passes through just one Sierpinski hole and one Mandelbrot set. Also, in a forthcoming paper by D. Cuzzocreo, it has been shown that each of the Sierpinski holes in the above result are known to be surrounded by infinitely many similar Mandelpinski sub-necklaces. And then this process continues recursively with each sub-necklace having its own sub-necklaces. See Figure 3.

These Mandelpinski necklaces arise from the following structure in the dynamical plane. Let $\gamma_0 = \gamma_0^\lambda$ be the circle given by $|z| = |\lambda|^{1/2n}$ in the dynamical plane. So γ_0 contains all of the critical points and prepoles which are arranged along γ_0 in alternating fashion. We call γ_0 the *critical circle.* Note that each involution H_λ preserves the critical circle and interchanges the interior and exterior of this circle. One checks easily that F_λ maps the critical circle $2n$-to-one onto the *critical value segment* which is the straight line segment connecting the two critical values and hence passing through 0.

The first necklace $\mathcal{C}_0$ then arises when the critical values lie on γ_0. This occurs when $|v^\lambda| = 2|\sqrt{\lambda}| = |\lambda|^{1/2n} = |c^\lambda|$. The necklace $\mathcal{C}_0$ is the circle in the parameter plane given by $|\lambda| = (1/4)^{n/n-1}$. We call this the *dividing circle.* As λ rotates one full turn around the dividing circle, each crtical value rotates one-half a turn along γ_0, while each of the prepoles and critical points rotates only $(1/2n)^{\text{th}}$ of a turn. This shows that there are exactly $n-1$ centers of Sierpinski and baby Mandelbrot sets on $\mathcal{C}_0$.

For the remaining necklaces, we assume that λ lies strictly inside the dividing circle in the parameter plane, so $|v^\lambda| < |c^\lambda|$. Therefore the critical circle is mapped strictly inside itself. As a consequence, the exterior of the critical circle is mapped as an n-to-one covering of the exterior of the critical value segment. Indeed, any circle outside γ_0 is mapped n-to-one onto an ellipse whose foci are $\pm v^\lambda$. Thus there is a preimage γ_1 of the critical circle that lies outside of γ_0. Then γ_1 contains $2n^2$ pre-critical points and the same number of pre-prepoles. Then the exterior of γ_1 is mapped as an n-to-one covering of the exterior of γ_0 and so there is another simple closed curve γ_2 that is mapped n-to-one onto γ_1. Thus γ_2 contains $2n^3$ points that are mapped to the critical points by F_λ^2 and the same number of points that are mapped to prepoles by F_λ^2. And this continues to produce a sequence of simple closed curves γ_k converging outward from γ_0 as $k \to \infty$, and each γ_k contains similar sets of $2n^{k+1}$ pre-critical points and prepoles.

By the H_λ symmetry, the interior of γ_0 is also mapped n-to-one onto the complement of the critical value segment. So we now have infinitely many other simple closed curves $\gamma_{-j}, j = 1, \ldots \infty$ lying inside γ_0. The curve γ_{-j} is mapped n-to-one onto γ_{j-1} and so these curves contain a similar number of points that are mapped to the critical points and prepoles by F_λ^j.

We then "transfer" this picture to the parameter plane as follows. As λ rotates once around the origin, a certain number of the preimages of the critical points and prepoles in $\mathcal{C}_k$ (namely, $(n-2)n^k + 1$ such points) can be shown to remain in the upper half plane. Consider the subset of the parameter plane consisting of an open annulus bounded by a circle strictly inside the McMullen domain and the dividing circle. Then remove the positive real axis from this set to give a simply connected set $\mathcal{O}$ in the parameter plane. Then we have two maps defined on $\mathcal{O}$. One map is given by $\lambda \mapsto v^\lambda$. The second is given by selecting one of the prepoles (or pre-critical points) on γ_{-k}^λ which remain in the upper half plane as λ rotates, excluding the critical points that lie on the real axis when λ is positive. Call this point z_λ^k where z_λ^k is chosen to vary analytically with λ. So this map is $\lambda \mapsto z_\lambda^k$.

The map $V(\lambda) = v^\lambda$ is invertible on $\mathcal{O}$, so we can consider the map $\lambda \mapsto V^{-1}(z_\lambda^k)$ from $\mathcal{O}$ to $\mathcal{O}$. Then the Schwarz Lemma implies that this map has a unique fixed point in $\mathcal{O}$. This fixed point is then the parameter λ^* for which $v^{\lambda^*} = z_{\lambda^*}^k$, so for the parameter λ^*, either one of the critical points c_{λ^*} returns to a critical point at iteration $k+1$, or else the orbit of a critical point lands on 0 at iteration $k+2$. When n is even, this implies that, in the first case, the orbit of some critical point lands on itself and hence is periodic of period $k+1$, whereas, if n is odd, this critical point may return to itself or to its negative and thus, by the $z \mapsto -z$ symmetry, this period is either $k+1$ or $2(k+1)$. This produces the parameters lying at the centers of the Sierpinski holes and the centers of the main cardioids of the baby Mandelbrot sets. Finding such a center of a Mandelbrot set in the case we excluded above when λ is positive is straightforward, since both v^λ and c^λ lie on the real axis in this case. Full details of this proof may be found in [**13**].

Producing the entire Sierpinski holes in these cases involves quasi-conformal surgery, while producing the entire baby Mandelbrot sets involves polynomial-like maps [**16**]. We shall illustrate how this works when we create the Mandelpinski spokes in a later section.

3. Dynamics on Sierpinski Curve Julia Sets

The Mandelbrot sets centered on the Mandelpinski necklaces $\mathcal{C}_k$ where $k > 0$ are all "buried" in the sense that they do not extend out to the external boundary of the connectedness locus. It is known that any parameter drawn from the main cardioid of such a buried Mandelbrot set is also a Sierpinski curve [**11**]. Thus we have a total of $2(n-2)n^k+2$ open disks along $\mathcal{C}_k$ for which the corresponding maps have Julia sets that are Sierpinski curves. Hence all of these Julia sets are homeomorphic to one another. So the natural question is: what about the dynamical behavior on these Julia sets? When is the dynamical behavior the same, i.e., when are the maps topologically conjugate to one another? And when is this behavior different?

Regarding these questions, first note that parameters drawn from the main cardioids of the Mandelbrot sets always have different dynamical behavior from those lying in Sierpinski holes, since, in the Mandelbrot set case, we have that the boundaries of the attracting basins are invariant under some iterate of F_λ, whereas only ∂B_λ is invariant under iterates of F_λ in the Sierpinski hole case.

It is known that any two parameters drawn from the same Sierpinski hole or main cardioid of a Mandelbrot set have conjugate behavior on their Julia sets [**14**], [**20**]. Regarding different Sierpinski holes (or main cardioids), it is also shown in [**14**] that only those that are symmetrically located in the parameter plane by a rotation by an $(n-1)^{\text{st}}$ root of unity or by complex conjugation contain parameters for which the dynamics are the same. The maps cannot be topologically conjugate on the Julia sets if they are drawn from different, non-symmetrically located disks.

So now the question is: What makes the dynamics different in the non-symmetrically located disks? This has been determined by Moreno Rocha in the Sierpinski hole case [**19**] where she produces a dynamical invariant for these maps. But this situation is still unresolved in the cardioid case.

As a remark, this conjugacy result holds for all Sierpinski holes (including those described in the following sections), not just those lying along the Mandelpinski necklaces. Indeed, it is known that there are exactly $(n-1)(2n)^{k-2}$ Sierpinski holes with escape time k for maps of the form $z^n + \lambda/z^n$ [**8**], [**20**]. As mentioned above, the escape time is the number of iterations for the critical orbits to enter the trap door. Then there are exactly $(2n)^{k-3}$ different conjugacy classes of these maps when n is even and $(2n)^{k-3}/2 + 2^{k-4}$ when n is odd [**14**]. This count of the number of Sierpinski holes arises from the fact that there is a unique center of each Sierpinski hole, i.e., a parameter for which $F_\lambda^k(c^\lambda) = 0$. Hence this count reduces to finding the number of roots of a polynomial equation, all of whose roots are known to be simple.

4. Mandelpinski Spokes

For the remainder of this paper, in order to keep the notation simple, we will restrict attention to the family of maps

$$F_\lambda(z) = z^2 + \frac{\lambda}{z^3}.$$

However, all of the following constructions go through with only minor changes for the more general family

$$F_\lambda(z) = z^n + \frac{\lambda}{z^d}$$

where $n \geq 2$ is even and $d \geq 3$ is odd.

4.1. The Initial Mandelpinski Arc. In this section, in preparation for the next two Mandelpinski structures, we shall construct a *Mandelpinski arc*. This will be an arc in the parameter plane that passes alternately along the spines of infinitely many baby Mandelbrot sets and through the centers of the same number of Sierpinski holes. By the spine of a Mandelbrot set we mean the analogue of the portion of the real axis lying in the usual Mandelbrot set associated with the quadratic family $z^2 + c$.

In this initial Mandelpinski arc, there will be infinitely many Mandelbrot sets $\mathcal{M}^k$ with $k \geq 2$ along this arc. Here k is the period of the attracting cycle for parameters drawn from the main cardioid of $\mathcal{M}^k$, i.e., the base period of $\mathcal{M}^k$. There will also be infinitely many Sierpinski holes $\mathcal{E}^k$ with $k \geq 1$ where k is the escape time in $\mathcal{E}^k$, i.e., the number of iterations it takes for the orbit of the critical points to enter T_λ. In this special case, the Mandelpinski arc will be the portion of the negative real axis in the parameter plane extending from the McMullen domain $\mathcal{E}^1$ down to the endpoint on the boundary of the connectedness locus in the left half plane. Then the Mandelbrot sets and Sierpinski holes will be arranged along this arc as follows:

$$\ldots \mathcal{M}^4 < \mathcal{E}^3 < \mathcal{M}^3 < \mathcal{E}^2 < \mathcal{M}^2 < \mathcal{E}^1.$$

In each case there will be an interval of nonzero length between any adjacent Mandelbrot set and Sierpinski hole lying along this arc. The Mandelpinski spokes we construct later will emanate from each of the $\mathcal{M}^k$, and the proof of the existence of these spokes will be similar in spirit to the proof that we sketch here.

In Figure 4 we display the parameter plane for $z^2 + \lambda/z^3$. Along the negative real axis, there are infinitely many red disks that are visible: these are the Sierpinski holes in the Mandelpinski arc. Between any two Sierpinski holes, there is then a (very small) Mandelbrot set as well as many more Sierpinski holes, as shown in the magnification in this figure.

There are now five critical points for the map F_λ that are given by $(3\lambda/2)^{1/5}$. We denote the critical point that lies in $\mathbb{R}^-$ when $\lambda \in \mathbb{R}^-$ by $c_0 = c_0^\lambda$ (and then c_0^λ varies analytically with λ). We denote the other critical points by $c_j = c_j^\lambda$ for $-2 \leq j \leq 2$ where the c_j are now arranged in the clockwise order as j increases. As λ moves half way around the origin from $\mathbb{R}^-$, c_0 rotates exactly one-tenth of a turn

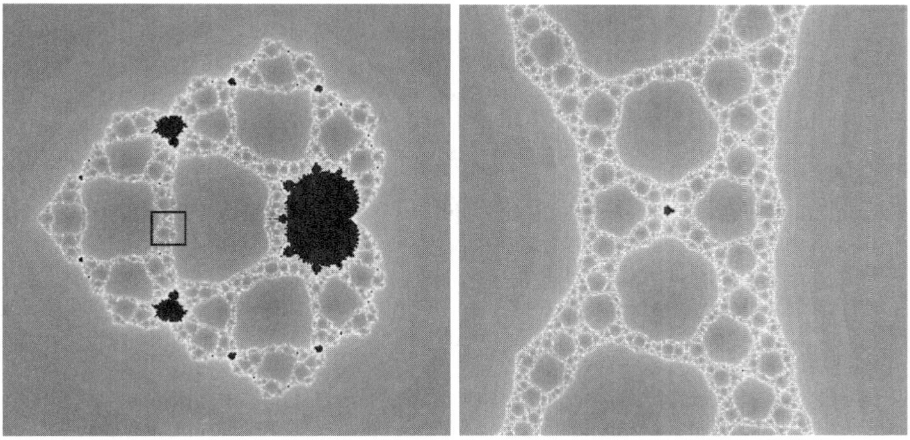

FIGURE 4. The parameter plane for $z^2 + \lambda/z^3$. The magnification shows a small Mandelbrot set between two of the large Sierpinski holes.

in the corresponding direction. Thus, when $\operatorname{Arg}\lambda$ decreases from π to 0, c_2 lies in $\mathbb{R}^+$ and when $\operatorname{Arg}\lambda$ increases from π to 2π, c_{-2} now lies in $\mathbb{R}^+$. The critical values of F_λ are then given by $v^\lambda = \kappa\lambda^{2/5}$ where κ is the constant given by $5/(2^{2/5}3^{3/5})$. One computes easily that $\kappa \approx 1.96$. We denote by v_j^λ the critical value that is the image of c_j^λ.

There are also five prepoles for F_λ given by $(-\lambda)^{1/5}$. We denote the prepole that lies in $\mathbb{R}^+$ when $\lambda \in \mathbb{R}^-$ by $p_2 = p_2^\lambda$. The other prepoles are denoted by $p_j = p_j^\lambda$ where again $-2 \le j \le 2$ and the p_j are arranged in the clockwise order as j increases. Note that, when $\lambda \in \mathbb{R}^-$, the critical point c_0 lies between the two rays starting at the origin and passing through p_0 and p_{-1}. Unlike the previous case where $n = d$, the critical points and prepoles no longer lie on the same circle.

The straight ray extending from the origin to ∞ and passing through the critical point c_j^λ is a *critical point ray*. This ray is mapped two-to-one onto the portion of the straight ray from the origin to ∞ that starts at the critical value v_j^λ and extends to ∞. A similar straight line extending from 0 to ∞ and passing through a prepole p_j^λ is a *prepole ray*, and this ray is mapped one-to-one onto the entire straight line passing through both the origin and the point $(-\lambda)^{2/5}$.

To construct the objects lying along this Mandelpinski arc, we will restrict attention at first to the λ-values lying in the annular region $\mathcal{O}$ in parameter plane given by $10^{-10} \le |\lambda| \le 2$. Also, let $\mathcal{A}$ be the annulus in the dynamical plane given by $\kappa 10^{-4} \le |z| \le \kappa 2^{2/5}$ where $\kappa \approx 1.96$ is defined as above. Then easy estimates show that, for any $\lambda \in \mathcal{O}$, all points on the outer circular boundary of $\mathcal{A}$ lie in B_λ, while all points on the inner circular boundary of $\mathcal{A}$ lie in T_λ. Moreover, F_λ maps each of these boundaries strictly outside the boundary of $\mathcal{A}$. Also, if λ lies on the inner circular boundary of $\mathcal{O}$, then each critical value lies on the inner circular boundary of $\mathcal{A}$ and so λ lies in the McMullen domain. And, if λ lies on the outer circular boundary of $\mathcal{O}$, then each critical value lies on the outer circular boundary of $\mathcal{A}$ and so λ lies in the Cantor set locus in the parameter plane. For a detailed proof of this, see [**6**].

We now restrict attention to a "smaller" subset of $\mathcal{O}$. Let $\mathcal{O}'$ be the subset of $\mathcal{O}$ containing parameters λ for which $0 \leq \operatorname{Arg}\lambda \leq 2\pi$. Despite the overlap of this region along the real axis, we will think of $\mathcal{O}'$ as being a closed disk (not an annulus) in the parameter plane with $\operatorname{Arg}\lambda = 0$ and $\operatorname{Arg}\lambda = 2\pi$ considered as different portions of the boundary. We do this because, as $\operatorname{Arg}\lambda$ increases from 0 to 2π, the critical point c_0 that we will be following rotates one-fifth of a turn in the dynamical plane. So this point will migrate to the position of a different critical point as $\operatorname{Arg}\lambda$ increases from 0 to 2π.

For any parameter in $\mathcal{O}'$, let L^λ be the closed "portion of the wedge" in the annulus $\mathcal{A}$ in the dynamical plane that is bounded by the two prepole rays through p_0 and p_{-1}. When $\lambda \in \mathbb{R}^-$, L^λ is thus bounded by the rays extending from 0 and passing through $\exp(2\pi i(2/5))$ and $\exp(2\pi i(3/5))$. So the critical point c_0 lies in the interior of L^λ. Next, let R^λ be the portion of the wedge in $\mathcal{A}$ that is bounded by the critical point rays passing through c_2 and c_{-2}. When $\lambda \in \mathbb{R}^-$, this wedge is bounded by the critical point rays extending from 0 and passing through $\exp(\pm 2\pi i/10)$. Note that R^λ is the symmetric image of L^λ under $z \mapsto -z$ for each $\lambda \in \mathcal{O}'$. See Figure 5.

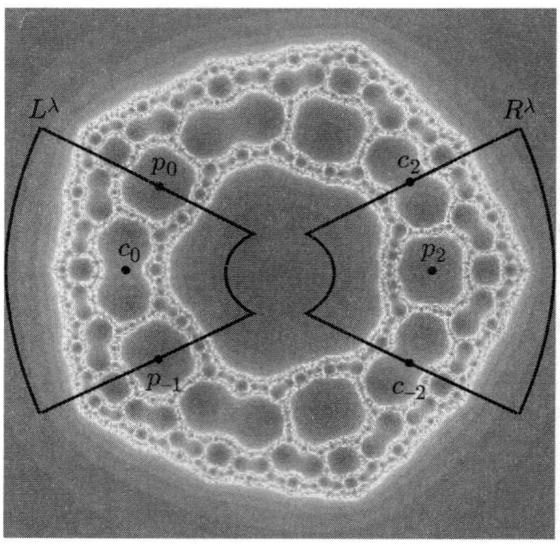

FIGURE 5. The wedges L^λ and R^λ for $\lambda = -0.09$.

Then it is easy to see that, for each $\lambda \in \mathcal{O}'$:

(1) F_λ maps the interior of R^λ in one-to-one fashion onto a region that contains the interior of $R^\lambda \cup L^\lambda$ together with a portion of T_λ that contains 0;
(2) F_λ maps the interior of L^λ two-to-one over a region that contains the interior of R^λ;
(3) As λ winds once around the boundary of $\mathcal{O}'$, the critical value $F_\lambda(c_0^\lambda) = v_0^\lambda$ winds once around the boundary of R^λ, (i.e., the winding index of the

vector connecting this critical value to the prepole p_2^λ lying in the interior of R^λ is one).

Again, the details of this proof may be found in [**6**].

Before constructing this Mandelpinski arc, we recall the concept of a polynomial-like map. Let G_μ be a family of holomorphic maps that depends analytically on the parameter μ lying in some open disk $\mathcal{D}$. Suppose each $G_\mu : U_\mu \to V_\mu$ where both U_μ and V_μ are open disks that also depend analytically on μ. G_μ is then said to be polynomial like of degree 2 if, for each μ:

- G_μ maps U_μ two-to-one onto V_μ and so there is a unique critical point in U_μ;
- V_μ contains U_μ;
- As μ winds once around the boundary of $\mathcal{D}$, the critical value winds once around U_μ in the region $V_\mu - U_\mu$.

As shown in [**16**], for such a family of polynomial-like maps, there is a homeomorphic copy of the Mandelbrot set in the disk $\mathcal{D}$. Moreover, for μ-values in this Mandelbrot set, $G_\mu \quad U_\mu$ is conjugate to the corresponding quadratic map given by this homeomorphism.

To show the existence of the Mandelpinski arc in the parameter plane, we will first observe a "similar" collection of sets in the dynamical plane, and then use polynomial-like maps and the argument principle to produce the analogous Mandelbrot sets $\mathcal{M}^k$ and Sierpinski holes $\mathcal{E}^k$ in the parameter plane.

We shall first consider the escape time case. By construction, for each $\lambda \in \mathcal{O}'$, there is a unique prepole p_2^λ in the interior of R^λ. Since F_λ maps R^λ one-to-one over itself, there is a unique preimage of this prepole, z_3^λ, in R^λ, so $F_\lambda^2(z_3^\lambda) = 0$. Continuing, for each $\lambda \in \mathcal{O}'$, there is a unique point z_k^λ in R^λ for which we have $F_\lambda(z_k^\lambda) = z_{k-1}^\lambda$ and so $F_\lambda^{k-1}(z_k^\lambda) = 0$. Now the points z_k^λ vary analytically with λ and are strictly contained in the interior of R^λ. So we may consider the function $H^k(\lambda)$ defined on $\mathcal{O}'$ by $H^k(\lambda) = v_0^\lambda - z_k^\lambda$ where $v_0^\lambda = F_\lambda(c_0^\lambda)$. When λ rotates once around the boundary of $\mathcal{O}'$, v_0^λ rotates once around the boundary of R^λ while z_k^λ remains in the interior of R^λ. Hence $H^k(\lambda)$ has winding number one along the boundary of $\mathcal{O}'$ and so there must be a unique zero in $\mathcal{O}'$ for each H^k. This is then the parameter that lies at the center of the escape time region $\mathcal{E}^k$. It is well known [**20**] that $\mathcal{E}^k$ is then an open disk in the parameter plane. Note that, as λ decreases along $\mathbb{R}^-$, both v_0^λ and z_k^λ increase along $\mathbb{R}^+$. It then follows that the portion of $\mathcal{E}^{k+1}$ in $\mathbb{R}^-$ lies to the left of $\mathcal{E}^k$ in the parameter plane.

To prove the existence of the Mandelbrot sets $\mathcal{M}^k$, recall that the orbit of the point z_k^λ under F_λ remains in R^λ before entering T_λ and landing at 0 at iteration $k-1$ (here $z_2^\lambda = p_2^\lambda$). For each $k \geq 2$, let E_λ^k be the open set surrounding z_k^λ in R^λ that is mapped to T_λ by F_λ^{k-1}. Let D_λ^k be the set in R^λ consisting of points whose first $k-2$ iterations lie in R^λ but whose $(k-1)^{\text{st}}$ iterate lies in the interior of L^λ. Since F_λ is univalent on R^λ, each D_λ^k is an open disk. Furthermore, the boundary of D_λ^k meets a portion of the boundaries of both E_λ^{k-1} and E_λ^k (where $E_\lambda^1 = T_\lambda$). Since F_λ^{k-1} maps D_λ^k one-to-one over the interior of L^λ and then F_λ maps L^λ two-to-one over a region that contains R^λ, we have that F_λ^k maps D_λ^k two-to-one over a region that completely contains R^λ. Moreover, the critical value for F_λ^k is just v_0^λ, which, as mentioned above, winds once around the exterior of R^λ as λ winds once around the boundary of $\mathcal{O}'$. Hence F_λ^k is a polynomial-like map of degree two on D^k_λ and this proves the existence of a baby Mandelbrot set $\mathcal{M}^k$ lying in $\mathcal{O}'$ for each $k \geq 2$. When λ is real and negative, we have that the centers of the escape regions $\mathcal{E}^k$ lie along $\mathbb{R}^-$ and, since the real line is invariant under F_λ when $\lambda \in \mathbb{R}^-$, both c_0^λ and v_0^λ also lie on the real axis. Then, by the $\lambda \mapsto \overline{\lambda}$ symmetry in the parameter plane, the spines of these Mandelbrot sets also lie in $\mathbb{R}^-$.

Next, since the E^k_λ and D^k_λ are arranged along the postive real axis in the following fashion:

$$T_\lambda = E^1_\lambda < D^2_\lambda < E^2_\lambda < D^3_\lambda < E^3_\lambda < \ldots$$

and, as shown above, the $\mathcal{E}^k$ decrease along $\mathbb{R}^-$ as k increases, we therefore have that the $\mathcal{E}^k$ and $\mathcal{M}^k$ are arranged along the negative real axis in the parameter plane in the opposite manner:

$$\ldots \mathcal{E}^3 < \mathcal{M}^3 < \mathcal{E}^2 < \mathcal{M}^2 < \mathcal{E}^1.$$

See Figure 6.

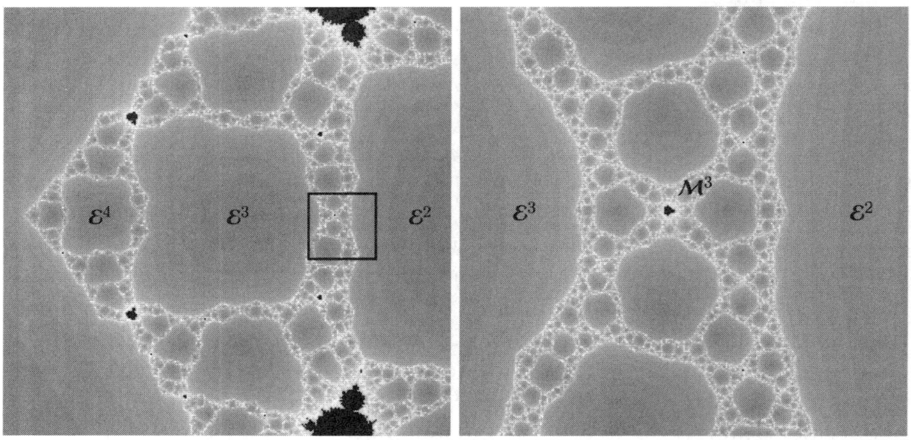

FIGURE 6. The Mandelpinski arc along the negative real axis. The $\mathcal{M}^k$ are so small that they are not visible in this picture. However, the magnification shows $\mathcal{M}^3$.

4.2. Construction of the Mandelpinski Spokes. In this section, we shall concentrate on a specific Mandelbrot set $\mathcal{M}^k$ and describe the infinite collection of Mandelpinski spokes emanating from this set. A *Mandelpinski spoke* is an arc in the parameter plane along which lie infinitely (or finitely) many Mandelbrot sets and Sierpinski holes in alternating fashion. With an eye toward how we shall proceed with this construction, note that, at this stage, we have already produced a single infinite Mandelpinski spoke extending to the left of $\mathcal{M}^k$ which contains the sets $\mathcal{M}^j$ with $j > k$ and $\mathcal{E}^j$ with $j \geq k$. And there is a finite Mandelpinski spoke lying on the other side of $\mathcal{M}^k$ which now contains finitely many sets $\mathcal{M}^j$ where $2 \leq j < k$ and $\mathcal{E}^j$ where $1 \leq j < k$. These will be the initial portions of two of the (eventually infinite) Mandelpinski spokes emanating from $\mathcal{M}^k$.

In this next phase of the construction, we shall show that, on each side of the Mandelbrot set $\mathcal{M}^k$ in the first spoke, there are a pair of infinite spokes, each extending over to one of the adjacent Sierpinski holes $\mathcal{E}^k$ and $\mathcal{E}^{k-1}$. We think of this as extending the two previously constructed spokes emanating from $\mathcal{M}^k$. In addition, we shall show that there are a pair of new finite spokes extending above and below each $\mathcal{M}^k$. These will be the initial portions of the first four spokes emanating from $\mathcal{M}^k$.

To begin this phase of the construction, let us assume that the critical value v_0^λ now lies in a particular open disk D_λ^k for some fixed $k \geq 2$. Let $\mathcal{O}_k \subset \mathcal{O}'$ denote the set of parameters for which this happens. Now the boundary of D_λ^k is mapped by F_λ^{k-1} one-to-one onto the boundary of L^λ, and the boundary of L^λ varies analytically with λ. So we can construct a natural parametrization of this boundary which also varies analytically with λ. Then we can pull back this parameterization to the boundary of each D_λ^k. Again, as we saw earlier, as λ rotates around the boundary of the original disk $\mathcal{O}'$ in the parameter plane, v_0^λ rotates once around the boundary of R^λ. Hence, arguing just as in the previous section, there must be a unique parameter λ for which v_0^λ lands on any given point in the parametrization of the boundary of D_λ^k. Hence we have that $\mathcal{O}_k$ is a disk contained inside $\mathcal{O}'$ and, as λ rotates once around the boundary of $\mathcal{O}_k$, the critical value has winding number one around the boundary of the disk D_λ^k.

Now consider the set of preimages in L^λ of all of the D_λ^j and E_λ^j under F_λ. Since we have assumed that v_0^λ lies in D_λ^k, it follows that there is a unique preimage of D_λ^k in L^λ which is a disk that contains c_0^λ and is mapped two-to-one onto D_λ^k. Call this special disk L_k^λ. For each other D_λ^j (with $j \neq k$), there are now two preimage disks lying in L^λ. Note that, when $\lambda \in \mathbb{R}^-$ and $j > k$, there are a pair of preimages of D_λ^j lying along $\mathbb{R}^-$, one to the right of L_k^λ and one to the left. These preimages tend away from D_λ^k in either direction as j increases. When $2 \leq j < k$, there are again two preimages of D_λ^j, but when $\lambda \in \mathbb{R}^-$, these preimages no longer lie on the negative axis; rather they branch out more or less perpendicularly above and below L_k^λ on this axis. As for the preimages of E_λ^j in L^λ, we have the same situation: there are infinitely many pairs of preimages of each E_λ^j lying along $\mathbb{R}^-$ on either side of the preimage of D_λ^k when $j \geq k$ and $\lambda \in \mathbb{R}^-$, and finitely many pairs extending above and below this preimage when $1 \leq j < k$. Thus we have a pair of infinite chains of alternating preimages of the disks D_λ^k and E_λ^k extending away from L_k^λ in the wedge L^λ and another pair of chains consisting of finitely many such preimages extending in a "perpendicular" direction away from L_λ^k.

Since F_λ^{k-1} maps D_λ^k one-to-one over L^λ, we thus have a similar collection of preimages that lie inside the disk D_λ^k. We denote by D_λ^{kj} each of the two disks in D_λ^k that are mapped onto D_λ^j by F_λ^k when $j \neq k$. And we let D_λ^{kk} denote the single preimage of D_λ^k under F_λ^k that is contained in D_λ^k. Points in D_λ^{kj} have orbits that remain in R^λ for the first $k-2$ iterations, then map into L^λ under the next iteration, and then map into D_λ^j under the next iteration. Then F_λ^{j-1} maps this set of points onto L^λ. So F_λ^{k+j-1} maps each D_λ^{kj} one-to-one onto all of L^λ (assuming $k \neq j$). Then the next iteration takes this set two-to-one onto all of R^λ. Now the critical value for F_λ^{k+j} is again v_0^λ, and, as we showed above, as λ rotates around the boundary of $\mathcal{O}_k$, v_0^λ circles around the boundary of D_λ^k. Hence F_λ^{j+k} is polynomial-like of degree two on each of the two disks D_λ^{kj} (where we again emphasize that we are assuming $j \geq 2$ and $j \neq k$). So this produces a pair of Mandelbrot sets $\mathcal{M}^{kj}$ with base period $k+j$ in $\mathcal{O}_k$. As in the previous construction, the Mandelbrot sets $\mathcal{M}^{kj}$ with $j > k$ all have spines lying along $\mathbb{R}^-$, one on each side of $\mathcal{M}^k$. The other Mandelbrot sets with $j < k$ now lie off the real axis, one above $\mathcal{M}^k$ and the other below $\mathcal{M}^k$.

Similar arguments as in the preceding section also produce a pair of Sierpinski holes $\mathcal{E}^{kj}$ on each side of $\mathcal{M}^k$ along the real axis where now $j \geq k$. And there are a pair of Sierpinski holes $\mathcal{E}^{kj}$, one above and one below $\mathcal{M}^k$, where now $1 \leq j < k$. As earlier, these Mandelbrot sets and Sierpinski holes alternate along each of these spokes. See Figure 7. For parameters in the Sierpinski hole $\mathcal{E}^{k1}$, the critical orbit $F_\lambda^i(c_0^\lambda)$ lies in R^λ for iterations $1 \leq i \leq k-1$. Then $F_\lambda^k(c_0^\lambda)$ returns to L^λ, and then $F_\lambda^{k+1}(c_0^\lambda)$ enters T_λ.

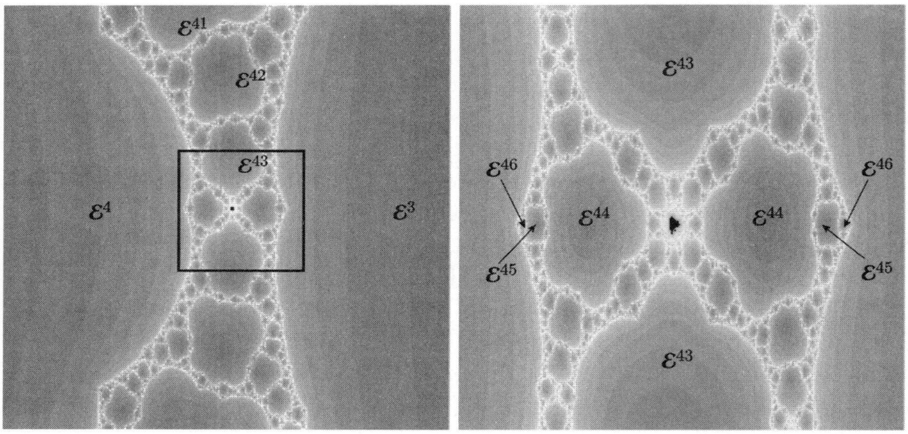

FIGURE 7. The finite spoke above and below $\mathcal{M}^4$ as well as a magnification showing the pair of infinite spokes along the real axis.

We now sketch the construction of the other Mandelpinski spokes by induction. For simplicity, we will only consider the next phase of the construction; all subsequent phases follow in exactly the same way. This time we will adjoin four infinite spokes that lie closer to $\mathcal{M}^k$ to those already in place, and then we will add four new finite spokes in betwen each of these infinite spokes.

To be precise, in the previous phase, we assumed that the critical value resided in a particular disk D_λ^k, and so there was a special disk $D_\lambda^{kk} \subset D_\lambda^k$ that was mapped two-to-one onto D_λ^k by F_λ^k. At this stage we make the further assumption that v_0^λ lies in D_λ^{kk}. Let $\mathcal{O}_{kk} \subset \mathcal{O}_k$ be the set of parameters for which this occurs. Note that $\mathcal{M}^k$ lies in $\mathcal{O}_{kk}$. We have that F_λ^k maps the boundary of D_λ^{kk} two-to-one onto the boundary of D_λ^k. Thus we may pull back the parametrization of ∂D_λ^k constructed earlier to produce a natural parametrization of ∂D_λ^{kk} which varies analytically with

λ. Thus there is a unique λ for which v_0^λ lands on a given point in the boundary of D_λ^{kk}, and so, as λ winds once around the boundary of $\mathcal{O}_{kk}$, v_0^λ winds once around ∂D_λ^{kk}.

By the prior construction, we have a pair of infinite chains each of which consists of the disks D_λ^{kj} with $j > k$ and E_λ^{kj} with $j \geq k$ lying in the annular region $D_\lambda^k - D_\lambda^{kk}$ as well as a pair of finite chains consisting of the disks D_λ^{kj} and E_λ^{kj} with $j < k$ lying in the same annulus. Since F_λ^k maps D_λ^{kk} two-to-one onto the entire disk D_λ^k, we therefore have four new infinite chains inside D_λ^{kk} that are the preimages of the two infinite chains in the annular region. These chains consist of disks that we denote by either D_λ^{kkj} with $j > k$ or E_λ^{kkj} with $j \geq k$. Each of these chains then connects to one of the two infinite or finite chains already constructed in the outer annular region. This follows since these outer chains were all mapped onto the left or right portion of the original chain by F_λ^k. We also have four finite chains in D_λ^{kk} consisting of disks D_λ^{kkj} and E_λ^{kkj} with $j < k$ that are preimages of the finite chains in the annular region. These chains do not connect to the previously constructed chains in the annular region.

Then the same arguments as above produce the corresponding spokes in the parameter plane. Each of the two finite and infinite spokes constructed earlier now have an added infinite spoke that lies in the region between $\mathcal{M}^k$ and that spoke. The Mandelbrot sets and Sierpinski holes in this new portion of the spoke are given by $\mathcal{M}^{kkj}$ where $j > k$ and $\mathcal{E}^{kkj}$ where $j \geq k$ and the four new finite spokes consist of similar sets with now $j < k$. These are all associated with rays of angle $\ell/8$ with ℓ even for the infinite spokes and ℓ odd for the finite spokes.

At this stage we now have eight Mandelpinski spokes emanating from $\mathcal{M}^k$, four finite spokes and four infinite spokes. Continuing inductively, at the next stage, we then add eight infinite spokes between each of these spokes and $\mathcal{M}^k$ as well as eight new finite spokes, one between each of these newly added infinite spokes. In the limit, we get an infinite collection of Mandelpinski spokes emanating from $\mathcal{M}^k$.

5. The Mandelpinski Maze

In this section we further elaborate on the construction of the Mandelbrot sets and Sierpinski holes in the parameter plane of F_λ by constructing what we call a *Mandelpinski maze*. In the previous sections, we actually constructed the first two portions of this maze. We first showed the existence of a string of interspersed Mandelbrot sets $\mathcal{M}^k$ with $k \geq 2$ and Sierpinski holes $\mathcal{E}^1$ with $k \geq 1$ along the real axis in the parameter plane. Next, we showed that, between each $\mathcal{E}^{k-1}$ and $\mathcal{E}^k$, there exist a pair of infinite spokes, each containing Mandelbrot sets $\mathcal{M}^{kj}$ where $j > k$ and Sierpinski holes $\mathcal{E}^{kj}$ where $j \geq k$ in the same alternating arrangement as earlier. One spoke extends from $\mathcal{M}^k$ to $\mathcal{E}^{k-1}$, the other from $\mathcal{M}^k$ to $\mathcal{E}^k$. The are also a pair of finite spokes extending away from $\mathcal{M}^k$ in opposite directions. These finite spokes contain the Mandelbrot sets $\mathcal{M}^{kj}$ where $2 \leq j < k$ and the Sierpinski holes $\mathcal{E}^{kj}$ where now $1 \leq j < k$. We think of this second collection of spokes emanating from $\mathcal{M}^k$ as a "plus sign" centered at $\mathcal{M}^k$.

In the previous construction we assumed that the critical value lies in the disk D^{kk}. Now we change this assumption so that the critical value now lies in either of the two disks D^{kj} where now $j \neq k$.

Recall that it is when $\lambda \in \mathcal{O}_k$ that the critical value v_0^λ lies in D_λ^k and that L_λ^k is the preimage of D_λ^k in L^λ that contains c_0^λ. In the previous construction, for each

$j > 1$ and $j \neq k$, we produced a pair of disks $D_\lambda^{kj} \subset D_\lambda^k$, and then we showed that F_λ^{k+j} was a polynomial-like map of degree two on D_λ^{kj}. This generated a pair of Mandelbrot sets that we called $\mathcal{M}^{kj}$. So fix k and j and concentrate on one of the two disks D_λ^{kj} and hence, in the parameter plane, on the corresponding Mandelbrot set $\mathcal{M}^{kj}$. As in the previous step, we now assume that v_0^λ lies in this disk D_λ^{kj}. This is possible since we have shown that v_0^λ winds once around the boundary of D_λ^k as λ winds around $\partial\mathcal{O}_k$, and $D_\lambda^{kj} \subset D_\lambda^k$. Let $\mathcal{O}_{kj}$ be the set of parameters for which this occurs. Then F_λ^{j+k-1} maps D_λ^{kj} one-to-one over L^λ. Thus we can pull back the earlier parametrization of the boundary of L^λ to construct an analytic parameterization of the boundary of D_λ^{kj}. Just as before, there then exists a unique λ in the boundary of $\mathcal{O}_{kj}$ for which v_0^λ lands on a given point on this parametrized boundary curve. Hence v_0^λ winds once around the boundary of the disk D_λ^{kj} as λ winds once around the boundary of $\mathcal{O}_{kj}$.

Since v_0^λ lies in D_λ^{kj}, there is then a preimage of the structure of all of the disks $D_\lambda^{k\ell}$ and $E_\lambda^{k\ell}$ contained in D_λ^k that now lies in the preimage of D_λ^k in L^λ, namely, L_λ^k. Each $D_\lambda^{k\ell}$ and $E_\lambda^{k\ell}$ now has two preimages in L_λ^k, with the exception of the chosen D_λ^{kj}, which has only one preimage that contains the critical point c_0^λ. Thus we again "duplicate" the preimage structure that we see in D_λ^k in the region L_λ^k, and center this duplication around the preimage of D_λ^{kj}. Then F_λ^{k+j-1} maps the disk D_λ^{kj} one-to-one onto L^λ since $j \neq k$. Hence there is a copy of this duplicated preimage structure that we see in L^λ that is now contained in the chosen disk D_λ^{kj}. Thus, for each $\ell > 1$, we now have four disks named $D^{kj\ell}$ that are contained in D_λ^{kj}. Each of the $D_\lambda^{kj\ell}$ is mapped one-to-one onto L^λ by $F_\lambda^{k+j+\ell-1}$ and hence two-to-one over themselves by $F_\lambda^{k+j+\ell}$. Then, arguing as before, this map is polynomial-like of degree two on each $D^{kj\ell}$ and this produces four new baby Mandelbrot sets $\mathcal{M}^{kj\ell}$ which are arranged in a similar pattern as the preimages of the disks in the dynamical plane. Similar arguments also yield four Sierpinski holes $\mathcal{E}^{kj\ell}$.

Note also that the maze structure in the small neighborhood of $\mathcal{M}^{kj}$ is now more complicated than before. For example, if we had chosen the disk D_λ^{kj} to be one of the disks in one of the two finite chains of disks emanating from D_λ^k, then the maze structure around $\mathcal{M}^{kj}$ would consist of a pair of finite spokes, one on each side of $\mathcal{M}^{kj}$, and also a pair of "plus signs," each again on opposite sides of $\mathcal{M}^{kj}$. On the other hand, had D_λ^{kj} been chosen to be one of the disks in the infinite string of disks, then there would now be a pair of infinite spokes emanating from $\mathcal{M}^{kj}$ and again a pair of "plus signs."

Then we may continue this process, each time selecting a previously constructed Mandelbrot set with itinerary $s_0 \dots s_n$. Assuming the sequence $s_0 \dots s_n$ is not a repeated finite sequence, i.e., not a repeating sequence of the form $s_0 \dots s_j \dots s_0 \dots s_j$, this inductive process then produces the more intricate maze structure around the given Mandelbrot set.

6. Open Questions

One open problem in this area is to determine the exact structure of the Mandelpinski maze. One should think of this at the mth stage as a graph with the vertices representing the Mandelbrot sets (and between each pair of connected vertices there is a unique Sierpinski hole). How this graph looks changes at each vertex at each stage. The question is how to sketch the graph at each stage and, then, how to describe the limiting "graph." Another open problem is to determine the exact structure of the entire spoke along, say, the negative real axis. We produced infinitely many Mandelbrot sets and Sierpinski holes along this axis, but what remains in the limit? Clearly, these are parameters for which the critical orbit has an infinite itinerary that is neither periodic nor escaping. Are all such parameters just singletons? We conjecture that the answer is yes.

References

[1] Cuzzocreo, D. and Devaney, R. L. Simple Mandelpinski Necklaces for $z^2 + C/z^2$. To appear.

[2] Robert L. Devaney, *Baby Mandelbrot sets adorned with halos in families of rational maps*, Complex dynamics, Contemp. Math., vol. 396, Amer. Math. Soc., Providence, RI, 2006, pp. 37–50, DOI 10.1090/conm/396/07392. MR2209085

[3] Robert L. Devaney, *Structure of the McMullen domain in the parameter planes for rational maps*, Fund. Math. **185** (2005), no. 3, 267–285, DOI 10.4064/fm185-3-5. MR2161407

[4] Robert L. Devaney, *Singular perturbations of complex polynomials*, Bull. Amer. Math. Soc. (N.S.) **50** (2013), no. 3, 391–429, DOI 10.1090/S0273-0979-2013-01410-1. MR3049870

[5] Devaney, R. L. A Mandelpinski Maze for Rational Maps of the Form $z^n + \lambda/z^d$. To appear in *Indagationes Math.*

[6] Devaney, R. L. Mandelpinski Spokes in a Family of Rational Maps. To appear in *JDEA*.

[7] Robert L. Devaney, *Cantor necklaces and structurally unstable Sierpinski curve Julia sets for rational maps*, Qual. Theory Dyn. Syst. **5** (2004), no. 2, 337–359, DOI 10.1007/BF02972685. MR2275444

[8] Robert L. Devaney, *The McMullen domain: satellite Mandelbrot sets and Sierpinski holes*, Conform. Geom. Dyn. **11** (2007), 164–190 (electronic), DOI 10.1090/S1088-4173-07-00166-X. MR2346215

[9] Robert L. Devaney, *Baby Mandelbrot sets adorned with halos in families of rational maps*, Complex dynamics, Contemp. Math., vol. 396, Amer. Math. Soc., Providence, RI, 2006, pp. 37–50, DOI 10.1090/conm/396/07392. MR2209085

[10] Robert L. Devaney and Daniel M. Look, *A criterion for Sierpinski curve Julia sets*, Topology Proc. **30** (2006), no. 1, 163–179. Spring Topology and Dynamical Systems Conference. MR2280665

[11] Robert L. Devaney and Daniel M. Look, *Buried Sierpinski curve Julia sets*, Discrete Contin. Dyn. Syst. **13** (2005), no. 4, 1035–1046, DOI 10.3934/dcds.2005.13.1035. MR2166716

[12] Robert L. Devaney, Daniel M. Look, and David Uminsky, *The escape trichotomy for singularly perturbed rational maps*, Indiana Univ. Math. J. **54** (2005), no. 6, 1621–1634, DOI 10.1512/iumj.2005.54.2615. MR2189680

[13] Robert L. Devaney and Sebastian M. Marotta, *The McMullen domain: rings around the boundary*, Trans. Amer. Math. Soc. **359** (2007), no. 7, 3251–3273 (electronic), DOI 10.1090/S0002-9947-07-04137-2. MR2299454

[14] Robert L. Devaney and Kevin M. Pilgrim, *Dynamic classification of escape time Sierpiński curve Julia sets*, Fund. Math. **202** (2009), no. 2, 181–198, DOI 10.4064/fm202-2-5. MR2506193

[15] Devaney, R. L. and Russell, E. Connectivity of Julia Sets of Singularly Perturbed Rational Maps. In *Chaos, CNN, Memristors and Beyond.* World Scientific Press (2013), 239-245.

[16] Adrien Douady and John Hamal Hubbard, *On the dynamics of polynomial-like mappings*, Ann. Sci. École Norm. Sup. (4) **18** (1985), no. 2, 287–343. MR816367

[17] Curt McMullen, *Automorphisms of rational maps*, Holomorphic functions and moduli, Vol. I (Berkeley, CA, 1986), Math. Sci. Res. Inst. Publ., vol. 10, Springer, New York, 1988, pp. 31–60, DOI 10.1007/978-1-4613-9602-4_3. MR955807

[18] John Milnor, *Dynamics in one complex variable*, 3rd ed., Annals of Mathematics Studies, vol. 160, Princeton University Press, Princeton, NJ, 2006. MR2193309

[19] Mónica Moreno Rocha, *A combinatorial invariant for escape time Sierpiński rational maps*, Fund. Math. **222** (2013), no. 2, 99–130, DOI 10.4064/fm222-2-1. MR3085235

[20] P. Roesch, *On capture zones for the family $f_\lambda(z) = z^2 + \lambda/z^2$*, Dynamics on the Riemann sphere, Eur. Math. Soc., Zürich, 2006, pp. 121–129, DOI 10.4171/011-1/6. MR2348958

[21] G. T. Whyburn, *Topological characterization of the Sierpiński curve*, Fund. Math. **45** (1958), 320–324. MR0099638

Department of Mathematics, Boston University, 111 Cummington Mall, Boston, Massachusetts 02215

Contemporary Mathematics
Volume **678**, 2016
http://dx.doi.org/10.1090/conm/678/13645

Substitutions and Möbius disjointness

S. Ferenczi, J. Kułaga-Przymus, M. Lemańczyk, and C. Mauduit

ABSTRACT. We show that Sarnak's conjecture on Möbius disjointness holds for all subshifts given by bijective substitutions and some other similar dynamical systems, e.g. those generated by Rudin-Shapiro type sequences.

1. Introduction

In 2010, Sarnak [**50**] formulated the following conjecture: for each zero entropy topological dynamical system (T, X) (X is a compact metric space and T is a homeomorphism of X), each $f \in C(X)$ and $x \in X$, we have

$$\frac{1}{N}\sum_{n\leqslant N} f(T^n x)\boldsymbol{\mu}(n) \to 0, \tag{1}$$

where $\boldsymbol{\mu}\colon \mathbb{N} \to \mathbb{C}$ is the Möbius function defined by $\boldsymbol{\mu}(1) = 1$, $\boldsymbol{\mu}(p_1 \cdot \ldots \cdot p_k) = (-1)^k$ for k different prime numbers p_i, and $\boldsymbol{\mu}(n) = 0$ in the remaining cases. The conjecture has already been proved in numerous cases, e.g. [**1, 2, 4–6, 10, 15, 19, 20, 33, 40, 42, 43, 51, 52**].

The aim of the present paper is to show that Sarnak's conjecture holds for some classes of dynamical systems of number theoretic origin: all systems given by bijective substitutions (a subclass of substitutions of constant length) [**48**] and other related systems given by some automata (e.g. by the sequences of the Rudin-Shapiro type). Our approach is purely ergodic and the main tool is the theory of compact group extensions of rotations. Throughout, we deal with zero entropy homeomorphisms which are additionally uniquely ergodic, i.e. homeomorphisms possessing exactly one invariant measure (which has to be ergodic).

Given a bijective substitution θ over a finite alphabet A, we define its group cover substitution $\overline{\theta}$ over a subgroup G of permutations of A, which hence carries an additional natural group structure. In Section 3, we establish that the dynamical system $(S, X(\theta))$[1] given by θ is a topological factor of the dynamical system $(S, X(\overline{\theta}))$ given by $\overline{\theta}$. Thus, it suffices to show that Sarnak's conjecture holds for $(S, X(\overline{\theta}))$. The group cover substitution $\overline{\theta}$ can be identified with a certain

The second and third author's research was supported by Narodowe Centrum Nauki grant UMO-2014/15/B/ST1/03736.

[1]Here and in what follows, S stands for the left shift on a closed S-invariant subset of the space of two-sided sequences, i.e. $(S, X(\theta))$ is an example of a *subshift*.

(generalized) Morse sequence x and the associated dynamical system $(S, X(x))$ is a Morse system. This is where compact group extensions come into play – each Morse dynamical system is (measure-theoretically) isomorphic to a compact group extension $(T_\psi, X \times G)$ given by a so called Morse cocycle $\psi\colon X \to G$ over a rotation T, more precisely over an odometer (T, X) (this and other relations between the dynamical systems described in this paragraph are illustrated in Figure 1). The main difficulty is that such compact group extensions have been studied so far mostly from the measure-theoretic point of view [**31, 38, 48**] (the dynamical systems under considerations are uniquely ergodic). The underlying reason and, at the same time, the main obstacle for us is that Morse cocycles are in general not continuous. Thus, we cannot deal directly with such models – Sarnak's conjecture requires topological systems. In order to bypass this difficulty, more tools are used. The Morse dynamical system $(S, X(x))$ turns out to have a Toeplitz dynamical system $(S, X(\widehat{x}))$ as a topological factor, which, in turn, is an almost 1-1-extension of the odometer (T, X). Moreover, the method of Toeplitz extensions [**38**] allows us to find a dynamical system topologically isomorphic to $(S, X(x))$, which also has a form of a compact group extension: $(S_\varphi, X(\widehat{x}) \times G)$ given by a continuous cocycle $\varphi\colon X(\widehat{x}) \to G$. If we denote the (natural) factoring map from $(S, X(\widehat{x}))$ to (T, X) by p, we have the following relation between the two cocycles: $\varphi = \psi \circ p$. Our goal will be to prove that Sarnak's conjecture holds for $(S_\varphi, X(\widehat{x}) \times G)$.

The passage to group substitutions seems to be unavoidable for our method. Indeed, we require that the substitution subshift has a topological factor, determined by a Toeplitz sequence, which is an almost 1-1 extension of the maximal equicontinuous factor (the underlying odometer) and is measure-theoretically isomorphic to it. However, in [**24**], Section 4.4, it is proved that a bijective substitution need not have a symbolic factor which is measure-theoretically isomorphic to the maximal equicontinuous factor (this anwers a question raised by Baake). For example, this surprising property holds for the substitution $a \mapsto aabaa$, $b \mapsto bcabb$ and $c \mapsto cbccc$ [**24**].

Recall that an arithmetic function $\boldsymbol{u}\colon \mathbb{N} \to \mathbb{C}$ is called *multiplicative* if $\boldsymbol{u}(m \cdot n) = \boldsymbol{u}(m) \cdot \boldsymbol{u}(n)$ whenever m, n are coprime. It is called *aperiodic* whenever $\frac{1}{N}\sum_{n \leqslant N} \boldsymbol{u}(an+b) \to 0$ for all $a, b \in \mathbb{N}$. The Möbius function $\boldsymbol{\mu}$ is multiplicative and aperiodic.

The first tool we use to deal with the continuous compact group extensions is the Katai-Bourgain-Sarnak-Ziegler criterion:

THEOREM 1.1 ([**6, 29**], see also [**23**]). *Assume that $(a_n) \subset \mathbb{C}$ is bounded and suppose that*

$$\frac{1}{N}\sum_{n \leqslant N} a_{nr}\overline{a}_{ns} \to 0$$

for all sufficiently large different prime numbers r, s. Then

$$\frac{1}{N}\sum_{n \leqslant N} a_n \boldsymbol{u}(n) \to 0$$

for each multiplicative function $\boldsymbol{u}\colon \mathbb{N} \to \mathbb{C}$, $|\boldsymbol{u}| \leqslant 1$.

This criterion is applied to sequences of the form $a_n = f(T^n x)$, $n \geqslant 1$, and, if satisfied, it yields a certain form of disjointness of different prime powers of the homeomorphism T; notice that "sufficiently large" in Theorem 1.1 may depend on f and x.

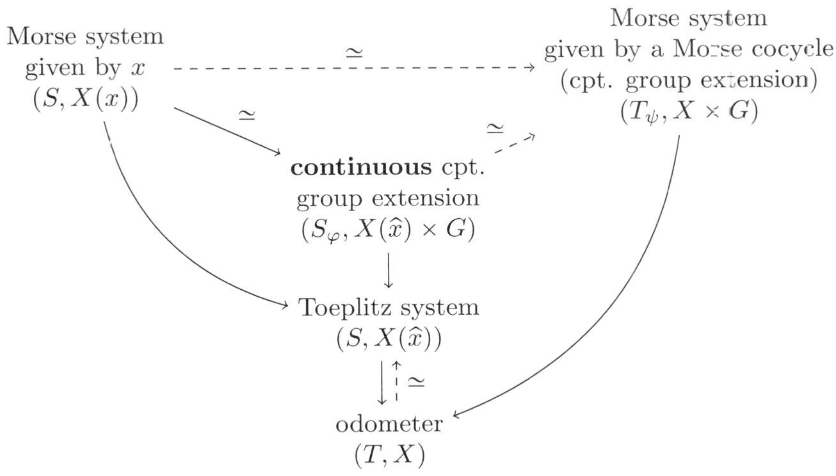

FIGURE 1. Morse and Toeplitz dynamical systems on a diagram. Plain and dashed lines denote topological and measure-theoretical maps, respectively (all depicted systems are uniquely ergodic).

Our second tool is based on the method of lifting generic points in the Cartesian products of different prime powers T^r and T^s to almost 1-1-extensions. It has already appeared in [**32**], where the almost 1-1-extensions are chosen in such a way that the original cocycle considered on the extended space becomes continuous. Moreover, we will study ergodic joinings of $(T_\psi)^r$ and $(T_\psi)^s$ and show that in our case this set consists only of the relatively independent extensions of isomorphisms between T^r and T^s for different sufficiently large primes r, s. This will allow us to control generic points in the Cartesian product of the continuous compact group extension $(S_\varphi)^r \times (S_\varphi)^s$.

In Section 5, we give the main application of our method – we prove Sarnak's conjecture for the dynamical systems given by:

- all bijective substitutions (for a relationship of our result with a recent result of Drmota [**12**], see Remark 5.7),
- certain subclass of regular (generalized) Morse sequences,
- certain sequences of the Rudin-Shapiro type.

In Section 6, we compare our results concerning generalized Morse sequences and the Rudin-Shapiro type sequences with some earlier results in which (1) has been proved only for $f(y) = (-1)^{y[0]}$. Using spectral approach, we prove, in some cases, that the validity of (1) for such f yields Sarnak's conjecture in its full form for the corresponding dynamical system. In particular, we show that Sarnak's conjecture holds for the dynamical systems given by Kakutani sequences [**4, 19**].[2] Sometimes, however, it seems that more than one function satisfying (1) is necessary for the validity of Sarnak's conjecture. E.g., this seems to be the case for the dynamical systems given by the Rudin-Shapiro type sequences. Here, (1) for f has been proved in [**43, 51**] by a purely combinatorial approach. The methods developed

[2]We have not been able to show the validity of Sarnak's conjecture for the dynamical systems given by Kakutani sequences by a use of the criterion proved in Section 4.3.

in [**43**] seem to be flexible enough to give (1) for finitely many functions described in 6.3, hence yields one more proof of Sarnak's conjecture for the corresponding dynamical system.

In Section 6.4, we compare our results with a recent work of Veech [**52**]. He provides a proof of Sarnak's conjecture for a class of dynamical systems given by some sequences over an arbitrary compact group. These sequences turn out to be generalizations of aforementioned Kakutani sequences. In particular, [**52**] gives an alternative proof of Sarnak's conjecture for Kakutani systems first proved in [**4, 19**].

2. Basic tools

2.1. Spectral theory. For an ergodic automorphism T of a standard probability Borel space $(X, \mathcal{B}, \mu)$, we consider the associated Koopman operator U_T on $L^2(X, \mathcal{B}, \mu)$ given by $U_T(f) = f \circ T$. Then there exist elements $f_n \in L^2(X, \mathcal{B}, \mu)$, $n \geqslant 1$, such that

$$L^2(X, \mathcal{B}, \mu) = \bigoplus_{n \geqslant 1} \mathbb{Z}(f_n) \text{ and } \sigma_{f_1} \gg \sigma_{f_2} \gg \dots, \tag{2}$$

where $\mathbb{Z}(f) = \overline{\text{span}}\{U_T^n(f) : n \in \mathbb{Z}\}$ is the *cyclic space* generated by f and σ_f denotes the only finite positive Borel measure on $\mathbb{T}$ such that $\int_X f \circ T^n \cdot \overline{f}\, d\mu = \int_{\mathbb{T}} z^n\, d\sigma_f(z)$ for each $n \in \mathbb{Z}$ (σ_f is called the *spectral measure of* f). The class of all measures equivalent to σ_{f_1} in the above decomposition is called *the maximal spectral type* of U_T and (2) is called a *spectral decomposition.* We say that the maximal spectral type is realized by $f \in L^2(X, \mathcal{B}, \mu)$ if σ_f is equivalent to σ_{f_1}.

If $L^2(X, \mathcal{B}, \mu) = \mathbb{Z}(f_1) \oplus \dots \oplus \mathbb{Z}(f_k)$ for some $f_i \in L^2(X, \mathcal{B}, \mu)$, we say that U_T has multiplicity at most k. If no such $k \geqslant 1$ exists, the multiplicity of U_T is infinite. If $k = 1$, U_T is said to have *simple spectrum.* All rank one automorphisms, e.g. [**14**], have simple spectrum [**3**].

Recall that T has *discrete spectrum* if the maximal spectral type of U_T is purely discrete. Equivalently, $L^2(X, \mathcal{B}, \mu)$ is generated by the eigenfunctions of U_T. Such transformations enjoy the rank one property [**8**]. By the Halmos-von Neumann theorem, T is, up to isomorphism, an ergodic rotation on a compact monothetic group. If, additionally, all eigenvalues of U_T are roots of unity, T is said to have *rational discrete spectrum.*

For more information on the spectral theory see, e.g., [**47**].

2.2. Joinings. Recall that if T and S are ergodic automorphisms on $(X, \mathcal{B}, \mu)$ and $(Y, \mathcal{C}, \nu)$, respectively, then by a *joining* between T and S we mean any $T \times S$-invariant measure κ on $(X \times Y, \mathcal{B} \otimes \mathcal{C})$ whose projections on X and Y are μ and ν, respectively. We denote by $J(T, S)$ the set of joinings between T and S and by $J^e(T, S)$ the subset of ergodic joinings. Clearly, $\mu \otimes \nu \in J(T, S)$. If $R\colon (X, \mathcal{B}, \mu) \to (Y, \mathcal{C}, \nu)$ establishes an isomorphism between T and S, then the measure μ_R determined by $\mu_R(B \times C) = \mu(B \cap R^{-1}C)$, $B \in \mathcal{B}, C \in \mathcal{C}$, belongs to $J^e(T, S)$. It is concentrated on the graph of R and is called a *graph* joining. When $T = S$, we speak about self-joinings of T and each graph self-joining is given by some element from the centralizer $C(T)$ of T. We remind that the centralizer $C(T)$ consists of automorphisms of $(X, \mathcal{B}, \mu)$ commuting with T.

Suppose that T and S are isomorphic, where the isomorphism is given by $R\colon X \to Y$, and have extensions to $\overline{T}$ on $(\overline{X}, \overline{\mathcal{B}}, \overline{\mu})$ and $\overline{S}$ on $(\overline{Y}, \overline{\mathcal{C}}, \overline{\nu})$, respectively.

The *relatively independent extension* of μ_R (to a joining of $\overline{T}$ and $\overline{S}$) is denoted by $\widetilde{\mu}_R$ and determined by

$$\int_{\overline{X}\times\overline{Y}} F\otimes G\, d\widetilde{\mu}_R = \int_X \mathbb{E}(F|X)\cdot\mathbb{E}(G|Y)\circ R\, d\mu$$

for $F \in L^2(\overline{X},\overline{\mu})$, $G \in L^2(\overline{Y},\overline{\nu})$.

2.3. Compact group extensions. Assume that T is an ergodic automorphism of a standard Borel probability space $(X,\mathcal{B},\mu)$. Let G be a compact metric group with Haar measure m_G.

DEFINITION 2.1. Any measurable map $\psi\colon X \to G$ is called a *cocycle*. The automorphism T_ψ of $(X\times G,\mathcal{B}\otimes\mathcal{B}(G),\mu\otimes m_G)$ defined by

$$T_\psi(x,g) := (Tx,\psi(x)g)$$

is called a *G-extension* of T (it is an example of a *compact group extension of* T). We say that ψ is *ergodic* if T_ψ is ergodic.

Compact group extensions enjoy the following relative unique ergodicity property.

LEMMA 2.1 ([**17**], the proof of Proposition 3.10). *If T_ψ is ergodic (i.e. if the product measure $\mu\otimes m_G$ is ergodic) then $\mu\otimes m_G$ is the only T_ψ-invariant measure projecting onto μ.*

Let τ_g be an automorphism of $(X\times G,\mu\otimes m_G)$ given by

$$\tau_g(x,g') = (x,g'\cdot g) \text{ for each } g'\in G.$$

Then $T_\psi\circ\tau_g = \tau_g\circ T_\psi$, that is, τ_g is an element of the centralizer $C(T_\psi)$ of T_ψ.

PROPOSITION 2.2 ([**45**]). *Assume that T is ergodic and $\psi\colon X\to G$ is ergodic as well. Assume additionally that T has discrete spectrum. Then each $\widetilde{S}\in C(T_\psi)$ is a lift of some $S\in C(T)$. More precisely, $\widetilde{S} = S_{f,v}$, where $S_{f,v}(x,g) = (Sx, f(x)v(g))$ for some $S\in C(T)$, some measurable $f\colon X\to G$ and some continuous group automorphism $v\colon G\to G$. Moreover, if $\widetilde{S}$ and $\overline{S}$ are two lifts of $S\in C(T)$ then $\widetilde{S} = \overline{S}\circ\tau_{g_0}$ for some $g_0\in G$.*

DEFINITION 2.2. We will say that T_ψ has *G-trivial centralizer* if

$$C(T_\psi) = \{T_\psi^k\circ\tau_g : k\in\mathbb{Z},\ g\in G\}.$$

DEFINITION 2.3. Let $H\subset G$ be a closed subgroup. The corresponding factor-automorphism $T_{\psi H}$ of $(X\times G/H,\mu\otimes m_{G/H})$ given by

$$T_{\psi H}(x,gH) := (Tx,\psi(x)gH)$$

is called a *natural factor* of T_ψ. It is called *nontrivial* if $H\neq G$, and it is called *normal* whenever H is normal.

REMARK 2.3. Notice that a power of a group extension is clearly a group extension: $(T_\psi)^r = T^r_{\psi^{(r)}}$[3] and the passage to natural factors is "commutative": $((T_\psi)^r)_H = (T_{\psi H})^r$.

We need some facts about joinings of compact group extensions.

[3] $\psi^{(r)}(x) := \psi(x)\psi(Tx)\ldots\psi(T^{r-1}x)$ for $r\geqslant 0$ and extends to $r\in\mathbb{Z}$, so that the cocycle identity $\psi^{(m+n)}(x) = \psi^{(m)}(x)\psi^{(n)}(T^m x)$ holds for every $m,n\in\mathbb{Z}$.

THEOREM 2.4 ([**45**]). *Assume that T is ergodic. Assume that $S \in C(T)$ and let $\psi_i : X \to G$ be an ergodic cocycle, $i = 1, 2$. Assume that $\kappa \in J^e(T_{\psi_1}, T_{\psi_2})$ and projects on the graph self-joining μ_S of T. Then there are two closed normal subgroups $H_1, H_2 \subset G$ and an isomorphism $\overline{S}$ (a lift of S) between the two normal natural factors $T_{\psi_1 H_1}$ and $T_{\psi_2 H_2}$ such that*

$$\kappa = \widetilde{(m_{G/H_1})_{\overline{S}}},$$

i.e. κ is the relatively independent extension of the graph joining $(m_{G/H_1})_{\overline{S}} \in J^e(T_{\psi_1 H_1}, T_{\psi_2 H_2})$ given by the isomorphism $\overline{S}$.

REMARK 2.5. Suppose that T has rational discrete spectrum and $(T_\psi)^r$ and $(T_\psi)^s$ are ergodic. Then T^r and T^s are isomorphic and the only ergodic joinings between them are the graph joinings. By Theorem 2.4, if there is no isomorphism between nontrivial normal natural factors of $(T_\psi)^r$ and $(T_\psi)^s$, then there are no ergodic joinings between $(T_\psi)^r$ and $(T_\psi)^s$, except for the "most independent" ones: the relatively independent extensions of isomorphisms between T^r and T^s. Notice also that if such a relative product is ergodic then automatically, by Lemma 2.1, it is the only invariant measure on $X \times G \times X \times G$ projecting on the graph of the isomorphism.

2.4. Generic points. Let T be a homeomorphism of a compact metric space X. Let μ be a T-invariant Borel probability measure on X.

DEFINITION 2.4. We say that $x \in X$ is *generic* for μ if $\frac{1}{N}\sum_{n\leqslant N} \delta_{T^n x} \to \mu$ weakly. If the convergence to μ takes place only along a subsequence (N_k) then x is called *quasi-generic* for μ.

REMARK 2.6. Notice that, by the compactness of X, the space of probability measures on X is also compact, hence each point is quasi-generic for some T-invariant measure (see the Krylov-Bogolyubov theorem, e.g. [**13**], p. 98).

3. Basic objects

3.1. Odometers, Morse cocycles and Toeplitz extensions.

Odometers. Assume that $(n_t)_{t\geqslant 0}$ satisfies $n_0 = 1$ and $n_t \mid n_{t+1}$ with $\lambda_t := n_{t+1}/n_t \geqslant 2$ for $t \geqslant 0$. Consider $X := \prod_{t\geqslant 0} \mathbb{Z}/\lambda_t\mathbb{Z}$ with the product topology and the group law given by addition mod λ_t, with carrying the remainder to the right. This makes X a compact metric Abelian group. We define the translation T by $(1, 0, 0, \ldots)$:

$$T(x_0, x_1, x_2, \ldots) = (x_0 + 1, x_1, x_2, \ldots)$$

to obtain $(X, \mathcal{B}, m_X, T)$ – an ergodic rotation.

DEFINITION 3.1. T defined above is called an *odometer*.

REMARK 3.1. Odometer T defined above has rational discrete spectrum given by the n_t-roots of unity, $t \geqslant 0$. For each $t \geqslant 0$, there is a Rokhlin tower $\mathcal{D}^t := \{D_0^t, D_1^t, \ldots, D_{n_t-1}^t\}$, i.e. a partition of X for which $T^i D_0^t = D_{i \bmod n_t}^t$ for each $i \geqslant 0$ (by ergodicity, such a tower is unique up to cyclic permutation of the levels). Indeed, $D_0^0 = X$ and we set

$$D_0^t := \{x \in X : x_0 = \ldots = x_{t-1} = 0\},\ t \geqslant 1.$$

Clearly, the partition $\mathcal{D}^{t+1}$ is finer than $\mathcal{D}^t$ and the sequence of such partitions tends to the partition into points.

REMARK 3.2 (cf. Remark 2.5). Notice that for each $r \geqslant 1$,

(3) T^r is isomorphic to T whenever T^r is ergodic.

Indeed, T^r has the same spectrum as T. To see the isomorphism more directly, notice that $\gcd(r, n_t) = 1$ and T^r permutes the levels of $\mathcal{D}^t$ – this extends to an isomorphism map between T and T^r.

REMARK 3.3 (cf. Remark 2.5). Since T has discrete spectrum, its only ergodic joinings are graph measures $(m_X)_W$, where $W \in C(T)$ is another rotation on X [**18**]. It easily follows that

(4) each point $(x, y) \in X \times X$ is generic for an ergodic self-joining of the form $(m_X)_W$.

Indeed, define W as the translation by $x - y$.

Morse cocycles. Assume that G is a compact metric group and (T, X) is an odometer.

DEFINITION 3.2 ([**21, 38**]). We say that $\psi\colon X \to G$ is a *Morse cocycle* if ψ is constant on each D_i^t, $t \geqslant 0$, $i = 0, 1, \ldots, n_t - 2$ ($\psi|_{D_i^t}$ may depend on i).

REMARK 3.4. To see what are the values of a Morse cocycle ψ on $D^t_{n_t-1}$, we first pass to the levels $D^{t+1}_{jn_t-1}$ for $j = 1, \ldots, \lambda_{t+1} - 1$, and read the values $\psi|_{D^{t+1}_{jn_t-1}}$. To read the values on $D^{t+1}_{n_{t+1}-1}$ ($n_{t+1} = \lambda_{t+1} n_t$), we pass to $\mathcal{D}^{t+2}$ etc. It is clear that ψ defined in this way is continuous everywhere (as the levels of the towers are clopen sets) except perhaps one point (given by the intersection of the top levels of all towers). Notice also that whenever G is finite then a Morse cocycle cannot be continuous unless it is constant on **each** level of the tower $\mathcal{D}^{t_0}$ for some t_0. In this case, T_ψ, if ergodic, is a direct product of T with a rotation on G. In particular, Sarnak's conjecture holds for T_ψ.

REMARK 3.5. The class of group extensions given by Morse cocycles is (up to measure-theoretic isomorphism) the same as the class of dynamical systems generated by generalized Morse sequences, see [**21, 31, 38, 49**], which we consider in the next section.

Toeplitz extensions. Morse cocycles yield extensions of odometers which are special cases of so called *Toeplitz extensions* studied in [**38**] (cf. earlier [**53, 54**] and the last section). Toeplitz extensions are also given by cocycles over odometers but in the definition of such cocycles we are letting more than one level have non-constant values. It is however required that the numbers of levels of $\mathcal{D}^t$ on which the cocycle is non-constant divided by n_t goes to zero. A reason for that is that we want to obtain a regular Toeplitz sequence which is behind such a construction, see [**38**] for more details. An example of a Toeplitz extension is given below.

EXAMPLE 3.1. Let $\lambda_t := 2$ for each $t \geqslant 0$ and $G := \mathbb{Z}/2\mathbb{Z}$. We define $\psi\colon X \to \mathbb{Z}/2\mathbb{Z}$ so that at stage t it is defined on each D_i^t, except for $i = 2^{t-1} - 1$ and $i = 2^t - 1$. Then, when we pass to $\mathcal{D}^{t+1}$, on the levels $D^{t+1}_{2^{t-1}-1}$ and $D^{t+1}_{2^t+2^{t-1}-1}$ (ψ must be defined here at this stage of the construction), we set the values 0 and 1 (or 1 and 0), respectively.

The class of Toeplitz extensions of the dyadic odometer described in Example 3.1 was considered in [**38, 41**]. The dynamical systems corresponding to the Rudin-Shapiro type sequences (see Section III.2 in [**38**]) are in this class.

3.2. Generalized Morse sequences. Let G be a compact metric group with the unit e.

DEFINITION 3.3. Let $b^t \in G^{\lambda_t}$ be a block over G of length $|b^t| = \lambda_t \geqslant 2$ and $b^t[0] = e$, $t \geqslant 0$. The associated *(generalized) Morse sequence* is defined by

$$x := b^0 \times b^1 \times \dots, \tag{5}$$

where $B \times C = (B \circ c[0])(B \circ c[1]) \dots (B \circ c[|C|-1])$ and $B \circ g := (b_0 g, \dots, b_{|B|-1} g)$ for B, C blocks over G and $g \in G$. By $(S, X(x))$ we denote the subshift corresponding to x ($X(x) \subset G^{\mathbb{Z}}$).

EXAMPLE 3.2. Generalized Morse sequences for $G = \mathbb{Z}/2\mathbb{Z}$ were first studied in [**31**]. If $b^t \in \{00, 01\}$, $t \geqslant 0$, we speak about *Kakutani sequences* [**34**].

DEFINITION 3.4 ([**27**]). We say that $u \in G^{\mathbb{N}}$ is a *Toeplitz sequence* whenever for each $n \in \mathbb{N}$ there exists $k_n \geqslant 1$ such that u is constant on the arithmetic progression $n + k_n\mathbb{N}$. (For the theory of dynamical systems given by Toeplitz sequences, see e.g. [**9**].)

LEMMA 3.6 (cf. Figure 1). *Let $x = b^0 \times b^1 \times \dots$ be a Morse sequence. The map*

$$y \mapsto \widehat{y}, \quad \widehat{y}[n] := y[n+1]y[n]^{-1} \tag{6}$$

yields an equivariant map between $(S, X(x))$ and $(S, X(\widehat{x}))$. Moreover, $\widehat{x}$ is a Toeplitz sequence.

PROOF. The first part is obvious. For the second, notice that for each $t \geqslant 1$, we have

$$x = c_t \times z_t, \text{ where } c_t = b^0 \times \dots \times b^{t-1}, \ z_t = b^t \times b^{t+1} \times \dots,$$

whence x is a concatenation of blocks of the form $c_t \circ g$. Moreover,

$$c_t[n+1]c_t[n]^{-1} = (c_t \circ g)[n+1](c_t \circ g)[n]^{-1} \text{ for } n = 0, \dots, |c_t| - 2.$$

It follows that

$$\widehat{x} = \widehat{c}_t * \widehat{c}_t * \widehat{c}_t * \dots, \tag{7}$$

where "$*$" stands for the unfilled place of $\widehat{x}$ at the stage $t \geqslant 1$. □

REMARK 3.7. Toeplitz sequence $\widehat{x}$ from (7) is regular in the sense of [**27**]. Hence $(S, X(\widehat{x}))$ is uniquely ergodic [**9**].

LEMMA 3.8 (cf. Figure 1). *$(S, X(x))$ is topologically isomorphic to $(S_\varphi, X(\widehat{x}) \times G)$, where $\varphi\colon X(\widehat{x}) \to G$ is the continuous cocycle given by*

$$\varphi(z) = z[0], \text{ for each } z \in X(\widehat{x}).$$

PROOF. The topological isomorphism is given by the (equivariant) map

$$y \mapsto (\widehat{y}, y[0]).$$

Indeed, this map is continuous as φ is continuous, it is onto since $\widehat{y} = \widehat{y \circ g}$, and finally it is 1-1 since $y[n+1]$ is determined by $y[n]$ and $\widehat{y}[n]$. □

REMARK 3.9 (cf. Figure 1). Let x be the Morse sequence given by (5) with $n_t := |c_t| = |b^0 \times \dots \times b^{t-1}|$, $t \geqslant 1$. Then the Toeplitz system $(S, X(\widehat{x}))$ has the $\{n_t\}$-odometer (T, X) as its topological factor. Moreover, the Morse dynamical system

$(S, X(x))$ is given by a Morse cocycle ψ over T. The values $(\psi|_{D_0^t}, \ldots, \psi|_{D_{n_t-2}^t})$ are determined by $\widehat{c}_t$:

$$\psi|_{D_i^t} = \widehat{c}_t[i] \text{ for } 0 \leqslant i \leqslant n_t - 2, t \geqslant 1.$$

It follows that

$$b^0[0] = e,\ b^0[i] = \prod_{j=0}^{i-1} \widehat{c}_1[j], 1 \leqslant i \leqslant \lambda_0 - 1$$

and then, inductively ($c_t = b^0 \times \ldots \times b^{t-1}$),

$$b^{t+1}[0] = e,\ b^{t+1}[i] = \prod_{j=i}^{1}(\widehat{c}_{t+1}[jn_t - 1]c_t[n_t - 1]), 1 \leqslant i \leqslant \lambda_{t+1} - 1$$

(for more details see, e.g. [**11**, **38**]). However, the Morse cocycle is not continuous. The passage to the Toeplitz dynamical system from Lemma 3.8 allows us to get its continuous version.

LEMMA 3.10. *The normal natural factors of Morse dynamical systems over G are Morse dynamical systems over G/H.*

PROOF. The assertion follows immediately from the equality $B \times C \bmod H = (B \bmod H) \times (C \bmod H)$ and Remark 3.9. □

3.3. Bijective substitutions. Fix a finite alphabet A with $|A| = r \geqslant 2$.

DEFINITION 3.5. A map $\theta \colon A \to A^\lambda$ ($\lambda \geqslant 1$) is called a *substitution on A of constant length λ* (in what follows, simply a *substitution*) if there exists $n \geqslant 1$ such that for each $a, a' \in A$ there exists k satisfying $\theta^n(a)[k] = a'$. We extend θ first to a map on blocks over A, then to a map $\theta \colon A^{\mathbb{N}} \to A^{\mathbb{N}}$. We will assume that $\theta(a_0)[0] = a_0$. By iterating θ at a_0, we obtain a fixed point for the map $\theta \colon A^{\mathbb{N}} \to A^{\mathbb{N}}$ and we denote by $(S, X(\theta))$ the corresponding subshift of $A^{\mathbb{Z}}$.

REMARK 3.11 (see Chapter 5 in [**48**]). For each substitution θ, $(S, X(\theta))$ is strictly ergodic.

REMARK 3.12. Let $\theta \colon A \to A^\lambda$ be a substitution such that $(S, X(\theta))$ is aperiodic. Recall that then for each $y \in X(\theta)$ there is a unique sequence $(i_t(y))_{t \geqslant 1} \subset \mathbb{Z}$ (*t-skeleton*) with $i_t(y) \in [-\lambda^t + 1, 0]$ such that $y[i_t + k\lambda^t, i_t + (k+1)\lambda^t - 1] = \theta^t(a_{k,t})$ for each $k \in \mathbb{Z}$ and some letters $a_{k,t} \in A$. This allows us to define the corresponding towers of height λ^t by setting the base of the t-tower

$$D_0^t := \{y \in X(\theta) : i_t(y) = 0\}$$

to obtain $\bigcup_{i=0}^{\lambda^t - 1} S^i D_0^t = X(\theta)$.

DEFINITION 3.6 ([**46**]). We say that substitution θ is *recognizable* if there exists a constant $M > 0$ such that if $y \in X(\theta)$, $t \geqslant 1$ and $i \in [-\lambda^t + 1, 0]$ satisfy

$$y[i, i + M\lambda^t - 1] = \theta^t(b_1) \ldots \theta^t(b_M)$$

for some $b_1, \ldots, b_M \in A$ then $i = i_t(y)$. We say that M is a *constant of recognizability.*

REMARK 3.13 ([**46**, **48**]). Each substitution θ such that $(S, X(\theta))$ is aperiodic, is recognizable. In what follows, we will tacitly assume that we deal with recognizable substitutions.

REMARK 3.14. Suppose that θ is recognizable. It follows that each function $\mathbb{1}_{D_0^t}$ depends on not more than $M\lambda^t$ coordinates.

DEFINITION 3.7 ([**48**]). Substitution θ is called *bijective* if

$$\overline{d}(\theta(a), \theta(a')) := \frac{|\{0 \leqslant k \leqslant \lambda - 1 : \theta(a)[k] \neq \theta(a')[k]\}|}{\lambda} = 1, \text{ whenever } a \neq a'$$

or, equivalently, the maps $\sigma_i(a) := \theta(a)[i]$ are bijections of A, $i = 0, \ldots, \lambda - 1$.

REMARK 3.15. We can assume (wlog) that $\sigma_0 = Id$ by considering, if necessary, its power.

DEFINITION 3.8. Let G be a finite group with the unit e. A substitution $\theta \colon G \to G^\lambda$ is called a *group substitution* whenever

$$\theta(g) = \theta(e) \circ g \text{ for each } g \in G.$$

REMARK 3.16. Each group substitution is bijective. Moreover, each group substitution can be identified with the Morse sequence $\theta(e) \times \theta(e) \times \ldots$

LEMMA 3.17 (cf. Lemma 3.10 and Remark 3.16). *The normal natural factors of dynamical systems given by group substitutions are determined by group substitutions.*

PROOF. Consider the group substitution given by $e \mapsto B$ (i.e. $g \mapsto B \circ g$). Then

$$gH \mapsto B \circ g \bmod H$$

yields a bijective substitution as in each column of the matrix corresponding to the group substitution we see all elements of G; in particular, by taking them mod H, we see all elements of G/H. $\square$

Denote by $\mathscr{S}_r$ the group of permutations of A. Define $\widetilde{\theta} \colon \mathscr{S}_r \to \mathscr{S}_r^\lambda$ by setting

$$\widetilde{\theta}(\tau) = (\sigma_0 \circ \tau, \sigma_1 \circ \tau, \ldots, \sigma_{\lambda-1} \circ \tau) = \widetilde{\theta}(Id) \circ \tau \tag{8}$$

for each $\tau \in \mathscr{S}_r$. Let $G \subset \mathscr{S}_r$ be the subgroup generated by $\sigma_0, \sigma_1, \ldots, \sigma_{\lambda-1}$ and define

$$\overline{\theta}(\tau) := \widetilde{\theta}(\tau) \text{ for } \tau \in G. \tag{9}$$

DEFINITION 3.9 (cf. Lemma 3.19 below). We call $\overline{\theta}$ the *group cover substitution* of θ.

LEMMA 3.18. *$\overline{\theta}$ is a (bijective) substitution.*

PROOF. Notice that if $\overline{\theta}^n(\sigma_0)[j] = \tau$ then, in $\overline{\theta}^{n+1}(\sigma_0)$, we can find the block $(\sigma_0 \circ \tau, \ldots, \sigma_{\lambda-1} \circ \tau)$. Since all elements in G are of finite order $(\sigma_i^{-1} = \sigma_i^{r!-1})$, it follows by induction that, for some n, we will see all symbols from G on $\overline{\theta}^n(\sigma_0)$. $\square$

LEMMA 3.19. *$(S, X(\theta))$ is a topological factor of its group cover substitution $(S, X(\overline{\theta}))$.*

PROOF. We define an equivalence relation on G by setting $\tau \equiv \tau'$ if $\tau(0) = \tau'(0)$.[4] For $y \in X(\overline{\theta})$, set

$$F(y)[n] := (y[n])(0).$$

[4]This relation is called *θ-consistent*, see [**28, 44**].

Notice that the image of F equals $X(\theta)$, F is equivariant and takes the same values on the equivalence classes of $\equiv$. Finally, notice that $\{\tau(0) : \tau \in G\} = A$ since θ is a substitution, whence G acts transitively on A. □

REMARK 3.20 (cf. Remark 3.16). The group cover substitution $\overline{\theta}$ can be identified with the Morse sequence $B \times B \times \ldots$ (over G), where $B = (\sigma_0, \sigma_1, \ldots, \sigma_{\lambda-1})$.

REMARK 3.21. Notice that, in order to prove Sarnak's conjecture for a bijective substitution θ, it suffices to prove it for $(S_\varphi, X(\widehat{x}) \times G)$, where $x = B \times B \times \ldots$ as $(S, X(\theta))$ is its topological factor (see Lemma 3.19 and Remark 3.20). Notice also that we do not claim that for $(S, X(\theta))$ the Toeplitz dynamical system $(S, X(\widehat{x}))$ is its topological factor (even though the odometer is its topological factor). In fact, there is a counterexample to such a claim due to Herning [**24**].

4. Sarnak's conjecture for finite group extensions

4.1. Lifting generic points for compact group extensions. We now recall a basic result on lifting generic points from [**32**]. Assume that $\overline{T}_i$ (T_i) is a uniquely ergodic homeomorphism, with a unique invariant measure $\overline{\mu}_i$ (μ_i), of a compact metric space $\overline{X}_i$ (X_i), $i = 1, 2$. Assume, moreover, that $\pi_i \colon \overline{X}_i \to X_i$ is continuous and yields (T_i, X_i) a topological factor of $(\overline{T}_i, \overline{X}_i)$.

PROPOSITION 4.1 ([**32**]). *Assume that $(\overline{T}_i, \overline{X}_i, \overline{\mu}_i)$ and (T_i, X_i, μ_i) are measure-theoretically isomorphic. Assume, moreover that (T_i, X_i, μ_i) is measure-theoretically coalescent*[5] *for $i = 1, 2$. Assume that $(x_1, x_2) \in X_1 \times X_2$ is generic for an ergodic $T_1 \times T_2$-invariant measure ρ. Then there exists a unique $\overline{T}_1 \times \overline{T}_2$-invariant measure $\overline{\rho}$, such that each pair $(\overline{x}_1, \overline{x}_2) \in (\pi_1 \times \pi_2)^{-1}(x_1, x_2)$ is generic for $\overline{\rho}$. Moreover, the system $(\overline{T}_1 \times \overline{T}_2, \overline{\rho})$ is isomorphic to $(T_1 \times T_2, \rho)$.*

Let T be an odometer acting on $(X, \mathcal{B}, m_X)$ and let $\overline{T}$ be a uniquely ergodic homeomorphism of $\overline{X}$ (with the unique invariant measure $\overline{m_X}$) such that $\pi \colon \overline{X} \to X$ is a topological factor map, and (T, m_X) and $(\overline{T}, \overline{m_X})$ are measure-theoretically isomorphic (then π is a.e. 1-1 as transformations with discrete spectrum are coalescent). Assume that $\psi \colon X \to G$ is ergodic and such that the cocycle $\overline{\psi} \colon \overline{X} \to G$ given by

$$\overline{\psi}(\overline{x}) := \psi(\pi(\overline{x})) \text{ is continuous} \tag{10}$$

(it is automatically ergodic as $(T_\psi, X \times G, m_X \otimes m_G)$ and $(\overline{T}_{\overline{\psi}}, \overline{X} \times G, \overline{m_X} \otimes m_G)$ are isomorphic). Assume that $r \neq s$ are such that $(T_\psi)^r$ and $(T_\psi)^s$ are ergodic, hence T^r and T^s are isomorphic (and they are isomorphic to T).

PROPOSITION 4.2. *Assume that the only ergodic joinings between $(T_\psi)^r$ and $(T_\psi)^s$ are the relatively independent extensions over the graphs of isomorphisms between T^r and T^s. Let $\overline{x} \in \overline{X}$ and let $\rho = (m_X)_R$ be the (ergodic) graph joining for which the point $(\pi(\overline{x}), \pi(\overline{x}))$ is generic. Then for each $g \in G$, the point $((\overline{x}, g), (\overline{x}, g))$ is generic for the $\widetilde{\overline{\rho}}$, where $\overline{\rho}$ comes from Proposition 4.1 ($\sim$ stands for the relatively independent extension). Moreover,*

$$((\overline{T}_{\overline{\psi}})^r \times (\overline{T}_{\overline{\psi}})^s, \widetilde{\overline{\rho}}) \text{ and } ((T_\psi)^r \times (T_\psi)^s, \widetilde{(m_X)_R}) \text{ are isomorphic.} \tag{11}$$

[5]An automorphism T of $(X, \mathcal{B}, \mu)$ is called *coalescent* [**22**] if each endomorphism commuting with T is invertible.

PROOF. The point $((\overline{x},g),(\overline{x},g))$ is quasi-generic for a $(\overline{T}_{\overline{\psi}})^r \times (\overline{T}_{\overline{\psi}})^s$-invariant measure κ. By Proposition 4.1, $(\overline{x},\overline{x})$ is generic for $\overline{\rho}$. Therefore, the projection of κ on $\overline{X}\times\overline{X}$ is equal to $\overline{\rho}$. Using Lemma 2.1 (applied to $\overline{T}^r\times\overline{T}^s$, $\overline{\psi}^{(r)}\times\overline{\psi}^{(s)}$ and $\overline{\rho}$), to conclude, we only need to prove that $((\overline{T}_{\overline{\psi}})^r\times(\overline{T}_{\overline{\psi}})^s,\widetilde{\overline{\rho}})$ is ergodic. Notice that (11) is obvious since $\rho=(m_X)_R$ and $\overline{\rho}$ yield isomorphic systems. This gives immediately that $((\overline{T}_{\overline{\psi}})^r\times(\overline{T}_{\overline{\psi}})^s,\widetilde{\overline{\rho}})$ is ergodic, whence $\kappa=\widetilde{\overline{\rho}}$. □

4.2. Criterion for the validity of Sarnak's conjecture for finite group extensions. In this section, we assume that $(T,X,\mathcal{B},\mu)$ is an ergodic transformation with discrete spectrum and $\psi\colon X\to G$ is an ergodic cocycle with values in a finite group G.

LEMMA 4.3. *Let $m=|G|$. Assume that $r\geqslant 2$ is an integer such that $(T_\psi)^r$ is ergodic and $\gcd(r,m)=1$. Then $C(T_\psi)=C((T_\psi)^r)$.*

PROOF. Assume that $\widetilde{S}\in C((T_\psi)^r)$. Since $T_\psi\in C((T_\psi)^r)$, we have $(T_\psi)^{-1}\circ\widetilde{S}\circ T_\psi\in C((T_\psi)^r)$. Since $(T_\psi)^{-1}\circ\widetilde{S}\circ T_\psi\in C((T_\psi)^r)$ is a lift of S and $(T_\psi)^r$ is ergodic, it follows by Proposition 2.2 that

$$(T_\psi)^{-1}\circ\widetilde{S}\circ T_\psi=\widetilde{S}\circ\tau_g \text{ for some } g\in G.$$

Therefore

$$(T_\psi)^{-2}\circ\widetilde{S}\circ(T_\psi)^2=(T_\psi)^{-1}\circ\widetilde{S}\circ\tau_g\circ T_\psi=(T_\psi)^{-1}\circ\widetilde{S}\circ T_\psi\circ\tau_g=\widetilde{S}\circ\tau_{g^2}$$

and, in a similar way, $(T_\psi)^{-m}\circ\widetilde{S}\circ(T_\psi)^m=\widetilde{S}\circ\tau_{g^m}=\widetilde{S}$, i.e. $\widetilde{S}\in C((T_\psi)^m)$. Let $a,b\in\mathbb{Z}$ be such that $am+br=1$. We conclude that $\widetilde{S}$ commutes with $(T_\psi)^{am+br}=T_\psi$ which completes the proof. □

PROPOSITION 4.4. *Assume that T_ψ has continuous spectrum on the orthocomplement of $L^2(X,\mathcal{B},\mu)\otimes\mathbb{1}_G$. Suppose that $r\geqslant 2$ is such that T^r is ergodic and $\gcd(m,r)=1$. Then $C(T_\psi)=C((T_\psi)^r)$.*

PROOF. Since, by assumptions, $(T_\psi)^r$ is ergodic, the assertion follows from Lemma 4.3. □

We can now formulate a general criterion concerning the validity of Sarnak's conjecture for continuous finite group extensions.

PROPOSITION 4.5. *Let $\overline{T}$ be a uniquely ergodic homeomorphism which is a continuous extension of an odometer T, measure-theoretically isomorphic to T. Assume that its (rational discrete) spectrum is determined by finitely many prime numbers. Assume that $\psi\colon X\to G$ is a cocycle with G finite,* (10) *is satisfied, and T_ψ has continuous spectrum in the orthocomplement of $L^2(X,\mathcal{B},m_X)\otimes\mathbb{1}_G$. Assume moreover that the centralizers for all normal natural factors $T_{\psi H}$ of T_ψ are G/H-trivial whenever $H\neq G$. Then, for each $f\in C(X)$ and $j\in C(G)$ of zero mean,* (1) *is satisfied for $\overline{T}_{\overline{\psi}}$ and $(f\circ\pi)\otimes j\in C(\overline{X}\times G)$ at each point.*

PROOF. Fix r,s two different prime numbers sufficiently large (so that $(T_\psi)^r$ and $(T_\psi)^s$ are ergodic). Notice that T^r is then isomorphic to T^s. Following Theorem 2.4 (applied to T^r isomorphic to T^s, both isomorphic to T) and Remark 2.5, we first will prove that if H_1,H_2 are proper normal subgroups of G then $(T_{\psi H_1})^r$ is not isomorphic $(T_{\psi H_2})^s$. For this aim, it is enough to notice is that $(T_{\psi H_2})^s$

cannot have an r-th root. Indeed, using the fact that the centralizer of $(T_{\psi H_2})^s$ is G/H_2-trivial and Proposition 4.4, if $(T_{\psi H_2})^s$ an r-th root then

$$(T_{\psi H_2})^s = ((T_{\psi H_2})^k \circ \tau_{gH_2})^r = (T_{\psi H_2})^{kr} \circ \tau_{g^r H_2}.$$

It follows that $\tau_{g^r H_2} = (T_{\psi H_2})^{s-kr}$, which is an absurd as s, r are prime $(s \neq kr)$ and T_ψ is aperiodic.

Take any $(\overline{x}, g)$. By the first part of the proof and Proposition 4.2, we obtain

$$\frac{1}{N}\sum_{n\leqslant N} \delta_{(\overline{T_{\overline{\psi}}})^{rn}\times(\overline{T_{\overline{\psi}}})^{sn}((\overline{x},g),(\overline{x},g))} \to \tilde{\tilde{\rho}}.$$

Therefore

$$\frac{1}{N}\sum_{n\leqslant N}((f\circ\pi)\otimes j)(\overline{T_{\overline{\psi}}})^{rn}\times(\overline{T_{\overline{\psi}}})^{sn}((\overline{x},g),(\overline{x},g)) \to \int((f\circ\pi)\otimes j)\cdot\overline{((f\circ\pi)\otimes j)}\,d\tilde{\tilde{\rho}}$$

$$= \int(f\otimes j)\cdot\overline{(f\otimes j)}\,d\widetilde{\mu_R} = \int_X f\cdot\overline{f\circ R}\,dm_X\cdot\int_{G\times G} j\otimes\overline{j}\,dm_G\otimes m_G = 0,$$

where the last equality follows by the assumption on j. The result follows by Theorem 1.1. □

REMARK 4.6. The assertion of Proposition 4.5 remains true if in the orthocomplement of $L^2(X, \mathcal{B}, m_X) \otimes \mathbb{1}_G$ there are finitely many rational eigenvalues (in the proof we need to exclude finitely many r, s).

4.3. Special case: 2-point extensions of odometers. We now consider the special case when $G = \mathbb{Z}/2\mathbb{Z}$. As an immediate consequence of Theorem 2.4, we obtain the following:

COROLLARY 4.7. *Let T be an odometer and let $\phi, \psi\colon X \to \mathbb{Z}/2\mathbb{Z}$ be ergodic cocyles. Then, either T_ϕ and T_ψ are isomorphic or they are relatively disjoint over T, i.e. $J^e(T_\phi, T_\psi) = \{\widetilde{(m_X)}_R : R \in C(T)\}$.*

REMARK 4.8. We give now a direct proof of Corollary 4.7. Fix $\rho \in J^e(T_\phi, T_\psi)$. We have $(T_\phi \times T_\psi, \rho) \simeq (T_{\phi\times\psi\circ R}, \kappa)$ where κ projects on m_X and $R \in C(T)$. If $\phi \times \psi \circ R$ is ergodic, it follows by Lemma 2.1 that $\kappa = m_X \otimes (m_G \otimes m_G)$, so $\rho = \widetilde{(m_X)}_R$. If $\phi \times \psi \circ R$ is not ergodic, then

$$\phi - \psi \circ R = \xi - \xi \circ T \text{ for some measurable } \xi\colon X \to \mathbb{Z}/2\mathbb{Z}.$$

It follows that T_ϕ and T_ψ are isomorphic: $R_\xi \circ T_\phi = T_\psi \circ R_\xi$.

We also have the following (cf. Proposition 4.4).

COROLLARY 4.9. *Let T be an odometer and let $\psi\colon X \to \mathbb{Z}/2\mathbb{Z}$ be ergodic. Assume that T_ψ has continuous spectrum in the orthocomplement of $L^2(X, \mathcal{B}, m_X) \otimes \mathbb{1}_{\mathbb{Z}/2\mathbb{Z}}$ and $C(T_\psi)$ is $\mathbb{Z}/2\mathbb{Z}$-trivial. Assume, moreover, that $r \neq s$ are prime numbers such that T^r and T^s are ergodic. Then $(T_\psi)^r$ and $(T_\psi)^s$ are not isomorphic.*

Now, using Corollary 4.7, the corresponding part of Proposition 4.5 takes the following form.

COROLLARY 4.10. *Let $\overline{T}$ be a uniquely ergodic homeomorphism which is a continuous extension of an odometer T, measure-theoretically isomorphic to T. Assume that its (rational discrete) spectrum is generated by finitely many prime numbers. Assume that $\psi\colon X \to \mathbb{Z}/2\mathbb{Z}$ is a cocycle, (10) is satisfied, and T_ψ has continuous* *spectrum in the orthocomplement of* $L^2(X, \mathcal{B}, m_X) \otimes \mathbb{1}_{\mathbb{Z}/2\mathbb{Z}}$. *Assume that for sufficiently large prime numbers* $r \neq s$, *the automorphisms* $(T_\psi)^r$ *and* $(T_\psi)^s$ *are not isomorphic. Then for each* $f \in C(X)$ *and* $\mathbb{1} \neq j \in \widehat{\mathbb{Z}/2\mathbb{Z}}$, (1) *is satisfied for* $\overline{T}_{\overline{\psi}}$ *and* $(f \circ \pi) \otimes j$ *at each point.*

5. Applications

5.1. Bijective substitutions. Let $\theta\colon A \to A^\lambda$ be a bijective substitution with the corresponding bijections $\sigma_i \in \mathscr{S}_r$. Let $C(\theta)$ denote the centralizer of the set $\{\sigma_i : i = 0, \dots, \lambda - 1\}$ in $\mathscr{S}_r$. Assume that $\eta \in C(\theta)$. Then η induces a map $\widetilde{\eta}$ (both on finite blocks over A and on $A^{\mathbb{Z}}$) given by

$$\widetilde{\eta}(y)[n] := \eta(y[n]) \text{ for each } n \in \mathbb{Z}.$$

We claim that $\widetilde{\eta}(X(\theta)) = X(\theta)$. Indeed, since $\eta(\sigma_i(0)) = \sigma_i(\eta(0))$, it follows that

$$\widetilde{\eta}(\theta^n(0)) = \theta^n(\eta(0))$$

and we use the transitivity of the action of the group G generated by $\sigma_0, \dots, \sigma_{\lambda-1}$ on A. Since $\widetilde{\eta}$ commutes with the shift, $\widetilde{\eta} \in C(S, X(\theta))$ (indeed, $(S, X(\theta))$ is uniquely ergodic, so $\widetilde{\eta}$ must preserve the unique measure). Now, the result from [**39**] shows that this is the only way to get non-trivial elements in the centralizer of the (measure-theoretic) dynamical system determined by a bijective substitution:

THEOREM 5.1 ([**39**]). $C(S, X(\theta)) = \{S^i \circ \widetilde{\eta} : i \in \mathbb{Z}, \eta \in C(\theta)\}$.

Suppose now that $\theta\colon G \to G^\lambda$ is a group substitution, i.e. $\theta(g) = \theta(e) \circ g$ for $g \in G$. Notice first that in each column j of the matrix for θ, we have elements $\theta(e)[j] \cdot g$. Therefore

$$\sigma_j \text{ is the left translation on } G \text{ by } \theta(e)[j]$$

and the group generated by $\theta(e)[j]$, $j = 0, \dots, \lambda - 1$, is G. It follows that the group generated by $\sigma_0, \dots, \sigma_{\lambda-1}$ is the group of all left translations on G. Its centralizer $C(\theta)$ is equal to the group of all right translations. Thus, we obtain the following consequence of Theorem 5.1:

COROLLARY 5.2. *The centralizer of the group substitutions is G-trivial.*

REMARK 5.3. The discrete part of the spectrum of the dynamical system $(S, X(\theta))$, where θ is a substitution, consists of the spectrum of the underlying odometer and a cyclic group determined by the *height* h of the substitution [**7**]. It follows that when the height is equal to 1, then the spectrum is continuous in the orthocomplement of the L^2-space of the underlying odometer. Otherwise, in this orthocomplement we have the cyclic group of eigenvalues generated by $e^{2\pi i/h}$.

We are now ready to show that Sarnak's conjecture holds for dynamical systems given by bijective substitutions.

THEOREM 5.4. *For each bijective substitution* $\theta\colon A \to A^r$, *each function* $F \in C(X(\theta))$, *each aperiodic multiplicative function* $\boldsymbol{u}\colon \mathbb{N} \to \mathbb{C}$, $|\boldsymbol{u}| \leqslant 1$, *and each* $y \in X(\theta)$, *we have*

$$\frac{1}{N} \sum_{n \leqslant N} F(S^n y)\boldsymbol{u}(n) \to 0 \text{ when } N \to \infty. \tag{12}$$

In particular, each topological dynamical system determined by a bijective substitution satisfies Sarnak's conjecture.

PROOF. It follows by Lemma 3.19 that it is enough to prove (12) for the dynamical system $(S, X(\overline{\theta}))$ corresponding to the group cover substitution $\overline{\theta}$ of θ. Moreover, in view of Lemma 3.8, we can study instead its topologically isomorphic model $(S_\varphi, X(\widehat{x}) \times G)$.

Fix $f \otimes j$ with $f \in C(X(\widehat{x}))$, where $j \in C(G)$, $\int j\, dm_G = 0$. In view of Theorem 1.1, Lemma 3.17, Corollary 5.2 and Proposition 4.5, Remark 4.6 and Remark 5.3, for each $(y, g) \in X(\widehat{x}) \times G$, we have

$$\frac{1}{N} \sum_{n \leqslant N} (f \otimes j)((S_\varphi)^n(y, g))\boldsymbol{u}(n) \to 0 \tag{13}$$

for each multiplicative function $\boldsymbol{u}$, $|\boldsymbol{u}| \leqslant 1$. If we now fix $\boldsymbol{u}$, then we have the relevant convergence (against this fixed $\boldsymbol{u}$) for a linearly dense set of functions in $C(X(\widehat{x}) \times G)$, hence for all functions in $C(X(\widehat{x}) \times G)$ and the result follows. □

REMARK 5.5. Let $j := \mathbb{1}_G$. Using (13), for each $y \in X(\widehat{x})$, we have

$$\frac{1}{N} \sum_{n \leqslant N} f(S^n y)\boldsymbol{u}(n) \to 0$$

for each aperiodic multiplicative function $\boldsymbol{u}$, $|\boldsymbol{u}| \leqslant 1$. This can be also proved more directly. Notice that for each odometer (T, X) we have (1) true with $\boldsymbol{\mu}$ replaced by $\boldsymbol{u}$ since each finite system enjoys this property and (T, X) is a topological inverse limit of such systems. If $(\overline{T}, \overline{X})$ is a uniquely ergodic topological extension of (T, X), measure-theoretically isomorphic to (T, X, m_X), we can apply Lemma 7 and Proposition 3 in [**1**] to lift the orthogonality condition (12) from the odometer to $(\overline{T}, \overline{X})$.

REMARK 5.6. The proof of Sarnak's conjecture also gives the following: whenever $(S, X(\theta))$ is a subshift given by a bijective substitution, for each ergodic powers S^r and S^s, each point $(y, z) \in X(\theta) \times X(\theta)$ is generic (for an ergodic measure).

REMARK 5.7. In part 5 of [**12**], Drmota explains how the method developed by Mauduit and Rivat in the proof of Theorem 1 of [**43**] can be applied to any bijective substitution (Definition 4.1 in [**12**] of "invertible λ-automatic sequence" corresponds to our bijective substitution of constant length λ). In particular, Theorem 5.5 of [**12**] says that any bijective substitution θ on the alphabet A satisfies a Prime Number Theorem, i.e. that for any $a \in A$, $\lim_{N\to\infty} \frac{1}{\pi(N)} |\{1 \leqslant p < N : p \text{ is prime}, x_\theta[p] = a\}|$ exists (here x_θ stands for a fixed point given by θ). In [**43**], the proof of Theorem 2 can be deduced from the proof of Theorem 1 just by replacing the classical Vaughan identity by the similar result for the Möbius function (see (13.39) and (13.40) in [**26**]). It follows from this remark that the arguments given by Drmota in [**12**] in order to generalize Theorem 1 from [**43**] to any bijective substitution is still valid to generalize Theorem 2 from [**43**] to any bijective substitution. This means that if θ is a bijective substitution and $J \colon A \to \mathbb{C}$ then

$$\frac{1}{N} \sum_{n \leqslant N} J(x_\theta[n])\boldsymbol{\mu}(n) \to 0.$$

REMARK 5.8. In [**6**], it is noted that Theorem 1.1 is applicable to sequences produced by a subclass of substitutions of constant length for which the spectral types are of intermediate dimension by applying some results from [**25**].

5.2. Regular Morse sequences and the Rudin-Shapiro case. In [**36**], it has been proved that the centralizer of the dynamical systems given by so called *regular Morse sequences* [**35**] $x = b^0 \times b^1 \times \dots$ ($b^t \in \{0,1\}^{\lambda_t}$, $t \geqslant 0$) is $\mathbb{Z}/2\mathbb{Z}$-trivial.

COROLLARY 5.9. *If $x = b^0 \times b^1 \times \dots$ is a regular Morse sequence for which the set $\{p : p$ is prime and $p|\lambda_t$ for some $t\}$ is finite. Then* (12) *holds in the dynamical system given by x.*

PROOF. The result follows from Corollaries 4.9 and 4.10 and the proof of Theorem 5.4. □

In [**38**], the Rudin-Shapiro type sequences are considered. These are 0-1-sequences $x \in \{0,1\}^{\mathbb{N}}$ such that $x[n]$ is equal to the mod 2 frequency of the block $1 * \dots * 1$ (with fixed number of $*$) in the block given by the binary expansion on n.[6] As shown in [**38**], the corresponding subshift is given by a Toeplitz type $\mathbb{Z}/2\mathbb{Z}$-extension of the dyadic odometer, and the whole method applies.

COROLLARY 5.10. *If x is a Rudin-Shapiro type sequence then* (12) *holds in $(S, X(x))$. In particular, Sarnak's conjecture holds in the dynamical system given by the classical Rudin-Shapiro sequence.*

PROOF. Since $(S, X(x))$ has the Lebesgue component of multiplicity 2^k in the spectrum in the orthocomplement of the space generated by eigenfunctions [**38**], it follows that its sth and rth power also have Lebesgue components in the spectrum, of multiplicity $s2^k$ and $r2^k$, respectively. Thus, these powers cannot be isomorphic, unless $s = r$. The result follows from Corollary 4.10. □

6. Spectral approach and other methods

Let (S, X) with $X \subset A^{\mathbb{Z}}$ be a subshift over a finite alphabet A with $|A| = r \geqslant 2$.

6.1. Some remarks.

LEMMA 6.1. *Suppose that* (1) *holds for arbitrary $x \in X$, for each function $f = \mathbb{1}_B$, where $B \in A^k$ is a block of finite length ($k \geqslant 1$ is arbitrary) that appears on X. Then Sarnak's conjecture holds for (S, X).*

PROOF. It suffices to show (1) for a linearly dense family of functions in $C(X)$: e.g. functions which depend on a finite number of coordinates. The space of (continuous) functions depending on coordinates $[-k, k]$ in the full shift has dimension r^{2k+1}, which is at the same time the number of possible blocks of length $2k+1$. In a similar way, for a subshift, we just need to count the number of distinct $(2k+1)$-blocks appearing on X. Moreover, the family of their characteristic functions is linearly independent. □

REMARK 6.2. There are other choices of finite families of functions than those in Lemma 6.1 which also yield the validity of Sarnak's conjecture. For example, when $r = 2$ we can use the so called Walsh basis: for each $K \geqslant 1$, we consider the characters of the group $\{0,1\}^{2K+1}$: $f_C(x) = (-1)^{\sum_{i \in C} x[i]}$ for $C \subset \{-K, \dots, K\}$ and $x \in X \subset A^{\mathbb{Z}}$.

[6] We can also consider $1 * \dots * 0$.

REMARK 6.3. In [**4**, **19**] the convergence in (1) is proved at any point for $f(y) = (-1)^{y[0]}$ ($f = f_{\{0\}}$ in the notation from Remark 6.2) for Kakutani sequences.[7] A natural question arises whether this is sufficient to obtain Sarnak's conjecture for the corresponding dynamical system. In general, it does not seem to be automatic that (1) for functions depending on one coordinate implies (1) for functions depending on more coordinates. E.g., in [**1**], where Sarnak's conjecture is proved for the 0-1-subshift generated by the Thue-Morse sequence, (1) for $f(y) = (-1)^{y[0]}$ is proved by completely different methods than for continuous functions invariant under the map $y \mapsto \widetilde{y}$, where $\widetilde{y}[n] = 1 - y[n]$.[8] We note that the method from Corollary 4.7 does not apply to Kakutani systems since their centralizer can be uncountable: there are Kakutani sequences for which the corresponding dynamical systems are rigid [**37**]. However, in Section 6.2 we provide an argument which in Section 6.3 will be used to show that [**4**, **19**] yield Sarnak's conjecture for the dynamical systems given by Kakutani sequences.

REMARK 6.4. We have already shown that Sarnak's conjecture holds for the dynamical system given by the Rudin-Shapiro sequence, see Corollary 5.10. Recall also that in this case (1) was shown earlier in [**43**] for $f(y) = (-1)^{y[0]}$ (at any point). Here the situation is more delicate if we want to apply the method from Section 6.2: we need more functions, see Section 6.3 for more details.

6.2. Spectral approach.

LEMMA 6.5. *Assume that T is a uniquely ergodic homeomorphism of a compact metric space X. Denote the unique T-invariant measure by μ. Assume that the unitary operator $U_T\colon L^2(X,\mathcal{B},\mu) \to L^2(X,\mathcal{B},\mu)$, $U_T g := g \circ T$, has simple spectrum. Assume that the maximal spectral type of U_T is realized by $F \in C(X)$. If F satisfies* (1) *at each point $x \in X$ then (T, X) satisfies Sarnak's conjecture.*

PROOF. Observe first that if F satisfies (1) at each point then the same is true for each function $p(U_T)F$ of F (where $p(z) = \sum_{\ell=-K}^{K} a_\ell z^\ell$ is a trigonometric polynomial). By the simplicity of the spectrum of U_T, the set of functions of the form $p(U_T)F$ is dense in $L^2(X,\mathcal{B},\mu)$. We now repeat the argument from Lemma 7 in [**1**]. Fix $G \in C(X)$, $x \in X$ and $\varepsilon > 0$. Find a trigonometric polynomial p so that $\|p(U_T)F - G\|_2 < \varepsilon$. Let N_0 be such that for $N \geqslant N_0$, $|\frac{1}{N}\sum_{n\leqslant N} p(U_T)F(T^n x)\boldsymbol{\mu}(n)| < \varepsilon$ for all $N > N_0$. Then (since T is uniquely ergodic and $|\boldsymbol{\mu}| \leqslant 1$)

$$
\begin{aligned}
&\left|\frac{1}{N}\sum_{n\leqslant N} G(T^n x)\boldsymbol{\mu}(n)\right| \\
&\leqslant \frac{1}{N}\sum_{n\leqslant N} |(G - p(U_T)F)(T^n x)||\boldsymbol{\mu}(n)| + \left|\frac{1}{N}\sum_{n\leqslant N} (p(U_T)F)(T^n x)\boldsymbol{\mu}(n)\right| \\
&\leqslant \frac{1}{N}\sum_{n\leqslant N} |(G - p(U_T)F)(T^n x)| + \varepsilon \to \|G - p(U_T)F\|_1 + \varepsilon
\end{aligned}
$$

when $N \to \infty$. Since $\|G - p(U_T)F\|_1 \leqslant \|G - p(U_T)F\|_2 < \varepsilon$, the result follows. □

[7]The uniformity of estimates in these papers yields indeed (1) at any point $y \in X(x)$).

[8]E.g. $g(y) = (-1)^{y[0]+y[1]}$ is invariant under this map; notice that $g = f_{\{0,1\}}$ from Remark 6.2.

REMARK 6.6.

(A) The assertion of Lemma 6.5 remains true if we take any arithmetic function $\boldsymbol{u}\colon \mathbb{N} \to \mathbb{C}$, $|\boldsymbol{u}| \leqslant 1$, instead of $\boldsymbol{\mu}$ (both in (1) and in Sarnak's conjecture). The proof is the same.

(B) Frączek [**16**] showed that for each automorphism T on $(X, \mathcal{B}, \mu)$, where X is a compact metric space, the maximal spectral type of U_T is always realized by a continuous function. However, in order to prove Sarnak's conjecture using Lemma 6.5, we look for *natural* continuous functions realizing the maximal spectral type for which we can show that (1) holds.

We would like to mention also an open problem raised by Thouvenot in the 1980th whether each ergodic dynamical system has L^1-simple spectrum, i.e., for some $f \in L^1(X, \mathcal{B}, \mu)$, we have $\operatorname{span}\{f \circ T^k : k \in \mathbb{Z}\}$ dense in $L^1(X, \mathcal{B}, \mu)$. If the answer to Thouvenot's problem is positive and the L^1-simplicity can be realized by a continuous function f in each uniquely ergodic system, then (cf. the proof of Lemma 6.5) to prove Sarnak's conjecture we need to check (1) for f (at each point).

(C) Suppose that the continuous and discrete part of the maximal spectral type of U_T are realized by $f \in C(X)$ and $g \in C(X)$, respectively. Then, by elementary spectral theory, $F = f + g \in C(X)$ realizes the maximal spectral type of U_T and, clearly, it suffices to check that (1) holds both for f and g (at each point) to see that it holds for F (at each point).

(D) Lemma 6.5 has a natural extension to uniquely ergodic homeomorphisms T such that U_T has non-trivial multiplicity. All we need to know is that $L^2(X, \mathcal{B}, \mu)$ has a decomposition into cyclic spaces: $L^2(X, \mathcal{B}, \mu) = \bigoplus_{k \geqslant 1} \mathbb{Z}(f_k)$ with $f_k \in C(X)$ and check (1) for these generators.[9] We will find such functions in the next section in case of the dynamical systems given by the Rudin-Shapiro type sequences.[10]

[9] Recall however that it is open whether for an arbitrary automorphism T on $(X, \mathcal{B}, \mu)$, where X is a compact metric space there are continuous functions f_k, $k \geqslant 1$ such that $L^2(X, \mathcal{B}, \mu) = \bigoplus_{k \geqslant 1} \mathbb{Z}(f_k)$ and $\sigma_{f_{k+1}} \ll \sigma_{f_k}$, $k \geqslant 1$, see e.g. [**30**] for more details.

[10] Recall that in the general case of dynamical systems given by the Rudin-Shapiro type sequences, this multiplicity is of the form 2^k, $k \geqslant 1$ [**38**].

6.3. Applications.
Generalized Morse sequences over $A = \{0,1\}$.

PROPOSITION 6.7. *Let x be a generalized Morse sequence over $A = \{0,1\}$. Then Sarnak's conjecture holds for $(S, X(x))$ if and only if* (1) *holds (at each point) for* $f(y) = (-1)^{y[0]}$.

PROOF. Recall that $(S, X(x))$ is uniquely ergodic [**31**] (with the unique invariant measure μ_x) and has simple spectrum [**34**]. As proved in [**34**] (see also [**21**]) $f(y) = (-1)^{y[0]}$ realizes the continuous part of the maximal spectral type of U_S (on $L^2(X(x), \mu_x)$). Moreover, the discrete part is given by the equicontinuous factor of $(S, X(x))$, which is the odometer determined by λ_t, $t \geqslant 0$. It follows that the eigenfunctions g_i, $i \geqslant 1$, are continuous. If $g = \sum_{i\geqslant 1} a_i g_i$, each $a_i \neq 0$, $\sum_{i\geqslant 1} |a_i| < +\infty$, then g is a continuous function realizing the discrete part of the maximal spectral type of U_S. Since each odometer is a topological inverse limit of systems defined on finitely many points, and for finite systems Sarnak's conjecture holds because of the PNT in arithmetic progressions, therefore g satisifes (1). Thus, in view of Lemma 6.5 and Remark 6.6 (C), it suffices to prove (1) for f to obtain the validity of Sarnak's conjecture. □

COROLLARY 6.8. *Sarnak's conjecture holds for the dynamical systems given by Kakutani sequences.*

PROOF. In view of the above proposition, it suffices to prove (1) for f which was done in [**4**, **19**] (cf. Remark 6.3). □

Rudin-Shapiro type sequences. Recall that the classical Rudin-Sharpiro sequence $x \in \{0,1\}^{\mathbb{N}}$ is defined in the following way:

- take the fixed point of the substitution $a \mapsto ab$, $b \mapsto ac$, $c \mapsto db$ and $d \mapsto dc$,
- use the code $a, b \mapsto 0$ and $c, d \mapsto 1$ to pass to the space of 0-1-sequences (the map arising from this code yields a topological isomorphism of the relevant subshifts on four and two letters).

The multiplicity of the corresponding dynamical system on the continuous part of the spectrum is equal to 2 [**38**, **48**]. It follows from [**38**] and Remark 6.6 (D) that, in order to obtain Sarnak's conjecture for the corresponding subshift, we need to check (1) for two continuous functions $f \cdot \mathbb{1}_{D_0^2}$ and $f \cdot \mathbb{1}_{D_1^2}$ (cf. Remark 3.12). It follows immediately from the definition of D_0^2 and the recognizability of substitutions [**46**] that $\mathbb{1}_{D_0^2}$ is a continuous function depending on a finite number of coordinates. Therefore, to obtain Sarnak's conjecture for the subshift given by the Rudin-Shapiro sequence, we would have to check (1) for the elements of the Walsh basis of order $4M$, where M is the constant of recognizability, see Remark 3.14. Recall that (1) was already shown for f in [**43**, **51**]. Notice that this approach to prove Sarnak's conjecture is completely different from the one presented in the preceding sections (cf. Remark 5.10). The above applies to all Rudin-Shapiro type sequences.

6.4. Comparison with results of Veech [52]. In the recent preprint [**52**], Veech considers a class of systems for which Sarnak's conjecture holds. We will now briefly present his work and then compare it with our results. Assume that $\lambda_n \geqslant 2$

for $n \geqslant 0$, then set $n_0 := 1$ and $n_t := \prod_{k=0}^{t-1} \lambda_k$, $t \geqslant 1$ and define

$$X := \operatorname{liminv}_{t\to\infty} \mathbb{Z}/n_t\mathbb{Z} = \{x = (x_t) : 0 \leqslant x_t < n_t, x_{t+1} = x_t \bmod n_t, t \geqslant 1\}.$$

This is a compact, Abelian, monothetic group on which we consider $Tx = x + \theta$ with $\theta = (1, 1, \ldots)$. It is not hard to see that the systems obtained this way are naturally isomorphic to the odometers considered in Section 3.1. The sequence of towers $\mathcal{D}^t$, $t \geqslant 1$, in the new coordinates is determined by

$$D_0^t := \{x \in X : x_t = 0\},$$

and we obtain pairwise disjoint sets $D_0^t, TD_0^t, \ldots T^{n_t-1}D_0^t$ with $\bigcup_{j=0}^{n_t-1} T^j D_0^t = X$. Then define

$$\tau(x) := \min\{t \geqslant 1 : x_t \neq n_t - 1\}.$$

We have $\lim_{x\to-\theta} \tau(x) = \infty$ and τ is continuous on $X\backslash\{-\theta\}$.

Let K be a compact group and take $(\Psi(t))_{t\geqslant 1} \subset K$. Set

$$f(x) := \Psi(\tau(x)).$$

Then Ψ is locally constant on $X\backslash\{-\theta\}$ and $f \in C(X\backslash\{-\theta\}, K)$. There are some assumptions on the sequence Ψ made in [**52**]:

(i) $\lim_{t\to\infty} \Psi(t)$ does not exist,

(ii) $\{\Psi(t) : t \geqslant 1\}$ generates a dense subgroup of K, and so does the set

$$\{\Psi(t)\Psi(u)^{-1} : t, u \geqslant 1\}$$

,

(iii) $(\Psi(t))_{t\geqslant 1}$ is recurrent, that is, every initial block of Ψ repeats infinitely often.

As noticed in [**52**], if $K = \mathbb{Z}/2\mathbb{Z}$, the conditions (ii) and (iii) are not necessary.

Let $M_\Psi \subset K^{\mathbb{Z}}$ be the closure of all sequences $(f(x + n\theta)_{n\in\mathbb{Z}}$ for $x \in X\backslash\mathbb{Z}\theta$. On M_Ψ, we consider the usual shift S. Let

$$m\colon M_\Psi \to K \text{ be given by } m(y) = y[0].$$

Finally, let $S_m\colon M_\Psi \times K \to M_\Psi \times K$ be the skew product defined as $S_m(y, k) = (Sy, m(y)k)$. Then S_m is a homeomorphism of $M_\Psi \times K$.

THEOREM 6.9 ([**52**]). *Suppose additionally that the set* $\{\lambda_t : t \geqslant 0\}$ *is finite. Then, under the above assumptions,* $(S_m, M_\Psi \times K)$ *satisfies Sarnak's conjecture.*

REMARK 6.10. It is not hard to see that in the language of [**9**], the function f is a semicocycle over an odometer (that is, a function continuous on a residual subset of an odometer). It follows from [**9**] that the dynamical system given by (S, M_Ψ) is a Toeplitz dynamical system (cf. Section 3.1). The system is regular [**9**], hence uniquely ergodic and measure-theoretically isomorphic to the odometer (T, X).

Notice that if $x_t = n_t - 1$ then also $x_k = n_k - 1$ for $1 \leqslant k \leqslant t - 1$. It follows immediately that τ is constant on each D_i^t, $0 \leqslant i \leqslant n_t - 2$. Therefore, the cocycle $f\colon X \to K$ defined above is a Morse cocycle (cf. Section 3.1). It has the following additional property:

$$f \text{ is constant on } \bigcup_{1\leqslant j\leqslant \lambda_{t+1}-1} D_{jn_t-1}^{t+1}. \tag{14}$$

Condition (14) yields the class of Morse sequences $x = b^0 \times b^1 \times \ldots$, where each block b_t, $t \geqslant 0$, is of the form $b^t = ek\ldots k^{|b^t|-1}$ (cf. Remark 3.9). In particular, if $K = \mathbb{Z}/2\mathbb{Z}$, we have $x = b^0 \times b^1 \times \ldots$, where $b^t = 0\ldots 0$, $b^t = 01\ldots 01$ or

$b^t = 010\dots 10$, $t \geqslant 0$. Notice that Kakutani sequences are of this form. It follows that for bijective substitutions θ over the alphabet $\{0,1\}$, Theorem 6.9 can be applied only if $\theta(0) = 01\dots 01$.

Notice also that $(S_m, M_\Psi \times K)$ corresponds to $(S_\varphi, X(\widehat{x}) \times G)$ defined in Section 3.1, which is, in turn, topologically isomorphic to $(S, X(x))$.

It follows that Theorem 6.9 is a significant extension of the validity of Sarnak's conjecture for Kakutani systems (cf. Corollary 6.8) to general compact groups.

References

[1] El Houcein El Abdalaoui, Stanisław Kasjan, and Mariusz Lemańczyk, *0-1 sequences of the Thue-Morse type and Sarnak's conjecture*, Proc. Amer. Math. Soc. **144** (2016), no. 1, 161–176, DOI 10.1090/proc/12683. MR3415586

[2] El Houcein El Abdalaoui, Mariusz Lemańczyk, and Thierry de la Rue, *On spectral disjointness of powers for rank-one transformations and Möbius orthogonality*, J. Funct. Anal. **266** (2014), no. 1, 284–317, DOI 10.1016/j.jfa.2013.09.005. MR3121731

[3] J. R. Baxter, *A class of ergodic transformations having simple spectrum*, Proc. Amer. Math. Soc. **27** (1971), 275–279. MR0276440

[4] J. Bourgain, *Möbius-Walsh correlation bounds and an estimate of Mauduit and Rivat*, J. Anal. Math. **119** (2013), 147–163, DOI 10.1007/s11854-013-0005-2. MR3043150

[5] ———, *On the correlation of the Moebius function with rank-one systems*, J. Anal. Math., 120 (2013), pp. 105–130.

[6] J. Bourgain, P. Sarnak, and T. Ziegler, *Disjointness of Moebius from horocycle flows*, From Fourier analysis and number theory to Radon transforms and geometry, Dev. Math., vol. 28, Springer, New York, 2013, pp. 67–83, DOI 10.1007/978-1-4614-4075-8_5. MR2986954

[7] F. M. Dekking, *The spectrum of dynamical systems arising from substitutions of constant length*, Z. Wahrscheinlichkeitstheorie und Verw. Gebiete **41** (1977/78), no. 3, 221–239. MR0461470

[8] Andrés del Junco, *Transformations with discrete spectrum are stacking transformations*, Canad. J. Math. **28** (1976), no. 4, 836–839. MR0414822

[9] Tomasz Downarowicz, *Survey of odometers and Toeplitz flows*, Algebraic and topological dynamics, Contemp. Math., vol. 385, Amer. Math. Soc., Providence, RI, 2005, pp. 7–37, DOI 10.1090/conm/385/07188. MR2180227

[10] T. Downarowicz and S. Kasjan, *Odometers and Toeplitz subshifts revisited in the context of Sarnak's conjecture*. To appear in Studia Math.

[11] T. Downarowicz, J. Kwiatkowski, and Y. Lacroix, *Spectral isomorphisms of Morse flows*, Fund. Math. **163** (2000), no. 3, 193–213. MR1758324

[12] Michael Drmota, *Subsequences of automatic sequences and uniform distribution*, Uniform distribution and quasi-Monte Carlo methods, Radon Ser. Comput. Appl. Math., vol. 15, De Gruyter, Berlin, 2014, pp. 87–104. MR3287361

[13] Manfred Einsiedler and Thomas Ward, *Ergodic theory with a view towards number theory*, Graduate Texts in Mathematics, vol. 259, Springer-Verlag London, Ltd., London, 2011. MR2723325

[14] Sébastien Ferenczi, *Systems of finite rank*, Colloq. Math. **73** (1997), no. 1, 35–65. MR1436950

[15] S. Ferenczi and C. Mauduit, *On Sarnak's conjecture and Veech's question for interval exchanges*. To appear in J. d'Analyse Math.

[16] Krzysztof M. Frączek, *On a function that realizes the maximal spectral type*, Studia Math. **124** (1997), no. 1, 1–7. MR1444806

[17] H. Furstenberg, *Recurrence in ergodic theory and combinatorial number theory*, Princeton University Press, Princeton, N.J., 1981. M. B. Porter Lectures. MR603625

[18] Eli Glasner, *Ergodic theory via joinings*, Mathematical Surveys and Monographs, vol. 101, American Mathematical Society, Providence, RI, 2003. MR1958753

[19] Ben Green, *On (not) computing the Möbius function using bounded depth circuits*, Combin. Probab. Comput. **21** (2012), no. 6, 942–951, DOI 10.1017/S0963548312000284. MR2981162

[20] Ben Green and Terence Tao, *The Möbius function is strongly orthogonal to nilsequences*, Ann. of Math. (2) **175** (2012), no. 2, 541–566, DOI 10.4007/annals.2012.175.2.3. MR2877066

[21] Mélanie Guenais, *Morse cocycles and simple Lebesgue spectrum*, Ergodic Theory Dynam. Systems **19** (1999), no. 2, 437–446, DOI 10.1017/S0143385799126579. MR1685402

[22] Frank Hahn and William Parry, *Some characteristic properties of dynamical systems with quasi-discrete spectra*, Math. Systems Theory **2** (1968), 179–190. MR0230877

[23] A. J. Harper, *A different proof of a finite version of Vinogradov's bilinear sum inequality.* `https://www.dpmms.cam.ac.uk/~ajh228/FiniteBilinearNotes.pdf`.

[24] Joseph L. Herning, *Spectrum and Factors of Substitution Dynamical Systems*, ProQuest LLC, Ann Arbor, MI, 2013. Thesis (Ph.D.)–The George Washington University. MR3167382

[25] Michael Hochman, *Geometric rigidity of $\times m$ invariant measures*, J. Eur. Math. Soc. (JEMS) **14** (2012), no. 5, 1539–1563, DOI 10.4171/JEMS/340. MR2966659

[26] Henryk Iwaniec and Emmanuel Kowalski, *Analytic number theory*, American Mathematical Society Colloquium Publications, vol. 53, American Mathematical Society, Providence, RI, 2004. MR2061214

[27] Konrad Jacobs and Michael Keane, *$0-1$-sequences of Toeplitz type*, Z. Wahrscheinlichkeitstheorie und Verw. Gebiete **13** (1969), 123–131. MR0255766

[28] Teturo Kamae, *A topological invariant of substitution minimal sets*, J. Math. Soc. Japan **24** (1972), 285–306. MR0293611

[29] I. Kátai, *A remark on a theorem of H. Daboussi*, Acta Math. Hungar. **47** (1986), no. 1-2, 223–225, DOI 10.1007/BF01949145. MR836415

[30] A. Katok and M. Lemańczyk, *Some new cases of realization of spectral multiplicity function for ergodic transformations*, Fund. Math. **206** (2009), 185–215, DOI 10.4064/fm206-0-11. MR2576267

[31] M. Keane, *Generalized Morse sequences*, Z. Wahrscheinlichkeitstheorie und Verw. Gebiete **10** (1968), 335–353. MR0239047

[32] J. Kułaga-Przymus and M. Lemańczyk, *Ergodic sequences along which Sarnak's conjecture holds for uniquely ergodic flows.* In preparation.

[33] J. Kułaga-Przymus and M. Lemańczyk, *The Möbius function and continuous extensions of rotations*, Monatsh. Math. **178** (2015), no. 4, 553–582, DOI 10.1007/s00605-015-0808-6. MR3422903

[34] Jan Kwiatkowski, *Spectral isomorphism of Morse dynamical systems* (English, with Russian summary), Bull. Acad. Polon. Sci. Sér. Sci. Math. **29** (1981), no. 3-4, 105–114. MR638749

[35] ———, *Isomorphism of regular Morse dynamical systems*, Studia Math., 72 (1982), pp. 59–89.

[36] Mariusz Lemańczyk, *The centralizer of Morse shifts*, Ann. Sci. Univ. Clermont-Ferrand II Probab. Appl. **4** (1985), 43–56. MR826355

[37] ———, *The rank of regular Morse dynamical systems*, Z. Wahrsch. Verw. Gebiete, 70 (1985), pp. 33–48.

[38] ———, *Toeplitz Z_2-extensions*, Ann. Inst. H. Poincaré Probab. Statist., 24 (1988), pp. 1–43.

[39] Mariusz Lemańczyk and Mieczysław K. Mentzen, *On metric properties of substitutions*, Compositio Math. **65** (1988), no. 3, 241–263. MR932072

[40] Jianya Liu and Peter Sarnak, *The Möbius function and distal flows*, Duke Math. J. **164** (2015), no. 7, 1353–1399, DOI 10.1215/00127094-2916213. MR3347317

[41] J. Mathew and M. G. Nadkarni, *A measure preserving transformation whose spectrum has Lebesgue component of multiplicity two*, Bull. London Math. Soc. **16** (1984), no. 4, 402–406, DOI 10.1112/blms/16.4.402. MR749448

[42] Christian Mauduit and Joël Rivat, *Sur un problème de Gelfond: la somme des chiffres des nombres premiers* (French, with English and French summaries), Ann. of Math. (2) **171** (2010), no. 3, 1591–1646, DOI 10.4007/annals.2010.171.1591. MR2680394

[43] ———, *Prime numbers along Rudin–Shapiro sequences*, J. Eur. Math. Soc. (JEMS), 17 (2015), pp. 2595–2642.

[44] Mieczysław K. Mentzen, *Invariant sub-σ-algebras for substitutions of constant length*, Studia Math. **92** (1989), no. 3, 257–273. MR985556

[45] ———, *Ergodic properties of group extensions of dynamical systems with discrete spectra*, Studia Math., 101 (1991), pp. 19–31.

[46] Brigitte Mossé, *Puissances de mots et reconnaissabilité des points fixes d'une substitution* (French), Theoret. Comput. Sci. **99** (1992), no. 2, 327–334, DOI 10.1016/0304-3975(92)90357-L. MR1168468

[47] William Parry, *Topics in ergodic theory*, Cambridge Tracts in Mathematics, vol. 75, Cambridge University Press, Cambridge, 2004. Reprint of the 1981 original. MR2140546

[48] Martine Queffélec, *Substitution dynamical systems—spectral analysis*, 2nd ed., Lecture Notes in Mathematics, vol. 1294, Springer-Verlag, Berlin, 2010. MR2590264

[49] E. A. Robinson, Jr., *Spectral multiplicity for nonabelian Morse sequences*, in Dynamical systems (College Park, MD, 1986–87), vol. 1342 of Lecture Notes in Math., Springer, Berlin, 1988, pp. 645–652.

[50] P. Sarnak, *Three lectures on the Möbius function, randomness and dynamics*. http://publications.ias.edu/sarnak/.

[51] T. Tao. http://mathoverflow.net/questions/97261/mobius-randomness-of-the-rudin-shapiro-sequence.

[52] W. A. Veech, *Möbius orthogonality for generalized Morse-Kakutani flows*. Preprint.

[53] ———, *Strict ergodicity in zero dimensional dynamical systems and the Kronecker-Weyl theorem* mod 2, Trans. Amer. Math. Soc., 140 (1969), pp. 1–33.

[54] ———, *Ergodic theory and uniform distribution*, in Journées Arithmétiques de Luminy (Colloq. Internat. CNRS, Centre Univ. Luminy, Luminy, 1978), vol. 61 of Astérisque, Soc. Math. France, Paris, 1979, pp. 223–234.

Institut de Mathématiques de Marseille, CNRS - UMR 7373, Case 907 - 163, av. de Luminy, F13288 Marseille Cedex 9, France
E-mail address: ssferenczi@gmail.com

Institute of Mathematics, Polish Academy of Sciences, Śniadeckich 8, 00-956 Warszawa, Poland — and — Faculty of Mathematics and Computer Science, Nicolaus Copernicus University, Chopina 12/18, 87-100 Toruń, Poland
E-mail address: joanna.kulaga@gmail.com

Faculty of Mathematics and Computer Science, Nicolaus Copernicus University, Chopina 12/18, 87-100 Toruń, Poland
E-mail address: mlem@mat.umk.pl

Institut de Mathématiques de Marseille, CNRS - UMR 7373, Case 907 - 163, av. de Luminy, F13288 Marseille Cedex 9, France
E-mail address: mauduit@iml.univ-mrs.fr

Contemporary Mathematics
Volume **678**, 2016
http://dx.doi.org/10.1090/conm/678/13646

Subshifts with slowly growing numbers of follower sets

Thomas French, Nic Ormes, and Ronnie Pavlov

ABSTRACT. For any subshift, define $F_X(n)$ to be the collection of distinct follower sets of words of length n in X. Based on a similar result proved in a work of Ormes and Pavlov, we conjecture that if there exists an n for which $|F_X(n)| \leq n$, then X is sofic. In this paper, we prove several results related to this conjecture, including verifying it for $n \leq 3$, proving that the conjecture is true for a large class of coded subshifts, and showing that if there exists n for which $|F_X(n)| \leq \log_2(n+1)$, then X is sofic.

1. Introduction

Let X be a subshift, i.e. a closed, shift-invariant subset of $A^{\mathbb{Z}}$ where A is some finite set. In this paper, we consider *follower sets* for words w appearing in X. By the follower set of w we mean the set of all one-sided infinite sequences s which may follow w in some point of X; see Section 2 for a formal definition. It is well known that the number of distinct follower sets in X is finite if and only if the subshift X is sofic [**2**]. In this paper, we consider the question of whether a sufficiently slow growth rate in the number of distinct follower sets for words of length n in X implies that X is a sofic subshift.

More specifically, let $F_X(n)$ denote the set of distinct follower sets in X for words of length n. We make the following conjecture:

CONJECTURE 1.1. *For a subshift X, if there exists n for which $|F_X(n)| \leq n$, then X is sofic.*

We are unable to prove this conjecture presently, but prove some supporting results in this paper. Firstly, we prove that if $|F_X(n)| \leq \log_2(n+1)$ for some $n \geq 1$, then X is sofic (Theorem 4.7). We prove a version of the conjecture relating to $|\bigcup_{\ell \leq n} F_X(n)|$ rather than $|F_X(n)|$ (Theorem 4.6). We also prove Conjecture 1.1 for $n = 1, 2, 3$ (Theorems 4.5, 4.9, and 4.10).

Conjecture 1.1 is motivated by a number of results, among them the following classical theorem of Morse and Hedlund.

THEOREM 1.2 ([**3**]). *For a subshift X, if there exists an n such that the number of words of length n is less than or equal to n, then X is a finite collection of periodic points.*

2010 *Mathematics Subject Classification.* Primary 37B10; Secondary 37B05.

An equivalent formulation is that a bound of n on the number of words of length n implies a uniform bound on the number of words of length n. With this phrasing, Conjecture 1.1 is equivalent to a version of Theorem 1.2 with "words of length n" replaced by "follower sets of words of length n".

Another motivation is recent work of the three authors. In [**1**], building on an example of Delacourt, the first author showed that even when bounded, the sequence $\{|F_X(n)|\}$ can exhibit some surprising behavior. In particular, he showed that while the sequence $\{|F_X(n)|\}$ is always eventually periodic, it is not necessarily eventually constant. In fact, the gaps between consecutive terms in the periodic portion can be arbitrarily prescribed. Even more closely related, the second and third authors proved in [**4**] that Conjecture 1.1 holds if follower sets are replaced by so-called *extender sets*. For a word w appearing in X, the extender set of w is the set of all pairs (p, s) of a left-infinite sequence p and a right-infinite sequence s such that the concatenation pws forms a legal point in X. Let $E_X(n)$ denote the set of distinct extender sets for words of length n in X.

THEOREM 1.3 ([**4**]). *Let X be a subshift. The following are equivalent.*

(1) *X is sofic*
(2) *the sequence $\{|E_X(n)|\}$ is uniformly bounded*
(3) *for some $n \geq 1$, $E_X(n) \leq n$.*

It is then natural to ask whether the above holds for follower sets as well, though the question seems more difficult.

We remark that one obvious approach would be to attempt to use Theorem 1.3 above. In other words, one might attempt to prove that a small number of follower sets implies a small number of extender sets, and therefore soficity. Indeed, this still may be an avenue to a proof. However, in Example 3.1, we show that the sequence $\{|E_X(n)|\}$ may grow exponentially while $\{|F_X(n)|\}$ grows linearly, meaning that this approach may not be enough on its own.

2. Definitions and preliminaries

We begin with a list of definitions. Let A denote a finite set, which we will refer to as our alphabet, elements of A will be referred to as letters.

DEFINITION 2.1. A **subshift** X on an alphabet A is some subset of $A^{\mathbb{Z}}$ which is shift-invariant and closed in the product topology.

DEFINITION 2.2. A **word** over A is a member of A^n for some $n \in \mathbb{N}$, which we call the **length** of w. We use $\varnothing$ to denote the **empty word**, the word of length zero.

DEFINITION 2.3. For any words $v \in A^n$ and $w \in A^m$, we define the **concatenation** vw to be the word in A^{n+m} whose first n letters are the letters forming v and whose next m letters are the letters forming w.

DEFINITION 2.4. For a word $u \in A^n$, if u can be written as the concatenation of two words $u = vw$ then we say that v is a **prefix** of u and that w is a **suffix** of u.

DEFINITION 2.5. The **language** of a subshift X, denoted by $L(X)$, is the set of all words which appear in points of X. For any finite $n \in \mathbb{N}$, define $L_n(X) = L(X) \cap A^n$, the set of words in the language of X with length n.

DEFINITION 2.6. For any subshift X on an alphabet A, and any word w in the language of X, we define the **follower set of** w **in** X, $F_X(w)$, to be the set of all right-infinite sequences $s \in A^{\mathbb{N}}$ such that the infinite word ws occurs in some point of X. (Note that $F_X(\varnothing)$ is simply the set of all right-infinite sequences appearing in any point of X). Similarly, we define the **predecessor set of** w **in** X, written $P_X(w)$, to be the set of all left-infinite sequences $p \in A^{-\mathbb{N}}$ such that pw occurs in some point of X.

DEFINITION 2.7. For any word $w \in L(X)$, we say that w is **shortenable** if there exists $v \in L(X)$ with strictly shorter length than w such that $F_X(w) = F_X(v)$.

DEFINITION 2.8. For any subshift X over the alphabet A, and any word w in the language of X, we define the **extender set of** w **in** X, $E_X(w)$, to be the set of all pairs (p, s) where p is a left-infinite sequence of symbols in A, s is a right-infinite sequence of symbols in A, and pws is a point of X.

DEFINITION 2.9. For any positive integer n, define the set $F_X(n) = \{F_X(w) \mid w \in L_n(X)\}$. Thus the cardinality $|F_X(n)|$ is the number of distinct follower sets of words of length n in X. Similarly, define $E_X(n) = \{E_X(w) \mid w \in L_n(X)\}$ and $P_X(n) = \{P_X(w) \mid w \in L_n(X)\}$, so that $|P_X(n)|$ and $|E_X(n)|$ are the numbers of distinct extender sets of words of length n in X and predecessor sets of words of length n in X respectively.

DEFINITION 2.10. A subshift X is **sofic** if it is the image of a shift of finite type under a continuous shift-commuting map.

Equivalently, sofic shifts are those with only finitely many follower sets, that is, a shift X is sofic iff $\{F_X(w) \mid w \text{ in the language of } X\}$ is finite (See Theorem 3.2.10 of [**2**]). The same equivalence exists for extender sets: X is sofic iff $\{E_X(w) \mid w \text{ in the language of } X\}$ is finite. (See Lemma 3.4 of [**4**])

3. An example with many more predecessor sets than follower sets

EXAMPLE 3.1. There exists a subshift X such that for every n, $|F_X(n)| = 2n+1$ and for every $n > 6$, $|P_X(n)| \geq 2^{\lfloor n/4 \rfloor}$.

PROOF. Define a labeled graph G as follows: the vertex set is $V = \mathbb{Z}^+ = \{0, 1, 2, \ldots\}$. From any vertex n are two outgoing edges: one leads to $n+1$ and is labeled with U (for 'up'), and the other leads to $\lfloor n/2 \rfloor$ and is labeled with D (for 'down'), unless it is the lone self-loop in the graph from 0 to itself, in which case it is labeled with E (for 'equals.') Then, define a subshift X with alphabet $\{D, U, E\}$ whose language consists of all labels of finite paths on G. For example, since from 17 one could follow D (to 8), D (to 4), U (to 5), U (to 6), D (to 3), D (to 1), D (to 0), and then E (to 0), $DDUUDDDE$ would be a word in the language of X. An example of a word not in the language of X would be $EUUDDD$, since E must terminate at 0, and then following U would take you to 1, another U to 2, D to 1, D to 0, and another D is not legal from 0. Note that G is right-resolving, i.e. given an initial vertex n and label a, there is at most one edge with initial vertex n labeled by a.

We will need two auxiliary notations: for any $w \in L(X)$, denote by $T_G(w)$ the set of terminal states of paths in G labeled by w, and by $I_G(w)$ the set of initial states of paths in G labeled by w. Similarly, for any $n \in V$, denote by $F_G(n)$ the set of labels of right-infinite paths in G with initial state n, and by $P_G(n)$ the set of labels of left-infinite paths in G with terminal state n. It should be clear that for any w, $F_X(w) = \bigcup_{n \in T_G(w)} F_G(n)$, and $P_X(w) = \bigcup_{n \in I_G(w)} P_G(n)$.

Claim 1: For every n, $|F_X(n)| = 2n + 1$.

Fix any n, and consider any $w \in L_n(X)$ which contains an E. Since E can only terminate at 0, and G is right-resolving, there is only one possible terminal vertex of a path in G labeled by w, and so $T_G(w) = \{k\}$ for some k, and correspondingly, $F_X(w) = F_G(k)$. Since the largest vertex in V that can be reached from 0 via a path of length less than n is $n - 1$, $0 \leq k \leq n - 1$. We claim that for $k < k' \in [0, n - 1]$, $F_G(k) \neq F_G(k')$; if 2^m is the smallest power of 2 greater than k, then the reader may check that $U^{2^m - k - 1} D^m EEEE\ldots$ is in $F_G(k)$ but not $F_G(k')$. Finally, we note that for each $k \in [0, n - 1]$, the word $w = E^{n-k} U^k$ has $T_G(w) = \{k\}$, and so all n of the distinct follower sets $F_G(0), F_G(1), \ldots, F_G(n - 1)$ are in $F_X(n)$.

Now, consider any $w \in L_n(X)$ which does not contain an E. We will prove by induction on n that $T_G(w) = [k, \infty)$ for some $0 \leq k \leq n$. The hypothesis is easy for $n = 1$; $T_G(U)$ is clearly $[1, \infty)$, and $T_G(D)$ is similarly clearly $[0, \infty)$. Now, suppose that the inductive hypothesis is true for n, and consider $w \in L_{n+1}(X)$. We can of course represent $w = w'a$, where $w' \in L_n(X)$. By the inductive hypothesis, $T_G(w') = [k, \infty)$ for $0 \leq k \leq n$. The reader can verify that if $a = D$, then $T_G(w) = [\lfloor k/2 \rfloor, \infty)$, and if $a = U$, then $T_G(w) = [k + 1, \infty)$, completing the inductive step. The proof is then completed, and so $F_G(w) = \bigcup_{i \geq k} F_G(i)$ for some $k \in [0, n]$.

We claim that for $k < k' \in [0, n]$, $\bigcup_{i \geq k} F_G(i) \neq \bigcup_{i \geq k'} F_G(i)$; again, if 2^m is the smallest power of 2 greater than k, then $U^{2^m - k - 1} D^m EEE\ldots$ is in $F_G(k)$ but not $F_G(i)$ for any $i \geq k'$. We note that for each $k \in [0, n]$, the word $w = D^{n-k} U^k$ has $T_G(w) = [k, \infty)$, and so follower sets $\bigcup_{i \geq k} F_G(i)$ are in $F_X(n)$ for $k = 0, 1, 2, \ldots, n$.

There are therefore n distinct follower sets of words $w \in L_n(X)$ containing an E and $n + 1$ distinct follower sets of words $w \in L_n(X)$ not containing an E. To prove the claim that there are $2n + 1$ in total, we verify that for all $k, k' \in [0, n]$, $\bigcup_{i \geq k'} F_G(i) \neq F_G(k)$. If $k \neq k'$, we have already distinguished these sets by identifying an element of $F_G(\min(k, k'))$ that is not in $F_G(i)$ for any $i > \min(k, k')$. If $k' = k$ and $2^m > k$, then $D^{m+1} EEE\ldots$ is in $F_G(2^m) \subseteq \bigcup_{i \geq k} F_G(i)$, but not in $F_G(k)$. This completes the proof of the claim.

Claim 2: For every $n > 6$, $|P_X(n)| \geq 2^{\lfloor n/4 \rfloor}$.

We will consider the set S of all $w \in \{D, U, E\}^n$ which end with $D^{\lceil n/2 \rceil - 1} E$, contain no other E, and do not contain consecutive U symbols. Clearly $|S|$ is the number of ($\lfloor n/2 \rfloor$)-letter words on $\{D, U\}$ without consecutive U symbols, which is greater than or equal to $2^{\lfloor n/4 \rfloor}$ (simply freely choose the first letter, force the second to be D, freely choose the third letter, and so on). Therefore, it suffices to show that $S \subseteq L_n(X)$ and that for $w \neq w' \in S$, $P_X(w) \neq P_X(w')$. We will verify both claims by the auxiliary claim that for any $w \in S$, $I_G(w)$ is a nonempty finite interval of integers, and that if $w \neq w' \in S$, then $I_G(w) \neq I_G(w')$. Clearly the fact that $I_G(w) \neq \varnothing$ will imply that $w \in L_n(X)$. Also, if $I_G(w) \neq I_G(w')$, then one can choose $k \in I_G(w) \triangle I_G(w')$. The reader may check that $\ldots EEEU^k \in P_G(k)$ and is not in $P_G(k')$ for any $k' \neq k$, and so $P_X(w) = \bigcup_{i \in I_G(w)} P_G(i)$ and $P_X(w') = \bigcup_{i \in I_G(w')} P_G(i)$ are distinct.

It remains only to prove the auxiliary claim. Consider any $w \in S$. We will work backwards from the end of w to determine $I_G(w)$. First, write $w = vD^{\lceil n/2 \rceil - 1}E$. The reader may check that $I_G(D^{\lceil n/2 \rceil - 1}E) = [2^{\lceil n/2 \rceil - 2}, 2^{\lceil n/2 \rceil - 1})$, which we write as $[a, b)$ for brevity. Write $v = v_{\lfloor n/2 \rfloor} \dots v_2 v_1$. Then, we will work from the right and state how each v_i will alter this interval. For instance, if $v_1 = D$, then $I_G(v_1 D^{\lceil n/2 \rceil - 1}E)$ is the set of all vertices which lead to a vertex in $[a, b)$ via an edge labeled D, or $[2a, 2b)$. Similarly, if $v_1 = U$, then $I_G(v_1 D^{\lceil n/2 \rceil - 1}E)$ is the set of all vertices which lead to a vertex in $[a, b)$ via an edge labeled U, or $[a-1, b-1)$. In fact, it is simple to see in the same way that each v_i will either double the endpoints of the interval (if $v_i = D$) or subtract one from the endpoints of the interval (if $v_i = U$). Since $a = 2^{\lceil n/2 \rceil - 2} > \lfloor n/2 \rfloor$ (for $n > 6$), clearly neither endpoint will ever go below 0 in this procedure.

This allows us to give a closed form for $I_G(w)$. Since U does not change the length of the interval and D doubles it, clearly the length of $I_G(w)$ is $2^j(b-a)$, where j is the number of D symbols in v. The left endpoint of $I_G(w)$, call it c, is obtained from a via a sequence of either doubling or subtracting 1, determined by whether the letters v_1, v_2, etc. are D or U respectively. Since no two consecutive v_i can be U, two subtractions in a row are not permitted. For instance, if $a = 16$ and $v = UDDDUD$, then $c = (((((16 \cdot 2) - 1) \cdot 2) \cdot 2) \cdot 2) - 1 = 247$. It is not hard to check that this final answer could also be written as $c = 2^j a - (2^{n_1} + 2^{n_2} + \ldots + 2^{n_k})$, where j is again the number of D symbols in v, k is the number of U symbols in v, and n_i is the number of D symbols preceding the ith U symbol in v. (Note that since v does not contain consecutive U symbols, $\{n_i\}$ is strictly decreasing). For instance, for $v = UDDDUD$, there are three D symbols preceding the rightmost U and zero D symbols preceding the leftmost U, and so $n_1 = 3$ and $n_2 = 0$, yielding $c = 2^4 a - (2^{n_1} + 2^{n_2}) = 2^4 \cdot 16 - 8 - 1 = 247$. This yields the closed form

$$I_G(w) = \left[2^j a - \sum_{i=1}^{k} 2^{n_i}, 2^j b - \sum_{i=1}^{k} 2^{n_i} \right).$$

Then, if w and w' have different numbers of D symbols, then the lengths of $I_G(w)$ and $I_G(w')$ are different, clearly implying that $I_G(w) \neq I_G(w')$. If w and w' have the same number of D symbols, then the choices of n_i for w and w' are distinct (since the n_i uniquely determine v), and by uniqueness of binary representation, the sum $\sum_{i=1}^{k} 2^{n_i}$ would take different values for w and w', again implying that $I_G(w) \neq I_G(w')$. Therefore, all words in S have distinct predecessor sets in X, and we are done.

□

We note that since predecessor sets are just projections of extender sets, this example clearly has $|E_X(n)| \geq 2^{\lfloor n/4 \rfloor}$ for every $n > 6$ as well, illustrating that in general, the number of extender sets of words of length n may be much greater than the number of follower sets of words of length n.

4. Main Results

We begin with some simple facts about follower sets, which will repeatedly be useful in our analysis. The proofs are simple and left to the reader.

LEMMA 4.1. *For any subshift X, any $w \in L_n(X)$, and any $m \in \mathbb{N}$, $F_X(w) = \bigcup_v F_X(vw)$, where the union is taken over those $v \in L_m(X)$ for which $vw \in L_{m+n}(X)$.*

LEMMA 4.2. *For any subshift X, any $w \in L_n(X)$, and any $m < n$, there exists a $v \in L_m(X)$ for which $F(v) \supseteq F(w)$.*

LEMMA 4.3. *Let X be a subshift. If for two words $w, u \in L(X)$, $F_X(w) = F_X(u)$, then for any $v \in L(X)$, $F_X(wv) = F_X(uv)$.*

The following will be our main tool for proving soficity of a subshift via the sets $F_X(n)$.

THEOREM 4.4. *For any subshift X, if there exists $n \in \mathbb{N}$ such that $F_X(n) \subseteq \bigcup_{\ell \leq n-1} F_X(\ell)$, then X is sofic.*

PROOF. If there exists $n \in \mathbb{N}$ such that $F_X(n) \subseteq \bigcup_{\ell \leq n-1} F_X(\ell)$, then for any word $w \in L_n(X)$, the follower set $F_X(w)$ is also the follower set of a strictly shorter word, so w is shortenable to a word of length strictly less than n. Now, let $v \in L(X)$ of length greater than n, say $v = v_1v_2...v_nv_{n+1}...v_k$ where $k > n$. Then $v_1v_2...v_n \in L_n(X)$, and so is shortenable to some word $v' \in L(X)$ of length less than n. But $F_X(v_1v_2...v_n) = F_X(v')$ implies $F_X(v_1v_2...v_nv_{n+1}...v_k) = F_X(v'v_{n+1}...v_k)$ by Lemma 4.3, so v is shortenable to a word $v'v_{n+1}...v_k$. If $v'v_{n+1}...v_k$ has length less than n, stop, but if $v'v_{n+1}...v_k$ has length at least n, we may apply the above process again and shorten repeatedly, getting shorter and shorter words with the same follower set until we find one with length less than n. So v is shortenable to a word of length less than n. But this means that $\bigcup_{\ell \leq n-1} F_X(\ell)$ contains all follower sets in X, so X has only finitely many follower sets, and thus, X is sofic. □

We can now show that $|F_X(n)| = 1$ for any n always implies soficity of X.

THEOREM 4.5. *For any subshift X, if there exists n for which $|F_X(n)| = 1$, then X is a full shift.*

PROOF. We prove the contrapositive. Without loss of generality, assume that the alphabet A of X consists entirely of letters which actually appear in points of X, and assume that X is not the full shift on A. Then there exists a word $w = w_1w_2 \dots w_k \in A^k$ which is not in the language of X; suppose that the length k of w is minimal. It must be the case that k is at least 2, since we assumed that all letters of A are in $L(X)$. Then since we assumed k to be minimal, $w_2 \dots w_k \in L(X)$, so we can choose some one-sided infinite sequence s appearing in X which begins with $w_2 \dots w_k$. Similarly, w_1 is in $L(X)$, so for any $n \in \mathbb{N}$, we may choose an n-letter word v ending with w_1. Then vs contains $w \notin L(X)$, so $s \notin F_X(v)$. However, since s appears in X, there exists some n-letter word u which can be followed by s in X, and so $s \in F_X(u)$. Hence $F_X(u) \neq F_X(v)$, so $|F_X(n)| \geq 2$, and since n was arbitrary, this is true for all n. □

We can now prove a version of Conjecture 1.1 for unions of the sets $F_X(n)$, rather than the sets themselves.

THEOREM 4.6. *For any subshift X, if there exists $n \in \mathbb{N}$ so that* $\left|\bigcup_{\ell \leq n} F_X(\ell)\right| \leq n$*, then X is sofic.*

PROOF. We prove the contrapositive, and so assume that X is nonsofic. By Theorem 4.5, $|F_X(1)| \geq 2$. Then, by Theorem 4.4, for every $n > 1$, there exists $S \in F_X(n) \setminus \bigcup_{\ell<n} F_X(\ell)$, and so $\left|\bigcup_{\ell \leq n} F_X(\ell)\right| > \left|\bigcup_{\ell \leq n-1} F_X(\ell)\right|$. Therefore, by induction, for each n, $\left|\bigcup_{\ell \leq n} F_X(\ell)\right| \geq n+1$. □

We may now prove the following, which establishes a logarithmic lower bound for the growth rate of $|F_X(n)|$ for nonsofic shifts.

THEOREM 4.7. *For any subshift X, if there exists $n \in \mathbb{N}$ such that $|F_X(n)| \leq \log_2(n+1)$, then X is sofic.*

PROOF. Suppose that for some n, $F_X(n) = \{F_1, F_2, ..., F_k\}$ where $k \leq \log_2(n+1)$. By Lemma 4.1, for each length $\ell < n$, every follower set of a word in $L_\ell(X)$ is a union of follower sets of words of length n. Therefore, every element of $\bigcup_{\ell \leq n} F_X(\ell)$ is a non-empty union of elements of $F_X(n)$. There are at most $2^k - 1 \leq 2^{\log_2(n+1)} - 1 = n$ such unions, so $\left|\bigcup_{\ell \leq n} F_X(\ell)\right| \leq n$, which implies that X is sofic by Theorem 4.6. □

Our next result shows that under the additional assumption that some non-empty word w has the same follower set as the empty word, Conjecture 1.1 is true.

LEMMA 4.8. *For any subshift X, if there exists a non-empty word $w \in L(X)$ such that $F_X(w) = F_X(\varnothing)$ and $n \in \mathbb{N}$ such that $|F_X(n)| \leq n$, then X is sofic.*

PROOF. The follower set of the empty word is the set of all right-infinite sequences appearing in any point of X. If there exists a word w such that any legal right-infinite sequence may appear after w, then by Lemma 4.2, there is a letter with this property as well. So we may assume that $F_X(a) = F_X(\varnothing)$ where a is a single letter.

The fact that $F_X(\varnothing) = F_X(a)$ implies by Lemma 4.3 that for every $w \in L(X)$, $F_X(w) = F_X(aw) = F_X(aaw) = \ldots$. Therefore, every follower set of a word of length ℓ is also a follower set of a word of any length greater than ℓ. In other words, $F_X(1) \subseteq F_X(2) \subseteq F_X(3) \subseteq \ldots$. Then, for every n, $F_X(n) = \bigcup_{\ell \leq n} F_X(\ell)$, and so if $|F_X(n)| \leq n$ for some n, clearly $\left|\bigcup_{\ell \leq n} F_X(\ell)\right| \leq n$, implying that X is sofic by Theorem 4.6. □

THEOREM 4.9. *For any subshift X, if there exists $n \geq 2$ for which $|F_X(n)| \leq 2$, then X is sofic.*

PROOF. The case where $|F_X(n)| = 1$ is treated by Theorem 4.5, so we choose any $n \geq 2$ and suppose that there are exactly 2 follower sets in X of words of length n, say F_1 and F_2. We consider the sets in $F_X(1)$. By Lemma 4.1, every element of $F_X(1)$ is either F_1, F_2, or $F_1 \cup F_2$. If $|F_X(1)| = 1$, X is sofic by Theorem 4.5, so assume that $|F_X(1)| \geq 2$, that is, at least two of the above sets must appear in $F_X(1)$. Note that $F(\varnothing) = \bigcup_{w \in L_X(n)} F(w) = F_1 \cup F_2$, so by Lemma 4.8, if

$F_1 \cup F_2$ is an element of $F_X(1)$, then X is sofic. The only remaining case is that $F_X(1) = \{F_1, F_2\} = F_X(n)$, and then X is sofic by Theorem 4.4. □

We are now prepared to prove Conjecture 1.1 for $n \leq 3$. Our proof is much more complicated than the cases where $n = 1, 2$.

THEOREM 4.10. *Let X be a subshift. If $|F_X(n)| \leq n$ for any $n \leq 3$, then X is sofic.*

PROOF. Clearly, for $n < 3$, Theorems 4.5 and 4.9 imply this result. We can then restrict to the case where $n = 3$. If $|F_X(3)| < 3$, then X is again sofic by either Theorem 4.5 or Theorem 4.9. We therefore suppose that $|F_X(3)| = 3$, say $F_X(3) = \{F_1, F_2, F_3\}$. We also note that $F(\varnothing) = F_1 \cup F_2 \cup F_3$, and if any of $F_X(1)$, $F_X(2)$, or $F_X(3)$ contains $F_1 \cup F_2 \cup F_3$ as an element, then X is sofic by Lemma 4.8. Therefore, in everything that follows, we assume that $F_1 \cup F_2 \cup F_3$ is not contained in $F_X(i)$ for $i \leq 3$.

We first show that if any F_i is contained entirely within another, then X is sofic. Suppose for a contradiction that some F_i is contained in another, and so without loss of generality, we say that $F_2 \subseteq F_1$. By Lemma 4.1, all elements of $F_X(2)$ are nonempty unions of F_1, F_2, and F_3. However, $F_1 \cup F_3 = F_1 \cup F_2 \cup F_3$, and so $F_1 \cup F_3 = F_1 \cup F_2 \cup F_3$ are not in $F_X(2)$ as assumed above. Also, $F_1 \cup F_2 = F_1$. Therefore, the only possible elements of $F_X(2)$ are F_1, F_2, F_3, and $F_2 \cup F_3$. If fewer than three of these four sets are part of $F_X(2)$, then X is sofic by Theorem 4.9. Thus we may assume at least three of the four sets appear. If F_1, F_2, and F_3 are all in $F_X(2)$, then $F_X(3) \subseteq F_X(2)$, implying that X is sofic by Theorem 4.4. Therefore, $F_2 \cup F_3 \in F_X(2)$. We note that by Lemma 4.2, some element of $F_X(2)$ must contain F_1. If F_3 contained F_1, then $F_3 = F_1 \cup F_2 \cup F_3$ is in $F_X(3)$, which we assumed not to be the case above. Similarly, $F_2 \cup F_3$ cannot contain F_1. Therefore, F_1 is the only set of F_1, F_2, F_3, and $F_2 \cup F_3$ to contain F_1, and so $F_1 \in F_X(2)$. Therefore $F_X(2)$ consists of F_1, $F_2 \cup F_3$, and exactly one of F_2 and F_3. We note that if $F_2 \cup F_3$ is equal to any of F_1, F_2, or F_3, then either $|F_X(2)| = 2$ or $F_X(3) \subseteq F_X(2)$, in either case implying soficity by either Theorem 4.9 or Theorem 4.4. So from now on we assume $F_2 \cup F_3$ is not equal to F_1, F_2, or F_3.

Now, let us consider $F_X(1)$. By Lemma 4.1, $F_X(1)$ can only consist of unions of sets in $F_X(2)$. The set $F_X(1)$ cannot contain $F_1 \cup F_3 = F_1 \cup F_2 \cup F_3$, and since $F_1 \cup F_2 = F_1$ we see that $F_X(1) \subseteq F_X(2)$. There exists some word $ab \in L_X(2)$ such that $F_X(ab) = F_2 \cup F_3$. Clearly $F_X(a)$ is an element of $F_X(1)$ and therefore $F_X(a) = F_X(xy)$ for some $xy \in L_2(X)$. But then by Lemma 4.3, $F_X(xyb) = F_X(ab) = F_2 \cup F_3$, a contradiction since we above noted that $F_2 \cup F_3$ does not equal any of F_1, F_2, or F_3. We have then shown that if any of the follower sets F_1, F_2, and F_3 are contained in one another, X is sofic, and so for the rest of the proof assume that no such containments exist. Note that this also implies that if any of $F_1 \cup F_2$, $F_1 \cup F_3$, or $F_2 \cup F_3$ contain each other, then the containing set is $F_1 \cup F_2 \cup F_3$, which we have assumed is not in $F_X(1)$, $F_X(2)$, or $F_X(3)$.

We break the remainder of the proof into cases by how many of the sets F_1, F_2, and F_3 are elements of $F_X(2)$. If all three of the sets are elements of $F_X(2)$, then X is sofic by Theorem 4.4. We then have three remaining cases.

Case 1: none of F_1, F_2, F_3 are in $F_X(2)$. By Lemma 4.1, $F_X(2)$ consists of nonempty unions of F_1, F_2, and F_3, and we have assumed that $F_1 \cup F_2 \cup F_3$ is not in $F_X(2)$. If $|F_X(2)| \leq 2$, then X is sofic by Theorem 4.9. The only possibility is then that $F_X(2) = \{F_1 \cup F_2, F_1 \cup F_3, F_2 \cup F_3\}$. Then by Lemma 4.2, $F_X(1)$ must contain supersets of each of these sets, and it cannot contain $F_1 \cup F_2 \cup F_3$. This forces $F_X(1)$ to also be $\{F_1 \cup F_2, F_1 \cup F_3, F_2 \cup F_3\}$, meaning that $F_X(2) = F_X(1)$, and so X is sofic by Theorem 4.4.

Case 2: exactly one of F_1, F_2, F_3 is in $F_X(2)$. Without loss of generality, suppose that $F_1 \in F_X(2)$ and $F_2, F_3 \notin F_X(2)$. At least two other sets must be elements of $F_X(2)$ or else X is sofic by Theorem 4.9, and they must be unions of F_1, F_2, and F_3 by Lemma 4.1. Therefore, $F_X(2)$ contains at least two of the sets $F_1 \cup F_2$, $F_1 \cup F_3$, and $F_2 \cup F_3$. By Lemma 4.2, supersets of any such unions are also present in $F_X(1)$, which must be the sets themselves since we've assumed that $F_1 \cup F_2 \cup F_3 \notin F_X(1)$. If F_1 is also in $F_X(1)$, $F_X(2) \subseteq F_X(1)$, and X would be sofic by Theorem 4.4, so $F_1 \notin F_X(1)$.

Now, let abc be some word such that $F_X(abc) = F_2$. What, then, is the follower set of ab? If it is any set in $F_X(1)$, then there would exist d so that $F_X(ab) = F_X(d)$, and then $F_X(abc)$ would equal $F_X(dc)$ by Lemma 4.3, meaning that $F_2 \in F_X(2)$, a contradiction. So the only choice for $F_X(ab)$ is F_1. Since at least two of $F_1 \cup F_2$, $F_1 \cup F_3$, and $F_2 \cup F_3$ are in $F_X(2)$, $F_X(2)$ contains a set of the form $F_1 \cup F_i$. Say that $F_X(xy) = F_1 \cup F_i$. Then, $F_X(xy) \supseteq F_X(ab)$, meaning that $F_X(xyc) \supseteq F_X(abc) = F_2$. Since none of the F_i contain each other, this means that $F_X(xyc) = F_2$. But then since $F_1 \cup F_i$ also is a member of $F_X(1)$, there exists z so that $F_X(z) = F_1 \cup F_i$, and then by Lemma 4.3, $F_X(zc) = F_2$, a contradiction since $F_2 \notin F_X(2)$. Hence X is sofic in this case as well.

Case 3: exactly two of F_1, F_2, F_3 are in $F_X(2)$. Without loss of generality, suppose that $F_1, F_2 \in F_X(2)$ and $F_3 \notin F_X(2)$. By Lemma 4.2, $F_X(2)$ must contain some superset of F_3 which is not $F_1 \cup F_2 \cup F_3$, so it is of the form $F_3 \cup F_i$ for $i = 1$ or 2. As in Case 2, any of the sets $F_1 \cup F_2$, $F_1 \cup F_3$, or $F_2 \cup F_3$ which is an element of $F_X(2)$ must be in $F_X(1)$ as well. This means that if F_1 and F_2 are both in $F_X(1)$, then $F_X(2) \subseteq F_X(1)$ and X would be sofic by Theorem 4.4, so we restrict to the case where at least one of these sets is not in $F_X(1)$.

Now, let abc be some word such that $F_X(abc) = F_3$. As in Case 2, the follower set of ab must be some set which occurs in $F_X(2)$ but not $F_X(1)$, which must be either F_1 or F_2 (depending on which is not part of $F_X(1)$). Without loss of generality, we say that $F_X(ab) = F_2$. We now show that neither $F_1 \cup F_2$ nor $F_2 \cup F_3$ is in $F_X(2)$. Suppose for a contradiction that there is a word $xy \in L(X)$ for which $F_X(xy) = F_2 \cup F_i$, $i = 1$ or 3. Then, since $F(xy) \supseteq F(ab) = F_2$, $F(xyc) \supseteq F(abc) = F_3$. Again, since no F_i contains another, this implies that $F(xyc) = F_3$. Finally, we note that $F(y) \supseteq F(xy) = F_2 \cup F_i$, so $F(y) = F_2 \cup F_i$. Therefore, by Lemma 4.3, $F(yc) = F(xyc) = F_3$, but this is a contradiction since $F_3 \notin F_X(2)$. We now know that neither $F_1 \cup F_2$ nor $F_2 \cup F_3$ is in $F_X(2)$. By Lemma 4.1, all sets in $F_X(2)$ are nonempty unions of F_1, F_2, and F_3, and if $|F_X(2)| < 3$, then X is sofic by Theorem 4.9. The only remaining case is then that $F_X(2) = \{F_1, F_2, F_1 \cup F_3\}$.

We now consider the sets in $F_X(1)$. Recall that $F_2 \notin F_X(1)$ and that $F_1 \cup F_3 \in F_X(1)$ since $F_1 \cup F_3 \in F_X(2)$. If $|F_X(1)| = 1$, then X is sofic by Theorem 4.5, so we can assume that $F_X(1)$ contains at least one other set, which must be a nonempty union of the elements of $F_X(2)$ by Lemma 4.1. The only possibilities are F_1 and

$F_1 \cup F_3$, since we assumed earlier that $F_1 \cup F_2 \cup F_3 \notin F_X(2)$. Therefore, every set in $F_X(1)$ is a superset of F_1.

Our final step will involve considering what happens when a word with follower set F_1 is extended on the right by a letter. Suppose for a contradiction that there exists a word $w \in L(X)$ with $F_X(w) = F_1$ and a letter i for which $F_X(wi) = F_2$. Then, for any letter j, since $F_X(j) \in F_X(1)$, $F_X(j) \supseteq F_X(w) = F_1$. Therefore, $F_X(ji) \supseteq F_X(wi) = F_2$. However, the only superset of F_2 in $F_X(2)$ is F_2 itself, and so for every $j \in A$, $F_X(ji) = F_2$. Finally, note that, by Lemma 4.1, $F_X(i) = \bigcup_j F_X(ji) = F_2$, a contradiction since $F_2 \notin F_X(1)$.

Similarly, let's assume for a contradiction that there exists a word $w \in L(X)$ with $F_X(w) = F_1$ and a letter i for which $F_X(wi) = F_3$. Then, choose a letter j with $F_X(j) = F_1 \cup F_3$. Then, since $F_X(j) \supseteq F_X(w) = F_1$, $F_X(ji) \supseteq F_X(wi) = F_3$. However, the only superset of F_3 in $F_X(2)$ is $F_1 \cup F_3$, so $F_X(ji) = F_1 \cup F_3$. Then, since $F_X(j) = F_X(ji) = F_1 \cup F_3$, by Lemma 4.3, $F_X(jii) = F_X(ji) = F_1 \cup F_3$, a contradiction since $F_1 \cup F_3 \notin F_X(3)$.

This means that for every word $w \in L(X)$ with $F_X(w) = F_1$ and any letter a for which $wa \in L(X)$, $F(wa) = F_1$. But then, since the follower set of every letter contains F_1, the follower set of every legal 2-letter word contains F_1, a contradiction since $F_X(2)$ contains F_2, and we assumed that none of the F_i contains another. Every case has either led to a contradiction or to the conclusion that X is sofic, and so we've proved that X is sofic. □

Our final result is a version of Conjecture 1.1 for a class of coded subshifts. Recall the definition of coded subshifts below.

DEFINITION 4.11. Given a set $\mathcal{W}$ of finite words, the **coded subshift** with **code words** $\mathcal{W}$ is the subshift generated by taking the closure of the set of all biinfinite sequences made from concatenating words in $\mathcal{W}$.

THEOREM 4.12. *Given a sofic shift X, choose a subset $\mathcal{W} \subseteq L(X)$ with the property that for any finite word $v \in L(X)$, there exists some $w \in \mathcal{W}$ such that v is a suffix of w. Create a coded subshift Y with code words $\{wc \mid w \in \mathcal{W}\}$ where c is a letter not appearing in the alphabet of X. Then if $|F_Y(n)| \le n$ for any $n \in \mathbb{N}$, then Y is sofic.*

PROOF. We begin with two preliminary observations. Firstly, $X \subseteq Y$, since any point of X is a limit of finite words in $L(X)$, all of which are suffixes of code words, which are themselves in $L(Y)$. We also note that any word in $L(Y)$ without a c must be a subword of a code word, and therefore in $L(X)$.

Secondly, for any word $ucv \in L(Y)$, $F_Y(ucv) = F_Y(cv)$. Clearly $F_Y(ucv) \subseteq F_Y(cv)$. Let $s \in F_Y(cv)$. Because uc is the suffix of a concatentation of code words and vs is the beginning of a concatenation of code words in Y, $ucvs$ occurs in Y and therefore $F_Y(ucv) \supseteq F_Y(cv)$.

We begin our proof by claiming that there are only finitely many follower sets in Y of words not containing the letter c. Given any word $w \in L(Y)$, if w does not contain a c then $w \in L(X)$. There are only finitely many follower sets in X, so it is sufficient to show that for any $w, v \in L(X)$, $F_X(w) = F_X(v)$ implies $F_Y(w) = F_Y(v)$. To that end, let $F_X(w) = F_X(v)$ and consider any $s \in F_Y(w)$. If s does not contain the letter c, then ws is a limit of longer and longer words in

$\mathcal{W}$, and since all such words are in $L(X)$, ws occurs in X, i.e. $s \in F_X(w)$. Since $F_X(w) = F_X(v)$, $s \in F_X(v)$, i.e. vs also occurs in X. Since $Y \supseteq X$, vs occurs in Y as well, and so $s \in F_Y(v)$.

On the other hand, if s contains the letter c and $s \in F_Y(w)$, then $s = s'cs''$ for some s' not containing c (s' may be the empty word). By the same logic as above, $ws' \in L(X)$, therefore $vs' \in L(X)$, and so vs' occurs as a suffix of some word in $\mathcal{W}$. But then, $vs'c$ is a suffix of some code word, and so $vs'cs''$ occurs in Y.

We have shown that in both cases, $s \in F_Y(w)$ implies $s \in F_Y(v)$, and so $F_Y(w) \subseteq F_Y(v)$. By the same argument, $F_Y(v) \subseteq F_Y(w)$, giving $F_Y(w) = F_Y(v)$. Therefore there are only finitely many follower sets in Y of words not containing c.

Now, we assume that n is such that $|F_Y(n)| \leq n$. Partition $L_n(Y)$ into $n+1$ sets based on the last appearance of the letter c in the word–the first set S_0 consists of words with no c, the second set S_1 consists of words ending with c, the third S_2 consists of words ending with c followed by another letter that is not c, and so on, up to the final set S_n which consists of words beginning with a c followed by $n-1$ other symbols which are not c. Since $X \subseteq Y$, there exist words in $L(Y)$ of every length without any c symbols, implying that $S_0 \neq \varnothing$. Therefore, there must exist $k > 0$ so that all follower sets (in Y) of words in S_k are also follower sets (in Y) of some word in S_i for some $i < k$; else each of the $n+1$ sets S_i would contribute a follower set not in any previous one, contradicting $|F_Y(n)| \leq n$.

Let w be a word in $L(Y)$ of length at least k. Our goal is to show that $F_Y(w)$ is either equal to one of the finitely many follower sets of words without a c or to the follower set of a word of length less than k. Clearly, if w does not contain a c, we are done, so suppose w contains the letter c. As noted earlier, $F_Y(w)$ is unchanged if all letters before the last occurrence of c are removed from w. If this removal results in a word of length less than k, then again we are done. So let us proceed under the assumption that w begins with c, has length k or greater and contains no other c symbols.

Let p denote the k-letter prefix of w. Since p begins with c, p can be arbitrarily extended backwards in any way to yield an n-letter word p' which has the same follower set as p. Note that $p' \in S_k$, and so there exists $i < k$ and $p'' \in S_i$ so that $F_Y(p) = F_Y(p') = F_Y(p'')$. There are two cases. If $i \neq 0$, then we may again remove the letters of p'' before the final c symbol to yield a word p''' of length $i < k$ for which $F_Y(p) = F_Y(p''')$. Then, we replace the prefix p of w by p''' to yield a new word w' with strictly smaller length, which still begins with a c and contains no other c symbols, and for which $F_Y(w) = F_Y(w')$ by Lemma 4.3. We then repeat the above steps. If at each step, $i \neq 0$, then eventually w will be shortened to a word of length at most k with the same follower set in Y, of which there are clearly only finitely many.

The only other case is that at some point, the prefix of length k has the same follower set in Y as a word in S_0. Then, again by Lemma 4.3 that prefix can be replaced by the word in S_0, yielding a word with no c symbols with the same follower set in Y as w. There are only finitely many follower sets in Y of words not containing c. We have then shown that $F_Y(w)$ (for arbitrary w of length at least k) has follower set in Y from a finite collection (namely all follower sets in Y of words with no c and all follower sets in Y of words with length at most $k-1$), which implies that Y is sofic.

□

REMARK 4.13. The class of coded subshifts Y which may be created as in Theorem 4.12 includes all so-called S-gap shifts [**2**] (with $X = \{0^\infty\}$ and $c = 1$) and the reverse context-free shift of [**5**] (with $X = \{a, b\}^{\mathbb{Z}}$ and c as in the theorem).

References

[1] Thomas French, *Characterizing follower and extender set sequences*, Submitted, 2015.

[2] Douglas Lind and Brian Marcus, *An introduction to symbolic dynamics and coding*, Cambridge University Press, Cambridge, 1995. MR1369092

[3] Marston Morse and Gustav A. Hedlund, *Symbolic dynamics II. Sturmian trajectories*, Amer. J. Math. **62** (1940), 1–42. MR0000745

[4] Nic Ormes and Ronnie Pavlov, *Extender sets and multidimensional subshifts*, Ergodic Theory Dynam. Systems **36** (2016), no. 3, 908–923, DOI 10.1017/etds.2014.71. MR3480350

[5] Ronnie Pavlov, *A class of nonsofic multidimensional shift spaces*, Proc. Amer. Math. Soc. **141** (2013), no. 3, 987–996, DOI 10.1090/S0002-9939-2012-11382-6. MR3003690

DEPARTMENT OF MATHEMATICS, UNIVERSITY OF DENVER, 2280 S. VINE ST., DENVER, COLORADO 80208

E-mail address: tfrench9@du.edu

DEPARTMENT OF MATHEMATICS, UNIVERSITY OF DENVER, 2280 S. VINE ST., DENVER, COLORADO 80208

E-mail address: normes@du.edu

URL: www.math.du.edu/~ormes/

DEPARTMENT OF MATHEMATICS, UNIVERSITY OF DENVER, 2280 S. VINE ST., DENVER, COLORADO 80208

E-mail address: rpavlov@du.edu

URL: www.math.du.edu/~rpavlov/

Contemporary Mathematics
Volume **678**, 2016
http://dx.doi.org/10.1090/conm/678/13647

Haar measures and Hausdorff dimensions of p-adic Julia sets

Joanna Furno

ABSTRACT. We investigate the Haar measures and Hausdorff dimensions of balls in finite extensions of the p-adic numbers. From this foundation, we calculate the Haar measures and Hausdorff dimensions of p-adic Julia sets. We include a broad class of p-adic polynomials and specific standard examples to which the results apply.

1. Introduction and statement of main result

Fix a prime p. Let $\mathbb{Q}_p$ be the field of p-adic numbers with the p-adic absolute value $|\cdot|_p$. The action of a p-adic polynomial or rational function on its Julia set is often described using the dynamics of a shift on a product space, such as in [**1**, **2**, **7**, **8**]. This paper describes the size of some Julia sets inside of finite extensions of $\mathbb{Q}_p$, using Haar measure and Hausdorff dimension.

In the rest of Section 1, we review definitions and outline the results in this paper. In particular, we use the context and Julia set definition from [**2**]. In Section 2, we build an understanding of the Haar measure and Hausdorff dimension of a ball in a finite extension of $\mathbb{Q}_p$. In Section 3, we use this understanding for balls to calculate the Haar measures and Hausdorff dimensions of Julia sets. In Section 4, we apply these results to calculate the Haar measures and Hausdorff dimensions of the Julia sets for a broad class of polynomials and for some specific standard examples.

Before stating our main result, we recall some facts about finite extensions of $\mathbb{Q}_p$ and set notation. See [**4**] for further details and proofs. Let K be a finite extension of $\mathbb{Q}_p$, and let $n = [K : \mathbb{Q}_p]$ be the degree of the extension. The p-adic metric $|\cdot|_p$ extends uniquely to K. Let $K^\times$ be the set of invertible elements in K. (Since K is a field, $K^\times = K\setminus\{0\}$.) Let

$$\nu_p(K^\times) = \left\{\log_p(|x|_p) : x \in K^\times\right\}$$

be the value group of K. There exists a positive integer e, called the *ramification index*, such that

$$\nu_p(K^\times) = \frac{1}{e}\mathbb{Z}.$$

Fix an element $\pi \in K$ such that $|\pi|_p = p^{-1/e}$. The set

$$\mathcal{O}_K = \{x \in K : |x|_p \leq 1\}$$

2010 *Mathematics Subject Classification.* Primary 37P40, 37P05, 37F35.

is the valuation ring and

$$\mathfrak{p}_K = \{x \in K : |x|_p < 1\}$$

is the maximal ideal of $\mathcal{O}_K$. Recall that the residue field $\mathbb{Z}_p / p\,\mathbb{Z}_p$ is isomorphic to $\mathbb{F}_p$, the field with p elements. Similarly, there exists a positive integer f such that the residue field $\mathcal{O}_K / \mathfrak{p}_K$ is isomorphic to $\mathbb{F}_{p^f}$. Fix a set of p^f coset representatives $A = \{0, c_1, \ldots, c_{p^f-1}\}$. Then every $a \in K$ has a unique expansion

$$a = \sum_{i=-m}^{\infty} a_i \pi^i,$$

where m is an integer and $a_i \in A$ for all $i \geq -m$.

The topology on K has a basis consisting of balls of the form

$$B_{p^r}(a) = \{x \in K : |x - a|_p \leq p^r\},$$

where $r \in \nu_p(K^\times)$ and $a \in K$. These balls are both open and closed.

Since K is a locally compact abelian group, there is a Haar measure on K that is unique up to scaling. Let μ be Haar measure on K, normalized so that $\mu(\mathcal{O}_K) = 1$. The Haar measure of a ball in $\mathbb{Q}_p$ is equal to its radius. In Section 2, we use the invariance of Haar measure under translation to prove a similar statement for the Haar measure of a ball in a finite extension of $\mathbb{Q}_p$.

PROPOSITION 1.1. *Let K be an extension of $\mathbb{Q}_p$ of degree n. Let μ be Haar measure on K, normalized so that $\mu(\mathcal{O}_K) = 1$. For all $r \in \nu(K^\times)$ and $a \in K$,*

$$\mu(B_{p^r}(a)) = (p^r)^n.$$

Proposition 1.1 says that the Haar measure of a ball is equal to the radius of the ball, raised to a power equal to the degree of the extension over $\mathbb{Q}_p$. Although stated differently, this formulation agrees with a more general statement in the preliminaries of [**6**] for the Haar measure of a ball in a non-Archimedean local field. For completeness and to demonstrate some counting arguments and finite covers used in later proofs, we include a proof of Proposition 1.1 in Section 2.

Since the p-adic absolute value induces a distance under which K is a separable metric space, it is possible to define Hausdorff measures and the Hausdorff dimension for subsets of K. Section 2 continues with these definitions and a proof of Proposition 1.2.

PROPOSITION 1.2. *Let K be an extension of $\mathbb{Q}_p$ of degree n. For all $r \in \nu(K^\times)$ and $a \in K$, the Hausdorff dimension of $B_{p^r}(a)$ is n.*

In Section 3, we calculate Haar measures and Hausdorff dimensions for Julia sets of p-adic maps. Following [**2**], let $f : X \to K$ be a map on a compact open $X \subset K$ satisfying the following two assumptions:

(i) $f^{-1}(X) \subset X$,

(ii) there exist $r, r_i \in \nu(K^\times)$ and $c_i \in K$ such that $X = \bigsqcup_{i=1}^{k} B_{p^r}(c_i)$ is a disjoint union of balls and for all $1 \leq i \leq k$

$$|f(x) - f(y)|_p = p^{r_i}|x - y|_p \text{ for all } x, y \in B_{p^r}(c_i). \tag{1.1}$$

In [**6**], a transformation satisfying condition (ii) is called a *locally scaling transformation.* Unlike [**6**], we do not assume that $f(X) \subset X$. Moreover, this inclusion will be false if $f^{-1}(X) \subsetneq X$, which is the case of main interest in this paper. For this reason, most of the results in [**6**] do not apply to our setting.

As in [**2**], the *Julia set* of a map $f : X \to K$ satisfying conditions (i) and (ii) is

$$\mathcal{J}(f, X) = \bigcap_{j=0}^{\infty} f^{-j}(X). \tag{1.2}$$

Theorem 1.3 is our main theorem and gives the Haar measure and Hausdorff dimension of Julia sets for a subclass of maps satisfying conditions (i) and (ii).

THEOREM 1.3. *Let K be an extension of $\mathbb{Q}_p$ of degree n, and let μ be Haar measure on K. Let $k \in \mathbb{N}$ and $f : X \to K$ be a map on a compact-open $X \subset K$ such that*

$$X = \bigcup_{i=1}^{k} B_{p^{r_1}}(c_i),$$

where

(a) $f^{-1}(X) \subset X$,
(b) $f(B_{p^{r_1}}(c_i)) \bigcap X \neq \emptyset$ *for all* $1 \leq i \leq k$,
(c) $\max_{1\leq i\leq k} \{|c_i|_p\} = p^{r_2}$ *for some* $r_2 > r_1$, *and*
(d) for all $1 \leq i \leq k$, $|f(x) - f(y)|_p = p^{r_2 - r_1}|x - y|_p$ *for all* $x, y \in B_{p^{r_1}}(c_i)$.

Then the following two statements hold:

1. *If* $k = p^{n(r_2 - r_1)}$, *then* $\mu(\mathcal{J}(f, X)) = p^{nr_2}$ *and* $Hdim(\mathcal{J}(f, X)) = n$.
2. *If* $k < p^{n(r_2 - r_1)}$, *then* $\mathcal{J}(f, X) \subsetneq X$, $\mu(\mathcal{J}(f, X)) = 0$, *and*

$$Hdim(\mathcal{J}(f, X)) = \frac{\ln(k)}{(r_2 - r_1)\ln(p)}.$$

If $f^{-1}(X) = X$, then the definition of the Julia set in equation (1.2) implies that $J_f = X$. Hence $\mu(J_f) = \mu(X)$ and $\mathrm{Hdim}(J_f) = n$. The case of $f^{-1}(X) \subsetneq X$ is more complicated. Hence, Theorem 1.3 has some hypotheses in addition to conditions (i) and (ii). Using the terminology of [**6**], Condition (d) says that f is locally scaling on X. Since $r_2 - r_1 > 0$, f is a *p-adic repeller* on $\mathcal{J}(f, X)$, using the terminology of [**2**]. Since the scale is the same for every ball, we could also say that f is $p^{r_2 - r_1}$-Lipschitz on X.

The assumption that $f(B_{p^{r_1}}(c_i)) \bigcap X \neq \emptyset$ for all $1 \leq i \leq k$ eliminates extraneous balls that would not contribute to the Julia set. As observed in [**2**], the Julia set $\mathcal{J}(f, X)$ is totally invariant, so $f(\mathcal{J}(f, X)) = \mathcal{J}(f, X) \subset X$. If $f(B_{p^{r_1}}(c_i)) \bigcap X = \emptyset$ for some $1 \leq i \leq k$, then $B_{p^{r_1}}(c_i) \bigcap \mathcal{J}(f, X) = \emptyset$. In this case, we can remove the ball from X and end up with the same Julia set as before.

In Section 4, we conclude with Proposition 4.1, which gives examples of polynomial maps that satisfy the conditions of Theorem 1.3. We conclude with a discussion of some standard examples included in Proposition 4.1. One of these examples illustrates the dependence of the Julia set on X, which depends in turn on K. Calculations of the Haar measure and Hausdorff dimension for Example 4.2 and Example 4.3 were included in the author's Ph.D. dissertation, completed under the supervision of Jane Hawkins at the University of North Carolina at Chapel Hill [**3**]. The author would like to thank Jane Hawkins and Cesar Silva for their conversations and suggestions, and to thank the referee for the comments and suggestions for improving the presentation.

2. Haar measure and Hausdorff dimension in finite extensions

This section contains the proofs of Proposition 1.1 and Proposition 1.2. For a ball in a finite extension of $\mathbb{Q}_p$, these propositions give the Haar measure and Hausdorff dimension, respectively.

Since K is a locally compact abelian group, there is a Haar measure on K that is unique up to scaling. Proving Proposition 1.1, we use the translation-invariance of Haar measure to calculate the measure of balls in K.

PROOF OF PROPOSITION 1.1. Let K be an extension of $\mathbb{Q}_p$ of degree n. Let μ be Haar measure on K, normalized so that $\mu(\mathcal{O}_K) = 1$. Let $r \in \nu(K^\times)$ and $a \in K$. The invariance of Haar measure under translation implies that $\mu(B_{p^r}(a)) = \mu(B_{p^r}(0))$, so we focus our attention on balls centered at 0. Moreover, a number $r \in \nu_p(K^\times)$ has the form k/e, where $k \in \mathbb{Z}$ and e is the ramification index.

If $k = 0$, then it follows from our normalization that

$$\mu(B_1(0)) = \mu(\mathcal{O}_K) = 1 = (p^0)^n.$$

If $k > 0$, then

$$B_{p^{k/e}}(0) = \bigsqcup_{(a_{-k},\dots,a_{-1})\in A^k} B_1(a_{-k}\pi^{-k} + \cdots + a_{-1}\pi^{-1}).$$

Recall that A contains p^f distinct elements, where $n = ef$. Then $B_{p^{k/e}}(0)$ is a disjoint union of p^{kf} translates of $\mathcal{O}_K$, so

$$\begin{aligned}\mu(B_{p^r}(0)) &= \mu(B_{p^{k/e}}(0))\\ &= p^{kf} = (p^r)^n.\end{aligned}$$

For the case of negative exponents, consider $-k$ with $k > 0$. Then

$$B_1(0) = \bigsqcup_{(a_0,\dots,a_{k-1})\in A^k} B_{p^{-k/e}}(a_0 + a_1\pi + \cdots + a_{k-1}\pi^{k-1}).$$

Now $\mathcal{O}_K$ is a disjoint union of p^{kf} translates of $B_{p^{-k/e}}(0)$, so

$$\begin{aligned}\mu(B_{p^r}(0)) &= \mu(B_{p^{-k/e}}(0))\\ &= p^{-kf} = (p^r)^n.\end{aligned}$$

□

Since K is a separable metric space under the p-adic metric, it is possible to define Hausdorff measures and the Hausdorff dimension for subsets of K. In order to set notation, we begin by reviewing these definitions. First, define an outer measure using countable covers of a set by balls with diameter less than $\delta > 0$. Since every point in a ball is a center of the ball, the diameter of a ball in K is equal to its radius. For $A \subset \mathbb{Q}_p$ and $s \geq 0$, define the outer measures by

$$h^*_{s,\delta}(A) = \inf\left\{\sum_{i=1}^{\infty}(r_i)^s : A \subset \bigcup_{i=1}^{\infty} B_{p^{r_i}}(a_i), r_i \in \nu(K^\times), p^{r_i} \leq \delta, a_i \in K\right\}$$

and

$$h^*_s(A) = \lim_{\delta\to 0} h^*_{s,\delta}(A).$$

The restrictions of $h^*_{s,\delta}$ and h^*_s to their measurable sets are the measure $h_{s,\delta}$ and h_s, respectively. Finally, the *Hausdorff dimension* of a measurable set $A \subset K$ is

$$\mathrm{Hdim}(A) = \sup\{s \geq 0 : h_s(A) > 0\}.$$

PROOF OF PROPOSITION 1.2. For simplicity of notation, we present the proof that the Hausdorff dimension of $\mathcal{O}_K = B_1(0)$ is equal to the degree of the finite extension. A similar proof gives the Hausdorff dimension for $B_{p^r}(a)$, where $r \in \nu(K^\times)$ and $a \in K$. The only change is the number of balls in the covers. For any $k \in \mathbb{N}$ such that $-k < r$, the ball $B_{p^r}(a)$ is equal to a disjoint union of $p^{n(r+k)}$ balls of radius p^{-k}.

Let K be an extension of $\mathbb{Q}_p$ of degree n. To show that $\mathrm{Hdim}(\mathcal{O}_K) = n$, we specify covers of $\mathcal{O}_K$ to give an upper bound on the s-dimensional Hausdorff measure $h_s(\mathcal{O}_K)$. Let $k \in \mathbb{N}$. Adjusting notation from the proof of Proposition 1.1, we have the cover

$$\mathcal{O}_K = \bigsqcup_{(a_0,\ldots,a_{ke-1})\in A^{ke}} B_{p^{-k}}(a_0 + a_1\pi + \cdots + a_{ke-1}\pi^{ke-1}).$$

This cover consists of $(p^f)^k e = p^{nk}$ balls of radius p^{-k}. For all $k \in \mathbb{N}$ such that $p^{-k} < \delta$, we have the upper bound

$$\begin{aligned} h_{s,\delta}(\mathcal{O}_K) &\leq p^{nk}\left(\frac{1}{p^k}\right)^s \\ &= p^{k(n-s)}, \end{aligned}$$

which yields the upper bound

$$\begin{aligned} h_s(\mathcal{O}_K) &= \lim_{\delta\to 0} h_{s,\delta}(\mathcal{O}_K) \\ &\leq \lim_{k\to\infty} p^{k(n-s)} \\ &= \begin{cases} 0 & \text{if } s > n \\ 1 & \text{if } s = n \\ \infty & \text{if } 0 \leq s < n \end{cases} \end{aligned} \tag{2.1}$$

By this upper bound on the Hausdorff measure, $h_s(\mathcal{O}_K) = 0$ for all $s > n$.

To complete the proof that $\mathrm{Hdim}(\mathcal{O}_K) = n$, we show that $h_n(\mathcal{O}_K) = 1$. If $\{B_{p^{t_i}}(a_i)\}_{i\geq 1}$ is a countable cover of $\mathcal{O}_K$ by balls such that $t_i \in \nu(K^\times)$ and $a_i \in K$, then

$$\begin{aligned} \sum_{i=1}^{\infty}(p^{t_i})^n &= \sum_{i=1}^{\infty}\mu(B_{p^{t_i}}(a_i)) \\ &\geq \mu\Big(\bigcup_{i=1}^{\infty} B_{p^{t_i}}(a_i)\Big) \\ &\geq \mu(\mathcal{O}_K) = 1. \end{aligned}$$

Since these inequalities hold for any such cover of $\mathcal{O}_K$ by balls, we have $h_{n,\delta}(\mathcal{O}_K) \geq 1$ for all $\delta > 0$. This lower bound and the upper bound from equation (2.1) combine to give $h_n(\mathcal{O}_K) = 1$. Since $h_s(\mathcal{O}_K) = 0$ for all $s > n$ and $h_s(\mathcal{O}_K) > 0$ for $s = n$, we have $\mathrm{Hdim}(\mathcal{O}_K) = n$. □

3. Haar measure and Hausdorff dimension of Julia sets

Before beginning the proof of Theorem 1.3, we give Lemma 3.1, which says that locally scaling maps are bijective on balls of sufficiently small radius.

LEMMA 3.1. *Let $f : X \to K$ be a map on a compact-open $X \subset K$ satisfying conditions* (i) *and* (ii)*. For each $1 \leq i \leq k$, for all $a \in B_{p_i^r}(c_i)$ and $r' \in \nu(K^\times)$ such that $r' < r_i$, the restricted map*

$$f : B_{p^{r'}}(a) \to B_{p^{r'+r_i}}(f(a))$$

is a bijection.

Lemma 3.1 follows from the proof of Lemma 5.1 in [**6**]. (Even though Kingsbery et al. implicitly assume that $T(X) \subset X$, they do not use this fact in the proof of Lemma 5.1.) Note that Lemma 2.1 in [**2**] is also a special case of Lemma 3.1.

Recall that $\mathcal{J}(f, X) = \bigcap_{i=0}^{\infty} f^{-i}(X)$. We prove Theorem 1.3 by using the open (and closed) sets $f^{-i}(X)$ as covers of $\mathcal{J}(f, X)$. Lemma 3.1 allows us to determing the size and number of the balls in $f^{-(i+1)}(X)$, given the size and number of the balls in $f^{-i}(X)$.

PROOF OF THEOREM 1.3. Let K be an extension of $\mathbb{Q}_p$ of degree n, and let μ be Haar measure on K. Let $k \in \mathbb{N}$ and $f : X \to K$ be a map on a compact-open $X \subset K$ such that

$$X = \bigcup_{i=1}^{k} B_{p^{r_1}}(c_i),$$

where $f^{-1}(X) \subset X$ and $\max_{1 \leq i \leq k} \{|c_i|_p\} = p^{r_2}$ for some $r_2 > r_1$. For each $1 \leq i \leq k$, suppose that $f(B_{p^r}(c_i)) \bigcap X \neq \emptyset$ and that

$$|f(x) - f(y)|_p = p^{r_2 - r_1}|x - y|_p$$

for all $x, y \in B_{r_1}(c_i)$. Having fixed f and X, we simplify notation in this proof with $\mathcal{J} = \mathcal{J}(f, X)$.

Since $r_2 > r_1$ and $\max_{1 \leq i \leq k} \{|c_i|_p\} = p^{r_2}$, the set X is contained in the ball $B_{p^{r_2}}(0)$. Since K is an extension of $\mathbb{Q}_p$ of degree n, $B_{p^{r_2}}(0)$ contains $p^{n(r_2 - r_1)}$ balls of radius p^{r_1}.

To prove the first statement of Theorem 1.3, we assume that $k = p^{n(r_2 - r_1)}$. In this case, $X = B_{p^{r_2}}(0)$, and f maps $B_{p^{r_1}}(c_i)$ bijectively onto $B_{p^{r_2}}(0)$ for each $1 \leq i \leq k$. Hence $f^{-1}(X) = X$, so $\mathcal{J} = X$. Thus, Proposition 1.1 implies that $\mu(\mathcal{J}) = k(p^{r_1})^n = p^{nr_2}$, and Proposition 1.2 implies that $\mathrm{Hdim}(\mathcal{J}) = n$.

To prove the second statement of Theorem 1.3, we assume that $k < p^{n(r_2 - r_1)}$. Since $\mathcal{J} = \bigcap_{j=0}^{\infty} f^{-j}(X)$, we use the $f^{-i}(X)$ as covers of $\mathcal{J}$. We use Lemma 3.1 and an induction proof to count the number of disjoint balls in $f^{-i}(X)$. The induction proof not only shows that there are k^{i+1} balls in $f^{-i}(X)$, but it also shows that these balls are evenly distributed through $f^{-(i-1)}(X)$. More precisely, there are k balls of $f^{-i}(X)$ in each ball of $f^{-(i-1)}(X)$. This distribution of balls is important when applying Lemma 3.1 to find the preimages of the balls in the induction step.

For the base case, we show that $f^{-1}(X)$ is a disjoint union of k^2 balls of radius $p^{r_1 - (r_2 - r_1)}$, such that k of these balls are contained in each ball of radius p^{r_1} in $f^{-0}(X) = X$. By Lemma 3.1, the restricted map

$$f : B_{p^{r_1}}(c_i) \to B_{p^{r_2}}(f(c_i)) \tag{3.1}$$

is a bijection. Then

$$f(B_{p^{r_1}}(c_i))\bigcap X = B_{p^{r_2}}(f(c_i))\bigcap X \neq \emptyset.$$

Since two balls of the same radius in K are either disjoint or equal, we conclude that $B_{p^{r_2}}(f(c_i)) = B_{p^{r_2}}(0)$. Then the bijection in (3.1) implies that for all $1 \leq i' \leq k$, there exists $a_{i'} \in B_{p^{r_1}}(c_i)$ such that $f(a_{i'}) = c_i$. By Lemma 3.1, $f^{-1}(X)\bigcap B_{p^{r_1}}(c_i)$ is a union of the k disjoint balls $B_{p^{r_1-(r_2-r_1)}}(a_{i'})$, for $1 \leq i' \leq k$. Since this argument holds for each $1 \leq i \leq k$, $f^{-1}(X)$ is a disjoint union of k^2 balls of radius $p^{r_1-(r_2-r_1)}$, such that k of these balls are contained in each ball of radius p^{r_1} in $f^{-0}(X)$.

For an induction hypothesis, we assume that there exists an integer $i \geq 1$ such that $f^{-i}(X)$ is a disjoint union of k^{i+1} balls of radius $p^{r_1-i(r_2-r_1)}$ such that k of these balls are in each ball of radius $p^{r_1-(i-1)(r_2-r_1)}$ in $f^{-(i-1)}(X)$. Suppose $B_{p^{r_1-(i-1)(r_2-r_1)}}(a)$ is a ball in $f^{-(i-1)}(X)$. By Lemma 3.1, the restricted map

$$f : B_{r_1-i(r_2-r_1)}(a) \to B_{p^{r_1-(i-1)(r_2-r_1)}}(f(a)) \tag{3.2}$$

is a bijection. Since $B_{p^{r_1-(i-1)(r_2-r_1)}}(f(a))$ is a ball in $f^{-(i-1)}(X)$, it contains k balls of radius $p^{r_1-i(r_2-r_1)}$ from $f^{-i}(X)$. The preimages of these k balls under (3.2) are k disjoint balls of radius $p^{r_1-(i+1)(r_2-r_1)}$. Hence, $f^{-(i+1)}(X)$ is a disjoint union of k^{i+2} balls of radius $p^{r_1-(i+1)(r_2-r_1)}$ such that k of these balls are in each ball of radius $p^{r_1-i(r_2-r_1)}$ in $f^{-i}(X)$.

Next, we use the number of balls in $f^{-i}(X)$ to calculate the Haar measure of $\mathcal{J}$. Since $f^{-i}(X)$ is a disjoint union of k^{i+1} balls of radius $p^{r_1-i(r_2-r_1)}$,

$$\begin{aligned}\mu(f^{-i}(X)) &= k^{i+1}(p^{r_1-i(r_2-r_1)})^n \\ &= kp^{r_1}\left(\frac{k}{p^{n(r_2-r_1)}}\right)^i.\end{aligned}$$

Under our assumption that $k < p^{n(r_2-r_1)}$, we have $\lim_{i\to\infty}\mu(f^{-i}(X)) = 0$. Since $\mathcal{J} \subset f^{-i}(X)$ for all $i \in \mathbb{N}$, we have $\mu(\mathcal{J}) = 0$.

Finally, we use the $f^{-i}(X)$ as covers of $\mathcal{J}$ to calculate the Hausdorff dimension of $\mathcal{J}$. As in the proof of Proposition 1.2, we first show that $h_s(\mathcal{J}) = 0$ for all $s > \ln(k)/(r_2-r_1)\ln(p)$. Then we show that $h_s(\mathcal{J}) > 0$ for $s = \ln(k)/(r_2-r_1)\ln(p)$.

If $p^{r_1-i(r_2-r_1)} < \delta$, then the k^{i+1} disjoint balls of radius $p^{r_1-i(r_2-r_1)}$ making up $f^{-i}(X)$ and covering $\mathcal{J}$ yield the inequality

$$\begin{aligned}h_{s,\delta}(\mathcal{J}) &\leq k^{i+1}(p^{r_1-i(r_2-r_1)})^s \\ &= kp^{sr_1}\left(\frac{k}{p^{s(r_2-r_1)}}\right)^i.\end{aligned}$$

Thus,

$$\begin{aligned}h_s(\mathcal{J}) &= \lim_{\delta\to 0} h_{s,\delta}(\mathcal{J}) \\ &\leq \lim_{i\to\infty} kp^{sr_1}\left(\frac{k}{p^{s(r_2-r_1)}}\right)^i \\ &= \begin{cases} 0 & \text{if } s > \ln(k)/(r_2-r_1)\ln(p) \\ k^{r_2/(r_2-r_1)} & \text{if } s = \ln(k)/(r_2-r_1)\ln(p) \\ \infty & \text{if } 0 \leq s < \ln(k)/(r_2-r_1)\ln(p) \end{cases} \end{aligned} \tag{3.3}$$

By this upper bound on the Hausdorff measure, $h_s(\mathcal{J}) = 0$ for all $s > \ln(k)/(r_2 - r_1)\ln(p)$.

Let $s = \ln(k)/(r_2 - r_1)\ln(p)$. Fix $\delta > 0$ and a cover $\{B_{p^{t_i}}(a_i)\}_{i=1}^{\infty}$ of $\mathcal{J}$ such that $t_i \in \nu(K^\times)$, $p^{t_i} \leq \delta$, and $a_i \in K$. Since $p^{r_1 - i(r_2-r_1)} \to 0$ as $i \to \infty$ and each $f^{-i}(X)$ covers $\mathcal{J}$ using only finitely many balls, $\mathcal{J}$ is a totally bounded subset of the metric space K. Hence, the Julia set is relatively compact. Thus, a countable cover of $\mathcal{J}$ by balls contains a finite subcover. Reindexing if necessary, let $\{B_{p^{t_i}}(a_i)\}_{i=1}^{j}$ be a finite subcover. Let $m = \min_{1\leq i \leq j} r_i$ and $i^* \in \mathbb{N}$ such that $r_1 - i^*(r_2 - r_1) \leq m$. Since balls in K are either disjoint or one ball is contained in the other, $f^{-i^*}(X) \subset \bigcup_{i=1}^{j} B_{p^{t_i}}(a_i)$. For $s = \ln(k)/(r_2 - r_1)\ln(p)$,

$$\begin{aligned} h_{s,\delta}\left(\bigcup_{i=1}^{\infty} B_{p^{t_i}}(a_i)\right) &\geq h_{s,\delta}\left(\bigcup_{i=1}^{j} B_{p^{t_i}}(a_i)\right) \\ &\geq h_{s,\delta}(f^{-i^*}(X)) \\ &= kp^{sr_1}\left(\frac{k}{p^{s(r_2-r_1)}}\right)^{i^*} \\ &= k^{r_2/(r_2-r_1)} > 0. \end{aligned}$$

Since this lower bound holds for any $\delta > 0$ and any cover of $\mathcal{J}$ by a union of balls of radius at most δ, we have $h_s(\mathcal{J}) \geq k^{r_2/(r_2-r_1)} > 0$ for $s = \ln(k)/(r_2 - r_1)\ln(p)$.

Since $h_s(\mathcal{J}) = 0$ for all $s > \ln(k)/(r_2 - r_1)\ln(p)$ and $h_s(\mathcal{J}) > 0$ for $s = \ln(k)/(r_2 - r_1)\ln(p)$, we conclude that $\text{Hdim}(\mathcal{J}) = \ln(k)/(r_2 - r_1)\ln(p)$. □

4. Examples and discussion

Proposition 4.1 gives a broad class of polynomials f with corresponding compact-open sets X so that f on X satisfies Theorem 1.3, which then gives the Haar measure and Hausdorff dimension of $\mathcal{J}(f, X)$. After stating and proving Proposition 4.1, we discuss standard examples of p-adic maps that fit the hypotheses of the proposition.

PROPOSITION 4.1. *Let K be an extension of $\mathbb{Q}_p$ of degree n, and let μ be Haar measure on K. Fix $r_1, r_2 \in \nu(K^\times)$ such that $r_1 < r_2$. For $1 \leq i \leq k \leq p^{n(r_2-r_1)}$, let $c_i \in K$ such that $\max_{1\leq i\leq k}\{|c_i|_p\} = p^{r_2}$ and $|c_i - c_j|_p = p^{r_2}$ for all $i \neq j$. Let $a \in K$ such that $|a|_p = p^{(2-k)r_2 - r_1}$ Set $X = \bigcup_{i=1}^{k} B_{p^{r_1}}(c_i)$, and define $f : X \to K$ by*

$$f(x) = a(x - c_1)(x - c_2)\cdots(x - c_k).$$

The polynomial f on X satisfies the hypotheses of Theorem 1.3.

PROOF. By hypothesis, we have that

$$X = \bigcup_{i=1}^{k} B_{p^{r_1}}(c_i),$$

such that $\max_{1\leq i\leq k}\{|c_i|_p\} = p^{r_2}$ for some $r_2 > r_1$ and that $k \leq p^{n(r_2-r_1)}$.

First, for all $1 \leq i \leq k$, we show that $|f(x) - f(y)|_p = p^{r_2-r_1}|x - y|_p$ for all $x, y \in B_{p^{r_1}}(c_i)$. If $x, y \in B_{p^{r_1}}(c_i)$ for some $1 \leq i \leq k$, then there exist $\bar{x}, \bar{y} \in B_{p^{r_1}}(0)$

such that $x = \bar{x} + c_i$ and $y = \bar{y} + c_i$. If $\bar{x} \in B_{p^{r_1}}(0)$, then

$$\begin{aligned} f(\bar{x} + c_i) &= a(\bar{x} + c_i - c_1)(\bar{x} + c_i - c_2) \cdots (\bar{x} + c_i - c_n) \\ &= a\bar{x} \prod_{j \neq i} (\bar{x} + (c_i - c_j)). \end{aligned}$$

Since $|\bar{x}|_p \le p^{r_1} < p^{r_2}$ and $|c_i - c_j|_p = p^{r_2}$ for $i \neq j$, the leading term (coefficient in front of the smallest power of π) in $\prod_{j \neq i}(\bar{x} + (c_i - c_j))$ is the same as the leading term in $\prod_{j \neq i}(c_i - c_j)$, which does not depend on $\bar{x} \in B_{p^{r_1}}(0)$. Hence,

$$\begin{aligned} |f(x) - f(y)|_p &= |f(\bar{x} + c_i) - f(\bar{y} + c_i)|_p \\ &= |a|_p |\bar{x} - \bar{y}|_p \left| \prod_{j \neq i} (c_i - c_j) \right|_p \\ &= p^{(2-k)r_2 - r_1} |(\bar{x} + c_i) - (\bar{y} + c_i)|_p p^{(k-1)r_2} \\ &= p^{r_2 - r_1} |x - y|_p. \end{aligned}$$

Finally, we show that $f(B_{p^{r_1}}(c_i)) \bigcap X \neq \emptyset$ for all $1 \le i \le k$ and $f^{-1}(X) \subset X$. Note that $f(c_i) = 0$ for all $1 \le i \le k$. By Lemma 3.1, the restricted map

$$f : B_{p^{r_1}}(c_i) \to B_{p^{r_2}}(0) \tag{4.1}$$

is a bijection. Since $X \subset B_{p^{r_2}}(0)$, the bijection in (4.1) implies that

$$f(B_{p^{r_1}}(c_i)) \bigcap X = B_{p^{r_2}}(0) \bigcap X \neq \emptyset.$$

Moreover, the bijection in (4.1) implies that each point in $X \subset B_{p^{r_2}}(0)$ has a preimage in each of the disjoint balls $B_{p^{r_1}}(c_i)$, for $1 \le i \le k$. Since f is a polynomial of degree k, a point cannot have more than k preimages. Hence, $f^{-1}(X) \subset X$. □

We conclude the paper with a discussion of some specific examples. Example 4.2 and Example 4.3 are standard examples that are discussed in Chapter 5 of [**7**] and generalized in [**2**]. Example 4.2 is also discussed in [**1**], and Example 4.3 is also discussed in [**5, 6, 8**].

EXAMPLE 4.2. Consider $f(x) = (x^2 - x)/p = x(x-1)/p$ on $X = B_{1/p}(0) \bigcup B_{1/p}(1) \subset \mathbb{Q}_p$. Since $k = 2 < p$, the Haar measure of $\mathcal{J}(f, X)$ in $\mathbb{Q}_p$ is zero and the Hausdorff dimension is $\ln 2 / \ln p$.

EXAMPLE 4.3. For $p \ge 3$, consider $f(x) = (x^p - x)/p$. The roots are 0 and the $(p-1)$th roots of unity. The compact-open set $X = \bigcup_{i=0}^{p-1} B_{1/p}(i) \subset \mathbb{Q}_p$ has one root in each ball. Since $k = p$, the Haar measure of $\mathcal{J}(f, X)$ in $\mathbb{Q}_p$ is $\mu(\mathbb{Z}_p) = 1$, and the Hausdorff dimension is 1.

Finally, Example 4.4 is a special case from a family of examples in [**2**] that generalize Example 4.3. This example illustrates that the Julia set $\mathcal{J}(f, X)$ can depend on X, which depends in turn on K.

EXAMPLE 4.4. For $p = 3$, consider $f(z) = (z^3 - 2z)/3$. If we take the compact-open set $X_1 = B_{1/3}(0) \subset \mathbb{Q}_3$, as in [**2**], then $\mathcal{J}(f, X_1) = \{0\}$. In this case, the Haar measure and Hausdorff dimension of $\mathcal{J}(f, X_1)$ in $\mathbb{Q}_p$ are both 0. If we take the compact-open set $X_2 = \bigcup_{i=0}^{2} B_{1/3}(i\sqrt{2}) \subset \mathbb{Q}_p[\sqrt{2}]$, then Proposition 4.1 and Theorem 1.3 imply that the Haar measure of $\mathcal{J}(f, X_2)$ in $\mathbb{Q}_p[\sqrt{2}]$ is still 0, but the Hausdorff dimension is $\ln 3 / \ln 9 = 1/2$.

References

[1] Robert Benedetto, Jean-Yves Briend, and Hervé Perdry, *Dynamique des polynômes quadratiques sur les corps locaux* (French, with English and French summaries), J. Théor. Nombres Bordeaux **19** (2007), no. 2, 325–336. MR2394889

[2] Aihua Fan, Lingmin Liao, Yue Fei Wang, and Dan Zhou, *p-adic repellers in $\mathbb{Q}_p$ are subshifts of finite type* (English, with English and French summaries), C. R. Math. Acad. Sci. Paris **344** (2007), no. 4, 219–224, DOI 10.1016/j.crma.2006.12.007. MR2292990

[3] Joanna Furno, *Ergodic Theory of p-adic Transformations*, ProQuest LLC, Ann Arbor, MI, 2013. Thesis (Ph.D.)–The University of North Carolina at Chapel Hill. MR3153461

[4] Fernando Q. Gouvêa, *p-adic Numbers*, 2nd ed., Universitext, Springer-Verlag, Berlin, 1997. An introduction. MR1488696

[5] Liang-Chung Hsia, *A weak Néron model with applications to p-adic dynamical systems*, Compositio Math. **100** (1996), no. 3, 277–304. MR1387667

[6] James Kingsbery, Alex Levin, Anatoly Preygel, and Cesar E. Silva, *On measure-preserving C^1 transformations of compact-open subsets of non-Archimedean local fields*, Trans. Amer. Math. Soc. **361** (2009), no. 1, 61–85, DOI 10.1090/S0002-9947-08-04686-2. MR2439398

[7] Joseph H. Silverman, *The Arithmetic of Dynamical Systems*, Graduate Texts in Mathematics, vol. 241, Springer, New York, 2007. MR2316407

[8] Christopher F. Woodcock and Nigel P. Smart, *p-adic chaos and random number generation*, Experiment. Math. **7** (1998), no. 4, 333–342. MR1678087

Indiana University-Purdue University Indianapolis, Department of Mathematical Sciences, 402 N. Blackford, LD 270, Indianapolis, Indiana 46202

E-mail address: jfurno@iupui.edu

Contemporary Mathematics
Volume **678**, 2016
http://dx.doi.org/10.1090/conm/678/13648

Lebesgue measure theoretic dynamics of rational maps

Jane Hawkins

ABSTRACT. We survey the role of one and two dimensional Lebesgue measure in complex dynamics. Before computers, rational maps with Julia set the whole sphere, a circle, or an arc were the only accessible maps so we begin with these classical examples. We then discuss some more recently studied families of rational maps that preserve finite or infinite measures equivalent to one and two dimensional Lebesgue measure. We end with a brief look at the idea behind Julia sets of quadratic polynomials with positive two-dimensional Lebesgue measure.

1. Introduction

Ever since the advent of computers, interest in the measurable dynamics of rational maps whose Julia sets are either the sphere or a smooth arc has been overshadowed by the multitude of beautifully illustrated studies of measures supported on fractal Julia sets. The fractal measures considered are a natural generalization of Lebesgue measure to noninteger dimension. However Lebesgue measurable dynamics still play an important and interesting role in the field of complex dynamics. One of the deepest results in the field in the past decade was the construction and proof of a Julia set which has positive area, but is not the entire sphere [**12**].

In this short survey article we start with the classical examples and move to more unusual families of rational maps of the Riemann sphere that exhibit interesting Lebesgue measurable dynamics. Many of the results mentioned in this paper have appeared elsewhere but we bring them together here under this common theme of Lebesgue measurable dynamics of rational maps. One of the main points of this survey is to illustrate that in the midst of fractal Julia sets lie many parametrized families of maps with smooth Julia sets and chaotic Lebesgue measure theoretic behavior.

In Section 2 we lay some groundwork with basic definitions and the classical examples. We continue the discussion in this section with two families of maps with chaotic one dimensional behavior; the measure we use is the arc length measure of a one-dimensional subset on the sphere, which we denote by m_1. One family of maps has the property that it is a nonpolynomial family (except for one map) of S-unimodal maps all of which are ergodic, exact, and admit an equivalent invariant probably measure.

2010 *Mathematics Subject Classification.* 37F10, 37A10, 37F45.

The next family is parametrized by one real parameter, and each map preserves Lebesgue measure ℓ on $\mathbb{R}$, the Julia set of each map. We connect these maps to inner functions by showing they are square roots of them. Many of the results in this section were obtained by two of the author's Ph.D. students [**6, 17**].

We then turn to m_2, normalized surface area measure in Section 4; we begin with a quick view of the Lattès examples dating back to 1918 and we then move forward to the 1980s and look at postcritically finite rational maps more generally. Theorem 4.1 is a new result. In Section 4.4 we give a brief idea of how one can show a Julia set for a polynomial could have positive m_2 measure.

The author thanks the co-organizers and hosts at Bryn Mawr and Williams College for the Oxtoby Centennial Conferences. Work of Oxtoby inspired this measurable approach to complex dynamics. The author is also grateful to the referee for suggestions that greatly improved the exposition in this paper.

2. Preliminary definitions and the first examples

We start with the first two classical examples of rational maps that one studies; they have smooth Julia sets. They are $p_0(z) = z^2$ and the polynomial occurring at the tip of the Mandelbrot set: $p_{-2}(z) = z^2 - 2$. The second map is more often presented to us as the degree two Chebychev polynomial $\tau_2(z) = 2z^2 - 1$. The maps p_{-2} and τ_2 are related via the map $\phi(z) = 2z$, which maps $[-1, 1]$ linearly onto $[-2, 2]$, and satisfies $\phi \circ \tau_2(z) = p_{-2} \circ \phi(z)$.

Throughout this paper we let $\mathbb{C}_\infty$ denote the Riemann sphere. The maps of interest are rational maps of the form: $R(z) = \dfrac{p(z)}{q(z)}$, with p, q polynomials over $\mathbb{C}$ with no common factor, and such that the maximum degree of p and q is at least 2. Rational maps characterize the analytic maps of the Riemann sphere; by R^n we denote the n-fold composition of R with itself. Two rational maps $R, S : \mathbb{C}_\infty \to \mathbb{C}_\infty$ are *conformally conjugate* if there exists a linear fractional transformation ϕ on $\mathbb{C}_\infty$ such that

$$\phi \circ R = S \circ \phi. \tag{1}$$

We showed above that p_{-2} and τ_2 are conformally conjugate maps.

The notion of conjugacy has a few meanings in this setting. It often occurs that we have two rational maps R, S related by a map ϕ on $\mathbb{C}_\infty$ such that $\phi \circ R = S \circ \phi$, but ϕ is only continuous, in which case we say R and S are *topologically conjugate.*

We consider the circle $S^1 = \{z : |z| = 1\} \subset \mathbb{C}_\infty$; by m_1 we denote arc length measure with respect to the spherical metric on $\mathbb{C}_\infty$, normalized so that $m_1(S^1) = 1$. In most introductory dynamics courses, for the map $p_0(z) = z^2$ we see the following:

- $p_0(S^1) = S^1 = p_0^{-1}(S^1)$;
- $\lim_{n\to\infty} p_0^n(z) = 0$ for any z with $|z| < 1$;
- $\lim_{n\to\infty} p_0^n(z) = \infty$ for any z with $|z| > 1$.

The dynamics of p_0 restricted to S^1 are interesting and will be discussed further. Using the Euler identity: $e^{i\theta} = \cos\theta + i\sin\theta$, if we set $\psi(z) = \dfrac{1}{2}\left(z + \dfrac{1}{z}\right)$, restrict our attention to the circle, and write $z = e^{i\theta}$, then we see that $\psi(e^{i\theta}) = \cos(\theta)$, so the unit circle is mapped by ψ in a 2-to-one manner, except at the critical points ± 1, onto $[-1, 1]$. It is easy to show that for all $z \in \mathbb{C}$,

$$\psi(p_0(z)) = \tau_2(\psi(z)) = \cos(2\theta), \tag{2}$$

and in fact replacing 2 by any integer $d \geq 2$, we use the following to define the dth Chebychev polynomial τ_d:

$$\tau_d(\cos(z)) = \cos(dz), \tag{3}$$

One other useful and noteworthy property of the maps $\mathfrak{X}_d(z) = z^d$, $d \geq 2$ and τ_d, is that they are measure theoretically isomorphic to one-sided Bernoulli shifts with respect to their invariant probability measures $\sim m_1$. We turn to a brief review of some needed measure theoretic definitions, but first we recall the definition of a one-sided Bernoulli shift.

DEFINITION 2.1. We fix an integer $d \geq 2$ and consider $\mathcal{A} = \{1, \ldots, d\}$, a finite state space with the discrete topology. A vector $p = \{p_1, \ldots, p_d\}$ such that $p_k > 0$ and $\sum p_k = 1$ determines a measure on $\mathcal{A}$, namely $p(\{k\}) = p_k$. Let $\Omega = \prod_{i=0}^{\infty} \mathcal{A}$ be the product space endowed with the product topology and product measure ρ determined by $\mathcal{A}$ and p. The map σ is the one-sided shift to the left, $(\sigma x)_i = x_{i+1}$. We say σ is a *one-sided Bernoulli shift* and denote it by $(\Omega, \mathcal{D}, \rho; \sigma)$, where $\mathcal{D}$ denotes the Borel σ-algebra generated by the cylinder sets, completed with respect to ρ.

2.1. Measure theoretic preliminaries. We assume throughout that every space $(X, \mathcal{B}, \mu)$ under consideration is a Lebesgue probability space though sometimes we specify that μ is a σ-finite infinite measure. In our setting, $X \subset \mathbb{C}_\infty$ is always a closed set and $\mathcal{B}$ denotes the σ-algebra of Borel measurable sets. We assume that the measure space is complete with respect to μ (every subset of every null set for μ is measurable), and that T is a surjective *nonsingular endomorphism*; i.e., $T : X \to X$ satisfies: $\mu(A) = 0 \iff \mu(T^{-1}A) = 0$ for every $A \in \mathcal{B}$, and $\mu(T(X)\triangle X) = 0$. If ν is a σ-finite measure such that $\nu \sim \mu$, and $\nu(T^{-1}A) = \nu(A)$ for all $A \in \mathcal{B}$, we say that T is *measure-preserving*, or equivalently T *preserves* ν. Without loss of generality we can assume that T is forward measurable and forward nonsingular; i.e., for all measurable sets A, $T(A) \in \mathcal{B}$ and $\mu(A) = 0 \iff \mu(TA) = 0$. When we say that a property holds on X (μ mod 0) or μ a.e., we mean that there is a set $N \in \mathcal{B}$ with $\mu(N) = 0$, (N is possibly the empty set), such that the property holds for all $x \in X \setminus N$.

DEFINITION 2.2. Let $T_1 : (X_1, \mathcal{B}_1, \mu_1) \circlearrowleft$ and $T_2 : (X_2, \mathcal{B}_2, \mu_2) \circlearrowleft$ be two measure-preserving endomorphisms.

A measurable map $\phi : X_1 \to X_2$ is a *homomorphism* if there exists a set $Y_1 \in \mathcal{B}_1$ of full measure and a set $Y_2 \in \mathcal{B}_2$ of full measure in X_2 such that ϕ maps Y_1 onto Y_2.

If there exists a homomorphism ϕ such that $T_1(Y_1) = Y_1$, $T_2(Y_2) = Y_2$, $\phi \circ T_1 = T_2 \circ \phi$ on Y_1, and $\mu_2(A) = \mu_1(\phi^{-1}(A))$ for all $A \in \mathcal{B}_1$, then T_2 is called a *factor* of T_1 (w.r.t. the measures μ_1 and μ_2), *with factor map* ϕ.

If in addition ϕ is injective on Y_1 we say it is an *isomorphism*. If T_2 is a factor of T_1 and ϕ is an isomorphism, then we say that the endomorphisms T_1 and T_2 are *isomorphic* endomorphisms.

For any two sets $A, B \in \mathcal{B}$ we define $A\triangle B = (A \setminus B) \cup (B \setminus A)$. The map T is *ergodic* if T has a trivial field of invariant sets, or equivalently, if any measurable set T with the property that $\mu(B\triangle T^{-1}B) = 0$ has either zero or full measure.

A map is *exact* if it has a trivial tail field $\cap_{n\geq 0}T^{-n}\mathcal{B} \subset \mathcal{B}$, or equivalently, if any set B with the property $\mu(T^{-n} \circ T^n(B) \triangle B) = 0$ for all n has either zero or full measure. It is clear that every exact map is also ergodic.

Assume $T : (X, \mathcal{B}, \mu) \to (X, \mathcal{B}, \mu)$ preserves μ. We recall a condition which is strictly weaker than one-sided Bernoulli for endomorphisms [**16**], but equivalent to Bernoulli in the invertible case [**15**]. We refer the reader to [**11, 23, 24**] if more detail is needed.

We consider partitions $\zeta = \{P_1, P_2, \cdots\}$ and $\eta = \{Q_1, Q_2, \cdots\}$ of X; each set is measurable and $\cup_{i\geq 1}P_i = \cup_{j\geq 1}Q_j = X$ ($\mu \bmod 0$), with each union disjoint ($\mu \bmod 0$). The notation $\zeta \vee \eta$ denotes the partition such that each set is of the form $P_i \cap Q_j$ for some i, j. The partition ζ *is independent of* η if

$$|\mu(P_i \cap Q_j) - \mu(P_i)\mu(Q_j)| = 0 \text{ for all } i, j.$$

The partition ζ is defined to be $\varepsilon-$*independent of* η if

$$\sum_i \sum_j |\mu(P_i \cap Q_j) - \mu(P_i)\mu(Q_j)| \leq \varepsilon.$$

DEFINITION 2.3. For an ergodic measure-preserving endomorphism T on $(X, \mathcal{B}, \mu)$, (invertible or noninvertible) a partition ζ is *weak Bernoulli* if given $\varepsilon > 0$, there exists $N \in \mathbb{N}$ such that for all $m \geq 1$,

$$\bigvee_0^m T^{-i}\zeta \quad \text{is} \quad \varepsilon - \text{independent of} \bigvee_N^{N+m} T^{-i}\zeta.$$

DEFINITION 2.4. (1) We say that a noninvertible endomorphism T on $(X, \mathcal{B}, \mu)$ has the *weak Bernoulli property* or that T is *weak Bernoulli* if there exists a weak Bernoulli partition ζ for T such that $\bigvee_{i=0}^{\infty} T^{-i}(\zeta) = \mathcal{B}\,(\mu \bmod 0)$.

(2) An automorphism $\tilde{T}$ is *the natural extension* of the (noninvertible endomorphism) T if T is a measurable factor of $\tilde{T}$ and any other automorphism S which has T as a factor also has $\tilde{T}$ as a factor.

The first example of a noninvertible endomorphism with a weak Bernoulli generator, that is not one-sided Bernoulli is due to Furstenberg [**16**]. It was proved by Friedman and Ornstein in [**15**] that for an invertible transformation T, if there exists a weak Bernoulli partition ζ such that:

$$\bigvee_{i=-\infty}^{\infty} T^{-i}(\zeta) = \mathcal{B}\,(\mu \bmod 0),$$

(ζ is a two-sided generator), then T is isomorphic to an (invertible) Bernoulli shift. It is clear that a measure-preserving endomorphism T is one-sided Bernoulli if and only if there exists an independent generating partition. The construction of the natural extension leads to a straightforward proof that the a weakly Bernoulli endomorphism has a weakly Bernoulli, hence Bernoulli natural extension (see e.g. [**23**]). Weakly Bernoulli endomorphisms exhibit many highly mixing properties, and conditions under which piecewise smooth bounded-to-one interval maps are weakly Bernoulli were given by Ledrappier in [**23**].

When we endow the Riemann sphere $\mathbb{C}_\infty$ with the σ-algebra of Borel sets, we consider rational maps R and nonsingular Borel measures μ for R, supported on the Julia set $J(R)$ (see Defn 3.1) , such that R is a nonsingular d-to-1 endomorphism of $\mathbb{C}_\infty$, where d is the degree of the map. We only consider measures with respect to which R is d-to-one in the following measurable sense.

DEFINITION 2.5. A nonsingular map T on $(X, \mathcal{B})$ is d-to-one with respect to a measure μ if there exists a partition $\zeta = \{A_1, A_2, \ldots A_d\}$ of X into d disjoint atoms of positive measure, called a *Rohlin partition*, and satisfying:

(1) the restriction of T to each A_i, which we will write as T_i, is one-to-one (μ mod 0);
(2) each A_i is of maximal measure in $X \setminus \cup_{j<i} A_j$ with respect to property (1);
(3) T_i is one-to-one and onto X (μ mod 0) for $i = 1, \ldots, d$.

3. A family of ergodic and exact nonpolynomial maps with an invariant probability measure equivalent to ℓ

The stereographic projection map $\mathfrak{S}$ takes $\mathbb{C}_\infty \setminus \{\infty\}$ injectively onto $\mathbb{C}$, and circles containing $\{\infty\}$ and their arcs are mapped to lines and line segments under $\mathfrak{S}$. Moreover on these sets the measure $\mathfrak{S}_* m_1 \sim \ell$, therefore when we consider rational maps in $\mathbb{C}$ we use Lebesgue measure ℓ on $\mathbb{R}$ if the Julia set is a line or line segment. We know that the map p_0 preserves m_1, the map τ_2 preserves a probability measure $\mu \sim \ell$, and that both are one-sided $\{1/2, 1/2\}$ Bernoulli with respect to their invariant measures. Hence p_0 and τ_2 are measure theoretically isomorphic. We now give some examples of maps that are topologically conjugate to τ_2, but neither measure theoretically nor conformally conjugate to it. These examples were studied by the author and Barnes in [**5**] and their Bernoulli properties proved by the author and Bruin in [**11**].

We consider the family of maps on $([-2,2], \mathcal{B}, \ell)$:

$$I_a(x) = \frac{-8 + (2+8a)x^2}{4 + (4a-1)x^2}, \; a \in (0,1). \tag{4}$$

Proposition 3.1 below from [**5**], establishes that each map I_a, $a \in (0,1)$ is a unimodal map in the sense that it is piecewise monotone with one turning point (the critical point at 0), and is has a finite postcritical orbit. When $a = \frac{1}{4}$, the map I_a is exactly p_{-2}, so isomorphic to τ_2. From [**5**] we list some properties, and show the graph of a typical map in Fig. 1.

PROPOSITION 3.1. For each $a \in (0,1)$, and I_a given in (4),

- $I_a(-2) = I_a(2) = 2$, $I_a(0) = -2$, and I_a has one critical point at $x = 0$, is strictly decreasing on $[-2, 0)$ and strictly increasing on $(0, 2]$.
- $I_a''(0) = 8a \neq 0$.
- $I_a[-2,2] = I_a^{-1}[-2,2] = [-2,2]$.
- $I_a'(-2) = -1/a$ and $I_a'(2) = 1/a$, so $x = 2$ is a repelling fixed point.
- Each I_a is finite postcritical (has a finite forward critical orbit).
- There is one other fixed point in $[-2,2]$, namely $p = \frac{-2}{1+2\sqrt{a}} \in (-2, -2/3)$, with derivative $I'(p) = -(1 + 2\sqrt{a})$. That is, p is always repelling.

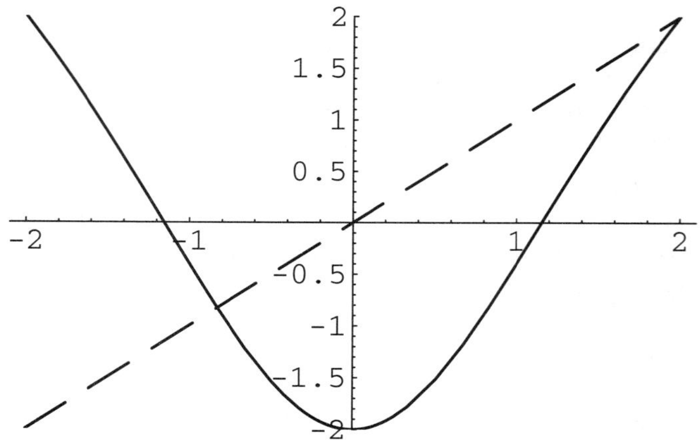

FIGURE 1. The graph of a non-Bernoulli S-unimodal map I_a with the dashed line $y = x$

These properties imply the existence of an invariant probability measure $\mu_a \sim \ell$ [**35**], and by [**23**] these maps are weakly Bernoulli, hence ergodic and exact with respect to ℓ (see [**5, 11**] for further details.)

It is also the case that the Schwarzian derivative of I_a is negative; namely,

$$S(I_a) := \frac{I_a'''}{I_a'} - \frac{3}{2}\left(\frac{I_a''}{I_a'}\right)^3 = \frac{-3}{2x^2} < 0,$$

but using ([**35**], Theorem D) the equivalent invariant probability measure follows from the fact that it is postcritically finite so this is not needed to prove the existence of an equivalent invariant measure.

In [**11**] it was proved that except for the $\{1/2, 1/2\}$ Bernoulli Chebyshev case, I_a is not one-sided Bernoulli. This is proved by showing that if I_a were $(\beta, 1-\beta)$ one-sided Bernoulli, then all periodic points of period k would need to have derivatives which are products of the form $\beta^i(1-\beta)^j$, with $i + j = k$, which cannot occur unless $\beta = 1/2$ and $a = 1/4$, the Chebychev case.

PROPOSITION 3.2. The map I_a is not one-sided Bernoulli except w.r.t. m if $a = 1/4$, the Chebyshev polynomial. For all $a \in (0, 1)$, I_a is ergodic, exact, and weak Bernoulli with respect to an invariant probability measure $\mu_a \sim \ell$.

Moreover, while the topological entropy of each map I_a, $a \in (0, 1)$ is $\log 2$, only the map $I_{\frac{1}{4}}$ has $h_\ell(I_{\frac{1}{4}}) = \log 2$, and for $a \in (0, 1) \setminus \frac{1}{4}$, preserving $\mu_a \sim \ell$, $h_{\mu_a} < \log 2$; this is discussed in [**5**]. The strict inequality on entropy follows from a result of Zdunik on rational maps ([**39**], Thm 1) in which it is shown that (when the Julia set is an interval) the unique measure of maximal entropy is singular with respect to ℓ except in the case of a Chebyshev polynomial.

COROLLARY 3.1. *The map $I_{\frac{1}{4}}$ is not measure theoretically isomorphic to I_a for any $a \in (0, 1) \setminus \{\frac{1}{4}\}$.*

I_a as a map of $\mathbb{C}_\infty$. Up to now we have not treated the maps I_a as maps of the sphere; a natural question is to ask is: what happens to points $z \in \mathbb{C}_\infty \setminus [-2, 2]$?

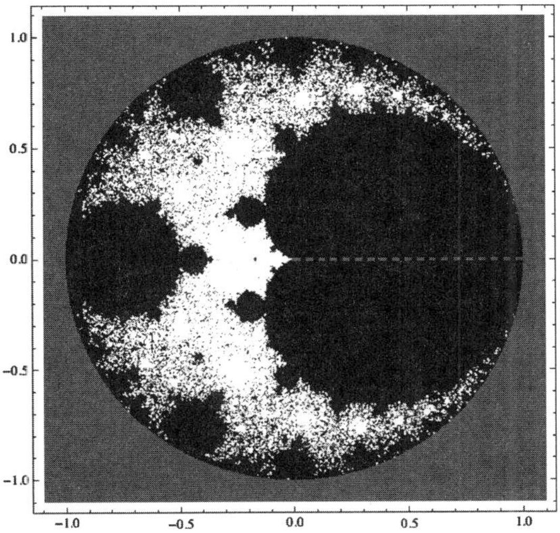

FIGURE 2. Parameter space for the maps I_a: the black parameters correspond to maps with attracting periodic orbits in $\mathbb{C}_\infty$; the dotted line corresponds to the parameters in Props. 3.1 and 3.2. Outside the unit disk $J(I_a)$ is a Cantor set.

It is not difficult to see that for parameters $a \in (0,1)$, there is an attracting fixed point outside the interval $[-2,2]$; in particular, $q = \frac{2}{(2\sqrt{a}-1)}$ is fixed, and has derivative $I_a'(q) = -1 + 2\sqrt{a}$. The derivative $I_a'(q)$ is an increasing function of a and is in $(-1,0)$, so is attracting. We can view the fixed point q as a decreasing function of a, with a pole at $a = \frac{1}{4}$. It follows then that if $a \in (0, 1/4]$, $q \in [-\infty, -2)$ and if $a \in (1/4, 1)$, $q \in (2, \infty)$

From this and the general theory of complex dynamics (see e.g. [**29**]), we can deduce:

$$\forall z \in \mathbb{C}_\infty \setminus [-2,2], \quad \lim_{n\to\infty} I_a^n(z) = q$$

It is also natural to ask what happens if a is a complex number not in $(0,1)$. For this we recall the definition of a Julia set and Fatou set.

DEFINITION 3.1. Let R be rational function of degree $d \geq 2$. The *Fatou set* of R is the open subset of $\mathbb{C}_\infty$ defined by:

$$\{z \in \mathbb{C}_\infty : \{R^n\} \text{ is equicontinuous at } z\}.$$

The *Julia set* of R is its complement in $\mathbb{C}_\infty$.

The terminology Julia set has been used for a long time - the first appearance in the literature seems to be around 1965 [**33**] - but the use of Fatou set for the set of normality only dates back to 1984 [**8**]. Earlier the Fatou set was usually referred to in the literature as the stable set and the set of normality. We write $F(R)$ and $J(R)$ for the Fatou and Julia sets respectively.

REMARK 3.1. There are other versions of the maps I_a discussed in [**5**]; we mention two here.

(1) The maps $f_a(z) = a(z + 1/z + 2)$, $a \in (0, 1)$ have $J(f_a) = [-\infty, 0]$. Each f_a is conformally conjugate to I_a via the linear fractional transformation $M(z) = \frac{2(z+1)}{z-1}$.
(2) The so-called modified Boole maps are defined by:

$$b_a(x) = \sqrt{a}(x - 1/x),$$

taking the positive square root of $a \in (0, 1)$. Each f_a is a factor of the modified Boole map b_a, via the two-fold branched covering map $\phi(x) = -x^2$.

In Fig. 2 we see that the parameters $a \in (0, 1)$ form the axis of a cardioid; this parameter space appeared in ([**28**], Fig. 11) and has been studied by the author in [**18**]. In particular the black areas of the parameter space represent open sets of parameters where an attracting periodic orbit exists. There are other areas of Fig. 2 that give maps of great interest in the study of Lebesgue dynamics. For example, we study a map corresponding to a non-real parameter, whose Julia set is the entire sphere, in Section 4. We now turn to the parameter $a = -1$, which generalizes to another parametrized family of quadratic maps, one whose only intersection with the space shown in Fig. 2 is that one point.

PROPOSITION 3.3. We define the map $\mathcal{I}(z) \equiv I_{-1}(z)$. We have $\mathcal{I} : \mathbb{R} \cup \{\infty\} \to \mathbb{R} \setminus (-2, \frac{6}{5})$, and $\mathcal{I}$ has the following properties as a rational map on $\mathbb{C}_\infty$:

(1) $\mathcal{I}(z) = \frac{-(8+6z^2)}{4-5z^2}$ has fixed points at 2 and $-\frac{2}{5} \pm \frac{4}{5}i$.
(2) $\mathcal{I}'(2) = -1$ and the other two fixed points are repelling.
(3) There are no period 2 cycles for $\mathcal{I}$; i.e., the fixed points of $\mathcal{I}^2$ are $2, 2, 2$, $-\frac{2}{5} + \frac{4}{5}i$ and $-\frac{2}{5} - \frac{4}{5}i$.

PROOF. One can check the fixed points and their derivatives by using Prop 3.1. Solving for period 2 points of $\mathcal{I}$ is equivalent to finding the roots of the polynomial $j(z) = 5z^5 - 26z^4 + 40z^3 - 16z^2 + 16z - 32$; using long division one can easily determine that $j(z) = (z - 2)^3(5z^2 + 4z + 4)$, from which the result follows. □

It is also stated in ([**28**], caption of Fig. 6) that a quadratic rational map has a fixed point with derivative -1 if and only if it has no period two orbits. This is proved in [**17**] along with a complete classification of which rational maps of degree d lack periodic orbits of period k; most of these maps result in nonsmooth Julia sets. However the quadratic family contains some examples of interest. The study of rational maps missing periodic orbits of certain periods goes back to Baker [**3**] and examples are given in [**7**]. Hagihara gave a complete list of such maps in her Ph.D. thesis [**17**].

COROLLARY 3.2. [**17**] *The map $\mathcal{I}$ is conformally conjugate to a map of the form:*

$$S_\alpha(z) = \frac{z^2 - z}{1 + \alpha z}, \tag{5}$$

for some $\alpha \in \mathbb{C} \setminus \{1\}$.

PROOF. It is proved in [**17**] that a quadratic map has no period 2 cycles if and only if it is conformally conjugate to a map of the form (5). □

While $J(\mathcal{I})$ is fractal and does not have nice Lebesgue measurable properties (meaning with respect to m_2, m_1 or ℓ), the family of maps S_α, which includes a conformally conjugate copy of $\mathcal{I}$, leads us to our discussion about infinite measure preserving maps.

3.1. A family of ergodic and exact rational maps with an infinite σ-finite invariant measure equivalent to ℓ. In this section we turn to infinite Lebesgue measure ℓ on $\mathbb{R}$. We have so far looked at families of finite measure preserving examples. There are maps with smooth one-dimensional Julia sets with the property that they preserve an infinite ergodic and exact σ-finite measure equivalent to m. Therefore they cannot preserve an equivalent ergodic finite measure so are fundamentally different from the family given in Eqn (4). We now return to the family of maps from Eqn (5): $S_\alpha(z) = \frac{z^2-z}{1+\alpha z}$, and we are interested only in $\alpha < -1$ for now.

The next modification we make is to conjugate S_α by the involution $\phi(z) = 1/z$, which gives us a map of the form:

$$U_{1+\alpha}(z) = -z - \frac{1+\alpha}{z-1} - (1+\alpha). \tag{6}$$

If $\alpha < -1$, then $U_{1+\alpha}$ is called a negative R-function; it can also be thought of as the square root of an inner function in the following sense. Recall that an inner function is a map f of the upper half plane satisfying: $\lim_{y\to 0} f(x+iy) = f(x)$, for ℓ a.e. $x \in \mathbb{R}$. The dynamics of inner functions have been studied extensively by Aaronson [**1**], along with the criteria supplied by Letac [**25**] on when they (and rational negative R-functions) preserve the infinite measure ℓ on $\mathbb{R}$. It is also directly verifiable that U_α preserves Lebesgue measure on $\mathbb{R}$. Rational maps that are negative R-functions flip the upper and lower half plane, but every even iteration is an inner function, so $U_{1+\alpha}^2$ is an inner function. To simplify notation, we write $b = 1+\alpha$, and we restrict to $\alpha \in [-3,-1)$ (see [**7, 17**]); for any $\alpha < -3$, S_α is conformally conjugate to $S_{\frac{3-\alpha}{1+\alpha}}$; then $b \in [-2, 0)$. For the next two theorems we need to recall the following concepts from infinite measure dynamics.

REMARK 3.2. (1) Since U_b preserves ℓ on $\mathbb{R}$, (for $b < 0$), and ℓ is infinite, for any fixed $c > 0$, U_b also preserves the scaled measure given by $(c\ell)(A) \equiv c \cdot \ell(A)$ for all measurable A. We say the maps U_{b_1} and U_{b_2} are *c-isomorphic* if for some $c > 0$, there exists a measurable isomorphism $\phi : \mathbb{R} \to \mathbb{R}$ with $\phi_*\ell \equiv \ell \circ \phi^{-1} = (c\ell)$, and such that $\phi \circ U_{b_1} = U_{b_2} \circ \phi$, ℓ -a.e.

(2) Assume $A \subset \mathbb{R}$ satisfies $0 < \ell(A) < \infty$ and $\ell(\mathbb{R} \setminus \cup_{i=0}^{\infty} U_b^{-i}(A)) = 0$. Then replacing U_b by U for simplicity of notation, by U_A we mean the map on A given by: $U_A(x) = U^{p(x)}(x)$ where $p(x) = \min\{i : U^i(x) \in A\}$.

(3) Since U preserves ℓ, (2) defines a measure preserving endomorphism $U_A : (A, \mathcal{B} \cap A, \ell|_A) \to (A, \mathcal{B} \cap A, \ell|_A)$ where $\ell|_A$ denotes the (non normalized) restriction of ℓ. We call U_A the *induced map on A*.

We have the following results from the Ph. D. thesis of Bayless [**6**]; (1) was shown in [**17**].

THEOREM 3.3. [**6**] *For* $b \in [-2,0)$, *for the map* $U_b(z) = -z - b + \frac{b}{1-z}$, *the following hold with respect to* ℓ:

(1) $J(U_b) = \mathbb{R} \cup \{\infty\}$;
(2) U_b *is conservative, exact, and ergodic;*
(3) *For each* U_b, *there exists a set* $A \in \mathcal{B}$ *of finite measure such that for* ℓ*-a.e.* $x \in \mathbb{R}$ *there exists a* $p(x) \in \mathbb{N}$ *such that* $U_b^{p(x)}(x) \in A$, *and such that the first return time partition (to* A*) of* $\mathbb{R}$ *has finite entropy (i.e.,* U_b *is said to be quasi-finite).*

3.1.1. *Entropy on infinite spaces.* We assume the reader has familiarity with the notion of measure theoretic entropy on probability spaces, and if not, Walters' book ([**37**], Chap 4) is a good source. When dealing with infinite invariant measures, there is an issue of scaling by a constant, which can change the value of invariants; here we use ℓ on $\mathbb{R}$, so $\ell([0,1]) = 1$. The earliest definition of entropy of an infinite measure preserving transformation is due to Krengel and is given as follows:

DEFINITION 3.2. [**22**] If $T : (X, \mathcal{B}, \mu) \to (X, \mathcal{B}, \mu)$ is a conservative measure-preserving system, then we define the Krengel entropy of T by:

$$h_{\text{kr}}(T) = \sup h(T_A, \mu_A),$$

where the supremum is taken over all $A \in \mathcal{B}$ satisfying $0 < \mu(A) < \infty$. We denote the Krengel entropy of U_b with respect to Lebesgue measure ℓ on $\mathbb{R}$, by $h_{\text{kr}}(U_b)$.

From Theorem 3.3 the Krengel entropy of each U_b is defined, and can be calculated for the maps U_b.

THEOREM 3.4. [**6**] *For each* $b \in [-2,0)$,

$$h_{\text{kr}}(U_b) = \int_{\mathbb{R}} \log |U_b'(t)| d\ell(t) = 2\pi\sqrt{-b}.$$

Moreover Bayless showed that every map of the form: $f(z) = -z - \beta - \dfrac{p}{t-z}$, with $\beta, t \in \mathbb{R}$, and $p > 0$ (i.e., any quadratic rational negative R-function), is c-isomorphic to exactly one map in the family U_b, and U_b and $U_{b'}$ are in turn mutually non-c-isomorphic for $b, b' \in [-2,0)$.

We mention that this family of mappings is embedded in a much more complicated parameter space studied by Hagihara in [**17**]. The parameter space resembles "the usual" ubiquitous Mandelbrot set missing the period 2 limb due to Corollary 3.2, and is difficult to produce due to the neutral fixed point. In Fig. 3 the dark unit disk centered at the origin shows parameters corresponding to maps S_α with an attracting fixed point at ∞, while the solid line from -3 to -1 shows parameters corresponding to maps S_α with Julia set the real line. There is a hole at $\alpha = -1$ where S_α does not have degree 2. We refer to a parameter space as *reduced* when no two points in parameter space represent conformally equivalent maps. Fig. 4 shows the reduced Mandelbrot set, arising for quadratic polynomials of the form $q_\kappa = \kappa z^2 + z$ (each of which is conformally conjugate to exactly one map p_c), for comparison with Fig. 3.

4. Two dimensional Lebesgue measure

We now turn to maps whose Julia set is two dimensional, and write m_2 for the normalized surface area measure on $\mathbb{C}_\infty$. We are interested in rational maps

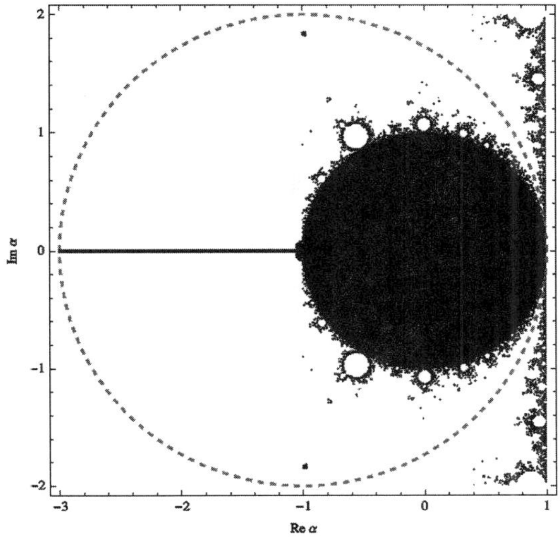

FIGURE 3. Parameter space for the maps S_α, equivalently U_b, with ℓ- preserving maps from Thm 3.3 marked by a solid (blue) line. The dashed (red) circle encloses the reduced parameter space.

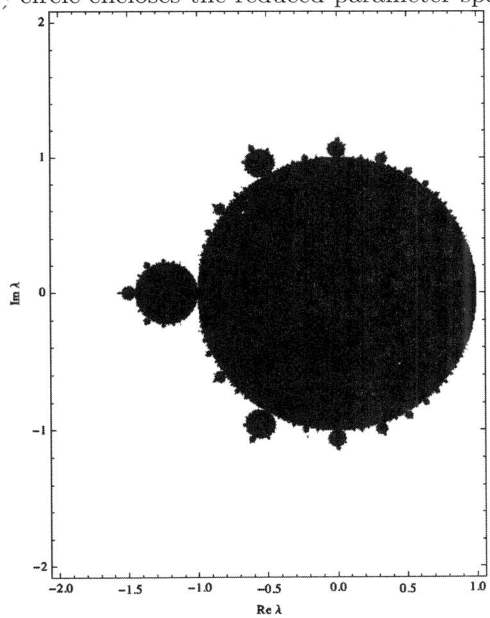

FIGURE 4. Reduced parameter space for the maps $q_\kappa(z) = \kappa z + z^2$, known as the Mandelbrot set. (Compare this with Fig. 3).

whose Julia set is all of $\mathbb{C}_\infty$ so that we can analyze their m_2 dynamical properties. Examples of this type are very old and it is natural they would predate computers. In fact there are many one-sided Bernoulli maps on the sphere that date back to Lattès in 1918. Milnor has written a clear and comprehensive exposition of Lattès' work along with updated examples, so we do not repeat that here, but refer the reader to the article [**31**]. Instead we illustrate the basic construction of these examples by giving a parametrized family of such maps (2 complex dimensions) that does not appear explicitly in [**31**] (though it can certainly be found there implicitly). Nonpolynomial Lattès maps are characterized by the property that the unique invariant probability measure of maximal entropy is equivalent to m_2 [**39**].

The idea behind a Lattès map in measure theoretic terms is to start with a uniformly expanding map of the torus, viewed as $\mathbb{C}/\Lambda$ with $\Lambda \subset \mathbb{C}$ a lattice; usually the map is as simple as $z \mapsto nz$, with $n \geq 2$ an integer. Such a map is easily seen to be ergodic, preserves ℓ^2, two dimensional Lebesgue measure on $\mathbb{C}/\Lambda$, and the generating partition to make it one-sided Bernoulli (i.e., isomorphic to the one-sided $\{\frac{1}{n^2}, \frac{1}{n^2}, \dots, \frac{1}{n^2}\}$ shift) is obvious. We then find a suitable factor map from the torus to the sphere, and look at the map generated on $\mathbb{C}_\infty$. If the factor map is finite-to-one, the pullback measure will inherit the Bernoulli measurable properties.

We also discuss postcritically finite maps that are not Lattès maps, and give the idea behind the result that such maps admit an ergodic finite invariant measure equivalent to m_2. There are many open questions about their measure theoretic entropy as well as the Hausdorff dimension of the maximal entropy measure whose support is necessarily the entire sphere.

4.1. A family of Lattès examples of degree 4. We begin with a complex torus: Let $\lambda_1, \lambda_2 \in \mathbb{C}\backslash\{0\}$ with $\lambda_2/\lambda_1 \notin \mathbb{R}$. A lattice is defined by $\Lambda = [\lambda_1, \lambda_2] := \{m\lambda_1 + n\lambda_2 \colon m, n \in \mathbb{Z}\}$; Λ is a discrete additive subgroup of $\mathbb{C}$ of rank 2, so $\mathbb{C}/\Lambda$ is a torus. Different sets of vectors can generate the same lattice Λ, but if $\Lambda = [\lambda_1, \lambda_2]$, and

$\Lambda = [\lambda_3, \lambda_4]$, the generators are related by

$$\begin{pmatrix} \lambda_3 \\ \lambda_4 \end{pmatrix} = \begin{pmatrix} a & b \\ c & d \end{pmatrix} \begin{pmatrix} \lambda_1 \\ \lambda_2 \end{pmatrix}$$

with $a, b, c, d \in \mathbb{Z}$ and $ad - bc = \pm 1$. Therefore given any lattice $\Lambda = [\lambda_1, \lambda_2]$, we interchange the order if needed, so that

$$\tau = \lambda_2/\lambda_1 \in \mathbb{H} = \{z : z = x + iy,\ y > 0\}.$$

It is useful to define an equally proportioned lattice $\Omega = [1, \tau]$, and set

$$\Lambda = k\Omega = k[1, \tau], \quad \tau \in \mathbb{H}, \quad k \neq 0. \tag{7}$$

A simple map such as $L(z) = \alpha z$, with the property that $\alpha\Lambda \subset \Lambda$ and $|\alpha| > 1$ defines a uniformly expanding map on $\mathbb{C}/\Lambda$ which gives rise via a meromorphic factor map to a rational map of $\mathbb{C}_\infty$ of degree α^2. We use $n = 2$ in the family of examples in Theorem 4.1, and integers $n \geq 2$ are the easiest to work with; however there are infinitely many noninteger examples, and in Fig. 2, at the parameter $a = -\frac{1}{4}$ we have a Lattès example obtained by using $\Lambda = [1, i]$ and $\alpha = \sqrt{2}$.

The factor map we use for these examples is an elliptic function which, by definition, is a meromorphic function in $\mathbb{C}$ periodic with respect to a lattice Λ. In our examples we use the Weierstrass elliptic $\wp$ function, defined by

$$\wp_\Lambda(z) = \frac{1}{z^2} + \sum_{w \in \Lambda\backslash\{0\}} \left(\frac{1}{(z-w)^2} - \frac{1}{w^2} \right), z \in \mathbb{C}.$$

Replacing every z by $-z$ in the definition we see that $\wp_\Lambda$ is an even function, $\wp_\Lambda$ is meromorphic, periodic with respect to Λ, and each pole, occurring exactly at lattice points, has order 2. The function $\wp$ exhibits some characteristics of trigonometric functions; it is periodic (with respect to a rank 2 subgroup of $\mathbb{C}$ rather than rank 1), it is related to its own derivative via a simple differential equation (see Eqn (8) below), and we have an "angle doubling" formula (see Eqn (12) below).

4.1.1. **ODEs for $\wp_\Lambda$ and its derivatives:** We have

$$\wp'_\Lambda(z)^2 = 4\wp_\Lambda(z)^3 - g_2\wp_\Lambda(z) - g_3, \tag{8}$$

where

$$g_2(\Lambda) = 60 \sum_{w\in\Lambda\setminus\{0\}} \frac{1}{w^4} \quad \text{and} \quad g_3(\Lambda) = 140 \sum_{w\in\Lambda\setminus\{0\}} \frac{1}{w^6}. \tag{9}$$

REMARK 4.1. We call (g_2, g_3) *invariants of* Λ ([**13**], Chap 2.22). In particular $g_2(\Lambda)$ and $g_3(\Lambda)$ are complete invariants of the lattice Λ since for any g_2 and g_3 such that $g_2^3 - 27g_3^2 \neq 0$ there exists a unique lattice having g_2 and g_3 as its sums in Eqn(9), ([**13**], Chap 2.11, and [**21**], Cor. 6.5.8), or equivalently, the pair (g_2, g_3) determines completely the value of $\wp_\Lambda(z)$ for all $z \in \mathbb{C}$. Moreover the invariants g_2, g_3 depend analytically on Λ in the sense that they vary analytically in $\tau \in \mathbb{H}$ when represented as in Eqn (7) (see [**21**], Thm 6.4.1), and from Eqn (9) it follows that

$$g_2(k\Omega) = k^{-4}g_2(\Omega), \text{ and } g_3(k\Omega) = k^{-6}g_3(\Omega). \tag{10}$$

Analogous to sine and cosine functions, we also have a second order ODE connecting the second derivative to the original function:

$$\wp''_\Lambda(z) = 6\wp^2_\Lambda(z) - \frac{g_2}{2}. \tag{11}$$

4.1.2. **Angle Doubling Formula for $\wp_\Lambda$:**

$$\wp_\Lambda(2z) = \frac{\wp''_\Lambda(z)^2}{4\wp'_\Lambda(z)^2} - 2\wp_\Lambda(z). \tag{12}$$

These classical identities and many others are worked out in detail in [**13**] or [**21**]. Putting together Eqns. (8) - (12), via the commuting diagram below, we obtain a rational map of the sphere of the form:

$$R(z) = \frac{z^4 + (g_2/2)z^2 + 2g_3z + g_2^2/16}{4z^3 - g_2z - g_3}, \tag{13}$$

depending on $\Lambda(g_2, g_3)$.

$$\begin{array}{ccc} (\mathbb{C}/\Lambda, \mathcal{B}, \ell^2) & \xrightarrow{2z} & (\mathbb{C}/\Lambda, \mathcal{B}, \ell^2) \\ \downarrow \wp_\Lambda & & \downarrow \wp_\Lambda \\ \mathbb{C}_\infty & \xrightarrow{R} & \mathbb{C}_\infty \end{array}$$

Since $\wp_\Lambda$ is meromorphic, the pullback measure on $\mathbb{C}_\infty$, $(\wp_\Lambda)_*\ell^2$, is equivalent to m_2. We are now ready to state our theorem, and give a brief proof.

THEOREM 4.1. *Given any* 3 *distinct complex numbers, a, b, c such that $a+b+c =$ 0, there exists a rational map of degree* 4 *of the form:*

$$R(z) = \frac{z^4 + (\alpha/2)z^2 + 2\beta z + \alpha^2/16}{4z^3 - \alpha z - \beta} \tag{14}$$

with $\alpha, \beta \neq 0$, such that:

(1) $J(R) = \mathbb{C}_\infty$
(2) *The critical values of R are exactly a, b, and c, each of multiplicity 2.*
(3) *The critical values get mapped to a fixed point at ∞, which is repelling.*
(4) *There are 6 distinct critical points for R, and they sum to 0.*
(5) *There exists a lattice Λ such that $\wp_\Lambda$ induces the map R from a toral endomorphism on $\mathbb{C}/\Lambda$;*
(6) *The map R is isomorphic to the one-sided $\{\frac{1}{4}, \frac{1}{4}, \frac{1}{4}, \frac{1}{4}\}$ Bernoulli shift with respect to the invariant probability measure $\mu = (\wp_\Lambda)_* \ell^2 \sim m_2$.*

PROOF. Given a, b, c we reorder them if necessary so that $a \neq 0, b \neq 0$; recall that at most one can be 0, so we assign that value to c if it occurs. We next set $p(z) = (z-a)(z-b)(z-c)$ and write it as:

$$p(z) = (z-a)(z-b)(z-(-a-b)) = z^3 - (a^2+ab+b^2)z + (a^2b+ab^2).$$

Using the angle doubling identities we get: $\alpha = 4(a^2+ab+b^2)$, $\beta = -4(a^2b+ab^2)$ to obtain

$$R(z) = \frac{z^4 + (\alpha/2)z^2 + 2\beta z + \alpha^2/16}{4z^3 - \alpha z - \beta}$$

While the statements about the critical values follow from some identities, in this case they can also be verified by hand. In particular, the 6 critical points for the map R are: $a \pm \sqrt{2a^2 - ab - b^2}$, $b \pm \sqrt{-a^2 - ab + 2b^2}$, and $c \pm \sqrt{2a^2 + 5ab + 2b^2}$, which sum to 0. The discriminant given for each critical point does not vanish because a, b, and c are distinct. Using $g_2 = \alpha$ and $g_3 = \beta$, we find a unique lattice $\Lambda(g_2, g_3)$ which we use for the covering torus. $\square$

It is not hard to vary the theorem a little in order to specify that $c_1, c_2 = \pm q \in \mathbb{C}$ be (distinct) critical points, instead of specifying the critical values. Choosing $g_2 = 4(\frac{q}{1+\sqrt{2}})^2$ and $g_3 = 0$ in (13) will work.

4.2. An example illustrating Theorem 4.1. In Theorem 4.1, we choose any nonzero $a \in \mathbb{C}, b = -a$, and $c = 0$. Following the proof, we see that $\alpha = 4(a^2 + ab + b^2) = 4a^2$ (which is g_2), and $\beta = -4(a^2b + ab^2) = 0$ (which is g_3), yielding the rational map:

$$F(z) = \frac{z^4 + 2a^2z^2 + a^4}{4z(z^2 - a^2)}. \tag{15}$$

It is a calculation to show that the 6 critical points for F are: $\pm ai, a(1 \pm \sqrt{2}), a(-1 \pm \sqrt{2})$. The critical values are as follows: $F(ai) = F(-ai) = 0$. One can substitute the other critical points and see that

$$F(a + a\sqrt{2}) = F(a - a\sqrt{2}) = a \text{ and } F(-a + a\sqrt{2}) = F(-a - a\sqrt{2}) = -a;$$

so we recover the three distinct critical values. It is then easy to see that each of the critical values has multiplicity 2 and is a pole, so is mapped under f to $\infty \in \mathbb{C}_\infty$.

The four fixed points for F that are in $\mathbb{C}$, are:

$$\pm ai\sqrt{-1 + \frac{2}{\sqrt{3}}}, \quad \pm a\sqrt{1 + \frac{2}{\sqrt{3}}},$$

and $F'(x) = -2$ for each of the fixed points. We note that ∞ is also a fixed point, and we calculate its multiplier, the analog of a derivative at ∞ when $\mathbb{C}_\infty$ is viewed as a compact surface. This is the derivative of: $z \mapsto 1/f(\frac{1}{z})$ at the origin, which can be shown to be 4.

Moreover every periodic point of period k, has derivative $\pm 2^m$ for some positive integer m, and in this example $m = k$ or $m = 2k$ for each k. This is a characteristic of Lattès examples, that their periodic points have derivatives of a particular form, but not verifiable unless you know how you arrived at the example (in our case, by multiplication by 2 on the torus, and then used $\wp$ as the factor map). This is discussed in ([**31**], Cor. 3.9).

Using Eqn.(10) we see that when $b = -a$, the underlying lattice Λ coming from Theorem 4.1 is of the form $\Lambda = [\lambda, \lambda i]$, $\lambda \neq 0$ satisfying

$$\lambda = \frac{1}{\sqrt{a}} \cdot \gamma,$$

where $\gamma > 0$ is the lemniscate constant which arises as the lattice side length for invariants $g_2 = 4, g_3 = 0$. Many details behind the constant γ and the invariants g_2, g_3 are discussed in [**19**].

4.3. Postcritically finite rational maps. The Lattès examples are special cases of postcritically finite rational maps. By definition, a rational map is *postcritically finite* if every critical point is preperiodic but not periodic. This excludes maps like $\mathfrak{X}_d(z) = z^d$, where 0 and ∞ are both fixed critical points on $\mathbb{C}_\infty$, but includes maps like the one in Example 4.2 and more general maps. We define the postcritical set of R by:

$$P(R) = \overline{\bigcup_{c \in C(R),\, n \in \mathbb{N}} R^n(c)}, \tag{16}$$

where $C(R)$ is the set of critical points of R; by definition $P(R)$ is closed and may or may not contain points from $C(R)$.

The reason a postcritically finite map R is of interest in Lebesgue ergodic theory is that $J(R) = \mathbb{C}_\infty$; a proof of this can be found in many places (see [**7**], Thm 9.4.4), but the basic idea is that any time there is a Fatou component, it maps eventually to a periodic component which must have a critical point associated to it (either in the Fatou set or with an infinite forward orbit, or both). This means that it is worth considering the measure theoretic structure of a postcritically finite map R with respect to the 2-dimensional probability measure m_2, as each such map is always topologically transitive (see e.g. [**7**], Thm 4.2.5). The measurable dynamics of postcritically finite maps have been extensively written about (e.g., [**4, 14, 18, 27, 34**]) and generalized to other settings (meromorphic postcritically finite, entire postcritically finite, and rational maps not postcritically finite but "close" to one). Here we review only a few of the basic measure theoretic results.

REMARK 4.2. (1) Assume that $P(R) = \{a_1, a_2, \ldots, a_k\}$ with each $a_j \notin C(R)$; it follows that $|P(R)| \geq 3$ ([**27**], Thm 3.6). To each a_j we assign the positive integer ν_j which is the least common multiple of the local degrees $\deg(R^k, y)$, for all $k > 0$ and y such that $y \in R^{-k}(a_j)$. We then have a set of ramification indices: $\mathcal{N} = \{\nu_1, \nu_2, \ldots, \nu_k\}$ with each $\nu_j \geq 2$. This leads to a metric which has a finite set of singularities for an otherwise smooth

Riemannian metric $\rho = \gamma(z)|dz|$ on $\mathbb{C}_\infty$, and the singularities are of the type $\rho = \dfrac{|dz|}{|z|^{1-1/\nu_j}}$ in local coordinates near each a_j.

(2) The pair $(\mathbb{C}_\infty \setminus P(R), \mathcal{N})$, with a complex atlas structure on $\mathbb{C}_\infty \setminus P(R)$, is called the *orbifold of* R and the metric ρ is its associated *orbifold metric.*

An application of the Schwarz lemma to the covering space of the orbifold $\mathbb{C}_\infty \setminus P(R)$, which is the disk with the hyperbolic metric since $|P(R)| \geq 3$, gives the following result. Details can be found in any one of these sources ([**14**, **27**, **30**])

THEOREM 4.2. [**14**] *A postcritically finite map R is expanding with respect to its orbifold metric and $\|R'(z)\| > \lambda > 1$ for all $z \in \mathbb{C}_\infty \setminus P(R)$.*

The next result is a compendium and a simplification of results by many authors. We cite a few places where the statements can be found, more or less as stated here; most of these results were shown by Rees [**34**].

THEOREM 4.3. *If R is a postcritically finite rational map then the following hold.*

(1) *R is ergodic, exact, and conservative with respect to m_2* [**4**, **27**, **34**]*;*
(2) *R admits an invariant probability measure $\nu \sim m_2$* [**34**].
(3) *Let $\mathfrak{M}$ be a connected 1 complex dimensional manifold, and suppose $\{R_a, a \in \mathfrak{M}\}$ is a parametrized family of rational mappings that vary holomorphically in a; assume there exists some $a_0 \in \mathfrak{M}$ such that R_{a_0} is postcritically finite. If R_a satisfies a nondegeneracy condition (given in Remark* 4.3 *below), then there exists a set of parameters of positive measure in $\mathfrak{M}$ such that (1) and (2) hold* [**34**].

REMARK 4.3. (1) Polynomials never satisfy the hypotheses of Theorem 4.3 since the point at ∞ is a fixed critical point for every polynomial. The examples and results from Sec. 3.1, S_α and U_b, can never be post critically finite either since a parabolic fixed point forces a map to have an infinite forward orbit (see [**7**], Thm 9.3.2).

(2) The nondegeneracy condition in Theorem 4.3 is described as follows. Fix the postcritically finite parameter a_0. By considering a higher iterate of R_a if necessary, assume each critical orbit terminates in a (necessarily repelling) fixed point. Assume the map R_{a_0} has critical points $c_1(a_0), \ldots, c_m(a_0)$, $m \leq 2d-2$. Write the fixed point at the end of the orbit of $c_i(a_0)$ as: $R^{s_i}(c_i(a_0)) = y_i(a_0) = R_{a_0}(y_i(a_0)) = z_i(a_0)$ for each $i = 1, \ldots, m$, where $s_i = \min\{k : R^k(c_i(a_0)) \text{ is fixed}\}$.

Each of these points moves holomorphically in $a \in \mathfrak{M}$. The nondegeneracy condition is that the function:

$$F_i(a) = R^{s_i+1}(c_i(a)) - R^{s_i}(c_i(a))$$

should satisfy:

$$\frac{DF_i}{Da}(a_0) \neq 0, \tag{17}$$

or equivalently,

$$\frac{Dz_i}{Da}(a_0) - \frac{Dy_i}{Da}(a_0) \neq 0.$$

(3) We give an example of a family of maps satisfying the nondegeneracy condition, from [**34**]. If $R_a(z) = a\dfrac{(z-2)^2}{z^2}$, with $a \in \mathbb{C} \setminus \{0\}$ then the critical points are $c_1 = 0$ and $c_2 = 2$, and the critical orbits look like:

$$2 \mapsto 0 \mapsto \infty \mapsto a \mapsto \frac{(a-2)^2}{a} \cdots$$

We set $R_a(a) = a$, and note that $R_a'(a) = \frac{4(a-2)}{a^2}$. Therefore choosing $a_0 = 1$, yields both critical orbits terminating at a repelling fixed point at 1. We can check the condition in Eqn (17) by brute force to see that both resulting derivatives yield $-4 \neq 0$.

(4) The family of maps shown in Fig. 2, which we write as: $f_a(z) = a(z + 1/z + 2)$, has a fixed critical orbit at $c_1 = -1$. Namely $-1 \mapsto 0 \mapsto \infty$ for all nonzero a. The multiplier of the fixed point at ∞ is $1/a$, so as long as a stays in the open unit disk D of parameters, this is a repelling fixed point. Therefore we only look at the second critical point, $c_2 = 1$ to check the nondegeneracy condition given by (17); the purpose of Condn (17) in [**34**] is to estimate the measure of the set of parameters on which both critical orbits "stay far enough away from" critical points, which is unaffected by only checking for c_2. If we set $a_0 = \frac{-3-\sqrt{7}i}{8}$, we see that under f_{a_0},

$$1 \mapsto \frac{-3-\sqrt{7}i}{2} \mapsto \frac{-3+\sqrt{7}i}{8},$$

and $z_0 = \frac{-3+\sqrt{7}i}{8}$ is fixed. Moreover $|f_{a_0}'(z_0)| = |1/a_0| = 2$.

Using the notation in (1),

$$F_2(a) = f_a^2(1) - f_a(1) = \frac{(1+4a)^2}{4} - 4a = \frac{(1-4a)^2}{4},$$

and $\frac{DF_2}{Da}(a_0) = -5 - \sqrt{7}i \neq 0$ when $a_0 = \frac{-3-\sqrt{7}i}{8}$. This implies that the family of map I_a, defined by Eqn (4), when parametrized over $a \in \mathbb{C}$ contains a set of parameters of positive Lebesgue measure for which $J(I_a) = \mathbb{C}_\infty$ and for which Thm 4.3 holds. This is mentioned in [**18**].

(5) Apart from the Lattès examples we do not look for parametrized families of rational maps whose Julia set is the whole sphere; they are very unstable mappings, and in a neighborhood of a parameter of a postcritically finite map one usually sees other similar maps as well as tiny Mandelbrot sets signaling the presence of maps with attracting orbits (see [**18, 27**] for example).

4.4. Recent results: fractal Julia sets of positive m_2 measure. In this section we review some recent results about the existence of quadratic polynomials whose Julia sets have positive m_2 measure [**12**]. We have a limited goal here of giving the reader an idea of how a positive Lebesgue measure Julia set could possibly arise. This result, Theorem 4.4 below, is quite remarkable and represents a creative and technical leap forward in the field. Like many good ideas, the basic construction can be explained in simple terms. And also like many significant breakthroughs the process of arriving at the construction involved many people, difficult technical tools (e.g., [**20**]), and a lot of groundwork before the key idea fell into place. The result appears in a paper by Buff and Cheritat [**12**] but their introduction mentions many contributions by others. Background material for this section can be found in [**29**].

We start by recalling the familiar definition of a filled Julia set of a polynomial map $p(z)$: namely,

$$K(p) = \{z \in \mathbb{C} : \{p^n(z)\}_{n \in \mathbb{N}} \text{ is bounded}\}. \tag{18}$$

We consider the family of quadratic polynomials parametrized as:

$$q_\kappa(z) = \kappa z + z^2, \quad \kappa \in \mathbb{C}. \tag{19}$$

In this form we see immediately that 0 is a fixed point with multiplier κ. For the construction we are only interested in $|\kappa| = 1$, and more precisely in $\kappa = e^{2\pi i \xi}$ with ξ an irrational number, so we replace the notation $q_{e^{2\pi i \xi}}$ with q_ξ. We refer to 0 as an irrationally indifferent fixed point, and these points are of two types: either 0 is a *Siegel point* if there exists a local holomorphic change of coordinate $z = \phi(w)$ such that:

$$q_\xi \circ \phi(w) = \phi \circ e^{2\pi i \xi} w$$

for all w near the origin, or it is a *Cremer point* if no such local map exists. While irrational numbers giving Siegel points for q_ξ have full Lebesgue measure in $[0, 1]$, Cremer point values of ξ form a dense G_δ set. The rotation numbers that occur for Siegel disks are relatively poorly approximable by rational numbers while the Cremer numbers are well approximable. Siegel points give rise to maps that have a Fatou component containing 0, so that q_ξ is locally conjugate to rotation through $2\pi\xi$ on a disk. In Fig. 5 we show a Siegel disk. By contrast, Cremer points are always in the Julia set of q_ξ.

In general, if $J(p)$ is connected, $\partial K(p) = J(p) \subsetneq K(p)$, but it can happen, as is shown in Fig. 6 that $K(p)$ has empty interior and is $J(p)$. If one starts with irrational numbers that have Siegel disks, and then takes a sequence ξ_n that has been carefully chosen so that the Lebesgue measure of $K(q_{\xi_{n+1}})$ stays close to the measure of $K(q_{\xi_n})$, then a limiting value ζ might appear which is not a Siegel number, but $m_2(K(q_\zeta)) > 0$ and $K(q_\zeta) = J(q_\zeta)$. This is roughly how the construction goes, after showing that the map $\xi \mapsto m_2(K(q_\xi))$ is upper semicontinuous. In other words, the authors are able to construct $\xi_n \to \zeta$ with these (and more) properties.

THEOREM 4.4. [**12**]

- *There exist quadratic polynomials that have a Cremer fixed point and a Julia set of positive m_2 measure.*
- *There exist quadratic polynomials which have a Siegel disk and a Julia set of positive m_2 measure.*

Some topological properties of the Julia sets resulting from this work have been published (see [**9**] for example.) Since there exists a unique invariant probability measure μ supported on the Julia set such that $h_\mu(p) = \log 2$, and it must be singular with respect to m_2 [**39**], the Lebesgue measure properties of these maps are still largely unknown.

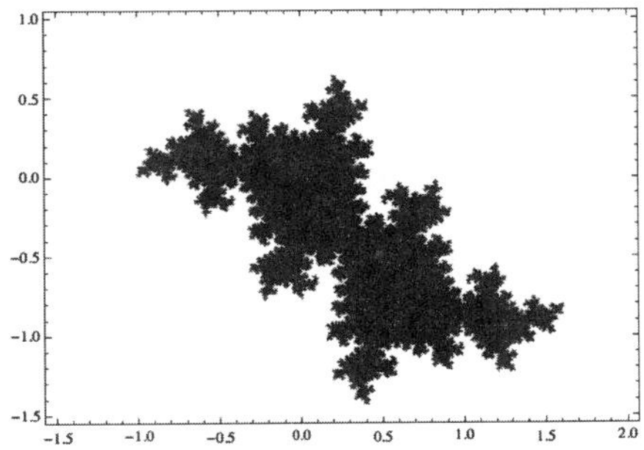

FIGURE 5. The map $q_{\sqrt{7}}(z) = e^{-2\pi i\sqrt{7}}z + z^2$ has a filled Julia set that is a Siegel disk centered at the fixed point at $z = 0$.

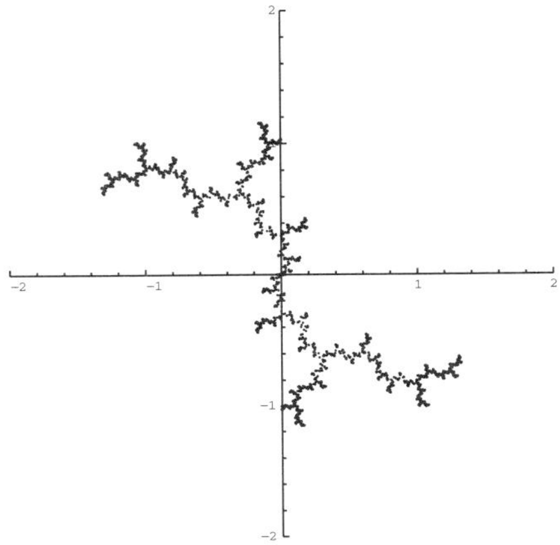

FIGURE 6. For the map $p_{-1}(z) = z^2 + i$, $K(p_i) = J(p_i)$

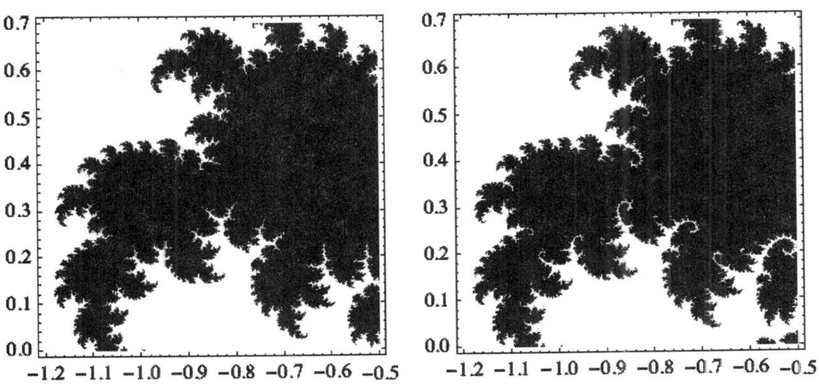

FIGURE 7. For the filled Julia set of the map $q_\xi(z) = e^{2\pi i\xi}z + z^2$, a small perturbation in ξ can cut deeper fjords in $K(q)$; on the left $\xi_1 = .131578\ldots$ and on the right, $\xi_2 = \xi_1 + \varepsilon$ with $\varepsilon < .00125$

References

[1] Jon Aaronson, *Ergodic theory for inner functions of the upper half plane* (English, with French summary), Ann. Inst. H. Poincaré Sect. B (N.S.) **14** (1978), no. 3, 233–253. MR508928

[2] Artur Avila, Xavier Buff, and Arnaud Chéritat, *Siegel disks with smooth boundaries*, Acta Math. **193** (2004), no. 1, 1–30, DOI 10.1007/BF02392549. MR2155030

[3] I. N. Baker, *Fixpoints of polynomials and rational functions*, J. London Math. Soc. **39** (1964), 615–622. MR0169989

[4] Julia A. Barnes, *Conservative exact rational maps of the sphere*, J. Math. Anal. Appl. **230** (1999), no. 2, 350–374, DOI 10.1006/jmaa.1998.6213. MR1672223

[5] Julia Barnes and Jane Hawkins, *Families of ergodic and exact one-dimensional maps*, Dyn. Syst. **22** (2007), no. 2, 203–217, DOI 10.1080/14689360600914730. MR2327993

[6] Bayless R (2013) Entropy of Infinite Measure-Preserving Transformations, PhD thesis, University of North Carolina at Chapel Hill.

[7] Alan F. Beardon, *Iteration of rational functions*, Graduate Texts in Mathematics, vol. 132, Springer-Verlag, New York, 1991. Complex analytic dynamical systems. MR1128089

[8] Paul Blanchard, *Complex analytic dynamics on the Riemann sphere*, Bull. Amer. Math. Soc. (N.S.) **11** (1984), no. 1, 85–141, DOI 10.1090/S0273-0979-1984-15240-6. MR741725

[9] Alexander Blokh and Lex Oversteegen, *The Julia sets of quadratic Cremer polynomials*, Topology Appl. **153** (2006), no. 15, 3038–3050, DOI 10.1016/j.topol.2006.02.001. MR2248408

[10] Henk Bruin and Jane Hawkins, *Exactness and maximal automorphic factors of unimodal interval maps*, Ergodic Theory Dynam. Systems **21** (2001), no. 4, 1009–1034, DOI 10.1017/S0143385701001481. MR1849599

[11] Henk Bruin and Jane Hawkins, *Rigidity of smooth one-sided Bernoulli endomorphisms*, New York J. Math. **15** (2009), 451–483. MR2558792

[12] Xavier Buff and Arnaud Chéritat, *Quadratic Julia sets with positive area*, Ann. of Math. (2) **176** (2012), no. 2, 673–746, DOI 10.4007/annals.2012.176.2.1. MR2950763

[13] Patrick Du Val, *Elliptic functions and elliptic curves*, Cambridge University Press, London-New York, 1973. London Mathematical Society Lecture Note Series, No. 9. MR0379512

[14] A. È. Erëmenko and M. Yu. Lyubich, *The dynamics of analytic transformations* (Russian), Algebra i Analiz **1** (1989), no. 3, 1–70; English transl., Leningrad Math. J. **1** (1990), no. 3, 563–634. MR1015124

[15] N. A. Friedman and D. S. Ornstein, *On isomorphism of weak Bernoulli transformations*, Advances in Math. **5** (1970), 365–394 (1970). MR0274718

[16] Harry Furstenberg, *Disjointness in ergodic theory, minimal sets, and a problem in Diophantine approximation*, Math. Systems Theory **1** (1967), 1–49. MR0213508

[17] Rika Hagihara, *Rational maps lacking certain periodic orbits*, ProQuest LLC, Ann Arbor, MI, 2007. Thesis (Ph.D.)–The University of North Carolina at Chapel Hill. MR2710025

[18] Jane Hawkins, *Lebesgue ergodic rational maps in parameter space*, Internat. J. Bifur. Chaos Appl. Sci. Engrg. **13** (2003), no. 6, 1423–1447, DOI 10.1142/S021812740300731X. MR1992056

[19] Jane Hawkins and Lorelei Koss, *Ergodic properties and Julia sets of Weierstrass elliptic functions*, Monatsh. Math. **137** (2002), no. 4, 273–300, DOI 10.1007/s00605-002-0504-1. MR1947915

[20] Inou H, Shishikura M (2006) The renormalization for parabolic fixed points and their perturbation, preprint.

[21] Gareth A. Jones and David Singerman, *Complex functions*, Cambridge University Press, Cambridge, 1987. An algebraic and geometric viewpoint. MR890746

[22] Ulrich Krengel, *Entropy of conservative transformations*, Z. Wahrscheinlichkeitstheorie und Verw. Gebiete **7** (1967), 161–181. MR0218522

[23] François Ledrappier, *Some properties of absolutely continuous invariant measures on an interval*, Ergodic Theory Dynamical Systems **1** (1981), no. 1, 77–93. MR627788

[24] François Ledrappier, *Quelques propriétés ergodiques des applications rationnelles* (French, with English summary), C. R. Acad. Sci. Paris Sér. I Math. **299** (1984), no. 1, 37–40. MR756305

[25] Gérard Letac, *Which functions preserve Cauchy laws?*, Proc. Amer. Math. Soc. **67** (1977), no. 2, 277–286. MR0584393

[26] M. Ju. Ljubich, *Entropy properties of rational endomorphisms of the Riemann sphere*, Ergodic Theory Dynam. Systems **3** (1983), no. 3, 351–385, DOI 10.1017/S0143385700002030. MR741393

[27] Curtis T. McMullen, *Complex dynamics and renormalization*, Annals of Mathematics Studies, vol. 135, Princeton University Press, Princeton, NJ, 1994. MR1312365

[28] John Milnor, *Geometry and dynamics of quadratic rational maps*, Experiment. Math. **2** (1993), no. 1, 37–83. With an appendix by the author and Lei Tan. MR1246482

[29] John Milnor, *On rational maps with two critical points*, Experiment. Math. **9** (2000), no. 4, 481–522. MR1806289

[30] John Milnor, *Dynamics in one complex variable*, Friedr. Vieweg & Sohn, Braunschweig, 1999. Introductory lectures. MR1721240

[31] John Milnor, *On Lattès maps*, Dynamics on the Riemann sphere, Eur. Math. Soc., Zürich, 2006, pp. 9–43, DOI 10.4171/011-1/1. MR2348953

[32] Michał Misiurewicz, *Absolutely continuous measures for certain maps of an interval*, Inst. Hautes Études Sci. Publ. Math. **53** (1981), 17–51. MR623533

[33] Toshio Nishino and Tsunéo Yoshioka, *Sur l'itération des transformations rationnelles entières de l'espace de deux variables complexes* (French), C. R. Acad. Sci. Paris **260** (1965), 3835–3837. MR0178162

[34] Mary Rees, *Positive measure sets of ergodic rational maps*, Ann. Sci. École Norm. Sup. (4) **19** (1986), no. 3, 383–407. MR870689

[35] Sebastian van Strien, *Hyperbolicity and invariant measures for general C^2 interval maps satisfying the Misiurewicz condition*, Comm. Math. Phys. **128** (1990), no. 3, 437–495. MR1045879

[36] V. A. Rohlin, *Exact endomorphism of a Lebesgue space* (Hungarian), Magyar Tud. Akad. Mat. Fiz. Oszt. Közl. **14** (1964), 443–474. MR0228654

[37] Peter Walters, *An introduction to ergodic theory*, Graduate Texts in Mathematics, vol. 79, Springer-Verlag, New York-Berlin, 1982. MR648108

[38] Yong Cheng Yin, *On the Julia sets of quadratic rational maps*, Complex Variables Theory Appl. **18** (1992), no. 3-4, 141–147. MR1157922

[39] Anna Zdunik, *Parabolic orbifolds and the dimension of the maximal measure for rational maps*, Invent. Math. **99** (1990), no. 3, 627–649, DOI 10.1007/BF01234434. MR1032883

DEPARTMENT OF MATHEMATICS, UNIVERSITY OF NORTH CAROLINA AT CHAPEL HILL, CB #3250, CHAPEL HILL, NORTH CAROLINA 27599-3250

E-mail address: jmh@math.unc.edu

Contemporary Mathematics
Volume **678**, 2016
http://dx.doi.org/10.1090/conm/678/13649

The inverse problem for canonically bounded rank-one transformations

Aaron Hill

Abstract. Given the cutting and spacer parameters for a rank-1 transformation, there is a simple condition which is easily seen to be sufficient to guarantee that the transformation under consideration is isomorphic to its inverse. Here we show that if the cutting and spacer parameters are canonically bounded, that condition is also necessary, thus giving a simple characterization of the canonically bounded rank-1 transformations that are isomorphic to their inverse.

1. Introduction

1.1. Background. A measure-preserving transformation is an automorphism of a standard Lebesgue space. Formally, it is a quadruple $(X, \mathcal{B}, \mu, T)$, where

(1) $(X, \mathcal{B}, \mu)$ is a measure space isomorphic to the unit interval with the Lebesgue measure on all Borel sets,
(2) T is a bijection from X to X such that T and T^{-1} are both μ-measurable and preserve the measure μ.

When the algebra of measurable sets is clear, we will refer to the transformation $(X, \mathcal{B}, \mu, T)$ by (X, μ, T). If $(X, \mathcal{B}, \mu, T)$ is a measure-preserving transformation, then so is its inverse, $(X, \mathcal{B}, \mu, T^{-1})$.

Two measure-preserving transformations $(X, \mathcal{B}, \mu, T)$ and $(X', \mathcal{B}', \mu', T')$ are isomorphic if there is a measure isomorphism ϕ from $(X, \mathcal{B}, \mu)$ to $(X', \mathcal{B}', \mu')$ such that μ almost everywhere, $\phi \circ T = T' \circ \phi$.

One of the central problems of ergodic theory, originally posed by von Neumann, is the isomorphism problem: How can one determine whether two measure-preserving transformations are isomorphic? The inverse problem is one of its natural restrictions: How can one determine whether a measure-preserving transformation is isomorphic to its inverse?

In the early 1940s, Halmos and von Neumann [**8**] showed that ergodic measure-preserving transformations with pure point spectrum are isomorphic iff they have the same spectrum. It immediately follows from this that every ergodic measure-preserving transformation with pure point spectrum is isomorphic to its inverse. About a decade later, Anzai [**2**] gave the first example of a measure-preserving

---

2010 *Mathematics Subject Classification.* Primary 37A05, 37A35.
Key words and phrases. rank-one transformation, isomorphic, canonically bounded.
The author acknowledges the US NSF grant DMS-0943870 for the support of his research.

transformation not isomorphic to its inverse. Later, Fieldsteel [**4**] and del Junco, Rahe, and Swanson [**9**] independently showed that Chacon's transformation–one of the earliest examples of what we now call rank-1 transformations–is not isomorphic to its inverse. In the late 1980s, Ageev [**1**] showed that a generic measure-preserving transformation is not isomorphic to its inverse.

In 2011, Foreman, Rudolph, and Weiss [**5**] showed that the set of ergodic measure-preserving transformations of a fixed standard Lebesgue space that are isomorphic to their inverse is a complete analytic subset of all measure-preserving transformations on that space. In essence, this result shows that there is no simple (i.e., Borel) condition which is satisfied if and only if an ergodic measure-preserving transformation is isomorphic to its inverse. However, in the same paper they show that the isomorphism relation becomes much simpler when restricted to the generic class of rank-1 transformations. It follows from their work that there exists a simple (i.e., Borel) condition which is satisfied if and only if a rank-1 measure-preserving transformation is isomorphic to its inverse. Currently, however, no such condition is known. In this paper we give a simple condition that is sufficient for a rank-1 transformation to be isomorphic to its inverse and show that for canonically bounded rank-1 transformations, the condition is also necessary.

1.2. Rank-1 transformations. In this subsection we state the definitions and basic facts pertaining to rank-1 transformations that will be used in our main arguments. We mostly follow the symbolic presentation in [**6**] and [**7**], but also provide comments that hopefully will be helpful to those more familiar with a different approach to rank-1 transformations. Additional information about the connections between different approaches to rank-1 transformations can be found in the survey article [**3**].

We first remark that by $\mathbb{N}$ we mean the set of all finite ordinals, including zero: $\{0, 1, 2, \ldots\}$.

Our main objects of study are symbolic rank-1 measure-preserving transformations. Each such transformation is a measure-preserving transformation $(X, \mathcal{B}, \mu, \sigma)$, where X is a closed, shift-invariant subset of $\{0,1\}^{\mathbb{Z}}$, $\mathcal{B}$ is the collection of Borel sets that X inherits from the product topology on $\{0,1\}^{\mathbb{Z}}$, μ is an atomless, shift-invariant (Borel) probability measure on X, and σ is the shift. To be precise, the shift σ is the bijection from $\{0,1\}^{\mathbb{Z}}$ to $\{0,1\}^{\mathbb{Z}}$, where $\sigma(x)(i) = x(i+1)$. Since the measure algebra of a symbolic measure-preserving transformation comes from the topology on $\{0,1\}^{\mathbb{Z}}$, we will omit the reference to that measure algebra and simply refer to a symbolic measure-preserving transformation as (X, μ, σ).

Symbolic rank-1 measure-preserving transformations are usually described by *cutting and spacer parameters.* The cutting parameter is a sequence $(r_n : n \in \mathbb{N})$ of integers greater than 1. The spacer parameter is a sequence of tuples $(s_n : n \in \mathbb{N})$, where formally s_n is a function from $\{1, 2, \ldots, r_n - 1\}$ to $\mathbb{N}$ (note that s_n is allowed to take the value zero). Given such cutting and spacer parameters, one defines the symbolic rank-1 system (X, σ) as follows. First define a sequence of finite words $(v_n : n \in \mathbb{N})$ by $v_0 = 0$ and

$$v_{n+1} = v_n 1^{s_n(1)} v_n 1^{s_n(2)} v_n \ldots v_n 1^{s_n(r_n-1)} v_n.$$

The sequence $(v_n : n \in \mathbb{N})$ is called a *generating sequence.* Then let

$$X = \{x \in \{0,1\}^{\mathbb{Z}} : \text{ every finite subword of } x \text{ is a subword of some } v_n\}.$$

It is straightforward to check that X is a closed, shift-invariant subset of $\{0,1\}^{\mathbb{Z}}$. These symbolic rank-1 systems are treated extensively–as topological dynamical systems–in [**6**]. In order to introduce a nice measure μ and thus obtain a measurable dynamical system, we make two additional assumptions on the cutting and spacer parameters.

(1) For every $N \in \mathbb{N}$ there exist $n, n' \geq N$ and $0 < i < r_n$ and $0 < i' < r_{n'}$ such that

$$s_n(i) \neq s_{n'}(i').$$

(2) $\sup_{n\in\mathbb{N}} \dfrac{\#\text{ of 1s in } v_n}{|v_n|} < 1$

It is straightforward to show that there is a unique shift-invariant measure on X which assigns measure 1 to the set $\{x \in X : x(0) = 0\}$. As long as the first condition above is satisfied, that measure is atomless. As long as the second condition above is satisfied, that measure is finite. Assuming both conditions are satisfied, the normalization of that measure is called μ and then (X, μ, σ) is a measure-preserving transformation. We call such an (X, μ, σ) a symbolic rank-1 measure-preserving transformation.

Below are several important remarks about symbolic rank-1 measure-preserving transformations that will be helpful in understanding the arguments in Section 2.

- *Bounded rank-1 transformations:* Suppose $(r_n : n \in \mathbb{N})$ and $(s_n : n \in \mathbb{N})$ are cutting and spacer parameters for (X, μ, σ). We say the cutting parameter is bounded if there is some $R \in \mathbb{N}$ such that for all $n \in \mathbb{N}$, $r_n \leq R$. We say that the spacer parameter is bounded if there is some $S \in \mathbb{N}$ such that for all $n \in \mathbb{N}$ and all $0 < i < r_n$, $s_n(i) \leq S$.

  Let (X, μ, σ) be a symbolic rank-1 measure-preserving transformation. We say that (X, μ, σ) is bounded if there are cutting and spacer parameters $(r_n : n \in \mathbb{N})$ and $(s_n : n \in \mathbb{N})$ that give rise to (X, μ, σ) that are both bounded.
- *Canonical cutting and spacer parameters:* There is an obvious bijective correspondence between cutting and spacer parameters and generating sequences, but there are many different generating sequences that give rise to the same symbolic rank-1 system. For example, any proper subsequence of a generating sequence will be a different generating sequence that gives rise to the same symbolic rank-1 system. There is a way, however, described in [**6**], to associate to each symbolic rank-1 system a unique canonical generating sequence, which in turn gives rise to the canonical cutting and spacer parameters of that symbolic system. The canonical generating sequence was used in [**6**] to fully understand topological isomorphisms between symbolic rank-1 systems; it was also used in [**7**] to explicitly describe when a bounded rank-1 measure-preserving transformation has trivial centralizer.

  There is only one fact about canonical generating sequences that is used in our argument. It is this: If $(r_n : n \in \mathbb{N})$ and $(s_n : n \in \mathbb{N})$ are the canonical cutting and spacer parameters for (X, μ, σ), then for all $n \in \mathbb{N}$, there is $0 < i < r_n$ and $0 < j < r_{n+1}$ such that $s_n(i) \neq s_n(j)$. (See the definition of canonical generating sequence in section 2.3.2 and 2.3.3 in [**6**])

- *Expected occurrences:* Let $(v_n : n \in \mathbb{N})$ be a generating sequence giving rise to the symbolic system (X, σ). Then for each $n \in \mathbb{N}$, there is a unique way to view each $x \in X$ as a disjoint collection of occurrences of v_n separated only by 1s. Such occurrences of v_n in x are called *expected* and the following all hold.
  (1) For all $x \in X$ and $n \in \mathbb{N}$, every occurrence of 0 in x is contained in a unique expected occurrence of v_n.
  (2) For all $x \in X$ and $n \in \mathbb{N}$, x has an expected occurrence of v_n beginning at position i iff $\sigma(x)$ has an expected occurrence of v_n beginning at position $(i-1)$.
  (3) If $x \in X$ has an expected occurrence of v_n beginning at position i, and $n' > n$, then the unique expected occurrence of $v_{n'}$ that contains the 0 at position i completely contains the expected occurrence of v_n that begins at i.
  (4) If $x \in X$ has expected occurrences of v_n beginning at positions i and j, with $|i - j| < |v_n|$, then $i = j$. In other words, distinct expected occurrences of v_n cannot overlap.
  (5) If $n > m$ and $x \in X$ has as expected occurrence of v_n beginning at i which completely contains an expected occurrence of v_m beginning at $i+l$, then whenever j is such that x has an expected occurrence of v_n beginning at j, that occurrence completely contains an expected occurrence of v_m beginning at $j + l$.

  For $n \in \mathbb{N}$ and $i \in \mathbb{Z}$ we define $E_{v_n,i}$ to be the set of all $x \in X$ that have an expected occurrence of v_n beginning at position i.
- *Relation to cutting and stacking constructions:* Let $(v_n : n \in \mathbb{N})$ be a generating sequence giving rise to the symbolic rank-1 measure-preserving system (X, μ, σ). One can take the cutting and spacer parameters associated to $(v_n : n \in \mathbb{N})$ and build, using a cutting and stacking construction, a rank-1 measure-preserving transformation. This construction involves a sequence of Rokhlin towers. There is a direct correspondence between the base of the nth tower in the cutting and stacking construction and the set $E_{v_n,0}$ in the symbolic system. The height of the nth tower in the cutting and stacking construction then corresponds to (i.e., is equal to) the length of the word v_n. If the reader is more familiar with rank-1 transformations as cutting and stacking constructions, one can use this correspondence to translate the arguments in Section 2 to that setting.
- *Expectedness and the measure algebra:* Let $(v_n : n \in \mathbb{N})$ be a generating sequence giving rise to the symbolic rank-1 measure-preserving system (X, μ, σ). If $\mathbb{M}$ is any infinite subset of $\mathbb{N}$, then the collection of sets $\{E_{v_n,i} : n \in \mathbb{M}, i \in \mathbb{Z}\}$ is dense in the measure algebra of (X, μ). Thus if A is any positive measure set and $\epsilon > 0$, there is some $n \in \mathbb{M}$ and $i \in \mathbb{Z}$ such that
$$\frac{\mu(E_{v_n,i} \cap A)}{\mu(E_{v_n,i})} > 1 - \epsilon$$
- *Rank-1 Inverses:* Let $(r_n : n \in \mathbb{N})$ and $(s_n : n \in \mathbb{N})$ be cutting and spacer parameters for the symbolic rank-1 measure-preserving transformation (X, μ, σ). It is straightforward to check that a simple modification of the parameters results in a symbolic rank-1 measure-preserving

transformation that is isomorphic to (X, μ, σ^{-1}). For each tuple s_n in the spacer parameter, let $\overline{s_n}$ be the reverse tuple, i.e., for $0 < i < r_n$, $\overline{s_n}(i) = s_n(r_n - i)$. It is easy to check that the cutting and spacer parameters $(r_n : n \in \mathbb{N})$ and $(\overline{s_n} : n \in \mathbb{N})$ satisfy the two measure conditions necessary to produce a symbolic rank-1 measure-preserving transformation. If one denotes that transformation by $(\overline{X}, \overline{\mu}, \sigma)$ and defines $\psi : X \to \overline{X}$ by $\psi(x)(i) = x(-i)$, then it is straightforward to check that ψ is an isomorphism between (X, μ, σ^{-1}) and $(\overline{X}, \overline{\mu}, \sigma)$. Thus to check whether a given symbolic rank-1 measure-preserving transformation (X, μ, σ) is isomorphic to its inverse, one need only check whether it is isomorphic to the symbolic rank-1 measure-preserving transformation $(\overline{X}, \overline{\mu}, \sigma)$.

1.3. The condition for isomorphism and the statement of the theorem. Let $(r_n : n \in \mathbb{N})$ and $(s_n : n \in \mathbb{N})$ be cutting and spacer parameters for the symbolic rank-1 measure-preserving transformation (X, μ, σ). Suppose that there is an $N \in \mathbb{N}$ such that for all $n \geq N$, $s_n = \overline{s_n}$. Let $\phi : X \to \overline{X}$ be defined so that $\phi(x)$ is obtained from x by replacing every expected occurrence of v_N by $\overline{v_N}$ (the reverse of v_N). It is straightforward to check that ϕ is an isomorphism between (X, μ, σ) and $(\overline{X}, \overline{\mu}, \sigma)$, thus showing that (X, μ, σ) is isomorphic to its inverse (X, μ, σ^{-1}).

As an example, Chacon2 is the rank-one transformation that can be defined by $v_{n+1} = v_n 1^n v_n$. (In the cutting and stacking setting, Chacon2 is usually described by $B_{n+1} = B_n B_n 1$, but that is easily seen to be equivalent to $B_{n+1} = B_n 1^n B_n$.) In this case $r_n = 2$ and $s_n(1) = n$, for all n. Since $s_n = \overline{s_n}$ for all n, Chacon2 is isomorphic to its inverse.

THEOREM 1.1. *Let $(r_n : n \in \mathbb{N})$ and $(s_n : n \in \mathbb{N})$ be the canonical cutting and spacer parameters for the symbolic rank-1 measure-preserving transformation (X, μ, σ). If those parameters are bounded, then (X, μ, σ) is isomorphic to (X, μ, σ^{-1}) if and only if there is an $N \in \mathbb{N}$ such that for all $n \geq N$, $s_n = \overline{s_n}$.*

We remark that in [**6**], the author and Su Gao have completely characterized when two symbolic rank-1 systems are *topologically* isomorphic, and as a corollary have a complete characterization of when a symbolic rank-1 system is *topologically* isomorphic to its inverse. A topological isomorphism between symbolic rank-1 systems is a homeomorphism between the underlying spaces that commutes with the shift. Since the underlying space of a symbolic rank-1 system admits at most one atomless, shift-invariant probability measure, every topological isomorphism between symbolic rank-1 systems is also a measure-theoretic isomorphism. On the other hand, there are symbolic rank-1 systems that are measure-theoretically isomorphic, but not topologically isomorphic.

We note here the main difference between these two settings. Suppose ϕ is an isomorphism–either a measure-theoretic isomorphism or a topological isomorphism–between two symbolic rank-1 systems (X, μ, σ) and (Y, ν, σ). Let $(v_n : n \in \mathbb{N})$ and $(w_n : n \in \mathbb{N})$ be generating sequences that gives rise to (X, μ, σ) and (Y, ν, σ), respectively. One can consider a set $E_{w_m,0} \subseteq Y$ and its pre-image, call it A, under ϕ. If ϕ is a measure-theoretic isomorphism then one can find some $E_{v_n,i}$ so that

$$\frac{\mu(E_{v_n,i} \cap A)}{\mu(E_{v_n,i})} > 1 - \epsilon.$$

However, if ϕ is in fact a topological isomorphism, then one can find some $E_{v_n,i}$ so that

$$E_{v_n,i} \subseteq A.$$

The stronger condition in the case of a topological isomorphism is what makes possible the analysis done by the author and Gao in [**6**]. In this paper, we are able to use the weaker condition, together with certain "bounded" conditions on the generating sequences $(v_n : n \in \mathbb{N})$ and $(w_n : n \in \mathbb{N})$ to achieve our results.

2. Arguments

We begin with a short subsection introducing two new pieces of notation. Then we prove a general proposition that can be used to show that certain symbolic rank-1 measure-preserving transformations are not isomorphic. Finally, we show how to use the general proposition to prove the non-trivial direction of Theorem 1.1.

2.1. New notation. The first new piece of notation is $*$, a binary operation on all finite sequences of natural numbers. The second is $\perp$, a relation (signifying incompatibility) between finite sequences of natural numbers that have the same length.

The notation $*$: We will first describe the reason for introducing this new notation. We will then then give the formal definition of $*$ and then illustrate that definition with an example. Suppose $(r_n : n \in \mathbb{N})$ and $(s_n : n \in \mathbb{N})$ are cutting and spacer parameters for the symbolic system (X, μ, σ) and that $(v_n : n \in \mathbb{N})$ is the generating sequence corresponding to those parameters. Fix $n_0 > 0$ and consider the generating sequence $(w_n : n \in \mathbb{N})$, defined as follows.

$$w_n = \begin{cases} v_n, & \text{if } n < n_0 \\ v_{n+1}, & \text{if } n \geq n_0 \end{cases}$$

It is clear that $(w_n : n \in \mathbb{N})$ is a subsequence of $(v_n : n \in \mathbb{N})$, missing only the element v_{n_0}; thus, $(w_n : n \in \mathbb{N})$ gives rise to the same symbolic system (X, μ, σ). We would like to be able to easily describe the cutting and spacer parameters that correspond to the generating sequence $(w_n : n \in \mathbb{N})$. Let $(r'_n : n \in \mathbb{N})$ and $(s'_n : n \in \mathbb{N})$ be those cutting and spacer parameters. It is clear that for $n < n_0$ we have $r'_n = r_n$ and $s'_n = s_n$. It is also clear that for $n > n_0$ we have $r'_n = r_{n+1}$ and $s'_n = s_{n+1}$. It is straightforward to check that $r'_{n_0} = r_{n_0+1} \cdot r_{n_0}$. The definition below for $*$ is precisely what is needed so that $s'_{n_0} = s_{n_0+1} * s_{n_0}$.

Here is the definition. Let s_1 be any function from $\{1, 2, \ldots, r_1 - 1\}$ to $\mathbb{N}$ and let s_2 be any function from $\{1, 2, \ldots, r_2 - 1\}$ to $\mathbb{N}$. We define $s_2 * s_1$, a function from $\{1, 2, \ldots, r_2 \cdot r_1 - 1\}$ to $\mathbb{N}$, as follows.

$$(s_2 * s_1)(i) = \begin{cases} s_1(k), & \text{if } 0 < k < r_1 \text{ and } i \equiv k \mod r_1 \\ s_2(i/r_1), & \text{if } i \equiv 0 \mod r_1 \end{cases}$$

It is important to note, and straightforward to check, that the operation $*$ is associative.

To illustrate, suppose that s_1 is the function from $\{1, 2, 3\}$ to $\mathbb{N}$ with $s_1(1) = 0$, $s_1(2) = 1$, and $s_1(3) = 0$ and that s_2 is the function from $\{1, 2\}$ to $\mathbb{N}$ such that $s_2(1) = 5$ and $s_2(2) = 6$; we abbreviate this by simply saying that $s_1 = (0, 1, 0)$ and $s_2 = (5, 6)$. Then $s_2 * s_1 = (0, 1, 0, 5, 0, 1, 0, 6, 0, 1, 0)$.

The notation $\perp$: Suppose s and s' are both functions from $\{1, 2, \ldots, r-1\}$ to $\mathbb{N}$. We say that s is *compatible* with s' if there exists a function c from $\{1\}$ to $\mathbb{N}$ so that s is a subsequence of $c * s'$. Otherwise we say that s is incompatible with s' and write $s \perp s'$.

To illustrate, consider $s = (0, 1, 0)$ and $s' = (0, 0, 1)$. Then s is compatible with s' because if $c = 0$, then $c * s' = (0, 0, 1, 0, 0, 0, 1)$ and $(0, 1, 0)$ does occur as a subsequence of $(0, 0, 1, 0, 0, 0, 1)$. If $s'' = (0, 1, 2)$, then s' is compatible with s'' (again let $c = 0$), but $s \perp s''$, because $(0, 1, 0)$ can never be a subsequence of $(0, 1, 2, c, 0, 1, 2)$.

Though not used in our arguments, it is worth noting, and is straightforward to check, that $s \perp s'$ iff $s' \perp s$. (It is important here that s and s' have the same length.)

We now state the main point of this definition of incompatibility. This fact will be crucial in the proof of Proposition 2.1. Suppose $(r'_n : n \in \mathbb{N})$ and $(s'_n : n \in \mathbb{N})$ are cutting and spacer parameters associated to the symbolic rank-1 measure-preserving transformation (Y, ν, σ) and that $(w_n : n \in \mathbb{N})$ is the generating sequence associated to those parameters. If n is such that $r_n = r'_n$ and $s_n \perp s'_n$, then no element of $y \in Y$ contains an occurrence of

$$w_n 1^{s_n(1)} w_n 1^{s_n(2)} \ldots 1^{s_n(r_n-1)} w_n$$

where each of the demonstrated occurrence of w_n is expected.

Indeed, suppose that beginning at position i, some $y \in Y$ did have such an occurrence of $w_n 1^{s_n(1)} w_n 1^{s_n(2)} \ldots 1^{s_n(r_n-1)} w_n$. The expected occurrence of w_n beginning at i must be completely contained in some expected occurrence of w_{n+1}, say that begins at position j. We know that the expected occurrence of w_{n+1} beginning at position j contains exactly r_n-many expected occurrences of w_n. Let $1 \le l \le r_n$ be such that the expected occurrence of w_n beginning at position i is the lth expected occurrence of w_n beginning at position j. If $l = 1$, then $s_n = s'_n$, which implies that s_n is a subsequence of $c * s'_n$ for any c. If, on the other hand, $1 < l \le r_n$, then letting $c = s_n(r_n - l + 1)$, we have that s_n is a subsequence of $c * s'_n$. In either case this would result in s_n being compatible with s'_n.

2.2. A general proposition guaranteeing non-isomorphism.

PROPOSITION 2.1. *Let $(r_n : n \in \mathbb{N})$ and $(s_n : n \in \mathbb{N})$ be the cutting and spacer parameters for a symbolic rank-1 system (X, μ, σ) and let $(r'_n : n \in \mathbb{N})$ and $(s'_n : n \in \mathbb{N})$ be the cutting and spacer parameters for a symbolic rank-1 system (Y, ν, σ). Suppose the following hold.*

(1) *For all n, $r_n = r'_n$ and* $\sum_{0<i<r_n} s_n(i) = \sum_{0<i<r_n} s'_n(i)$.

(2) *There is an $S \in \mathbb{N}$ such that for all n and all $0 < i < r_n$,*

$$s_n(i) \le S \text{ and } s'_n(i) \le S.$$

(3) *There is an $R \in \mathbb{N}$ such that for infinitely many n,*

$$r_n \le R \text{ and } s_n \perp s'_n.$$

Then (X, μ, σ) and (Y, ν, σ) are not measure-theoretically isomorphic.

PROOF. Let $(v_n : n \in \mathbb{N})$ be the generating sequence associated to the cutting and spacer parameters $(r_n : n \in \mathbb{N})$ and $(s_n : n \in \mathbb{N})$. Let $(w_n : n \in \mathbb{N})$ be the generating sequence associated to the cutting and spacer parameters $(r'_n : n \in \mathbb{N})$

and $(s'_n : n \in \mathbb{N})$. Condition (1) implies that for all n, $|v_n| = |w_n|$. Now suppose, towards a contradiction, that ϕ is an isomorphism between (X, μ, σ) and (Y, ν, σ).

First, choose $m \in \mathbb{N}$ so that $|v_m| = |w_m|$ is greater than the S from condition (2). Next, consider the positive μ-measure set

$$\phi^{-1}(E_{w_m,0}) = \{x \in X : \phi(x) \text{ has an expected occurrence of } w_m \text{ at } 0\}.$$

Let $\mathbb{M} = \{n \in \mathbb{N} : r_n \leq R \text{ and } s_n \perp s'_n\}$, where R is from condition (3), and note that $\mathbb{M}$ is infinite. We can then find $n \in \mathbb{M}$ and $k \in \mathbb{N}$ such that

$$\frac{\mu(E_{v_n,k} \cap \phi^{-1}(E_{w_m,0}))}{\mu(E_{v_n,k})} > 1 - \frac{1}{R}.$$

One can loosely describe the above inequality by saying: Most $x \in X$ that have an expected occurrence of v_n beginning at position k are such that $\phi(x)$ has an expected occurrence of w_m beginning at position 0.

We say that an expected occurrence of v_n in any $x \in X$ (say it begins at i) is a *good* occurrence of v_n if $\phi(x)$ has an expected occurrence of w_m beginning at position $i-k$. In this case we say the good occurrence of v_n beginning at i *forces* the expected occurrence of w_m beginning at position $i - k$. Note that an expected occurrence of v_n beginning at position i in $x \in X$ is good iff $\sigma^i(x) \in E_{v_n,k} \cap \phi^{-1}(E_{w_m,0})$, since ϕ commutes with σ. A simple application of the ergodic theorem shows that μ almost every $x \in X$ satisfies

$$\lim_{N \to \infty} \frac{|\{i \in [-N, N] : \ x \text{ has a good occurrence of } v_n \text{ at } i\}|}{|\{i \in [-N, N] : \ x \text{ has an expected occurrence of } v_n \text{ at } i\}|} > 1 - \frac{1}{R}.$$

Since $r_n \leq R$, this implies that μ almost every $x \in X$ contains an expected occurrence of v_{n+1} such that each of the r_n-many expected occurrences of v_n that it contains is good. We say that such an occurrence of v_{n+1} is *totally good*.

Let $x \in X$ and $i \in \mathbb{Z}$ be such that x has a totally good occurrence of v_{n+1} beginning at i. There are r_n-many expected occurrences of v_n in the expected occurrence v_{n+1} beginning at i and each of them forces an expected occurrence of w_m in $\phi(x)$. The first of these forced expected occurrences of w_m in $\phi(x)$ begins at position $i - k$ and must be part of some expected occurrence of w_n, say it begins at position i'. We claim that, in fact, $\phi(x)$ must have an occurrence of

$$w_n 1^{s_n(1)} w_n 1^{s_n(2)} \ldots 1^{s_n(r_n - 1)} w_n$$

beginning at i', where each of the demonstrated occurrence of w_n is expected. This will contradict the fact that $s_n \perp s'_n$.

Proving the claim involves an argument that is repeated $r_n - 1$ times. The next paragraph contains the first instance of that argument, showing that the expected occurrence of w_n beginning at i' in $\phi(x)$ is immediately followed by $1^{s_n(1)}$ and then another expected occurrence of w_n, this one containing the second forced occurrence of w_m. The next instance of the argument would show that the expected occurrence of w_n beginning at $i' + |w_n| + s_n(1)$ is immediately followed by $1^{s_n(2)}$ and then another expected occurrence of w_n, this one containing the third forced occurrence of w_m. After the $r_n - 1$ instances of that argument, the claim would be proven.

Here is the first instance of the argument: We know that $\phi(x)$ has an expected occurrence of w_n beginning at i' and that this expected occurrence of w_n contains the expected occurrence of w_m beginning at position

$$i - k = (i') + (i - k - i').$$

Thus, by point (5) of the remark about expected occurrences in Section 1.2, if $j \in \mathbb{Z}$ is such that $\phi(x)$ has an expected occurrence of w_n beginning at position j, that occurrence completely contains an expected occurrence of w_m beginning at position $j + (i - k - i')$.

The expected occurrence of w_n beginning at position i' must be followed by 1^t and then another expected occurrence of w_n, for some $0 \leq t \leq S$. The expected occurrence of w_n beginning at position $i' + |w_n| + t$ must contain an expected occurrence of w_m beginning at position

$$(i - k) + |w_n| + t = (i' + |w_n| + t) + (i - k - i').$$

But we also know that the expected occurrence of v_n beginning at position $i + |v_n| + s_n(1)$ in x forces an expected occurrence of w_m beginning at $i + |w_n| + s_n(1)$ in $\phi(x)$. Since $0 \leq s_n(1), t \leq S$ it must be that $0 \leq |t - s_n(1)| \leq S$ and thus, since $|w_m| > S$, the expected occurrences beginning at positions $i + |w_n| + s_n(1)$ and $i + |w_n| + t$ must overlap. Since distinct expected occurrences of w_m cannot overlap, it must be the case that $s_n(1) = t$ (See point (4) of the remark about expected occurrences in Section 1.2). Thus, the expected occurrence of w_n beginning at i' in $\phi(x)$ is immediately followed by $1^{s_n(1)}$ and then another expected occurrence of w_n, this one containing the second forced occurrence of w_m. □

2.3. Proving the theorem. We start this subsection with a comment and a simple lemma. The comment is that if $(r_n : n \in \mathbb{N})$ and $(s_n : n \in \mathbb{N})$ are the canonical cutting and spacer parameters for a symbolic rank-1 measure-preserving transformation (X, μ, σ), then for all $n \in \mathbb{N}$, $s_{n+1} * s_n$ is not constant. This is simply a restatement, using the notation $*$, of the last sentence in the remark about canonical generating sequences in Section 1.2.

LEMMA 2.2. *Let s_1 be a function from $\{1, 2, \ldots, r_1 - 1\}$ to $\mathbb{N}$ such that $s_1 \neq \overline{s_1}$. Let s_2 be function from $\{1, 2, \ldots, r_2 - 1\}$ to $\mathbb{N}$ that is not constant. Then $s_2 * s_1 \perp \overline{s_2 * s_1}$.*

PROOF. Suppose, towards a contradiction, that $s_2 * s_1$ is compatible with $\overline{s_2 * s_1}$. Then there is function c from $\{1\}$ to $\mathbb{N}$ so that $s_2 * s_1$ is a subsequence of

$$c * (\overline{s_2 * s_1}) = c * (\overline{s_2} * \overline{s_1}) = (c * \overline{s_2}) * \overline{s_1}.$$

In other words, there is some $0 \leq k \leq r_2 \cdot r_1$ such that for all $0 < l < r_2 \cdot r_1$, $(s_2 * s_1)(l) = ((c * \overline{s_2}) * \overline{s_1})(k + l)$. We now have two cases.

Case 1: $k \equiv 0 \mod r_1$. Then for all $0 < i < r_1$,

$$s_1(i) = (s_2 * s_1)(i) = ((c * \overline{s_2}) * \overline{s_1})(k + i) = \overline{s_1}(i).$$

Thus $s_1 = \overline{s_1}$, which is a contradiction.

Case 2: There is some $0 < m < r_1$ such that $k + m \equiv 0 \mod r_1$. For all $0 \leq d < r_2$, we have $(s_2 * s_1)(m + dr_1) = s_1(m)$. But also,

$$(s_2 * s_1)(m + dr_1) = ((c * \overline{s_2}) * \overline{s_1})(k + m + dr_1) = (c * \overline{s_2})\left(\frac{k - m}{r_1} + d\right).$$

This implies that the function $c * \overline{s_2}$ is constant (taking the value $s_1(m)$) for r_2-many consecutive inputs. This implies that $\overline{s_2}$ must be constant, which is a contradiction. □

We will now prove the non-trivial direction of the theorem.

PROOF. Let $(\tilde{r}_n : n \in \mathbb{N})$ and $(\tilde{s}_n : n \in \mathbb{N})$ be the canonical cutting and spacer parameters for a symbolic rank-1 measure-preserving transformation (X, μ, σ). Suppose both parameters are bounded; let $\tilde{R}$ be such that for all $n \in \mathbb{N}$, $\tilde{r}_n \leq \tilde{R}$ and let S be such that for all $n \in \mathbb{N}$ and all $0 < i < \tilde{r}_n$, $\tilde{s}_n(i) \leq S$. Also, assume that for infinity many n, $\tilde{s}_n \neq \overline{\tilde{s}_n}$. To prove the non-trivial direction of the theorem, we need to show that (X, μ, σ) is not isomorphic to its inverse.

Let $(u_n : n \in \mathbb{N})$ be the generating sequence corresponding to the parameters $(\tilde{r}_n : n \in \mathbb{N})$ and $(\tilde{s}_n : n \in \mathbb{N})$. We will now describe a subsequence $(v_n : n \in \mathbb{N})$ of $(u_n : n \in \mathbb{N})$ and let $(r_n : n \in \mathbb{N})$ and $(s_n : n \in \mathbb{N})$ be the cutting and spacer parameters corresponding to the generating sequence $(v_n : n \in \mathbb{N})$ (which also gives rise to (X, μ, σ)). First, let $v_0 = u_0 = 0$. Now, suppose v_{2n} has been defined as u_k. Let $m > k$ be as small as possible so that $\tilde{s}_m \neq \overline{\tilde{s}_m}$, and define $v_{2n+1} = u_m$ and $v_{2n+2} = u_{m+3}$. It is very important to note here that

$$r_{2n+1} = \tilde{r}_{m+2} \cdot \tilde{r}_{m+1} \cdot \tilde{r}_m$$

and that

$$s_{2n+1} = \tilde{s}_{m+2} * \tilde{s}_{m+1} * \tilde{s}_m.$$

This has two important consequences. First, we have that for $n \in \mathbb{N}$, $r_{2n+1} \leq \tilde{R}^3$. By the remark before Lemma 2.2, we also have that $\tilde{s}_{m+3} * \tilde{s}_{m+2}$ is not constant and thus, by Lemma 2.2, $\tilde{s}_{m+3} * \tilde{s}_{m+2} * \tilde{s}_{m+1} \perp \overline{\tilde{s}_{m+3} * \tilde{s}_{m+2} * \tilde{s}_{m+1}}$; put another way, $s_{2n+1} \perp \overline{s_{2n+1}}$.

Now for each n, let $r'_n = r_n$ and $s'_n = \overline{s_n}$. Let (Y, ν, σ) be the symbolic rank-1 transformation corresponding to the cutting and spacer parameters $(r'_n : n \in \mathbb{N})$ and $(s'_n : n \in \mathbb{N})$. As mentioned in the remark on rank-1 inverses at the end of Section 1.2, the transformation (Y, ν, σ) is isomorphic to the inverse of (X, μ, σ). Thus to show that (X, μ, σ) is not isomorphic to its inverse, we can show that (X, μ, σ) and (Y, ν, σ) are not isomorphic.

To do this we will apply Proposition 2.1. We need to check that the following three conditions hold.

(1) For all n, $r_n = r'_n$ and $\sum_{0<i<r_n} s_n(i) = \sum_{0<i<r_n} s'_n(i)$.

(2) There is an $S \in \mathbb{N}$ such that for all n and all $0 < i < r_n$,

$$s_n(i) \leq S \text{ and } s'_n(i) \leq S.$$

(3) There is an $R \in \mathbb{N}$ such that for infinitely many n,

$$r_n \leq R \text{ and } s_n \perp s'_n.$$

Condition (1) above follows immediately from the fact that for all $n \in N$, $r'_n = r_n$ and $s'_n = \overline{s_n}$. Condition (2) follows from the fact that each $s_n(i)$ is equal to some $\tilde{s}_m(j)$ which is less than or equal to S. Finally, to verify condition (3), let $R = \tilde{R}^3$ and note that, as remarked above, for all $n \in \mathbb{N}$ we have that $r_{2n+1} \leq \tilde{R}^3$ and $s_{2n+1} \perp \overline{s_{2n+1}}$. We now apply Proposition 2.1 and conclude that (X, μ, σ) and (Y, ν, σ) are not isomorphic. Thus (X, μ, σ) is not isomorphic to its inverse. $\square$

References

[1] O. N. Ageev, *Conjugacy of a group action to its inverse* (Russian), Mat. Zametki **45** (1989), no. 3, 3–11, 127. MR1001691

[2] Hirotada Anzai, *On an example of a measure preserving transformation which is not conjugate to its inverse*, Proc. Japan Acad. **27** (1951), 517–522. MR0047742

[3] Sébastien Ferenczi, *Systems of finite rank*, Colloq. Math. **73** (1997), no. 1, 35–65. MR1436950

[4] A. Fieldsteel, An uncountable family of prime transformations not isomorphic to their inverses, unpublished manuscript.

[5] Matthew Foreman, Daniel J. Rudolph, and Benjamin Weiss, *The conjugacy problem in ergodic theory*, Ann. of Math. (2) **173** (2011), no. 3, 1529–1586, DOI 10.4007/annals.2011.173.3.7. MR2800720

[6] Su Gao and Aaron Hill, *Topological isomorphism for rank-1 systems*, J. Anal. Math. **128** (2016), 1–49, DOI 10.1007/s11854-016-0001-4. MR3481169

[7] Su Gao and Aaron Hill, *Bounded rank-1 transformations*, J. Anal. Math. **129** (2016), 341–365, DOI 10.1007/s11854-016-0024-x. MR3540602

[8] Paul R. Halmos and John von Neumann, *Operator methods in classical mechanics. II*, Ann. of Math. (2) **43** (1942), 332–350. MR0006617

[9] A. del Junco, M. Rahe, and L. Swanson, *Chacon's automorphism has minimal self-joinings*, J. Analyse Math. **37** (1980), 276–284, DOI 10.1007/BF02797688. MR583640

[10] J. von Neumann, *Zur Operatorenmethode in der klassischen Mechanik* (German), Ann. of Math. (2) **33** (1932), no. 3, 587–642, DOI 10.2307/1968537. MR1503078

Department of Mathematics, University of Louisville, Louisville, Kentucky 40292
E-mail address: aaron.hill@louisville.edu

Contemporary Mathematics
Volume **678**, 2016
http://dx.doi.org/10.1090/conm/678/13650

Higher dimensional shift spaces with shear

Steve Kass and Kathleen Madden

ABSTRACT. We introduce a property of higher-dimensional shift spaces called *shear*, as well as a class of shift spaces with shear that we call *generalized shift products*. We investigate conditions under which a shift space with shear is a generalized shift product.

1. Background Definitions

Let $\mathcal{A}$ be a finite alphabet, and for $d \geq 1$, view $\mathcal{A}^{\mathbb{Z}^d}$ as a topological space under the product topology, using the discrete topology on $\mathcal{A}$. The d-dimensional full shift on $\mathcal{A}$ is the set $\mathcal{A}^{\mathbb{Z}^d}$ together with the $\mathbb{Z}^d$-action $\sigma : \mathbb{Z}^d \times \mathcal{A}^{\mathbb{Z}^d} \to \mathcal{A}^{\mathbb{Z}^d}$ defined by

$$\sigma(\vec{v}, x)_{\vec{u}} = \sigma_{\vec{v}}(x)_{\vec{u}} = x_{\vec{v}+\vec{u}}$$

for any $x \in \mathcal{A}^{\mathbb{Z}^d}$ and any $\vec{v}$, $\vec{u} \in \mathbb{Z}^d$. Here $x_{\vec{u}}$ denotes the symbol appearing in x at $\vec{u}$ for $x \in \mathcal{A}^{\mathbb{Z}^d}$ and $\vec{u} \in \mathbb{Z}^d$. Similarly, for $x \in \mathcal{A}^{\mathbb{Z}^d}$ and $D \subseteq \mathbb{Z}^d$, we will denote the configuration of symbols appearing in x in the locations in D as x_D.

A closed, shift-invariant subset X of $\mathcal{A}^{\mathbb{Z}^d}$ together with the restriction of the $\mathbb{Z}^d$-action to X is called a $\mathbb{Z}^d$-shift space and is denoted by $(X, \mathbb{Z}^d)$. When needed for clarity, we will denote the alphabet of a shift space $(X, \mathbb{Z}^d)$ by $\mathcal{A}_X$.

A continuous, shift commuting map $\phi : X \to Y$ between shift spaces $(X, \mathbb{Z}^d)$ and $(Y, \mathbb{Z}^d)$ is called a factor map. A symbol map $\Phi : \mathcal{A}_X \to \mathcal{A}_Y$ can be used to define a factor map $\phi : X \to Y$ via $\phi(x)_{\vec{u}} = \Phi(x_{\vec{u}})$ for all $\vec{u} \in \mathbb{Z}^d$.

All shift spaces can be defined via a set $\mathcal{F}$ of forbidden configurations. If $\mathcal{F}$ can be chosen to be finite, $(X, \mathbb{Z}^d)$ is called a shift of finite type. If the set of forbidden configurations defining a shift of finite type consists of pairs of adjacent symbols (i.e. $\mathcal{F} \subseteq \bigcup_{i=1}^{d} \mathcal{A}^{\{0,\vec{e_i}\}}$) then $(X, \mathbb{Z}^d)$ is called a nearest neighbor shift of finite type. A shift space $(Y, \mathbb{Z}^d)$ is said to be sofic if it is the image of a factor map $\phi : X \to Y$ where $(X, \mathbb{Z}^d)$ is a shift of finite type.

See [**LM**] for additional background on shift spaces, particularly in the one-dimensional case.

When $(X, \mathbb{Z}^d)$ is a shift space and $D \subseteq \mathbb{Z}^d$, we say a D-configuration $C \in \mathcal{A}^D$ *occurs* in X if there exists $x \in X$ with $x_D = C$. We say C is *allowed* if C does not contain any of the defining forbidden configurations. Note that all occurring

2010 *Mathematics Subject Classification.* Primary 37B50; Secondary 37B10.
Key words and phrases. Symbolic dynamics, higher-dimensional shifts, sofic shifts.

configurations are necessarily allowed, but allowed configurations might or might not occur.

We will be particularly interested in configurations that occur on the following subsets of $\mathbb{Z}^d$: for $j, k \in \mathbb{Z}$ with $j < k$. Let

$$\begin{aligned} B_j &= \{\vec{v} \in \mathbb{Z}^d \mid v_d < j\}, \text{ the } j^{th} \textit{ bottom half-space}, \\ T_j &= \{\vec{v} \in \mathbb{Z}^d \mid v_d > j\}, \text{ the } j^{th} \textit{ top half-space}, \\ R_j &= \{\vec{v} \in \mathbb{Z}^d \mid v_d = j\}, \text{ the } j^{th} \textit{ center slice}, \text{ and} \\ R(j,k) &= \{\vec{v} \in \mathbb{Z}^d \mid j \leq v_d \leq k\}, \text{ the } \textit{j-k slice}. \end{aligned} \tag{1.1}$$

In the definitions above, we represent $\vec{v} \in \mathbb{Z}^d$ as $\sum_{i=1}^{d} v_i\vec{e}_i$ where each v_i is an element of $\mathbb{Z}$ and $\{\vec{e}_1, \vec{e}_2, \ldots, \vec{e}_d\}$ is the set of standard unit basis vectors in $\mathbb{Z}^d$. We will denote B_0, T_0, and R_0 by B, T, and R, respectively.

We will also be interested in configurations on affine sublattices of $\mathbb{Z}^d$. By this we mean the following: $\mathbb{Z}^d$ is a free abelian group under vector addition generated by the standard unit basis vectors. If we let $\mathcal{B} \subseteq \{\vec{e}_1, \vec{e}_2, \ldots, \vec{e}_d\}$ and define

$$L = \mathcal{L}(\mathcal{B}) = \left\{ \sum_{\vec{e}_i \in \mathcal{B}} v_i\vec{e}_i : v_i \in \mathbb{Z} \right\}$$

then L is a subgroup (or sublattice) of $\mathbb{Z}^d$. Each sublattice $L = \mathcal{L}(\mathcal{B})$ has a complementary sublattice $L^\perp = \mathcal{L}(\mathcal{B}^\perp)$ where $\mathcal{B}^\perp = \{\vec{e}_1, \vec{e}_2, \ldots, \vec{e}_d\} \backslash \mathcal{B}$. If $\vec{w} \in L^\perp$, then we will call $L + \vec{w}$ an affine sublattice. We note that $\mathbb{Z}^d = \bigcup_{\vec{w} \in L^\perp}(L + \vec{w})$, and we can write any $\vec{v} \in \mathbb{Z}^d$ uniquely as $\vec{v} = \vec{v}_L + \vec{v}_{L^\perp}$, where $\vec{v}_L \in L$ and $\vec{v}_{L^\perp} \in L^\perp$.

There are many subgroups (sublattices) of $\mathbb{Z}^d$ that are not of the form $\mathcal{L}(\mathcal{B})$, for example $\{\sum_{i=1}^d 2v_i\vec{e}_i : v_i \in \mathbb{Z}\}$. We will only be interested in sublattices of the form $\mathcal{L}(\mathcal{B})$.

If $|\mathcal{B}| = k$, the set of configurations in X restricted to affine sublattices $L + \vec{w}$ where $L = \mathcal{L}(\mathcal{B})$ can be viewed as a $\mathbb{Z}^k$-shift space. This is called the *projection* of $(X, \mathbb{Z}^d)$ onto L. The dynamical properties of projections were considered in [**JKM**] and [**PS**].

2. Shift Spaces with Shear: Definitions and Examples

We can view the shift operator $\sigma_{\vec{v}}$ as acting on configurations in $\mathcal{A}^{\mathbb{Z}^d}$ by moving each symbol of the configuration by a displacement vector $-\vec{v}$. We now define a new operator, called a shear operator, that also acts on configurations in $\mathcal{A}^{\mathbb{Z}^d}$. This new operator only moves the symbols on one affine sublattice, shifting them by a displacement vector parallel to the affine sublattice:

DEFINITION 2.1. Let $L + \vec{w}$ be an affine sublattice of $\mathbb{Z}^d$. The **shear operator** $\sigma^{L+\vec{w}} : L \times \mathcal{A}^{\mathbb{Z}^d} \to \mathcal{A}^{\mathbb{Z}^d}$ is defined as follows:

$$\sigma^{L+\vec{w}}(\vec{v}, x)_{\vec{u}} = \sigma_{\vec{v}}^{L+\vec{w}}(x)_{\vec{u}} = \begin{cases} x_{\vec{u}+\vec{v}} & \vec{u} \in L + \vec{w} \\ x_{\vec{u}} & \vec{u} \notin L + \vec{w} \end{cases}$$

for $\vec{v} \in L$ and $x \in \mathcal{A}^{\mathbb{Z}^d}$.

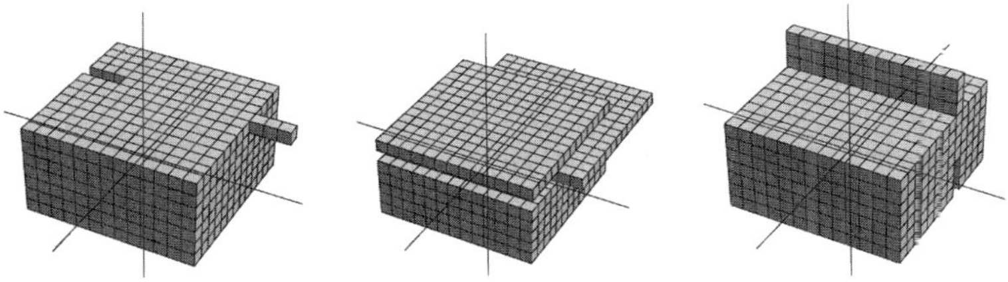

$H = L = \text{span}\langle \vec{e_1} \rangle$	$H = L = \text{span}\langle \vec{e_1}, \vec{e_2} \rangle$	$H = \text{span}\langle \vec{e_3} \rangle$
$\vec{w} = (0, 7, -1)$	$\vec{w} = (0, 0, -2)$	$L = \text{span}\langle \vec{e_2}, \vec{e_3} \rangle$
$\vec{v} = (3, 0, 0)$	$\vec{v} = (2, 4, 0)$	$\vec{w} = (0, 7, 0)$
		$\vec{v} = (0, 0, 3)$

FIGURE 1. Examples of shear in $\mathbb{Z}^3$

DEFINITION 2.2. A shift space $(X, \mathbb{Z}^d)$, $d \geq 2$, is said to have (L, H)**-shear** for sublattices $H \subseteq L \subseteq \mathbb{Z}^d$ if and only if for all $\vec{v} \in H$, $\sigma^L_{\vec{v}}(X) \subseteq X$. If $H = L$, we will say that $(X, \mathbb{Z}^d)$ has L**-shear**.

Figure 2 illustrates the shear operator action for affine sublattices of dimension one and two in $\mathbb{Z}^3$.

Notice that if $L = \mathbb{Z}^d$, then $\sigma^L_{\vec{v}} = \sigma_{\vec{v}}$, and all $\mathbb{Z}^d$ shift spaces have $\mathbb{Z}^d$-shear. We will only be interested in shear in cases where L is a proper sublattice of $\mathbb{Z}^d$.

We make two straightforward observations. The first describes the relationship between the shift and shear operators:

OBSERVATION 2.3. For $L + \vec{w}$ an affine sublattice of $\mathbb{Z}^d$, $\vec{v} \in L$, and $x \in \mathcal{A}^{\mathbb{Z}^d}$,

$$\sigma^{L+\vec{w}}_{\vec{v}}(x) = \sigma_{-\vec{w}} \sigma^L_{\vec{v}} \sigma_{\vec{w}}(x).$$

It follows from Observation 2.3 and from the fact that X is shift invariant that $(X, \mathbb{Z}^d)$ has (L, H)-shear if and only if $\sigma^{L+\vec{w}}_{\vec{v}}(X) \subseteq X$ for all $\vec{w} \in L^\perp$ and $\vec{v} \in H$.

Our second observation describes the implications of (L, H)-shear for sub- and super-lattices:

OBSERVATION 2.4. Suppose $0 \subset H_2 \subset H \subseteq L \subset L_2 \subset \mathbb{Z}^d$ are sublattices. Then (L, H)- shear implies (L, H_2)-, (L_2, H)-, and (L_2, H_2)-shear.

We close this section with several examples:

EXAMPLE 2.5. **Full Shifts**
A full shift $(\mathcal{A}^{\mathbb{Z}^d}, \mathbb{Z}^d)$ has L-shear for all sublattices $L \subseteq \mathbb{Z}^d$.

EXAMPLE 2.6. **Shift Products**
Let $(Y, \mathbb{Z}^r)$ be a shift space and let $L = \mathcal{L}(\mathcal{B}) \subseteq \mathbb{Z}^d$ where $\mathcal{B} = \{\vec{e}_1, \vec{e}_2, \ldots, \vec{e}_r\}$. Intuitively, the $\mathbb{Z}^{d-r}$ shift product of $(Y, \mathbb{Z}^r)$, denoted $(Y^{\mathbb{Z}^{d-r}}, \mathbb{Z}^d)$, consists of copies of the $\mathbb{Z}^{d-r}$ lattice with an element of Y at each lattice point of $\mathbb{Z}^{d-r}$. Formally, we define

$$Y^{\mathbb{Z}^{d-r}} = \{ f : \mathbb{Z}^{d-r} \to Y \}$$

and a $\mathbb{Z}^d$-action on $Y^{\mathbb{Z}^{d-r}}$ given by

$$\sigma_{\vec{v}}(f)(\vec{z}) = \sigma_{\vec{v}_L}\left(f(\vec{z} + \vec{v}_{L^\perp})\right)$$

for all $\vec{v} \in \mathbb{Z}^d$ and $\vec{z} \in \mathbb{Z}^{d-r}$. (Note that in an abuse of notation we are viewing $\vec{v}_L$ as being in both $\mathbb{Z}^r$ and in $\mathbb{Z}^d$. A similar comment applies to $\vec{v}_{L^\perp}$.) Clearly the shift product $(Y^{\mathbb{Z}^{d-r}}, \mathbb{Z}^d)$ has L-shear.

We can reconcile the definition of a shift product with our intuition regarding $Y^{\mathbb{Z}^{d-r}}$ as a shift space by viewing each $f \in Y^{\mathbb{Z}^{d-r}}$ as a $\mathbb{Z}^d$ array of symbols in $(\mathcal{A}_Y)^{\mathbb{Z}^d}$. For each $\vec{v} \in \mathbb{Z}^d$, the symbol found in the $\vec{v}$ location of f is $(f(\vec{v}_{L^\perp}))_{\vec{v}_L}$.

There is only one kind of shear of interest in a $\mathbb{Z}^2$ space, namely L-shear when L is a one-dimensional sublattice of $\mathbb{Z}^2$. For $\mathbb{Z}^3$ shift spaces, there are three kinds of shear: L-shear when L has dimension one; L-shear when L has dimension two; and (L, H)-shear when L has dimension two and H has dimension one. (These appear left-to-right in Figure 2.) Let $(Y_1, \mathbb{Z})$ be a one-dimensional shift space and let $(Y_2, \mathbb{Z}^2)$ be a two-dimensional shift space without shear. The shift products $(Y_1^{\mathbb{Z}^2}, \mathbb{Z}^3)$ and $(Y_2^{\mathbb{Z}}, \mathbb{Z}^3)$ are examples, respectively, of the first two of these three kinds of shear. The following example illustrates the third kind of shear possible for $\mathbb{Z}^3$ shift spaces.

EXAMPLE 2.7. (L, H)**-Shear in $\mathbb{Z}^3$ with L, H Dimension Two, One.** Define a space $X \subseteq \{0,1\}^{\mathbb{Z}^3}$ as follows: $x \in X$ if and only if

(1) For a fixed j, $x_{(i,j,k)} = 1$ for at most one value of (i, j, k) (The symbol 1 appears at most once in each plane orthogonal to $\vec{e}_2$.);
(2) For a fixed k, $x_{(i,j,k)} = 1$ for at most two values of (i, j, k) (The symbol 1 appears at most twice in each plane orthogonal to $\vec{e}_3$.); and
(3) If $x_{(i,j,k)} = x_{(i',j',k)} = 1$, then $|i - i'| + |j - j'| \leq 10$. (If the symbol 1 appears twice with the same z-coordinate, the appearances must be "close" together.)

Then $(X, \mathbb{Z}^3)$ is a shift space with (L, H)-shear for $L = \mathcal{L}(\vec{e}_1, \vec{e}_2)$ and $H = \mathcal{L}(\vec{e}_1)$.

EXAMPLE 2.8. **Three Ones Example**
Shift space $X \subseteq \{0,1\}^{\mathbb{Z}^d}$ has the property that each $x \in X$ has at most three occurrences of the symbol 1. Then $(X, \mathbb{Z}^d)$ has L-shear for all sublattices $L \subseteq \mathbb{Z}^d$.

EXAMPLE 2.9. **A Two Dimensional Context Free Type Shift**
Let $\mathcal{A} = \{0, \mathrm{a}, \mathrm{b}, \mathrm{c}, \mathrm{A}, \mathrm{B}, \mathrm{R}, \mathrm{L}\}$. Define $X \subseteq \mathcal{A}^{\mathbb{Z}^2}$ by a set of 2×2 forbidden blocks, where a block is forbidden if it does not appear in Figure 2. Then $(X, \mathbb{Z}^2)$ is a shift of finite type. Let Y be the image of X under the symbol map defined by $\Phi(s) = s$ if $s \in \{a, b, c\}$ and $\Phi(s) = 0$ otherwise. The space $(Y, \mathbb{Z}^2)$ has L-shear where $L = \mathcal{L}(\vec{e}_1)$.

Every shift space is contained in a "minimal" space with shear. We call this space the shear closure:

EXAMPLE 2.10 (**Shear Closure**). Let $(X, \mathbb{Z}^d)$ be any shift space and let $H \subseteq L \subset \mathbb{Z}^d$ be sublattices. Define $(\hat{X}, \mathbb{Z}^d)$, the (L, H)-shear closure of $(X, \mathbb{Z}^d)$, via $\hat{X} = \bigcap \tilde{X}$, where the intersection is over all $\tilde{X} \supseteq X$ for which $(\tilde{X}, \mathbb{Z}^d)$ is a shift space with (L, H)-shear. It is clear that by construction, that (i) $(\hat{X}, \mathbb{Z}^d)$ has (L, H)-shear, (2) $\mathcal{A}_X = \mathcal{A}_{\hat{X}}$, and (3) $X \subseteq \hat{X}$. Note also that if $(X, \mathbb{Z}^d)$ has (L, H)-shear, then $\hat{X} = X$.

0	0	0	0	0	0	0	0	0	0	0	0	0	0	0	0	0	0	0	0	0
0	0	0	0	L	R	0	0	0	0	0	0	0	0	0	0	0	0	0	0	0
0	0	0	L	A	B	R	0	0	0	0	0	0	0	0	0	0	0	0	0	0
0	0	L	A	A	B	B	R	0	0	0	L	R	0	0	0	0	0	0	0	0
0	L	A	A	A	B	B	B	R	0	L	A	B	R	0	0	0	0	L	R	0
c	a	a	a	a	b	b	b	b	c	a	a	b	b	c	c	c	c	a	b	c
0	0	0	0	0	0	0	0	0	0	0	0	0	0	0	0	0	0	0	0	0

FIGURE 2. Allowed blocks for Example 2.9.

3. Generalized Shift Products

In this section, we describe the construction of an important class of examples, the generalized shift products. As the name suggests, generalized shift products are a generalization of Example 2.6. In Section 4, we will show that many shift spaces with shear are in fact generalized shift products.

Let $L = \mathcal{L}(\mathcal{B})$, $L^\perp \subseteq \mathbb{Z}^d$ be complementary sublattices of $\mathbb{Z}^d$ with $|\mathcal{B}| = r$. Let $(W, \mathbb{Z}^{d-r})$ be any $(d-r)$-dimensional shift space. For each $w \in \mathcal{A}_W$, let $(Y_w, \mathbb{Z}^r)$ be an r-dimensional shift space. We will create a new shift space, denoted by

$$(X(W, \{Y_w\}_{w\in\mathcal{A}_W}), \mathbb{Z}^d).$$

Intuitively, we think of this space as copies of the $\mathbb{Z}^{d-r}$ lattice with both a symbol $w_{\vec{z}} \in \mathcal{A}_W$ and a point $y_{\vec{z}}$ (where $y_{\vec{z}} \in Y_{w_{\vec{z}}}$) at each lattice point $\vec{z} \in \mathbb{Z}^{d-r}$, where the symbols $w_{\vec{z}}$ form a point of W: $(w_{\vec{z}})_{\vec{z}\in\mathbb{Z}^{d-r}} \in W$. We define $(X(W, \{Y_w\}_{w\in\mathcal{A}_W}), \mathbb{Z}^d)$ formally as follows. First let

$$\mathcal{S} = \{\, (w, y) \mid w \in \mathcal{A}_W \text{ and } y \in Y_w\}.$$

Then let

$$X(W, \{Y_w\}_{w\in\mathcal{A}_W}) = \{\, f : \mathbb{Z}^{d-r} \to \mathcal{S} \mid (\pi_1(f(\vec{z})))_{\vec{z}\in\mathbb{Z}^{d-r}} \in W\,\} \tag{3.1}$$

where $\pi_i \circ f$ is the ith coordinate of the image of f for $i = 1, 2$. We define a $\mathbb{Z}^d$-action on $X(W, \{Y_w\}_{w\in\mathcal{A}_W})$ via

$$\sigma_{\vec{v}}(f)(\vec{z}) = (\pi_1(f(\vec{z} + \vec{v}_{L^\perp})), \sigma_{\vec{v}_L}\pi_2(f(\vec{z} + \vec{v}_{L^\perp})))$$

for all $\vec{v} \in \mathbb{Z}^d$ and $\vec{z} \in \mathbb{Z}^{d-r}$. It is not difficult to verify that $(X(W, \{Y_w\}_{w\in\mathcal{A}_W}), \mathbb{Z}^d)$ has L-shear.

We can reconcile the definition of $X(W, \{Y_w\}_{w\in\mathcal{A}_W})$ with our intuition of $X(W, \{Y_w\}_{w\in\mathcal{A}_W})$ as a shift space by viewing each $f \in X(W, \{Y_w\}_{w\in\mathcal{A}_W})$ as a $\mathbb{Z}^d$ array of symbols in $\mathcal{A}^{\mathbb{Z}^d}$ where $\mathcal{A} = \{\, (w, a) \mid w \in \mathcal{A}_W \text{ and } a \in \mathcal{A}_{Y_w}\,\}$. For each $\vec{v} \in \mathbb{Z}^d$, the symbol found in the $\vec{v}$ location of f is the ordered pair

$$(\pi_1(f(\vec{v}_{L^\perp})), \pi_2(f(\vec{v}_{L^\perp}))_{\vec{v}_L}).$$

We will call a shift space a **generalized shift product** if it can be obtained from a shift space of the form $X(W, \{Y_w\}_{w\in W})$ by "forgetting" the information given by W.

DEFINITION 3.1. The shift space $(Y, \mathbb{Z}^d)$ is a **generalized shift product** if there exists $(X(W, \{Y_w\}_{w\in W}), \mathbb{Z}^d)$ and a factor map $\phi : X(W, \{Y_w\}_{w\in\mathcal{A}_W}) \to Y$ where ϕ is defined via $(\phi(f))_{\vec{v}} = \pi_2(f(\vec{v}_{L^\perp}))_{\vec{v}_L}$ for all $\vec{v} \in \mathbb{Z}^d$.

Figure 3 gives an intuitive picture of a generalized shift product for $d = 2$ and $r = 1$. In the figure, shading patterns represent the symbols in Y_w. The figure shows, from left to right: a point $w \in W$, a point in $X(W, \{Y_a, Y_b, Y_c\})$, and the corresponding point in the generalized shift product Y obtained by "forgetting" the information given by W.

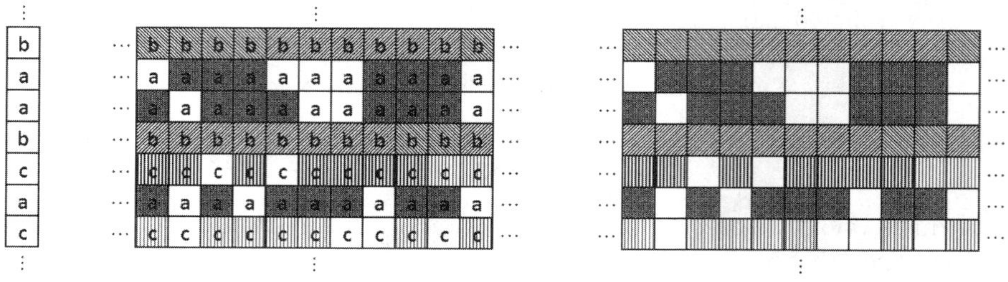

FIGURE 3. A point in Y from $X(\{a, b, c\}^{\mathbb{Z}}, \{Y_a, Y_b, Y_c\})$

Shift spaces that are shift products are also generalized shift products. To see this, let $(X, \mathbb{Z}^d) = (Y^{\mathbb{Z}^{d-r}}, \mathbb{Z}^d)$ where $(Y, \mathbb{Z}^r)$ is a shift space. Then $(X, \mathbb{Z}^d)$ is the image of $(X(W, Y_a), \mathbb{Z}^d)$ where W is the single-point shift $(\{a\}^{\mathbb{Z}^{d-r}}, \mathbb{Z}^{d-r})$ and $Y_a = Y$.

The definition of decouplable nearest-neighbor shifts of finite type [**OP**] helped motivate the idea of a generalized shift product. Any higher-dimensional *decouplable* nearest-neighbor shift of finite type is a generalized shift product, though not all generalized shift products are decouplable. Briefly, a generalized shift product is a decouplable nearest-neighbor shift of finite type if it is a nearest-neighbor $\mathbb{Z}$-shift or can be obtained from a shift space of the form $X(W, \{Y_w\}_{w \in W})$ where W is a nearest-neighbor $\mathbb{Z}$-shift and each Y_w is decouplable.

4. $\mathbb{Z}^{d-1}$-Shear in $\mathbb{Z}^d$ Shift Spaces

We will now investigate shift spaces with L-shear in the special case where $L \subseteq \mathbb{Z}^d$ is $(d-1)$-dimensional. For simplicity, we will assume that $L = \mathcal{L}(\mathcal{B})$ where $\mathcal{B} = \{\vec{e}_1, \vec{e}_2, \ldots, \vec{e}_{d-1}\}$, although we could take $\mathcal{B}$ to be any $d-1$ basis vectors. We will call this type of shear $\mathbb{Z}^{d-1}$-shear.

The main theorem of this section provides a criterion to determine when a $\mathbb{Z}^d$ shift space with $\mathbb{Z}^{d-1}$-shear is a generalized shift product. In the proof, we will construct generalized shift products with a one-dimensional shift space $(W, \mathbb{Z})$. We restrict our attention to $\mathbb{Z}^{d-1}$–shear specifically to avoid some of the higher-dimensional complications that arise in constructing generalized shift products entirely with shift spaces of dimension greater than one.

We begin with the following useful definition:

DEFINITION 4.1. Let $(Y, \mathbb{Z}^d)$ be a shift space, and let C be configuration that occurs in Y on the bottom-half space $B \subseteq \mathbb{Z}^d$. The **bottom half-space extender set** of C, denoted $\mathcal{E}(C)$, is given as follows:

$$\mathcal{E}(C) = \{y_{R \cup T} \,|\, y \in Y \text{ and } y_B = C\}.$$

Higher dimensional extender sets were first defined in [**KM**], where they were used to find a sufficient condition for soficness in higher dimensional shift spaces.

In one dimension, bottom half-space extender sets are the same as follower sets, and a shift space is sofic if and only if it has a finite number of follower sets. Theorem 4.2 shows that for higher dimensional shift spaces with Z^{d-1}-shear, having a finite number of bottom half-space extender sets implies that the space is a generalized shift product.

THEOREM 4.2. *Let $(Y, \mathbb{Z}^d)$ be a shift space with $\mathbb{Z}^{d-1}$-shear. The set of bottom half-space extender sets of $(Y, \mathbb{Z}^d)$ is finite if and only if $(Y, \mathbb{Z}^d)$ is a generalized shift product where the space $(W, \mathbb{Z})$ is a nearest neighbor shift of finite type.*

Before proving this theorem, we will describe the shift spaces $\{(Y_w, \mathbb{Z}^{d-1})\}_{w \in \mathcal{A}_W}$ in the case where the set of bottom half-space extender sets is finite Let $(Y, \mathbb{Z}^d)$ be a shift space with $\mathbb{Z}^{d-1}$-shear and suppose there are finitely many bottom half-space extender sets, $\mathcal{E}_i$, for $1 \le i \le N$. Define

$$Y_{i,j} = \left\{ r \in (\mathcal{A}_Y)^{\mathbb{Z}^{d-1}} \;\middle|\; \begin{array}{l} \text{for any } y \in Y \text{ with } \mathcal{E}(y_B) = \mathcal{E}_i \text{ and any} \\ \text{configuration } E \in \mathcal{E}_j, \text{ the point } x \in (\mathcal{A}_Y)^{\mathbb{Z}^d} \text{ with} \\ x_B = y_B,\ x_R = r, \text{ and } x_T = E \text{ is in } Y \end{array} \right\}.$$

It will be convenient to denote the point $x \in (\mathcal{A}_Y)^{\mathbb{Z}^d}$ where $x_B = y_B$, $x_R = r$, and $x_T = E$ as $y_B \cdot r \cdot E$.

LEMMA 4.3. *Let $(Y, \mathbb{Z}^d)$ be a shift space with $\mathbb{Z}^{d-1}$-shear and bottom half-space extender sets $\mathcal{E}_i$ for $1 \le i \le N$. Then for each $1 \le i, j \le N$, $(Y_{i,j}, \mathbb{Z}^{d-1})$ as defined above is a shift space.*

PROOF. If $Y_{i,j}$ is empty, the claim is trivial, so assume $Y_{i,j}$ is non-empty. The shift invariance of $Y_{i,j}$ follows immediately from the fact that $(Y, \mathbb{Z}^d)$ has $\mathbb{Z}^{d-1}$-shear, so we need only show that $(Y_{i,j}, \mathbb{Z}^{d-1})$ is closed.

Let $\{r_\ell\}_{\ell=1}^\infty \subseteq Y_{i,j}$ with $\lim_{\ell \to \infty} r_\ell = r$. We first note that because each $r_\ell \in Y_{i,j}$, that given any occurring configuration H on B with $\mathcal{E}(H) = \mathcal{E}_i$ and any configuration $E \in \mathcal{E}_j$, the points $y_\ell \in \mathcal{A}^{\mathbb{Z}^d}$ given by $y_\ell = H \cdot r_\ell \cdot E$ are in Y. In order to show closure, we need to show that for any such E and H, $H \cdot r \cdot E \in Y$. So let occurring configurations H on B with $\mathcal{E}(H) = \mathcal{E}_i$ and $E \in \mathcal{E}_j$ be given, and let $y_\ell = H \cdot r_\ell \cdot E$. Because Y is closed and $y_\ell \in Y$, by taking a subsequence if necessary, we may assume that $\lim_{\ell \to \infty} y_\ell = y \in Y$. But clearly, since $(y_\ell)_B = H$ and $(y_\ell)_T = E$ for all $\ell \in \mathbb{N}$, $y_B = H$ and $y_T = E$. Furthermore, because $(y_\ell)_R = r_\ell$ for all $\ell \in \mathbb{N}$ with $\lim_{\ell \to \infty} r_\ell = r$, we have $y_R = r$. Thus $y = H \cdot r \cdot E \in Y$ as desired. □

We are now ready to prove Theorem 4.2.

PROOF. First, let $(Y, \mathbb{Z}^d)$ be a generalized shift product given by a nearest neighbor shift of finite type $(W, \mathbb{Z})$ and shift spaces $\{(Y_w, \mathbb{Z}^{d-1})\}_{w \in \mathcal{A}_W}$ with $\phi : X(W, \{Y_w\}_{w \in \mathcal{A}_W}) \to Y$. Let $y \in Y$ and note that

$$\mathcal{E}(y_B) = \bigcup \{\, \phi(\mathcal{E}(f_B)) \mid f \in X(W, \{Y_w\}_{w \in \mathcal{A}_W}) \text{ with } \phi(f) = y \,\}.$$

We will show that $X(W, \{Y_w\}_{w \in \mathcal{A}_W})$ has at most $|\mathcal{A}_W|$-many bottom half-space extender sets. Then there are at most $|\mathcal{A}_W|$ distinct sets $\phi(\mathcal{E}(f_E))$, and at most

$2^{|\mathcal{A}_W|}$ unions of one or more of the $\phi(\mathcal{E}(f_B))$. Every bottom-half space extender set for $(Y, \mathbb{Z}^d)$ is one of these $2^{|\mathcal{A}_W|}$ unions.

Let $f \in X(W, \{Y_w\}_{w \in \mathcal{A}_W})$. Then $\mathcal{E}(f_B)$ is determined by $\pi_1(f(-1)) \in \mathcal{A}_W$. To see this, note that if $g \in X(W, \{Y_w\}_{w \in \mathcal{A}_W})$, and if the transition $\pi_1(f(-1))\pi_1(g(0))$ is allowed in the nearest neighbor shift $(W, \mathbb{Z})$, then the function $h : \mathbb{Z} \to \mathcal{S}$ for $\mathcal{S} = \{(w, y) \mid w \in W \text{ and } y \in Y_w\}$ defined via

$$h(n) = \begin{cases} f(n) & n < 0 \\ g(n) & n \geq 0 \end{cases}$$

is in $X(W, \{Y_w\}_{w \in \mathcal{A}_W})$. Thus, the number of bottom half-space extender sets for $X(W, \{Y_w\}_{w \in \mathcal{A}_W})$ is at most the number of different values of $\pi_1(f(-1))$, which is $|\mathcal{A}_W|$, as desired.

For the other direction of the proof, suppose that $(Y, \mathbb{Z}^d)$ is a shift space with $\mathbb{Z}^{d-1}$-shear and only finitely many bottom half-space extender sets $\mathcal{E}_i$ for $i = 1, \ldots, N$. Let $\mathcal{A}_W = \{(i, j) \mid 1 \leq i, j \leq N \text{ and } Y_{i,j} \neq \emptyset\}$ where the $(Y_{i,j}, \mathbb{Z}^{d-1})$ are the shift spaces defined for Lemma 4.3. We define a one-dimensional shift of finite type $(W, \mathbb{Z})$ with alphabet $\mathcal{A}_W$ using the transition rules $(i, j) \to (\ell, k)$ if and only if $j = \ell$. For each $(i, j) \in \mathcal{A}_W$, we associate the shift space $(Y_{i,j}, \mathbb{Z}^{d-1})$.

We construct $X(W, \{Y_{i,j}\}_{(i,j) \in \mathcal{A}_W})$ as defined in Equation 3.1. Then

$$X(W, \{Y_{i,j}\}_{(i,j) \in \mathcal{A}_W}) = \{f : \mathbb{Z} \to \mathcal{S} \mid \pi_1(f(n))_{n \in \mathbb{Z}} \in W\}$$

where $\mathcal{S} = \{((i, j), r) \mid (i, j) \in \mathcal{A}_W \text{ and } r \in Y_{i,j}\}$. We can view elements in $X\left(W, \{Y_{i,j}\}_{(i,j) \in \mathcal{A}_W}\right)$ as being of the form $(f(n))_{n \in \mathbb{Z}} = ((i_n, j_n), r_n)_{n \in \mathbb{Z}}$ where $r_n \in Y_{i_n, j_n}$ and $j_n = i_{n+1}$. That is:

$$(f(n))_{n \in \mathbb{Z}} = \begin{array}{lcl} \vdots & & \vdots \\ ((i_2, i_3) & , & r_2) \\ ((i_1, i_2) & , & r_1) \\ ((i_0, i_1) & , & r_0) \\ ((i_{-1}, i_0) & , & r_{-1}) \\ ((i_{-2}, i_{-1}) & , & r_{-2}) \\ \vdots & & \vdots \end{array}$$

where $r_n \in Y_{i_n, i_{n+1}}$ Note that because $r_n \in Y_{i_n, i_{n+1}}$, for any occurring configurations H on B with $\mathcal{E}(H) = \mathcal{E}_{i_n}$ and $E \in \mathcal{E}_{i_{n+1}}$, the point $y = H \cdot r_n \cdot E$ constructed via $y_B = H$, $y_R = r_n$, and $y_T = E$ will be in Y.

We first show that if $y \in Y$, then there exists $f \in X(W, \{Y_{i,j}\}_{(i,j) \in \mathcal{A}_W})$ with $\phi(f) = y$. So, let $y \in Y$ be given. For each $j \in \mathbb{Z}$, let $r_j = y_{R_j}$. That is, r_j is the $\mathbb{Z}^{d-1}$-configuration on the jth center slice in y. Choose (i_j, i_{j+1}) as follows: if $\mathcal{E}(y_{B_j}) = \mathcal{E}_m$ and $\mathcal{E}(y_{B_j} \cdot r_j) = \mathcal{E}_n$, then $(i_j, i_{j+1}) = (m, n)$.

We first note that $r_j \in Y_{i_j, i_{j+1}}$. To see this, let H be any configuration on B with $\mathcal{E}(H) = \mathcal{E}_{i_j}$ and let E be any element of $\mathcal{E}_{i_{j+1}}$. To show that $r_j \in Y_{i_j, i_{j+1}}$, we must show that $H \cdot r_j \cdot E$ is in Y. We have $\mathcal{E}(H) = \mathcal{E}_{i_j} = \mathcal{E}(y_{B_j})$, so $H \cdot r_j \cdot E \in Y$ if $y_{B_j} \cdot r_j \cdot E \in Y$. But $E \in \mathcal{E}_{i_{j+1}} = \mathcal{E}(y_{B_j} \cdot r_j)$, so $y_{B_j} \cdot r_j \cdot E \in Y$. Thus, $H \cdot r_j \cdot E \in Y$ for all H and E with $\mathcal{E}(H) = \mathcal{E}_{i_j}$ and $E \in \mathcal{E}_{i_{j+1}}$, and therefore $r_j \in Y_{i_j, i_{j+1}}$.

We next note that $\mathcal{E}(y_{B_j} \cdot r_j) = \mathcal{E}_{i_{j+1}} = \mathcal{E}(y_{B_{j+1}})$, so $(i_j, i_{j+1})_{j \in \mathbb{Z}} \in W$. So, $f = ((i_j, i_{j+1}), r_j)_{j \in \mathbb{Z}}$ is a point in $X(W, \{Y_{i,j}\}_{(i,j) \in \mathcal{A}_W})$ whose image under ϕ is y. Thus, $Y \subseteq \phi(X(W, \{Y_{i,j}\}_{(i,j) \in \mathcal{A}_W}))$.

Next let $f = ((i_j, i_{j+1}), r_j)_{j\in\mathbb{Z}} \in X(W, \{Y_{i,j}\}_{(i,j)\in\mathcal{A}_W})$. We need to show that

$$\phi(f) = \phi\left(((i_j, i_{j+1}), r_j)_{j\in\mathbb{Z}}\right) = (r_j)_{j\in\mathbb{Z}} \in Y.$$

We will show instead that for all $M \in \mathbb{Z}$ and $k \in \mathbb{N}$, the $(M, M+k)$-slice configuration

$$\phi\left((i_j, i_{j+1}), r_j\right)_{R(M,M+k)} = (r_j)_{M\le j\le M+k}$$

occurs in Y. A limiting argument will then give us that $\phi(f) \in Y$ as desired.

Let $M \in \mathbb{Z}$ and $k \in \mathbb{N}$. First, let H_0 be any half-space configuration on B occurring in Y with $\mathcal{E}(H_0) = \mathcal{E}_{i_M}$. By definition of $X(W, \{Y_{i,j}\}_{(i,j)\in\mathcal{A}_W})$, $r_M \in Y_{i_M, i_{M+1}}$ and thus for any $E \in \mathcal{E}_{i_{M+1}}$, the point $H_0 \cdot r_M \cdot E$ occurs in Y.

Now let H_M denote the configuration on B given by $H_0 \cdot r_M$. That is,

$$(H_M)_{B_{-1}} = H_0 \text{ and } (H_M)_{R_{-1}} = r_M.$$

(Informally, H_M is H_0 "topped" with r_M.) As noted above, for any $E \in \mathcal{E}_{i_{M+1}}$, the point $H_0 \cdot r_M \cdot E$ occurs in Y. This tells us both that H_M occurs in Y and that $\mathcal{E}(H_M) \supseteq \mathcal{E}_{i_{M+1}}$.

Consider r_{M+1}. By definition of X, $r_{M+1} \in Y_{i_{M+1}, i_{M+2}}$, and for any half-space configuration H occurring in Y with $\mathcal{E}(H) = \mathcal{E}_{i_{M+1}}$ and any $E \in \mathcal{E}_{i_{M+2}}$, the point $H \cdot r_{M+1} \cdot E$ occurs in Y. In particular, for H_M with $\mathcal{E}(H_M) \supseteq \mathcal{E}_{i_{M+1}}$ and any $E \in \mathcal{E}_{i_{M+2}}$, the point $H_M \cdot r_{M+1} \cdot E$ occurs in Y.

Now let H_{M+1} denote the configuration on B given by $H_M \cdot r_{M+1} = H_0 \cdot r_M \cdot r_{M+1}$:

$$(H_{M+1})_{B_{-1}} = H_M \text{ and } (H_{M+1})_{R_{-1}} = r_{M+1}$$

which means

$$(H_{M+1})_{B_{-2}} = H_0 \text{ and } (H_{M+1})_{R_{-2}} = r_M \text{ and } (H_{M+1})_{R_{-1}} = r_{M+1}.$$

Then H_{M+1} occurs in Y and $\mathcal{E}(H_{M+1}) \supseteq \mathcal{E}_{i_{M+2}}$.

Continue in this way to construct H_{M+k} on B given by

$$H_{M+k} \cdot r_{M+k} = H_0 \cdot r_M \cdot r_{M+1} \cdots r_{M+k}$$

with

$$(H_{M+k})_{B_{-k-1}} = H_0 \text{ and } (H_{M+k})_{B_{-k-1+j}} = r_{M+j-1} \text{ for } 0 < j \le k.$$

Then H_{M+k} occurs in Y and $\mathcal{E}(H_{M+k}) \supseteq E_{i_{M+k+1}}$. Choose any $E \in \mathcal{E}(H_{M+k})$ and create $y = H_{M+k} \cdot E \in Y$ and then

$$y_{R(-M-k+1,-1)} = (r_j)_{M\le j\le M+k}$$

as desired.

□

It is not difficult to check that Examples 2.5, 2.6, and 2.8 all have both $\mathbb{Z}^{d-1}$-shear and a finite number of bottom half-space extender sets; thus, they are all generalized shift products. Example 2.9 has $\mathbb{Z}^{d-1}$-shear and infinitely many bottom half-space extender sets; thus, it can not be a generalized shift product. Example 2.7 does not have $\mathbb{Z}^{d-1}$-shear, so Theorem 4.2 does not apply.

5. $\mathbb{Z}^{d-1}$–Shear in $\mathbb{Z}^d$ Shifts of Finite Type

Shifts of finite type are a restrictive class of general shift spaces, so we might hope to sharpen Theorem 4.2 in this situation. We begin with the following observation:

LEMMA 5.1. *If $(X, \mathbb{Z}^d)$ is a shift of finite type with $\mathbb{Z}^{d-1}$-shear, the set of bottom half-space extender sets is finite.*

PROOF. Let $(X, \mathbb{Z}^d)$ be a shift of finite type with $\mathbb{Z}^{d-1}$-shear. Without loss of generality, we may assume that X is defined via a set of forbidden configurations $\mathcal{F} \subseteq (\mathcal{A}_X)^D$ where

$$D = \left\{ \sum_{i=1}^{d} a_i e_i \,|\, 0 \leq a_i \leq m-1 \right\}$$

is the $[0, m-1]^d$ cube in $\mathbb{Z}^d$.

Let x, $x' \in X$ and suppose that the same set of configurations in $(\mathcal{A}_X)^D$ occur on $x_{R(-m+1,0)}$ and $x'_{R(-m+1,0)}$, where the sets $R(j,k)$ are as defined in Equation 1.1. Then $\mathcal{E}(x_B) = \mathcal{E}(x'_B)$; that is, the configurations x_B and x'_B can be extended in exactly the same ways.

To see this, suppose for some $y \in X$, $y_{R \cup T} = C \in \mathcal{E}(x_B)$. We will show that $C \in \mathcal{E}(x'_B)$ as well. Suppose not. Then $x_B \cdot C \in X$ but $x'_B \cdot C \notin X$. This means that $x'_B \cdot C$ must contain a forbidden configuration F from $\mathcal{F}$. The configurations x'_B and C both occur in X, so the forbidden configuration F is not contained entirely in x'_B or entirely in C. So F must be found in

$$(x'_B \cdot C)_{R(-m+2,m-2)} = (x'_B)_{R(-m+2,-1)} \cdot C_{R(0,m-2)}.$$

Say $F = A_1 \cdot A_2$ where A_1 is contained in $(x'_B)_{R(-m+2,-1)}$ and A_2 is contained in $C_{R(0,m-2)}$.

Since x, $x' \in X$ have the same set of configurations on $R(-m+2,-1)$, A_1 is contained in $(x_B)_{R(-m+2,-1)}$ as well. But, using the fact that $x_B \cdot C \in X$ and X has shear, this implies that $F = A_1 \cdot A_2$ occurs in X, contradicting the fact that F is forbidden. Thus, no such F exists and $x'_B \cdot C \in X$. Thus, $C \in \mathcal{E}(x'_B)$ and $\mathcal{E}(x_B) \subseteq \mathcal{E}(x'_B)$.

Reversing the roles of x and x' in the previous argument gives $\mathcal{E}(x'_B) \subseteq \mathcal{E}(x_B)$, and $\mathcal{E}(x_B) = \mathcal{E}(x'_B)$ as desired.

Thus the bottom half-space extender sets for X are determined by subsets of $(\mathcal{A}_X)^D$. Because there are only finitely many such subsets, the set of bottom half-space extender sets for X is finite.

□

As a result of Lemma 5.1 and Theorem 4.2, we know that shifts of finite type with $\mathbb{Z}^{d-1}$-shear are all generalized shift products where the space $(W, \mathbb{Z})$ is a nearest neighbor shift of finite type. In fact, we can say more:

THEOREM 5.2. *Let $(X, \mathbb{Z}^d)$ be a shift of finite type with $\mathbb{Z}^{d-1}$-shear. Then $(X, \mathbb{Z}^d)$ is a generalized shift product given by a nearest neighbor shift of finite type $(W, \mathbb{Z})$ and shifts of finite type $\{(Y_w, \mathbb{Z}^{d-1})\}_{w \in \mathcal{A}_W}$.*

PROOF. Let $(X, \mathbb{Z}^d)$ be a shift of finite type with $\mathbb{Z}^{d-1}$-shear. As in the previous proof, we assume that the set of forbidden configurations, $\mathcal{F}$, is a subset of $(\mathcal{A}_X)^D$

where D is the $[0, m-1]^d$ cube in $\mathbb{Z}^d$. Let $\{\mathcal{E}_1, \mathcal{E}_2, \ldots, \mathcal{E}_N\}$ denote the collection of bottom-half space extender sets which, via Lemma 5.1, is finite.

We begin by recalling the construction found in the proof of Theorem 4.2. First, for each pair (i,j) with $1 \leq i, j \leq N$ let

$$Y_{i,j} = \left\{ r \in (\mathcal{A}_X)^{\mathbb{Z}^{d-1}} \;\middle|\; \begin{array}{l} \text{for any } y \in X \text{ with } \mathcal{E}(y_B) = \mathcal{E}_i \text{ and any} \\ \text{configuration } E \in \mathcal{E}_j, \text{ the point } x \in (\mathcal{A}_X)^{\mathbb{Z}^d} \text{ with} \\ x_B = y_B,\ x_R = r, \text{ and } x_T = E \text{ is in } X \end{array} \right\}.$$

Lemma 4.3 tells us that $(Y_{i,j}, \mathbb{Z}^{d-1})$ is a shift space.

Next, define shift of finite type $(W, \mathbb{Z})$ with alphabet

$$\mathcal{A}_W = \{(i,j) \,|\, 1 \leq i, j \leq N \text{ and } Y_{i,j} \neq \emptyset\}$$

and transition rules $(i,j) \to (\ell, k)$ if and only if $j = \ell$. Theorem 4.2 tells us that $(X, \mathbb{Z}^d)$ is a generalized shift product with $X(W, \{Y_{i,j}\}_{(i,j)\in\mathcal{A}_W})$ mapping onto X by a map on symbols. So, we need only show that the shift spaces $\{(Y_{i,j}, \mathbb{Z}^{d-1})\}_{(i,j)\in\mathcal{A}_W}$ are themselves shifts of finite type. We will do this by finding a finite set of forbidden blocks defining each $(Y_{i,j}, \mathbb{Z}^{d-1})$.

Fix $1 \leq i, j \leq N$. Let $\{A_\ell\}_{\ell=1}^L \subseteq (\mathcal{A}_X)^D$ be all configurations that occur on $x_{R(-m,-1)}$ for some $x \in X$ with $\mathcal{E}(x_B) = \mathcal{E}_i$. Similarly, let $\{C_k\}_{k=1}^K \subseteq (\mathcal{A}_X)^D$ be all configurations that occur on $\mathcal{E}_{R(1,m)}$ for some $E \in \mathcal{E}_j$.

We define a set of forbidden configurations $\mathcal{F}_{i,j} \subseteq (\mathcal{A}_X)^{D'}$ where

$$D' = \left\{ \sum_{i=1}^{d-1} a_i e_i \,|\, 0 \leq a_i \leq m-1 \right\}$$

is the $[0, m-1]^{d-1}$ cube in $\mathbb{Z}^{d-1}$ as follows: $F \in (\mathcal{A}_X)^{D'}$ is in $\mathcal{F}_{i,j}$ if and only if $A_\ell \cdot F \cdot C_k$ contains an element of $\mathcal{F}$ for some $1 \leq \ell \leq L$, $1 \leq k \leq K$.

Now let $(X_{i,j}, \mathbb{Z}^{d-1})$ be the $\mathbb{Z}^{d-1}$ shift of finite type given by forbidden blocks $\mathcal{F}_{i,j}$. We will show that $X_{i,j} = Y_{i,j}$.

First let $r \in X_{i,j}$. In order to show that $r \in Y_{i,j}$ as well, we need to show that given any $y \in X$ with $\mathcal{E}(y_B) = \mathcal{E}_i$ and any $E \in \mathcal{E}_j$, the point $x = y_B \cdot r \cdot E \in (\mathcal{A}_X)^{\mathbb{Z}^d}$ with $x_B = y_B$, $x_R = r$, and $x_T = E$ is in X. Clearly if $y \in X$ with $\mathcal{E}(y_B) = \mathcal{E}_i$, then the D-configurations occurring on $y_{R(-m,-1)}$ are all elements of the set $\{A_\ell\}_{\ell=1}^L$. Similarly, if $E \in \mathcal{E}_j$, then the D-configurations occurring on $\mathcal{E}_{R(1,m)}$ are all elements of the set $\{C_k\}_{k=1}^K$. But then $x = y_B \cdot r \cdot E$ can contain no forbidden D-blocks and $x \in X$ as desired. So, we have $X_{i,j} \subseteq Y_{i,j}$.

Next let $r \in Y_{i,j}$. This means that for any $y \in X$ with $\mathcal{E}(y_B) = \mathcal{E}_i$ and any $E \in \mathcal{E}_j$, the point $x \in (\mathcal{A}_X)^{\mathbb{Z}^d}$ with $x_B = y_B$, $x_R = r$, and $x_T = E$ is in X. We'd like to show that r cannot contain any of the forbidden configurations in $\mathcal{F}_{i,j}$. Suppose to the contrary that $F \in \mathcal{F}_{i,j}$ occurs in r. Because $F \in \mathcal{F}_{i,j}$, we can find some $1 \leq \ell \leq L$, $1 \leq k \leq K$ so that $A_\ell \cdot F \cdot C_k$ contains a forbidden configuration from $\mathcal{F}$. But some $y \in X$ with $\mathcal{E}(y_B) = \mathcal{E}_i$ must contain A_ℓ among the D-configurations occurring on $y_{R(-m,-1)}$ and some $E \in \mathcal{E}_j$ must contain C_k among the D-configurations occurring on $E_{R(1,m)}$, and so $y_B \cdot r \cdot E$ does not occur in X. This contradicts the fact that $r \in Y_{i,j}$. Thus, no such $F \in \mathcal{F}_{i,j}$ occurs in r, and $r \in X_{i,j}$. So $Y_{i,j} \subseteq X_{i,j}$ and $X_{i,j} = Y_{i,j}$ as desired. ∎

6. $\mathbb{Z}^{d-1}$–Shear in $\mathbb{Z}^d$ Sofic Shifts

Based on Theorem 5.2, we might conjecture that if $(Y, \mathbb{Z}^d)$ is a sofic shift with $\mathbb{Z}^{d-1}$-shear, then it must be a generalized shift product in which all of the shift spaces $\{(Y_w, \mathbb{Z})\}_{w\in W}$ are sofic. However, as noted earlier, Example 2.9 is a sofic $\mathbb{Z}^d$ shift space with $\mathbb{Z}^{d-1}$-shear that does not have a finite number of bottom-half space extender sets. Thus, it is not even a generalized shift product. So, we might refine our conjecture about $\mathbb{Z}^d$ sofic shift spaces with $\mathbb{Z}^{d-1}$-shear as follows:

CONJECTURE 6.1. *Let $(Y, \mathbb{Z}^d)$ be a sofic shift space with $\mathbb{Z}^{d-1}$-shear. Then $(Y, \mathbb{Z}^d)$ is a generalized shift product given by a nearest neighbor shift of finite type $(W, \mathbb{Z})$ and sofic shift spaces $\{(Y_w, \mathbb{Z}^{d-1})\}_{w\in\mathcal{A}_W}$ if and only if the set of bottom half-space extender sets is finite.*

It is easy to see that the shift product $(Y^{\mathbb{Z}}, \mathbb{Z}^d)$ has a single bottom half-space extender set, and therefore Conjecture 6.1 generalizes the following existing conjecture [**GJ**].

CONJECTURE 6.2. *$(Y^{\mathbb{Z}}, \mathbb{Z}^2)$ is sofic only if $(Y, \mathbb{Z})$ is sofic.*

The converse of Conjecture 6.2 and one direction of Conjecture 6.1 are trivial. While the other direction of Conjecture 6.1 remains open, Theorem 6.3 tells us that if a counterexample exists, it will be a a sofic shift space with $\mathbb{Z}^{d-1}$-shear for which there is no shift of finite type cover with $\mathbb{Z}^{d-1}$-shear.

THEOREM 6.3. *Let $(Y, \mathbb{Z}^d)$ be a sofic shift space with $\mathbb{Z}^{d-1}$-shear. Then $(Y, \mathbb{Z}^d)$ is a generalized shift product given by a nearest neighbor shift of finite type $(W, \mathbb{Z})$ and sofic shift spaces $\{(Y_w, \mathbb{Z}^{d-1})\}_{w\in\mathcal{A}_W}$ if and only if it is the image under a factor map of a shift of finite type $(X, \mathbb{Z}^d)$ with $\mathbb{Z}^{d-1}$-shear.*

PROOF. First if $(Y, \mathbb{Z}^d)$ is a generalized shift product given by a nearest neighbor shift of finite type $(W, \mathbb{Z})$ and sofic shift spaces $\{(Y_w, \mathbb{Z}^{d-1})\}_{w\in\mathcal{A}_W}$, we simply find shifts of finite type $(X_w, \mathbb{Z}^{d-1})$ for each sofic $(Y_w, \mathbb{Z}^{d-1})$ so that Y_w is the image of X_w under a factor map. Notice that without loss of generality, the $(X_w, \mathbb{Z}^{d-1})$ can be assumed to be nearest neighbor shifts of finite type. Then the generalized shift product $(X(W, \{X_w\}_{w\in\mathcal{A}_W}), \mathbb{Z}^d)$ is a nearest neighbor shift of finite type with $\mathbb{Z}^{d-1}$-shear type that factors onto $(Y, \mathbb{Z}^d)$.

Conversely, let $\alpha : X \to Y$ be a factor map where $(X, \mathbb{Z}^d)$ is a shift of finite type with $\mathbb{Z}^{d-1}$-shear. By Theorem 5.2, $(X, \mathbb{Z}^d)$ is a generalized shift product given by a nearest neighbor shift of finite type $(W, \mathbb{Z})$ and shifts of finite type $\{(Y_w, \mathbb{Z}^{d-1})\}_{w\in\mathcal{A}_W}$. Thus there exists a factor map from $(X(W, \{X_w\}_{w\in\mathcal{A}_W}), \mathbb{Z}^d)$ to Y as desired. □

7. Future Research Questions

In this paper, we have focused on Z^d-shift spaces with Z^{d-1}-shear. Do our results, or similar ones, hold when different kinds of shear are present?

Many foundational theorems for one-dimensional shift spaces, such as those describing entropy, have no higher-dimensional analogs but can be applied to shift products of one-dimensional shift spaces. We might hope to extend some one-dimensional theorems to generalized shift products and/or higher-dimensional shift spaces with shear.

Finally, we have only considered shear and generalized shift products with respect to sublattices $\mathcal{L}(\mathcal{B})$ generated by standard unit vectors $\vec{e_i}$. An even more general setting, allowing arbitrary sublattices, would require us to work with shift spaces on groups other than $\mathbb{Z}^d$. While the notion of shear might not generalize to a more group-theoretical (as opposed to geometric) setting, this might be a fruitful avenue of research. For example, *interleavings of full shifts*, a class of one-dimensional shifts defined in [**BE**], resemble generalized shift products using the sublattice $L = k\mathbb{Z} \subset \mathbb{Z}$.

References

[BE] Burke, N. and J. Enns, *Coincidence of 1- and 2-dimensional entropies*, 2014 IEEE International Symposium on Information Theory, 166-170.

[GJ] Guillon, P. and E. Jeandel, *Infinite Communication Complexity*, arXiv preprint, 2015, arXiv:1501.05814v1.

[JKM] Aimee Johnson, Steve Kass, and Kathleen Madden, *Projectional entropy in higher dimensional shifts of finite type*, Complex Systems **17** (2007), no. 3, 243–257. MR2373706

[LM] Douglas Lind and Brian Marcus, *An introduction to symbolic dynamics and coding*, Cambridge University Press, Cambridge, 1995. MR1369092

[KM] Steve Kass and Kathleen Madden, *A sufficient condition for non-soficness of higher-dimensional subshifts*, Proc. Amer. Math. Soc. **141** (2013), no. 11, 3803–3816, DOI 10.1090/S0002-9939-2013-11646-1. MR3091770

[OP] Nic Ormes and Ronnie Pavlov, *Extender sets and multidimensional subshifts*, Ergodic Theory Dynam. Systems **36** (2016), no. 3, 908–923, DOI 10.1017/etds.2014.71. MR3480350

[PS] Ronnie Pavlov and Michael Schraudner, *Classification of sofic projective subdynamics of multidimensional shifts of finite type*, Trans. Amer. Math. Soc. **367** (2015), no. 5, 3371–3421, DOI 10.1090/S0002-9947-2014-06259-4. MR3314811

Department of Mathematics and Computer Science, Drew University, Madison, New Jersey 07940

E-mail address: skass@drew.edu

School of Natural Sciences, Mathematics and Engineering, California State University Bakersfield, Bakersfield, California 93311

E-mail address: kmadden2@csub.edu

Contemporary Mathematics
Volume **678**, 2016
http://dx.doi.org/10.1090/conm/678/13651

Hereditary subshifts whose simplex of invariant measures is Poulsen

Joanna Kułaga-Przymus, Mariusz Lemańczyk, and Benjamin Weiss

ABSTRACT. We give a sufficient condition for the simplex of invariant measures for a hereditary system to be Poulsen. In particular, we show that this simplex is Poulsen in case of positive entropy $\mathscr{B}$-free systems. We also give an example of a positive entropy hereditary system whose simplex of invariant measures is not Poulsen.

1. Introduction

Recall that given a homeomorphism T of a compact metric space X, the set $\mathcal{P}(T,X)$ of probability Borel T-invariant measures on X is a Choquet simplex, with the set of extremal points equal to $\mathcal{P}^e(T,X)$ – the subset of ergodic measures (see e.g. [**24**], [**19**] or [**10**], page 95). Recall also that the set of invariant measures of a minimal flow may have the affine-topological structure of an arbitrary metrizable Choquet simplex [**7**].[1]

DEFINITION 1.1. We say that a simplex is *Poulsen* if it is non-trivial and the set of its extreme points is dense.

Recall that every Choquet simplex is affinely homomorphic to a face of a Poulsen simplex [**16**]. Moreover, up to affine homeomorphism, there is only one Poulsen simplex [**15**]. A natural question arises:

QUESTION 1. When is $\mathcal{P}(T,X)$ a Poulsen simplex?

We deal with Question 1 in case of *hereditary* subshifts, i.e. subshifts $X \subset \{0,1\}^{\mathbb{Z}}$ with the additional property that for $x,y \in \{0,1\}^{\mathbb{Z}}$ such that $x \in X$ and $y \le x$ (coordinatewise), we have $y \in X$ (recall that $X \subset \{0,1\}^{\mathbb{Z}}$ is called a *subshift* if it is closed and invariant under the left translation S).[2]

Given a subshift $X \subset \{0,1\}^{\mathbb{Z}}$, we define $\widetilde{X}$ as the smallest hereditary subshift containing X, i.e.

$$\widetilde{X} := \{y \in \{0,1\}^{\mathbb{Z}} : y \le x \text{ coordinatewise for some } x \in X\}.$$

2010 *Mathematics Subject Classification.* Primary 37B10, 37A05; Secondary 37B40, 37A35.
The first author's research was supported by National Science Centre, Poland, grant UMO-2014/15/B/ST1/03736.

[1] Moreover, all such simplices are obtained for the class of zero entropy Toeplitz subshifts.

[2] In other words, the heredity of X means that this set is closed under coordinatewise multiplication by arbitrary 0-1-sequences.

Let $M\colon X \times \{0,1\}^{\mathbb{Z}} \to \{0,1\}^{\mathbb{Z}}$ be given by

$$M(x,y)(n) = x(n) \cdot y(n) \text{ for each } n \in \mathbb{Z}.$$

Clearly $M(X \times \{0,1\}^{\mathbb{Z}}) = \widetilde{X}$. Suppose now additionally that there exists $\nu \in \mathcal{P}^e(S,X)$ such that

$$(1.1) \qquad \mathcal{P}^e(S,\widetilde{X}) = \{M_*(\rho) : \rho \in \mathcal{P}^e(S \times S, X \times \{0,1\}^{\mathbb{Z}}), \rho|_X = \nu\}.$$

Our main result is the following:

THEOREM 1.2. *Suppose that* (1.1) *holds. Then $\mathcal{P}(S,\widetilde{X})$ is Poulsen provided that $h_{top}(S,\widetilde{X}) > 0$.*

Notice also that in the zero entropy case, i.e., when $h_{top}(S,\widetilde{X}) = 0$, we have $\mathcal{P}(S,\widetilde{X}) = \{\delta_{(\dots,0,0,0,\dots)}\}$ (see [**14**]), in particular, $\mathcal{P}(S,\widetilde{X})$ is not Poulsen.

We apply Theorem 1.2 in the following two situations:

(i) for hereditary systems $(S,\widetilde{X})$ with (S,X) being uniquely ergodic,
(ii) for $\mathscr{B}$-free systems.

More precisely, in case (i), we have:

COROLLARY 1. *Suppose that (S,X) is uniquely ergodic. Then $\mathcal{P}(S,\widetilde{X})$ is Poulsen, provided that $h_{top}(S,\widetilde{X}) > 0$.*

PROOF. Given $\mu \in \mathcal{P}^e(S,\widetilde{X})$, let x be a generic point for μ and let $y \in X$ be such that $x \le y$. Clearly, $(y,x) \in X \times \{0,1\}^{\mathbb{Z}}$ is quasi-generic for some measure ρ with $\rho|_X = \nu$. Moreover, $M_*(y,x) = x$ and it follows immediately that $M_*(\rho) = \mu$. It follows by considering the ergodic decomposition of ρ that (1.1) holds and we can apply Theorem 1.2. □

To state the result in case (ii), we need to recall first some basic notions. Let $\mathscr{B} \subset \mathbb{N}$. The set $\mathcal{F}_{\mathscr{B}} := \mathbb{Z} \setminus \bigcup_{b\in\mathscr{B}} b\mathbb{Z}$ is called the *$\mathscr{B}$-free set*. Let $\eta := \mathbb{1}_{\mathcal{F}_{\mathscr{B}}} \in \{0,1\}^{\mathbb{Z}}$ and consider (S,X_η) with

$$X_\eta = \overline{\{S^k\eta : k \in \mathbb{Z}\}} \subset \{0,1\}^{\mathbb{Z}}.$$

Note that in this case $\widetilde{X}_\eta \subset X_{\mathscr{B}}$, where $(S,X_{\mathscr{B}})$ is the *$\mathscr{B}$-admissible subshift* [**2**]: $z \in X_{\mathscr{B}}$ if and only if we have $|\{n \in \mathbb{Z} : z(n) = 1\} \bmod b| < b$ for each $b \in \mathscr{B}$. There is a natural S-invariant measure ν_η on X_η, called the Mirsky measure, for which η is quasi-generic [**2**]. This means that there exists an increasing sequence $(N_k)_{k\ge1} \subset \mathbb{N}$ such that

$$\frac{1}{N_k}\sum_{n\le N_k} f(S^n\eta) = \int f\,d\nu_\eta$$

for any continuous function f on X_η. Moreover, ν_η is of zero entropy and we have:

THEOREM 1.3 ([**2**], for $\mathscr{B}$ pairwise coprime with $\sum_{b\in\mathscr{B}} 1/b < \infty$, see [**12**]). *For any $\mathscr{B} \subset \mathbb{N}$,*

$$\mathcal{P}^e(S,\widetilde{X}_\eta) = \{M_*(\rho) : \rho \in \mathcal{P}^e(S \times S, X_\eta \times \{0,1\}^{\mathbb{Z}}), \rho|_{X_\eta} = \nu_\eta\}.$$

We also have:

THEOREM 1.4 ([**2**]). *For any $\mathscr{B} \subset \mathbb{N}$, $h_{top}(S,\widetilde{X}_\eta) = h_{top}(S,X_{\mathscr{B}}) = \overline{d}(\mathcal{F}_{\mathscr{B}})$.*[3]

[3]For $A \subset \mathbb{Z}$, $\overline{d}(A)$ stands for the upper density of $A \cap \mathbb{N}$.

As an immediate consequence of Theorem 1.2, Theorem 1.3 and Theorem 1.4, we obtain:

COROLLARY 2. *Let $\mathscr{B} \subset \mathbb{N}$ be such that $\overline{d}(\mathcal{F}_{\mathscr{B}}) > 0$. Then $\mathcal{P}(S, \widetilde{X}_\eta)$ is Poulsen.*

In many cases, the $\mathscr{B}$-free system (S, X_η) itself is hereditary, i.e. $\widetilde{X}_\eta = X_\eta$. In particular, using results from [**2**], we obtain the following:

COROLLARY 3. *Let $\mathscr{B} \subset \mathbb{N}$ be such that $\overline{d}(\mathcal{F}_{\mathscr{B}}) > 0$, with $\overline{d}(\bigcup_{b \geq K} b\mathbb{Z}) \to 0$ when $K \to \infty$. Assume moreover that $\mathscr{B}$ contains an infinite pairwise coprime subset. Then (S, X_η) is hereditary and the simplex $\mathcal{P}(S, X_\eta)$ is Poulsen.*

In particular, $\mathcal{P}(S, X_\eta)$ is Poulsen in the following classical cases:

- when $\mathscr{B}$ is infinite, pairwise coprime and $\sum_{b \in \mathscr{B}} 1/b < \infty$; for example, the result holds for the square-free subshift given by $\mathscr{B} = \{p^2 : p \text{ is prime}\}$;
- when $\mathscr{B} = \mathscr{B}_{\mathbb{A}}$, where $\mathscr{B}_{\mathbb{A}}$ stands for the set of primitive abundant numbers [**8**].

Finally, we give an example of a hereditary system of positive entropy whose simplex of invariant measures fails to be Poulsen.

REMARK 1.5. For $\mathscr{B}$ pairwise coprime with $\sum_{b \in \mathscr{B}} 1/b < \infty$ another proof of Corollary 2 has been presented in [**11**]. This proof is using a different method than ours.

REMARK 1.6. Our original motivation for Theorem 1.2 was to study the simplex of invariant measures for $\mathscr{B}$-free systems, where the Mirsky measure ν_η that plays the role of ν in condition (1.1) is of zero entropy. For this reason, we include a complete proof of Theorem 1.2 under the extra assumption that ν has zero entropy and then explain the necessary changes to obtain the full version of our result.

REMARK 1.7. While the name *Poulsen simplex* comes from [**20**], where a simplex with a dense set of extreme points was constructed, this is historically not the first such example. The most basic dynamical system with the simplex of invariant measures being Poulsen is the full shift. For the 2-shift, the fact that any invariant measure can be approximated by measures concentrated on periodic orbits (such measures are of course ergodic) follows from [**23**] (see the proof of Theorem 3 therein, in particular, the comments on page 13). For the full shift over any Polish space, the fact that ergodic measures are dense in the space of all invariant measures was proved by Parthasarathy [**18**]. There is a further discussion of this with a proof that measures concentrated on periodic orbits are dense in the paper of Oxtoby [**17**]. Moreover, Sigmund [**21, 22**] gave a condition on (T, X) (so-called periodic specification property) that implies that $\mathcal{P}(T, X)$ is Poulsen. His results were applied in many situations, e.g., in [**3, 4, 6, 13**]. See [**9**] for more details.

2. Proof of Theorem 1.2 for ν of zero entropy

LEMMA 2.1. *Assume that $(X_i, \mathcal{B}_i, \mu_i)$, $i = 1, 2$, are standard probability Borel spaces with automorphisms T_i. Assume that $A \subset X_1 \times X_2$ is a Borel set and let $\rho \in J(T_1, T_2)$.[4] Let $\pi_{X_1} : X_1 \times X_2 \to X_1$ stand for the projection onto the first coordinate. Then $\mu_1(\pi_{X_1}(A)) \geq \rho(A)$.*

[4] $\rho \in J(T_1, T_2)$ is a *joining* of T_1 and T_2, i.e. $\rho \in \mathcal{P}(T_1 \times T_2, X_1 \times X_2)$ and has projections μ_1 and μ_2 respectively.

PROOF. Since π_{X_1} is Borel measurable and A is Borel, the set $\pi_{X_1}(A)$ is μ_1-measurable. Moreover, $A \subset \pi_{X_1}(A) \times X_2$ and the result follows. □

Given a subshift $Y \subset \{0,1\}^{\mathbb{Z}}$, let $\mathcal{L}(Y)$ stand for the family of all blocks appearing in a $y \in Y$.

LEMMA 2.2. *Given $n \geq 1$, $\delta > 0$, $\mu \in \mathcal{P}^e(S, \widetilde{X})$, and $A_n \subset \{0,1\}^n$ with $\mu(A_n) > 1 - \delta$, let*

$$\mathcal{C}_n := \{u \in \mathcal{L}(X) : |u| = n \text{ and } u \geq w \text{ for some } w \in A_n\}.$$

Then $\nu(\mathcal{C}_n) > 1 - \delta$, where ν is as in (1.1).

PROOF. Let $\rho \in \mathcal{P}(S \times S, X \times \{0,1\}^{\mathbb{Z}})$ be such that $M_*(\rho) = \mu$ and $\rho|_X = \nu$. It is not hard to see $C_n \supset \pi_X(M^{-1}(A_n))$. The result follows from Lemma 2.1. □

Fix $\nu_1, \nu_2 \in \mathcal{P}^e(S, \widetilde{X})$. All we need to show is that the measure $\frac{1}{2}(\nu_1 + \nu_2)$ can be approximated by ergodic measures.[5] By the definition of weak topology on measures, it follows that, given $k_0 \geq 1, \varepsilon_0 > 0$, we need to find $\overline{\eta} \in \widetilde{X}$ such that:

(2.1) $\overline{\eta}$ is generic for an ergodic measure,

(2.2) the empirical distribution of k_0-blocks on $\overline{\eta}$ is, up to $\varepsilon_0 > 0$, equal to $\frac{1}{2}(\nu_1|_{\{0,1\}^{k_0}} + \nu_2|_{\{0,1\}^{k_0}})$.

Fix $k_0 \geq 1$ and $\varepsilon_0 > 0$. Fix also $\varepsilon > 0$ much smaller than ε_0. Using the ergodic theorem for ν_1 and ν_2 respectively, we can find $n_0 \geq k_0$, $\mathcal{F}_i \subset \{0,1\}^{n_0}$, $\nu_i(\mathcal{F}_i) > 1 - \varepsilon/2$, $i = 1, 2$ such that

(2.3) the empirical k_0-distribution in any $w \in \mathcal{F}_i$ is ε-close to $\nu_i|_{\{0,1\}^{k_0}}$.

Let

$$\mathcal{G}_{n_0} := \{u \in \mathcal{L}(X) : |u| = n_0 \text{ and } u \geq w_i \text{ for some } w_i \in \mathcal{F}_i, i = 1, 2\}. \tag{2.4}$$

Note that, for $i = 1, 2$, we have

$$u \in \mathcal{L}(X) \text{ and } u \geq w_i \text{ for some } w_i \in \mathcal{F}_i \iff u \in \pi_X(M^{-1}(\mathcal{F}_i)),$$

i.e.

$$\mathcal{G}_{n_0} = \pi_X(M^{-1}(\mathcal{F}_1)) \cap \pi_X(M^{-1}(\mathcal{F}_2)). \tag{2.5}$$

Therefore, applying Lemma 2.2 to $n = n_0$ and $A_n = \mathcal{F}_i$, $i = 1, 2$, we obtain

$$\nu(\mathcal{G}_{n_0}) > 1 - \varepsilon.$$

Based on (2.4), we define two maps $R_i \colon \mathcal{G}_{n_0} \to \{0,1\}^{n_0}$, so that

$$R_i(u) \in \mathcal{F}_i, \ i = 1, 2. \tag{2.6}$$

Consider now the Markov chain with the states $\{1, \ldots, n_0, n_0 + 1\}$ and the transition probabilities, i.e. the stochastic matrix $P' = (p'_{ij})$, given by:

$$p'_{i,i+1} = 1 \text{ for } i = 1, \ldots, n_0 - 1, \ p'_{n_0,n_0+1} = 1/2, \ p'_{n_0,1} = 1/2, \ p'_{n_0+1,1} = 1.$$

Let $p' = (p'_1, \ldots, p'_{n_0}, p'_{n_0+1})$ be the probabilistic vector in which

$$p'_i = \frac{1}{n_0 + 1/2} \text{ for } i = 1, \ldots, n_0 \text{ and } p'_{n_0+1} = \frac{1}{2(n_0 + 1/2)}.$$

[5] It is enough to show that the closure of the ergodic measures is a convex set and for this it suffices to verify the midpoint condition for ergodic measures.

We have $p' \cdot P' = p'$ whence the formula

$$\kappa'(a_0, a_1, \ldots, a_m) = p'_{a_0} p'_{a_0,a_1} \cdots p'_{a_{m-1},a_m}$$

for $a_k \in \{1, \ldots, n_0, n_0+1\}$ yields an S-invariant (Markov) measure κ' on $W' := \{1, \ldots, n_0, n_0+1\}^{\mathbb{Z}}$. It is not hard to see that $(P')^{n_0+1}$ has all entries positive, that is, P' is aperiodic, and therefore the Markov measure κ' yields a mixing Markov shift.

Let $\eta_\nu \in \{0,1\}^{\mathbb{Z}}$ be a generic point for ν.

LEMMA 2.3. *If $z' \in \{0,1\}^{\mathbb{Z}}$ is a generic point for κ', then (η_ν, z') is a generic point for the product measure $\nu \otimes \kappa'$ with the latter measure being ergodic.*

PROOF. Each mixing Markov subshift is a K-system, so the result follows directly from disjointness of zero entropy systems with K-systems, see, e.g., [**10**]. □

Notice that z' above consists of consecutive blocks

$$(1, 2, \ldots, n_0) \text{ or } (1, 2, \ldots, n_0, n_0+1).$$

Moreover, by Lemma 2.3,

$$(2.7)\quad (1-\varepsilon)\kappa'([1]) \le \nu \otimes \kappa'(\mathcal{G}_{n_0} \times [1]) = \lim_{N\to\infty} \frac{1}{N} \sum_{s\le N} \mathbb{1}_{\mathcal{G}_{n_0}\times[1]}((S\times S)^s(\eta_\nu, z')).$$

REMARK 2.4. The natural representation of z' as a concatenation of blocks of length n_0 and n_0+1 induces the corresponding representation of η_ν as concatenation of block of the same lengths. An interpretation of (2.7) is that when we look at η_ν as a concatenation of n_0- and (n_0+1)-blocks, then for "most" of the blocks, we see that the either the n_0-block belongs to $\mathcal{G}_{n_0}$ or the beginning n_0-block in case of length n_0+1, belongs to $\mathcal{G}_{n_0}$. Note also that we cannot simplify this argument by representing η_ν as concatenation only of n_0-blocks. Indeed, we do not know whether (S^{n_0}, X, ν) is ergodic, hence we cannot be sure that for most of the blocks in such a concatenation, we are in $\mathcal{G}_{n_0}$. For example, if we consider the square-free case, (S^m, X_η, ν_η) is not ergodic for any $m \ge 2$ as the spectrum contains $e^{2\pi i/p^2}$, $p \in \mathscr{P}$, hence the roots of all prime degrees.

Remembering that due to the natural representation of z' as a concatenation of n_0- and (n_0+1)-blocks, the sequence η_ν is represented as a concatenation of blocks of length n_0 or n_0+1, let us now define a new sequence $\eta' = \eta'(\eta_\nu, z') \in \{*, 0\}^{\mathbb{Z}}$ in the following way:

(a) if in the above concatenation the block in η_ν is of length n_0 and belongs to $\mathcal{G}_{n_0}$, we replace it by the all $*$ n_0-block;
(b) if in the above concatenation the block in η_ν is of length n_0+1 and the starting n_0-block is in $\mathcal{G}_{n_0}$, we replace it by the all $*$ n_0-block adding 0 at the end to obtain a block of length n_0+1;
(c) if in the above concatenation the block in η_ν is of length n_0 and does not belong to $\mathcal{G}_{n_0}$ or it is of length n_0+1 but the starting n_0-block does not belong to $\mathcal{G}_{n_0}$, we replace it by the all 0 n_0- or (n_0+1)-block.

LEMMA 2.5. *The sequence $\eta' \in \{*,0\}^{\mathbb{Z}}$ is generic for an ergodic measure.*

PROOF. We simply show that η' is obtained from (η_ν, z') by a (finite) code.[6] Indeed, if we want to determine $\eta'(i)$ (i.e. $*$ or 0), we first look at $z'(i)$ and seek the first symbol 1 on the left (we do not check more than n_0+1 positions); we determined a position j in this way, and we check now whether $\eta_\nu(j, j+n_0-1)$ does or does not belong to $\mathcal{G}_{n_0}$. Now the $\eta'(i)$ is determined by (a)-(c). □

Consider now the following Markov shift: the set of states consists of

$$\{1,\ldots,n_0,n_0+1\} \text{ and its disjoint copy } \{\overline{1},\ldots,\overline{n_0},\overline{n_0+1}\}.$$

Let

$$p = (p_1,\ldots,p_{n_0},p_{n_0+1},p_{\overline{1}},\ldots,p_{\overline{n_0}},p_{\overline{n_0+1}})$$

be the probabilistic vector for which $p_i = p_{\overline{i}} = \frac{1}{2n_0+1}$ for $i = 1,\ldots,n_0$ and $p_{n_0+1} = p_{\overline{n_0+1}} = \frac{1}{2(2n_0+1)}$. The matrix P of transition probabilities is given by the following:

$$p_{i,i+1} = p_{\overline{i},\overline{i+1}} = 1 \text{ for } i = 1,\ldots,n_0-1,$$

$$p_{n_0,n_0+1} = p_{\overline{n_0},\overline{n_0+1}} = 1/2,$$

$$p_{n_0,1} = p_{\overline{n_0},\overline{1}} = p_{n_0,\overline{1}} = p_{\overline{n_0},1} = 1/4$$

and

$$p_{n_0+1,1} = p_{n_0+1,\overline{1}} = p_{\overline{n_0+1},1} = p_{\overline{n_0+1},\overline{1}} = 1/2.$$

We have $p\cdot P = p$ whence the formula

$$\kappa(a_0,a_1,\ldots,a_m) = p_{a_0}p_{a_0,a_1}\cdots p_{a_{m-1},a_m}$$

for $a_k \in \{i,\overline{i} : i = 1,2,\ldots,n_0\}$ yields an S-invariant (Markov) measure κ on $W := \{1,\ldots,n_0+1,\overline{1},\ldots,\overline{n_0+1}\}^{\mathbb{Z}}$. It is not hard to see that $P^{2(n_0+1)}$ has all entries positive, that is, P is aperiodic, and therefore the Markov measure κ yields a mixing Markov shift.

Fix z a generic point for κ. Then z is a concatenation of 4 types of blocks (of length either n_0 or n_0+1): $(1,\ldots,n_0)$, $(1,\ldots,n_0,n_0+1)$, $(\overline{1},\ldots,\overline{n_0})$ and $(\overline{1},\ldots,\overline{n_0},\overline{n_0+1})$. Moreover, on a sufficiently long initial part of z, the non-barred-blocks and the barred-blocks are equally probable (as $\kappa(1) = \kappa(\overline{1})$). Similarly, to Lemma 2.5, we obtain the following.

LEMMA 2.6. *If $z \in \{0,1\}^{\mathbb{Z}}$ is a generic point for κ, then (η_ν, z) is a generic point for the product measure $\nu\otimes\kappa$ with the latter measure being ergodic.*

Moreover, the map which to i and $\overline{i}$ associates i, for $i = 1,\ldots,n_0,n_0+1$ yields a continuous factor map Λ between (S,W,κ) and (S,W',κ'); in particular, if z is generic for κ then $z' := \Lambda(z)$ is generic for κ'.

We now fix z a generic point for κ and repeat the construction of $\eta' \in \{0,1\}^{\mathbb{Z}}$ for z'. Then, we transform η' into $\overline{\eta}$ by replacing the n_0-$*$-block w by either $R_1(w)$

[6] Formally, we should define a block map. This can be done for example as follows. Consider all blocks (over the double alphabet) of length $4n_0+1$ that appear in (η_ν, z'). Due to the special form of z', there will be at least 3 symbols 1 on the second coordinates of this block. Look for the first 1 on the left of the middle of the block. This gives a certain position j. Look at the block on η_ν of length n_0 (or n_0+1 if , as the symbol(!), n_0+1 appears (on the second coordinate) on the right to the middle position before 1 reappears), read whether the corresponding block on η_ν belongs or does not to $\mathcal{G}_{n_0}$ and code the whole $4n_0+1$-block (over the double alphabet) by $*$ or 0, respectively.

or $R_2(w)$ (cf. (2.6)) depending on the fact whether on z we considered the non-barred- or the barred-block. Up to density $\varepsilon_1 > 0$, we have now replaced half of the $*$-blocks by blocks from $\mathcal{F}_1$ and the second half by blocks from $\mathcal{F}_2$. Clearly, $\overline{\eta} \leq \eta_\nu$. Moreover, by (2.3), the empirical distribution of k_0-blocks on $\overline{\eta}$ is ε-close to the distribution of k_0-blocks for the measure $\frac{1}{2}(\nu_1 + \nu_2)$. Finally, following the proof of Lemma 2.5, we obtain that $\overline{\eta}$ is generic for an ergodic measure (it is obtained by a finite code from (η_ν, z)). We have proved (2.1) and (2.2), so Theorem 1.2 follows.

3. Sketch of the proof of Theorem 1.2 for general ν

We explain now how to modify the proof from Section 2 after dropping the additional assumption that ν has zero entropy. Instead of the Markov shift (S, W, κ), we consider an arbitrary ergodic aperiodic automorphism (T, W, κ) disjoint from (S, X, ν) (the existence of such an automorphism follows from [**5**], in fact, a generic automorphism is disjoint from (S, X, ν)). By the Alpern's Lemma [**1**], W can be decomposed into two towers, say, $\mathcal{W}_0$ and $\mathcal{W}_1$, of height n_0 and $n_0 + 1$, respectively. We split each of them into two further towers of equal measure and of the same height as the original tower: $\mathcal{W}_0 = \mathcal{W}_0^L \cup \mathcal{W}_0^R$, $\mathcal{W}_1 = \mathcal{W}_1^L \cup \mathcal{W}_1^R$. By assigning symbols $1, \dots, n_0$ and $1, \dots, n_0 + 1$ to the consecutive levels of $\mathcal{W}_0^L$ and $\mathcal{W}_1^L$, and symbols $\overline{1}, \dots, \overline{n_0}$ and $\overline{1}, \dots, \overline{n_0 + 1}$ to the consecutive levels of $\mathcal{W}_0^L$ and $\mathcal{W}_1^L$, we obtain a coding of points from W by two-sided sequences over the alphabet $\{1, \dots, n_0, n_0 + 1, \overline{1}, \dots, \overline{n_0}, \overline{n_0 + 1}\}$. The remaining part of the proof of Theorem 1.2 stays the same as in Section 2.

4. Hereditary system of positive entropy whose simplex of invariant measures is not Poulsen

In this section we will show that there are hereditary systems of positive entropy whose simplex of invariant measures is not Poulsen. For this, we recall an example, which was used in [**12**] to show that there are hereditary systems which are not intrinsically ergodic.

Given a block $C \in \{0, 1\}^n$, let $x_C \in \{0, 1\}^{\mathbb{Z}}$ be the infinite concatenation of C and let $X_C \subset \{0, 1\}^{\mathbb{Z}}$ stand for the orbit closure of x_C (equal to its orbit). Let $\widetilde{X}_C$ be the smallest hereditary subshift containing X_C. Finally, let ν_C be the periodic measure for which x_C is a generic point. Notice that if $\operatorname{supp} C \neq \emptyset$ then $\nu_C \neq \delta_{(\dots,0,0,0,\dots)}$.

Let $A := 101001000$, $B := 101000100$ and consider $X := X_A \cup X_B$. Then the smallest hereditary subshift $\widetilde{X}$ containing X equals $\widetilde{X}_A \cup \widetilde{X}_B$. Since both (S, X_A) and (S, X_B) are uniquely ergodic (with zero entropy), it follows by Corollary 1 that both $\mathcal{P}(S, \widetilde{X}_A)$ and $\mathcal{P}(S, \widetilde{X}_B)$ are Poulsen simplices (they are both non-trivial as $\nu_A, \nu_B \neq \delta_{(\dots,0,0,0,\dots)}$).

Suppose that $\mathcal{P}(S, \widetilde{X})$ is also Poulsen and take its arbitrary element ν. Since ν can be approximated by ergodic measures and $\mathcal{P}^e(S, \widetilde{X}) = \mathcal{P}^e(S, \widetilde{X}_A) \cup \mathcal{P}^e(S, \widetilde{X}_B)$, it follows that

$$\nu \in \mathcal{P}(S, \widetilde{X}_A) \cup \mathcal{P}(S, \widetilde{X}_B). \tag{4.1}$$

Let now $\nu := \frac{1}{2}(\nu_A + \nu_B)$. Since $\nu_A(X_A) = \nu_B(X_B) = 1$ and $X_A \cap X_B = \emptyset$, it follows that $\nu(\widetilde{X}_A) = \nu(\widetilde{X}_B) = 1/2$. This contradicts (4.1) and we conclude that $\mathcal{P}(S, \widetilde{X})$ cannot be Poulsen.

Finally, notice that it is easy to modify the above example, so that $(S, \widetilde{X})$ becomes intrinsically ergodic (i.e. it has exactly one measure of maximal entropy). E.g., we can take $A' := 111001000$ instead of A and consider $X' := X_{A'} \cup X_B$. Then $M_*(\nu_{A'} \otimes B(1/2, 1/2))$ (where $B(1/2, 1/2)$ stands for the Bernoulli measure $(1/2, 1/2)$ on $\{0,1\}^{\mathbb{Z}}$) is the unique measure of maximal entropy for $(S, \widetilde{X'})$, see [**12**] for more details.

References

[1] Steve Alpern, *Return times and conjugates of an antiperiodic transformation*, Ergodic Theory Dynamical Systems **1** (1981), no. 2, 135–143. MR661814

[2] A. Bartnicka, S. Kasjan, J. Kułaga-Przymus, and M. Lemańczyk, *$\mathscr{B}$-free sets and dynamics*, Preprint.

[3] A. M. Blokh, *Decomposition of dynamical systems on an interval* (Russian), Uspekhi Mat. Nauk **38** (1983), no. 5(233), 179–180. MR718829

[4] Jérôme Buzzi, *Specification on the interval*, Trans. Amer. Math. Soc. **349** (1997), no. 7, 2737–2754, DOI 10.1090/S0002-9947-97-01873-4. MR1407484

[5] Andrés del Junco, *Disjointness of measure-preserving transformations, minimal self-joinings and category*, Ergodic theory and dynamical systems, I (College Park, Md., 1979–80), Progr. Math., vol. 10, Birkhäuser, Boston, Mass., 1981, pp. 81–89. MR633762

[6] Manfred Denker, Christian Grillenberger, and Karl Sigmund, *Ergodic theory on compact spaces*, Lecture Notes in Mathematics, Vol. 527, Springer-Verlag, Berlin-New York, 1976. MR0457675

[7] Tomasz Downarowicz, *The Choquet simplex of invariant measures for minimal flows*, Israel J. Math. **74** (1991), no. 2-3, 241–256, DOI 10.1007/BF02775789. MR1135237

[8] P. Erdös, *On the Density of the Abundant Numbers*, J. London Math. Soc. **9** (1934), no. 4, 278–282. MR1574879

[9] K. Gelfert and D. Kwietniak, *The (Poulsen) simplex of invariant measures*, `http://arxiv.org/abs/1404.0456`, 04 2014.

[10] Eli Glasner, *Ergodic theory via joinings*, Mathematical Surveys and Monographs, vol. 101, American Mathematical Society, Providence, RI, 2003. MR1958753

[11] J. Konieczny and D. Kwietniak, *Invariant measures of B-free shifts.*

[12] Joanna Kułaga-Przymus, Mariusz Lemańczyk, and Benjamin Weiss, *On invariant measures for $\mathcal{B}$-free systems*, Proc. Lond. Math. Soc. (3) **110** (2015), no. 6, 1435–1474, DOI 10.1112/plms/pdv017. MR3356811

[13] Jaroslaw Kwapisz, *Cocyclic subshifts*, Math. Z. **234** (2000), no. 2, 255–290, DOI 10.1007/s002099900107. MR1765882

[14] Dominik Kwietniak, *Topological entropy and distributional chaos in hereditary shifts with applications to spacing shifts and beta shifts*, Discrete Contin. Dyn. Syst. **33** (2013), no. 6, 2451–2467, DOI 10.3934/dcds.2013.33.2451. MR3007694

[15] J. Lindenstrauss, G. Olsen, and Y. Sternfeld, *The Poulsen simplex* (English, with French summary), Ann. Inst. Fourier (Grenoble) **28** (1978), no. 1, vi, 91–114. MR500918

[16] Wolfgang Lusky, *Separable Lindenstrauss spaces*, Functional Analysis: surveys and recent results (Proc. Conf., Paderhorn, 1976), Notas Mat., vol. 63, North-Holland, Amsterdam-New York, 1977, pp. 15–28. MR481894

[17] John C. Oxtoby, *On two theorems of Parthasarathy and Kakutani concerning the shift transformation*, Ergodic Theory (Proc. Internat. Sympos., Tulane Univ., New Orleans, La., 1961), Academic Press, New York, 1963, pp. 203–215. MR0160875

[18] K. R. Parthasarathy, *On the category of ergodic measures*, Illinois J. Math. **5** (1961), 648–656. MR0148850

[19] Robert R. Phelps, *Lectures on Choquet's theorem*, 2nd ed., Lecture Notes in Mathematics, vol. 1757, Springer-Verlag, Berlin, 2001. MR1835574

[20] Ebbe Thue Poulsen, *A simplex with dense extreme points* (English, with French summary), Ann. Inst. Fourier. Grenoble **11** (1961), 83–87, XIV. MR0123903

[21] Karl Sigmund, *Generic properties of invariant measures for Axiom* A *diffeomorphisms*, Invent. Math. **11** (1970), 99–109. MR0286135

[22] Karl Sigmund, *On dynamical systems with the specification property*, Trans. Amer. Math. Soc. **190** (1974), 285–299. MR0352411

[23] J. Ville, *Étude critique de la notion de collectif*, vol. 218, Thèses françaises de l'entre-deux-guerres, Paris, 1939 (fre).

[24] Peter Walters, *An introduction to ergodic theory*, Graduate Texts in Mathematics, vol. 79, Springer-Verlag, New York-Berlin, 1982. MR648108

Institute of Mathematics, Polish Acadamy of Sciences, Śniadeckich 8, 00-956 Warszawa, Poland

Current address: Faculty of Mathematics and Computer Science, Nicolaus Copernicus University, Chopina 12/18, 87-100 Toruń, Poland

E-mail address: joanna.kulaga@gmail.com

Faculty of Mathematics and Computer Science, Nicolaus Copernicus University, Chopina 12/18, 87-100 Toruń, Poland

E-mail address: mlem@mat.umk.pl

Institute of Mathematics, Hebrew University of Jerusalem, Jerusalem, Israel

E-mail address: weiss@math.huji.ac.il

Contemporary Mathematics
Volume **678**, 2016
http://dx.doi.org/10.1090/conm/678/13652

Substitutions, symbolic codings, and discrepancy

David Ralston

ABSTRACT. We provide an overview of the use of symbolic codings in the study of discrepancy, the supremal deviation of ergodic sums from their expected value. We specifically utilize the interplay between induced maps and substitutions, progressing from the classical study of Sturmian sequences to more recent results in self-similar dynamics. We then present a generalization of the techniques used in those scenarios and conclude with discussion about the new machinery.

1. Introduction

In this survey we will discuss the use of substitutions to study the growth rate of ergodic sums in dynamical systems. We will begin in 2 with a brief overview of common notation, terminology, and definitions which are necessary to discuss symbolic dynamical systems, growth rates of ergodic sums in (not necessarily symbolic) dynamical systems, and the *codings* which may be used to relate one to the other. In 3 we will briefly discuss the history of symbolic codings, as well as developing our introductory example: a coding of irrational circle rotations derived through substitutions. Then, in 4 we will discuss *self-similar systems*, which provide a natural application of symbolic codings through primitive substitutions. In 5 we will generalize the notion of a substitution system to account for symbolic dynamical systems which are derived from applications of multiple different substitutions, and in 6 we see how such systems might encode a wider range of systems than self-similar systems, including a proof of the well-known fact that Sturmian sequences have bounded discrepancy. Finally, we conclude in 7 with a broad discussion and possible directions for future research.

2. Terminology

We assume that the reader is familiar with basic terms from dynamical systems (*e.g. measure preserving, isomorphic, ergodic*), pausing briefly only to define *induced transformation.* If (X, T, μ) is a measure preserving system, $\mu(X) = 1$, and $A \subset X$ such that $0 < \mu(A) < 1$, then by the Poincaré recurrence theorem we have for almost every $x \in A$ a finite positive integer $n(x)$, the least positive integer such that $T^{n(x)}(x) \in A$. This quantity $n(x)$ is called the *return time* to A, and it defines the *induced system* $(A, T|_A, \mu|_A)$ given by $T|_A(x) = T^{n(x)}(x)$. If there is some choice of A such that the induced system is isomorphic to the original system, then

2010 *Mathematics Subject Classification.* Primary 37-06, Secondary 37C35, 37B10.

(X, T, μ) is called *self-similar*. We will now define terminology specific to symbolic dynamical systems.

Let $\mathcal{A}$ be a compact set (the *alphabet*) and $\mathcal{A}^*$ be the free monoid on $\mathcal{A}$, with the identity element denoted by ϵ. Elements of $\mathcal{A}^*$ will be called *words*. If $\omega \in \mathcal{A}^*$ and $\omega = \omega_0\omega_1 \ldots \omega_{n-1}$, where each $\omega_i \in \mathcal{A}$, then ω is said to be a *finite word* of *length* n; we will use $|\omega|$ to represent the length of a word. We use the same terminology to include sequences in $\mathcal{A}$ indexed by $\mathbb{N}$, which are called *infinite* words. If $\alpha, \beta, \gamma, \omega$ are all words such that

$$\omega = \alpha\gamma\beta,$$

then α is called a *left factor* (or *prefix*) of ω, β is called a *right factor* (or *postfix*), and γ is simply called a *factor* (or *infix*). The set of all finite-length factors of ω is called the *language*, denoted $\mathcal{L}(\omega)$; typically one only discusses languages of infinite words, and one may analogously define for a set of words X the language $\mathcal{L}(X)$ to be the union of all $\mathcal{L}(\omega)$ for $\omega \in X$. The empty word (consisting of no symbols) is denoted ϵ, and is a factor of all words.

We may map one infinite word to another by the *shift map*, denoted S and defined by $(S(\omega))_i = \omega_{i+1}$. Defining a topology on $\mathcal{A}^{\mathbb{N}}$ through cylinder sets results in S being a continuous map. Let $\Omega \subseteq \mathcal{A}^{\mathbb{N}}$ be closed and $S(\Omega) \subseteq \Omega$: then (Ω, S) is a *symbolic dynamical system*. As we assumed $\mathcal{A}$ to be compact, $\mathcal{A}^{\mathbb{N}}$ is compact, so then by the theorem of Krylov and Bogolyubov, there exists a finite Borel measure μ on Ω which is S-invariant [**20**]. The derivation of measure-theoretic properties of symbolic systems is an enormous field; we refer specifically to [**10, 27**] as comprehensive references. More specific results may be found in other sources as well, such as [**6, 23**].

The *forward orbit closure* of an infinite word ω is the closure (with respect to the same cylinder topology) of the set $\{\omega, S(\omega), S^2(\omega), S^3(\omega), \ldots\}$, which we denote X_ω, and which defines a symbolic dynamical system (X_ω, S).

A homomorphism $\sigma : \mathcal{A}^* \mapsto \mathcal{A}^*$ such that for all $a \in \mathcal{A}$, $\sigma(a) \neq \epsilon$, is called a *substitution*. Note that a substitution is completely defined by the image of each $a \in \mathcal{A}$, may be naturally extended to act on infinite words as well, and since no letter maps to the empty word, for any word ω, $\sigma(\omega)$ cannot be of shorter length. From this point forward we assume that $\mathcal{A}$ is finite, but with at least two elements, as is standard. If we order the elements of $\mathcal{A}$ as $\{a_1, a_2, \ldots, a_k\}$, we define the *incidence matrix* M_σ of the substitution σ through

$$(M_\sigma)_{i,j} = \sum_{\ell=0}^{|\sigma(a_j)|-1} \chi_i(\sigma(a_j)_\ell),$$

where χ_i is the characteristic function of $\{a_i\}$. The entry at i, j counts the number of appearances of a_i in the word $\sigma(a_j)$. If there is some $n \geq 1$ such that every entry of the matrix $(M_\sigma)^n$ is positive, then σ is called *primitive*. Note that $(M_\sigma)^n = M_{\sigma^n}$, so equivalently: there is some $n \in \mathbb{N}$ so that for any choice of $a, b \in \mathcal{A}$, $\sigma^n(a)$ contains b as a factor. The principal application of primitivity in studying substitutions is:

THEOREM 2.1 (see *e.g.* Proposition 5.4.3 in [**10**]). *Suppose that σ is a primitive substitution on $\mathcal{A}$, and ω is a word such that $\sigma(\omega) = \omega$. Then the symbolic dynamical system (X_ω, S) is uniquely ergodic.*

DEFINITION 1. Suppose that (Ω, S) is a symbolic dynamical system and (X,T) is a topological dynamical system. Further suppose that there are subsets $\Omega' \subseteq \Omega$ and $X' \subseteq X$ and a map $\phi : \Omega' \to X'$ such that

- ϕ is bicontinuous and bijective,
- $\Omega \setminus \Omega'$ and $X \setminus X'$ are at most countably infinite, and
- $T \circ \phi = \phi \circ S$.

Then we say that (Ω, S) is a *coding* of (X,T).

A coding of a dynamical system is simply a second, symbolic, dynamical system which is *semi-topologically conjugate* to the first. The typical interpretation, however, is as follows: with the subset X' partitioned into sets A_1 through A_k, the map ϕ is a way to describe each point $x \in X'$ through its orbit, where $\phi^{-1}(x)_i = a_j$ if and only if $T^i(x) \in A_j$. Results pertaining to (Ω, S) therefore give information about the distribution of orbits in (X,T) with respect to the partition by A_i.

The definition above comes from [**10**, Definition 5.2.14], but is subject to variations in practice. For example, in [**22**, §2.4] the term *coding* is used for a finite-to-one measure-theoretic conjugacy on a set of full measure (even though the complement is uncountable). Whether results in the symbolic system push to (X,T) to hold on sets of full measure, on residual sets, or on all but countably many points will depend on the specific type of conjugacy employed. Informally, codings are the use of some form of conjugacy to relate symbolic dynamical systems to others.

If $\omega, \omega', \psi, \rho$ are all words on some alphabet $\mathcal{A}$ such that

$$\omega = \rho\psi, \quad \omega' = \psi\rho,$$

then we say that ω' is *conjugate* to ω. If a word is not equal to any of its non-trivial conjugates (*i.e.* neither ψ nor ρ is the empty word), we say the word is *primitive* (this definition is equivalent to saying that the word ω cannot be expressed as the power of a smaller word).

A *Lyndon word* is a word which is not larger than any of its conjugates with respect to a lexicographic ordering derived from some total ordering on the set of letters (see e.g. [**24**, §5.1]). If the word ω is primitive then there must be a unique conjugate of ω which is a Lyndon word. A related notion is that of a *heavy word* [**28**, Def. 6]. We update the definition with the following:

DEFINITION 2 (Heavy Word). Suppose ω is a word of length n on the alphabet $\mathcal{A}$, and $f : \mathcal{A}^* \to \mathbb{R}$ is a homomorphism with respect to addition (and therefore defined by its value on each $a_i \in \mathcal{A}$). If for some α and for every nonempty ρ a prefix of ω we have $f(\rho) \geq \alpha$, then ω is called a *α-heavy word.* If ω is a finite word which is $f(\omega)$-heavy, then ω is simply called *heavy*: the minimal sum over all nonempty left factors of ω is the total value $f(\omega)$. In the event that ω is an infinite word, if μ is a measure on X_ω preserved by S, then we call the infinite word ω *heavy* if for any prefix ρ we have that $f(\rho)$ is at least as large as its expected value:

$$f(\rho) \geq |\rho| \sum_{i=1}^{k} f(a_i)\mu(a_i).$$

Finding factors ψ of ω which are heavy are useful when computing deviation of ergodic sums across ω, as they provide a fixed, known lower bound on f evaluated on all left factors of ψ.

Finally, we introduce the definition of and notation for *discrepancy*. Let (Ω, S, μ) be a symbolic dynamical system and $\omega \in \Omega$. Define for any $\psi \in \mathcal{L}(\omega)$

$$D_n(\omega, \psi) = \left(\sum_{j=0}^{n-1} \chi_\psi \left(\omega_j \cdots \omega_{j+|\psi|-1} \right) \right) - n\mu(\psi). \tag{2.1}$$

This function counts the number of times that ψ appears as a factor of ω in the first n indices, and compares that quantity to the expected total. Setting ψ to be a single letter is common. We then define

$$D_n(\omega) = \sup_{a \in \mathcal{A}} |D_n(\omega, a)|, \tag{2.2}$$

$$D_n(X_\omega) = \sup_{\omega' \in X_\omega} \sup_{\psi \in \mathcal{L}(\omega)} |D_n(\omega', \psi)|. \tag{2.3}$$

These are called the *symbolic discrepancy* of ω and the *discrepancy function* of X_ω, respectively. We have not included an absolute value in (2.1): this choice preserves the useful cocycle structure

$$D_{n+m}(\omega, \psi) = D_m(\omega, \psi) + D_n(S^m\omega, \psi).$$

Discrepancy of sequences is a classical subject, measuring how far certain sequences deviate from their expected distributions; see [**21**] for the standard reference on the subject, or [**8**] for a more modern update. The typical task is to establish for some sequence f_n whether

$$0 < \liminf_{n\to\infty} \frac{D_n(X_\omega)}{f_n} \quad \text{and/or} \quad \limsup_{n\to\infty} \frac{D_n(X_\omega)}{f_n} < \infty.$$

A common theme in the study of discrepancy which we will observe is to first consider the simpler function $D_n(\omega, \psi)$, which allows a specific choice of the word ω, such as one generated by a primitive substitution. If one may understand $D_n(\omega, \psi)$, then it is hopeful to generalize slightly to any $D_n(S^m\omega, \psi)$. If the symbolic dynamical system (X_ω, S) is uniquely ergodic (as in the case of words generated by primitive substitutions, by 2.1) then this improvement leads to an understanding of $D_n(X_\omega)$.

3. History

The use of symbolic maps to study the systems conjugate to them goes back to [**11**], in which Hedlund used a symbolic coding (which appeared at least as early as [**26**]) to prove that the geodesic flow on a surface of constant negative curvature is ergodic with respect to the Liouville measure. The classical context for developing the theory is coding geodesic flow on the modular surface $SL_2(\mathbb{R})/SL_2(\mathbb{Z})$. See [**13, 14**] for expositions on the use of symbolic dynamics for coding geodesics on surfaces of constant negative curvature. See also the work of Adler and Series: [**34–36**]. We will not present this line of development, but the interested reader can investigate how properties of Fuchsian groups lead to a development of standard continued fractions [**32, 33**] with connections to other types of continued fractions in the work of S. Katok and Ugarcovici [**15–17**].

The introductory scenario for our purposes is to construct a coding for the system $(\mathbb{S}^1, R_\theta)$: rotation of the unit circle by some $\theta \notin \mathbb{Q}$. In contrast to the classical systems mentioned above, we will realize our coding through *substitutions*. In the following section we somewhat follow the presentation of [**10**], with the difference

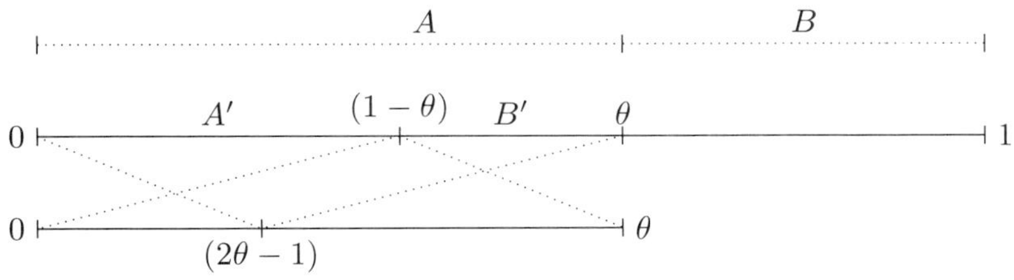

FIGURE 1. The induced transformation on the interval $[0, \theta)$.

that we will not be studying *substitution systems*, in which a single substitution is used to define a dynamical system, but rather systems which may be defined using an infinite sequence of possibly distinct substitutions. We mention first the following basic lemma:

LEMMA 1. *Suppose that (X, T) is a topological dynamical system, x is a point whose orbit under T is dense in X, and that the space X is partitioned up to a countable set by the open sets $\{A_1, \ldots, A_k\}$. Then if we construct an infinite word ω in the alphabet $\mathcal{A} = \{a_1, \ldots, a_k\}$ by*

$$\omega_i = a_j \quad \Longleftrightarrow \quad T^i(x) \in A_j,$$

then the symbolic system (X_ω, S) is a coding of (X, T).

PROOF. The proof is straightforward and omitted. The factor map ϕ is constructed by mapping each $T^i(x)$ to $\sigma^i(\omega)$ and extending continuously to all other $y \in X$ such that every $T^i(y)$ belongs to some element of the original partition. The countably infinite set of exclusions allowed in 1 is exactly the set of points not included in our partition along with their orbits. □

REMARK 3.1. The condition that the A_i partition X up to a countable set may be adjusted if different versions of conjugacy are to be used (*e.g.* 'up to measure zero for some preserved measure μ' is a common replacement).

EXAMPLE 1. Partition $\mathbb{S}^1 = \mathbb{R}/\mathbb{Z}$ into the sets $A_0 = (0, \theta)$, $A_1 = (\theta, 1)$, where

$$\theta = \frac{-1 + \sqrt{5}}{2}.$$

Note that $(2\theta - 1)/\theta = (1 - \theta)$. Since we have $0 < 1 - \theta < \theta$, we observe that the induced transformation on A is given according to 1. We observe two things. First, the induced map on A is the system $([0, \theta), R_{2\theta-1})$, which by rescaling by θ is isomorphic to $(\mathbb{S}^1, R_{1-\theta})$, which is isomorphic to our original transformation $(\mathbb{S}^1, R_\theta)$. Second, the return map naturally defines two intervals within A, according to whether the return time was two (the interval we have labeled A') or one (the interval we have labeled B'). Following these sets A' and B' through the isomorphisms mentioned above, A' exactly maps to A and B' to B.

Not only does every $x \in B'$ return to A in one application of R_θ, but each such x satisfies $T^0(x) \in A$. Similarly, each $x \in A'$ has a return time of two and satisfies $T^0(x) \in A$, $T^1(x) \in B$. We therefore have a labeling scheme that we can systematically use to describe the orbit points in our original transformation. The orbit of any point in A' is described by the sequence of values AB, while the orbit

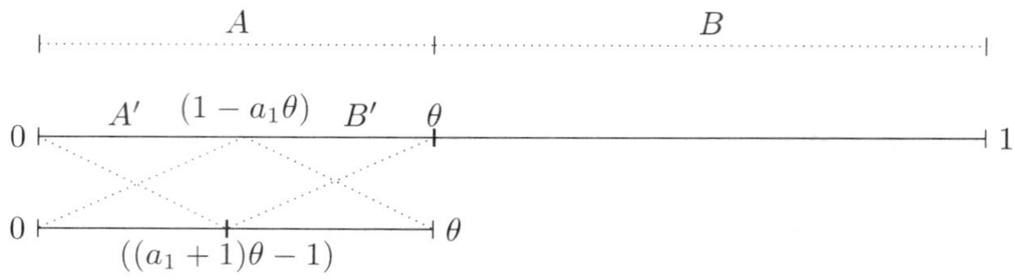

FIGURE 2. The induced transformation on the interval $[0, \theta)$.

of any point in B' may be described by the sequence A. Therefore, if ω were a finite word in $\{0, 1\}^N$ which described the orbit of some point under the induced map, then $\sigma(\omega)$ would correctly describe the orbit of the same point under the original map, where

$$\sigma : \begin{array}{l} A \mapsto AB \\ B \mapsto A \end{array}$$

Since the induced transformation is isomorphic to the original, and the sets A' and B' are correspond to A and B through this isomorphism, we can repeat this process. Consider the induced transformation on A', which defines the intervals A'' and B'', and points in these sets have their orbits in the original transformation coded by σ^2, the square of the same substitution as above. By repeating this process, we create an infinite word given by the limit of the words

$$A, \quad \sigma(A) = AB, \quad \sigma^2(A) = ABA, \quad \sigma^3(A) = ABAAB, \quad \dots$$

Each finite word encodes the orbit of some interval through the time given by the length of the word. The infinite limiting word ω, then, encodes the orbit of some point $x \in \mathbb{S}^1$, albeit with a small catch: if the intersection of the closure of the intervals A, A', A'', etc. is some preimage of θ it is possible that our coding has at most two errors in it. This small adjustment is an artifact of making a choice as to whether the endpoints 0 and θ belong to the intervals A and B. Clearly it may only affect the encoding of any trajectory in at most two locations, an effect which is generally either ignored or bypassed by working in the totally disconnected space in which the rotation by θ is continuous, but both A and B are clopen.

The particular properties of our choice of θ were convenient, but for general $\theta \notin \mathbb{Q}$ we get a similar situation.

EXAMPLE 2. We will again induce on the interval $[0, \theta)$. If we represent θ by its *standard continued fraction* representation:

$$\theta = [a_1, a_2, a_3, \dots] = \cfrac{1}{a_1 + \cfrac{1}{a_2 + \cfrac{1}{a_3 + \cdots}}}, \qquad (a_i \in \mathbb{Z},\ a_i \geq 1)$$

then $(a_1 + 1)\theta \bmod 1$ is the first solution to $n\theta \mod 1 < \theta$, so if we again look at the induced map on $[0, \theta)$, we obtain 2.

The induced transformation is again a rotation, isomorphic to rotation by

$$\frac{(a_1 + 1)\theta - 1}{\theta} = 1 + a_1 - \frac{1}{\theta} = 1 - \gamma(\theta),$$

where γ is the *Gauss map*, which acts on irrational $\theta \in (0,1)$ by shifting entries in the continued fraction expansion by one index:

$$\gamma([a_1, a_2, a_3, \ldots]) = [a_2, a_3, \ldots].$$

Rotation by $1 - \gamma(\theta)$ is isomorphic to rotation by $\gamma(\theta)$. Furthermore, again by observing the orbits of points in A' and B' through their return to A, we obtain the substitution

$$\sigma : \begin{array}{l} A \mapsto AB^{a_1} \\ B \mapsto AB^{a_1 - 1} \end{array}$$

which perfectly matches our observation when θ was the golden mean, with $a_1 = 1$ and $\gamma(\theta) = \theta$.

In this general scenario our induced transformation is almost certainly not isomorphic to the original transformation, and so repeated applications of one substitution cannot be used to encode the orbit of any point. In this example, however, we see that we will obtain a sequence of substitutions σ_1, σ_2, etc., where

$$\sigma_i : \begin{array}{l} A \mapsto AB^{a_i} \\ B \mapsto AB^{a_i - 1} \end{array}$$

Since $\sigma_i(A)$ begins with A and is of length larger than one for all i, there is necessarily a limiting word

$$\omega = \lim_{n\to\infty} \sigma_1 \circ \sigma_2 \circ \cdots \circ \sigma_{n-1}(A), \tag{3.1}$$

which is a sequence of words which encodes the orbit of smaller and smaller intervals (given by nesting choices of intervals on which to consider induced transformations) in the original rotation. Note the order of composition; σ_1, which comes from a_1, is the final substitution applied, and therefore gives rise to the 'most localized' properties of the limiting sequence. Larger-scale properties arise from σ_i with larger i.

As before, this infinite word encodes the orbit of some point, again with at most two errors. We will revisit this example in 6.1 to show that $D_n(X_\omega)$ is bounded. We will postpone the statement and proof of this theorem until we develop the machinery to represent the general scenario where a sequence of substitutions gives rise to a symbolic coding. First we will discuss the simpler situation, where a single substitution is used to encode a system.

4. Self-Similar Systems

In 1 we observed that if we may find a subset A on which the induced transformation $(A, T|_A)$ is isomorphic to the original system, a primitive substitution arose such that repeated applications of that substitution generated a limiting word ω which encoded the orbit of some point (with perhaps up to two errors), and therefore (X_ω, S) was a coding of (X, T) (modulo these possible errors, which affect a finite number of indices for at most countably many points).

Symbolic dynamical systems arising from fixed points of primitive substitutions are uniquely ergodic (2.1), so if we let μ be the preserved measure, we may therefore compute for each $a \in \mathcal{A}$ the cylinder measure

$$\mu(a) = \lim_{N \to \infty} \frac{1}{N} \sum_{i=0}^{N-1} \chi_a(\omega_i),$$

where $\sigma(\omega) = \omega$, and χ_a is the characteristic function of the set $\{a\}$.

For self-similar systems, the discrepancy is governed by spectral properties of the incidence matrix. The reader should refer to [**1**] for a full treatment, which we will briefly summarize. Given two positive sequences a_n and b_n, we say (following the notation of [**1**]) $a_n = (O \cap \Omega)(b_n)$ if the ratio a_n/b_n is both bounded and bounded away from zero.

THEOREM 4.1 (Theorem 1 of [**1**]). *Let ω be a fixed point of a primitive substitution σ, with $D_n(\omega)$ as given by (2.2). Let θ be the principal eigenvalue of M_σ, and let θ_2 be the secondary eigenvalue with multiplicity α_2. Then*

(1) *if $|\theta_2| < 1$, then $D_n(\omega)$ is bounded,*
(2) *if $|\theta_2| > 1$, then*

$$D_n(\omega) = (O \cap \Omega)((\log n)^{\alpha_2} n^{\log_\theta \theta_2},$$

(3) *if $|\theta_2| = 1$ and θ_2 is not a root of unity, then*

$$D_n(\omega) = (O \cap \Omega)((\log n)^{\alpha_2 + 1}),$$

(4) *if $|\theta_2| = 1$ and θ_2 is a root of unity, then there is an explicit computable constant $A(\sigma, \omega)$ such that if $A(\sigma, \omega) \neq 0$ we have*

$$D_n(\omega) = (O \cap \Omega)(\log n)^{\alpha_2},$$

and otherwise if $A(\sigma, \omega) = 0$

$$D_n(\omega) = (O \cap \Omega)(\log n)^{\alpha_2 + 1}.$$

See also [**3, 4**] for further works along these lines. Self-similar systems have been studied in their own right without invoking substitution systems which encode them as well, as in [**2**]. We emphasize that these bounds are very precise, with a single function $f(n)$ acting as both lower and upper bound on D_n.

The above result characterizes the discrepancy of single letters within the single word ω. However, the result may be generalized to cover arbitrary factors of arbitrary sequences in the space:

THEOREM 4.2 (Theorem 3 of [**1**]). *Let ω be a fixed point of a primitive substitution σ, with $D_n(X_\omega)$ as in (2.3). Let θ be the principal eigenvalue of M_σ, and let $\theta_{2,2}$ and $\alpha_{2,2}$ be the second eigenvalue of the* second-order substitution *(derived from σ, but which we do not define herein). Then*

(1) *if $|\theta_{2,2}| < 1$, then $D_n(X_\omega)$ is bounded,*
(2) *if $|\theta_{2,2}| > 1$, then*

$$D_n(X_\omega) = (O \cap \Omega)((\log n)^{\alpha_{2,2}} n^{\log_\theta \theta_{2,2}},$$

(3) *if $|\theta_{2,2}| = 1$ and $\theta_{2,2}$ is not a root of unity, then*

$$D_n(X_\omega) = (O \cap \Omega)\left((\log n)^{\alpha_{2,2}+1}\right),$$

(4) *if $|\theta_{2,2}| = 1$ and $\theta_{2,2}$ is a root of unity, then*

$$D_n(X_\omega) = O\left((\log n)^{\alpha_{2,2}+1}\right), \quad D_n(X_\omega) = \Omega\left((\log n)^{\alpha_{2,2}}\right).$$

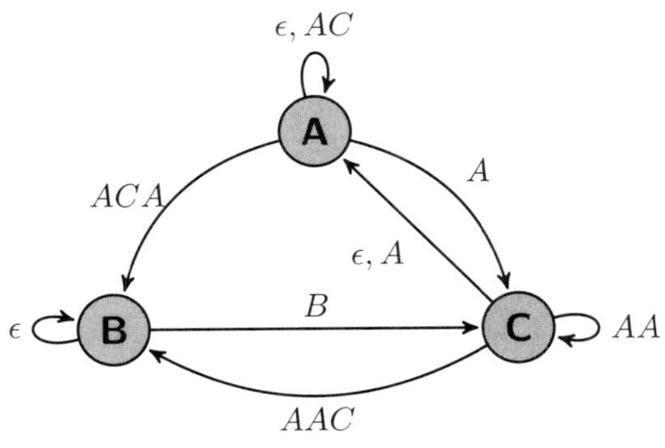

FIGURE 3. The prefix automaton defined by the substitution $A \mapsto ACAB$, $B \mapsto BC$, $C \mapsto AACB$.

Note the importance of spectral properties of the adjacency matrix. For a survey specific to the interplay of substitution dynamical systems and spectral theory, see [**27**].

Similar to the notion of the adjacency matrix, one may also associate to each substitution a *prefix automaton*, a crucial tool in the proofs of the above two theorems: see *e.g.* [**5, 9, 10, 12, 31**]. Suppose that σ is a substitution on some (finite) alphabet $\mathcal{A}$. Then we may create a labeled directed graph as follows. First, create a vertex set labeled by the elements of $\mathcal{A}$. Suppose that for $i, j \in \mathcal{A}$ we have $\sigma(i) = \psi j \xi$ for some words (possibly empty) ψ and ξ. Then create an arrow from i to j labeled with ψ.

Suppose that $\sigma(A)$ begins with A, then as σ is primitive, it follows that the length of $\sigma^n(A)$ tends to infinity, so there is a natural word ω defined as the limit of $\sigma^n(A)$, which is a fixed point in $\mathcal{A}^{\mathbb{N}}$ under σ. The principal observation about these automata is the following:

THEOREM 4.3 (see [**9, 31**]). *Let ω be a fixed point under σ generated as a limit of $\sigma^n(A)$. Then we have:*

(1) *For every positive integer N, there is a unique nonnegative integer n_N and a sequence of edges $E_0, E_1, \dots E_{n_N}$ which define a path in the prefix automaton beginning at vertex A, and such that ω_N, the prefix of ω of length N, is given by*
$$\omega_N = \sigma^{n_N}(E_0)\sigma^{n_N-1}(E_1)\sigma^{n_N-2}(E_2)\dots\sigma^{n_N-(n_N-1)}(E_{n_N-1})\sigma^0(E_N).$$
(2) *Conversely, to any path in the graph beginning at the vertex labeled A and beginning with $E_0 \neq \epsilon$, we may create a word according to the above which gives a prefix of ω.*
(3) *The quantities n_N are given by the inequality $|\sigma^{n_N}(1)| \leq N < |\sigma^{n_N+1}(1)|$.*

The proof is direct once properly interpreted; rather than replicate it, we will give an example using the automaton in 3. Using the substitution on $\{A, B, C\}$
$$\sigma : \begin{array}{l} A \mapsto ACAB \\ B \mapsto BC \\ C \mapsto AACB \end{array}$$
$$\omega = ACABAACBACABBCACABACABAACBBCACABAACB\dots$$

Consider the prefix of length 21: $\omega_{21} = ACABAACBACABBCACABACA$. First, we find the number n_{21} by finding the maximal $\sigma^n(A)$ which is a left prefix of this word: $\sigma^2(A) = ACABAACBACABBC$, while $\sigma^3(A)$ is much longer. We conclude that $n_{21} = 2$. We then find the longest ξ, a prefix of $\sigma(A)$, such that $\sigma^2(\xi)$ is a prefix of ω_{21} and we see that $\sigma^2(A)$ is a prefix of ω_{21} while $\sigma^2(AC)$ is not, so we set $E_0 = A$.

We now have $\omega_{21} = \sigma^2(A)ACABACA$. Recall that we are interpreting E_0 as a prefix of $\sigma(A)$, and this $E_0 = A$ is followed by C (within $\sigma(A)$). So we now consider prefixes of C and apply $\sigma^{n_{21}-1} = \sigma^1$ to these prefixes, finding the longest we can which is a prefix of our 'remainder' $ACABACA$. Within the word $\sigma(C) = AACB$, we first check the prefix A and see that $\sigma(A) = ACAB$ is a prefix of our remainder, while $\sigma(AA)$ is not. So we set $E_1 = A$.

At this point we have reduced our exponent to zero, and we see that we have represented $\omega_{21} = \sigma^2(A)\sigma(A)ACA$, and the new remainder ACA is itself a prefix of $\sigma(A)$, so $E_2 = ACA$.

Alternately, let's consider a path in the prefix automaton:

$$E_0 = AC, \quad E_1 = \epsilon, \quad E_2 = A.$$

This corresponds to a path beginning at vertex A, traversing the edge labeled AC which ends back at vertex A, then traversing the edge labeled ϵ towards the vertex labeled A, then taking the edge labeled A. Since we have again set $n_N = 2$, we compute the word

$$\sigma^2(AC)\sigma(\epsilon)A = \sigma^2(A)A = ACABAACBACABBCACABACABAACBBC,$$

which we see is ω_{28}.

5. Dynamically Defined Substitution Schemes

In this section we generalize the techniques of self-similar systems to generate codings of certain systems which are not self-similar, but which nonetheless have subsets whose induced systems are related to (X, T) through a separate action on some sort of parameter space. What follows is a generalization of techniques developed in [**29, 30**].

Let (Σ, ν) be a probability measure space, and let $g : \Sigma \to \Sigma$ be some measurable function on that space. Assume that there is a function σ which assigns to each $y \in \Sigma$ a substitution σ_y on a fixed alphabet $\mathcal{A}$ such that $\sigma^{-1}(\pi)$ is measurable for each substitution π. For convenience we denote the adjacency matrix of σ_y simply by M_y. By $\|M\|$ we mean the modulus of the principal eigenvalue of M, and for a column vector $\vec{u}$ with real entries, by $\|\vec{u}\|$ we mean the sum of the absolute values of the entries. There are various assumptions which one may or may not make beyond these, which are listed for convenient reference in 1.

If we were primarily concerned with topological dynamical properties, we could similarly assume only that Σ is a topological space, and g is continuous, with appropriate alterations to the other definitions. Since our questions will be phrased in a measure-theoretic context, however, we will not formalize this and similar assumptions.

In the assumption of primitivity note the order of compositions: the first composition applied is the last one generated in the g-orbit of y, and the last substitution

Assumption	Definition
Ergodicity	g is ergodic.
Regularity	There is some $a_1 \in A$ so that for almost every y, $\sigma_y(a_1)$ begins with a_1.
Primitivity	For almost every y, there is some $m = m(y)$ such that for all i, j: $\left(M_{\sigma_y} \cdot M_{\sigma_{g(y)}} \cdots M_{\sigma_{g^{m-1}(y)}}\right)_{i,j} \geq 1$.
Strong Primitivity	In addition to primitivity, we have $\int_\Sigma m(y)d\nu < \infty$.
Invertibility	For almost every y, the adjacency matrix M_y is invertible.
Integrability	$\log\left(\|M_y\|\right) \in L^1(\Sigma, \nu)$, and if invertibility is assumed, then $\log\left(\|M_y^{-1}\|\right) \in L^1(\Sigma, \nu)$ as well.

TABLE 1. Various optional assumptions which may be used when discussing dynamically defined substitution schemes

applied is σ_y. With this composition in mind, we define

$$\sigma_y^{(i)} = \sigma_y \circ \sigma_{g(y)} \circ \cdots \circ \sigma_{g^{i-1}(y)},$$

(recall (3.1)) an analogue to an ergodic sum or product which we term an *ergodic composition.*

The key observation is that for any $m, n \in \mathbb{N}$ we have a cocycle equation:

$$\sigma_y^{(n+m)} = \sigma_y^{(n)}\left(\sigma_{g^n(y)}^{(m)}\right). \tag{5.1}$$

Regularity is a natural consequence when substitutions arise from consideration of induced transformations, so long as the set on which we induce is a subset of one of the partition elements. When studying substitutions in their own right, this assumption is frequently made (for one fixed substitution) in order to ensure that $\sigma^n(a_1)$ has a limit as $n \to \infty$; our stronger assumption of regularity (see 1) is the appropriate condition to maintain this property:

LEMMA 2. *Assuming regularity, ergodicity, and that*

$$\int_\Sigma |\sigma_y(a_1)|\, d\nu > 1, \tag{5.2}$$

we have for generic $y \in \Sigma$

$$\lim_{n\to\infty} \left|\sigma_y^{(n)}(a_1)\right| = \infty.$$

Furthermore, primitivity implies (5.2).

PROOF. Recall that substitutions never map letters to the empty word, so we always have $|\sigma(\omega)| \geq |\omega|$ for any word ω. Denote by $N(\omega, a_1)$ the quantity of a_1 appearing in the word ω. By our assumption of regularity, we may assume generically that

$$N(\sigma_y(\omega), a_1) \geq N(\omega, a_1),$$

and whenever $\sigma_y(a_1)$ is of length at least two, which is true on a set of positive measure by (5.2), we have

$$|\sigma_y(\omega)| \geq |\omega| + N(\omega, a_1).$$

Between both of these facts we see that the lengths of the words

$$a_1, \sigma_y(a_1), \sigma_y^{(2)}(a_1), \sigma_y^{(3)}(a_1), \ldots$$

increases by at least one whenever $\sigma_{g^i(y)}(a_1)$ is of length at least two; our integral assumption and ergodicity guarantee that for generic y this happens infinitely many times.

For the final claim, assume that (5.2) fails. Then regularity implies that $\sigma_y(a_1) = a_1$ for almost every y, and this immediately gives that the system is not primitive, since we now have $\sigma_y^{(n)}(a_1) = a_1$ for all $n \in \mathbb{N}$ for almost every y. $\square$

So under the assumptions of 2 a limiting word ω_y may be found (recall that $\sigma_y(a_1)$ begins with a_1 for almost every y). It is worth emphasizing that just because there is a limiting word ω_y does not at all imply that $\sigma_y(\omega_y) = \omega_y$, but rather (via (5.1)): $\omega_y = \sigma_y(\omega_{g(y)})$.

Lemma 2 can in fact be used to show that for generic y not only does $\left|\sigma_y^{(n)}(a_1)\right|$ diverge, but it does so at least linearly-fast: $|\sigma_y^{(n)}(a_1)| \geq \delta n$ for some $\delta > 0$ (we may set, for example, δ to be the measure of the set on which $|\sigma_y(a_1)| > 1$). This linear growth is not fast enough for most purposes, however, so we will show that the strong primitivity condition guarantees exponential growth in the length of $\sigma_y^{(n)}(a_1)$.

Let $M_y^{(n)} = M_y \cdot M_{g(y)} \cdots M_{g^{n-1}(y)}$ be the adjacency matrix of $\sigma_y^{(n)}$. Given a word $\omega \in \mathcal{A}^*$, we may create a vector $\vec{u}$ by letting $\vec{u}_i$ be the quantity of a_i which appear in ω. Note, then, that $\|\vec{u}\|$ is the length of ω, and

$$\left\| M_y^{(n)} \begin{pmatrix} 1 \\ 0 \\ \vdots \\ 0 \end{pmatrix} \right\| = \left|\sigma_y^{(n)}(a_1)\right|. \tag{5.3}$$

LEMMA 3. *Assuming regularity, ergodicity, invertibility, and integrability, there is some $\lambda \geq 1$ such that for almost every $y \in \Sigma$ we have*

$$\lim_{n\to\infty} \left|\sigma_y^{(n)}(a_1)\right|^{1/n} = \lambda \geq 1.$$

If furthermore we assume strong primitivity, then $\lambda > 1$.

PROOF. This lemma is a generalization of [**30**, Lemma 4.2 and preceding commentary]. The existence of λ is a direct application of the Oseledec ergodic theorem: ergodicity, integrability, invertibility, and (5.3) imply that this limit λ almost surely exists. Since substitutions cannot decrease the length of words, we see that $\lambda \geq 1$.

Assume, then, the condition of strong primitivity. Recall that $k \geq 2$ has been used to denote the (finite) cardinality of the alphabet. Then since $\sigma_y^{(m(y))}(a_i)$ is of length at least k for all a_i, we see that

$$\sigma_{g^{m(y)}}^{\left(m(g^{m(y)}(y))\right)} \circ \sigma_y^{(m(y))}(a_1)$$

is of length at least k^2, and so forth for successive compositions of length determined by the function $m(y)$: each 'block' of consecutive compositions increases the word length by a factor of at least k. As we have assumed ergodicity and strong primitivity, for any $\hat{m} > \int_\Sigma m(y)d\nu$, for almost every y we have for sufficiently large n

$$\left|\sigma_y^{(\lceil n\hat{m}\rceil)}(a_1)\right| > k^n,$$

where $\lceil n\hat{m} \rceil$ is the least integer greater than $n\hat{m}$. That is, $\hat{m}$ consecutive substitutions is on average sufficient to increase word length by a factor of k, from which we see that

$$\lambda \geq k^{\left(\int_\Sigma m(y)d\nu\right)^{-1}} > 1. \qquad \square$$

With only strong regularity and ergodicity we could derive an exponential lower bound on the growth of $\left|\sigma_y^{(n)}(a_1)\right|$; the extra assumptions of invertibility and integrability allow us to apply the Oseledec theorem to derive the existence of the limit λ.

For convenience we denote the forward orbit closure X_{ω_y} as simply X_y. So under the conditions of 2 we therefore have a mapping

$$y \longmapsto (X_y, S),$$

which associates to each $y \in \Sigma$ the symbolic dynamical system generated by the limiting word $\omega_y = \sigma_y^{(n)}(a_1)$ as $n \to \infty$.

It is possible to generalize the notion of the prefix automaton to these words generated by the sequence of substitutions $\sigma_y^{(n)}$, but we cannot do so with a fixed, finite graph, as our substitutions are changing with each successive iteration of the map g. Rather, we coin the term *prefix Bratelli diagram.* We have a sequence of vertex sets V_n for all $n = 0, 1, 2, \ldots$; for our purposes, each $V_n = \mathcal{A}$. From V_n to V_{n+1} we draw an arrow from i to j labeled by ψ if ψ is a prefix of $\sigma_{g^n(i)}$, and again we allow ψ to be the empty word ϵ. Note that each $y \in \Sigma$ will give rise to its own diagram!

EXAMPLE 3. Consider the vertex set $\mathcal{A} = \{A, B\}$, and suppose that we have some Σ and g so that for some $y \in \Sigma$ we have

$$\sigma_y : \begin{array}{l} A \mapsto ABB \\ B \mapsto AB \end{array} \quad \sigma_{g(y)} : \begin{array}{l} A \mapsto AAB \\ B \mapsto BBA \end{array} \quad \sigma_{g^2(y)} : \begin{array}{l} A \mapsto AB \\ B \mapsto ABA \end{array} \tag{5.4}$$

Then we would create a prefix Bratelli diagram which would begin as given by Figure 4. Similarly to Theorem 4.3 we obtain:

THEOREM 5.1. *For some $y \in \Sigma$, assume that $\omega_y = \lim_{n\to\infty} \sigma^{(n)}(y)$ exists and is of infinite length. Let $\omega_{y,N}$ denote the prefix of ω_y of length N.*

- *For every $N = 0, 1, 2, \ldots$, there is some $n_N \in \mathbb{N}$ (possibly zero) and a unique path in the prefix Bratelli diagram, beginning from the vertex labeled A in the initial vertex set V_0, given by the edges $E_0, E_1, \ldots, E_{n_N}$, so that*

$$\omega_{y,N} = \sigma_y^{n_N}(E_0)\sigma_{g(y)}^{n_N-1}(E_1)\sigma_{g^2(y)}^{n_N-2}(E_2)\cdots\sigma^1_{g^{n_N-1}(y)}(E_{n_N-1})\sigma^0_{g^{n_N}(y)}(E_{n_N}),$$

  and $E_0 \neq \epsilon$.
- *Conversely, to every such finite path in the prefix Bratelli diagram we may obtain a unique prefix of ω_y.*
- *The quantity n_N is obtained through the inequality*

$$\left|\sigma^{(n_N)}(A)\right| \leq N < \left|\sigma^{(n_N+1)}(A)\right|.$$

PROOF. This theorem is not much different from 4.3, so we do not present a separate proof. Recall (5.1) to keep track of the proper subscripts and superscripts. Note also that in the first item we have maintained the substitution $\sigma_{g^{n_N}(y)}$ in the final entry, but as the exponent on this term is zero, this substitution acts as identity on the final edge E_{n_N}. $\square$

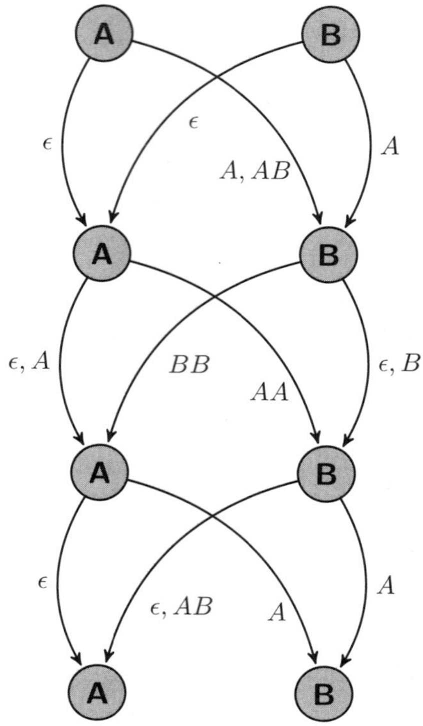

FIGURE 4. The first four levels of the prefix Bratelli diagram corresponding to a hypothetical dynamically defined substitution which begins as given by (5.4).

These prefix Bratelli diagrams are dependent on the choice of $y \in \Sigma$, and the manner in which elements of $\mathcal{A}$ are connected to each other changes as we move through the orbit of y. This 'malleable connection state' phenomenon appears similar to *random walks in random environments.*

6. Systems Coded by Dynamically Defined Substitutions

Observe that if σ is a primitive substitution on $\mathcal{A}$ such that $\sigma(a)$ begins with a for some $a \in \mathcal{A}$, then if we define the space Σ to be a single point with the identity transformation, then we see that the symbolic dynamical system associated to a primitive substitution is a trivial example of dynamically defined substitution.

Let us develop a more complicated example, with $\mathcal{A} = \{A, B\}$. Let $\Sigma = \mathbb{S}^1 = [0, 1)$, and for each $y \in \Sigma$ define the substitution

$$\sigma_y : \begin{array}{l} A \mapsto AB^{a_1} \\ B \mapsto AB^{a_1 - 1} \end{array} \tag{6.1}$$

where a_1 is the first partial quotient in the standard continued fraction expansion of y. Let the map g be given by the Gauss map γ:

$$\gamma(y) = \frac{1}{y} - a_1.$$

This is an ergodic transformation with respect to the Gauss measure on $[0, 1)$, which is mutually absolutely continuous with Lebesgue measure. Then as we saw in 2, the associated symbolic system

$$(X_y, S)$$

is, in fact, a coding for the system $(\mathbb{S}^1, R_y)$, with A corresponding to the interval $[0, y)$ and B corresponding to $[y, 1)$. Sequences which code the orbit of some $x \in \mathbb{S}^1$ with respect to this partition under this rotation (for any fixed irrational y) are called *Sturmian* sequences, and they form a classical setting for the study of symbolic systems, originally developed in [**25**]. See also [**10, 28**].

Through these substitutions it is possible to derive the following well-known result (in greatest generality, see [**18**, Theorem 4], and for the original symbolic discrepancy equivalent, see [**25**, Theorem 2.3]):

THEOREM 6.1. *The discrepancy function*

$$D_N(x,y) := \left(\sum_{i=0}^{N-1} \chi_{[0,y)}(x+iy \mod 1)\right) - Ny, \tag{6.2}$$

is uniformly bounded in all of $N \in \mathbb{N}$, $x \in \mathbb{S}^1$, and $y \in (\mathbb{S}^1 \setminus \mathbb{Q})$.

Note the similarity to the discrepancy functions in 4; here the length of an interval is used in place of the measure of a cylinder. We will prove this theorem in three basic steps. First, we will establish that it is sufficient to prove that the symbolic discrepancy of the associated symbolic coding is bounded. Next, since this coding may be realized through a dynamically defined substitution scheme, we establish how these substitutions interact with symbolic discrepancy (Lemma 4), which we will then use to show that the symbolic discrepancy remains bounded (Corollary 1).

We have not taken an absolute value in (6.2) as is standard; we wish to preserve the cocycle structure of D_N, as they are ergodic sums:

$$D_N(x,y) = \sum_{i=0}^{N-1} f(x+iy \mod 1),$$

where the function $f(x)$ is given by

$$f(x) = \begin{cases} (1-y) & (x<y) \\ -y & (x \geq y). \end{cases}$$

As irrational rotations of the circle are minimal, any prefix in the orbit of any x will appear somewhere as a factor of the word ω_y, where ω_y is limiting word generated by the dynamically defined substitution scheme outlined at the beginning of this section. It therefore suffices to show that for $N \geq M$

$$D_{M,N}(y) := \left(\sum_{i=M}^{N-1} \chi_A(\omega_y)_N\right) - (N-M)y$$

is uniformly bounded in N, M, y. We need only to find a uniform bound in N when we set $M = 0$, as we have $D_{M,N}(y) = D_{0,N}(y) - D_{0,M}(y)$ (this fact follows from the definition of $D_{M,N}(y)$ and the cocycle structure of the sums involved). With $M = 0$ fixed, note that we are in fact computing the symbolic discrepancy $D_N(\omega_y, A)$ as given in (2.1); as the symbol A corresponds in the coding to an interval of length

y in the uniquely ergodic irrational circle rotation, we certainly have $\mu(A) = y$. So we may approach 6.1 through the associated symbolic coding.

Our next step is to see how the substitutions σ_y (6.1) interact with this discrepancy function through the Gauss map γ. For convenience, for finite words ω we denote

$$d_y(\omega) = D_{|\omega|}(\omega, A);$$

we seek to show, then that $d_y(\omega)$ is bounded for all irrational $y \in \mathbb{S}^1$ and ω a left factor of the limiting word ω_y. We begin by determining how symbolic discrepancy is affected by our substitutions:

LEMMA 4. *For any irrational $y \in \mathbb{S}^1$ and any word $\psi \in \{A, B\}^*$,*

$$d_y(\sigma_y(\psi)) = -y d_{\gamma(y)}(\psi).$$

PROOF. Suppose that ψ is of length n, with k appearances of A and $(n-k)$ of B. Then for any $z \in (0,1)$:

$$d_z(\psi) = k - nz. \tag{6.3}$$

Recall that

$$\sigma_y : \begin{array}{l} A \mapsto AB^{a_1} \\ B \mapsto AB^{a_1 - 1} \end{array},$$

where a_1 is the first partial quotient of y. Then we see that the word $\sigma_y(\psi)$ has exactly n appearances of the letter A (one for each letter in ψ, which was of length n), and a total of

$$ka_1 + (n-k)(a_1 - 1) = n(a_1 - 1) + k$$

appearances of the letter B (for a total word length of $na_1 + k$), which allows us to compute (using (6.3), and also that $\gamma(y) = (1/y) - a_1$):

$$\begin{aligned} d_y(\sigma_y(\psi)) &= n - (na_1 + k)y \\ &= -yk + n - na_1 y \\ &= -yk + ny\left(\frac{1}{y} - a_1\right) \\ &= -y\left(k - n\gamma(y)\right) \\ &= -y d_{\gamma(y)}(\psi). \end{aligned}$$

□

Finally, we see that the above result shows that the symbolic discrepancy $D_n(\omega_y, A)$ is bounded:

COROLLARY 1. *For any y, for any n, for ψ any left-factor of $\sigma_y^{(n)}(A)$, we have $-1 < d_y(\psi) < 1$.*

PROOF. For $n = 1$, simply consider the substitutions σ_y and see that the discrepancy of any left-factor of $\sigma_y(A)$ has discrepancy no larger than $1 - y < 1$ (which is achieved by the initial letter A) and no smaller than $1 - (a_1 + 1)y > -1$ (which is achieved by the complete word $\sigma_y(A)$.

Now suppose that for some $n > 1$ we have a left factor ψ of $\sigma_y^{(n)}(A)$ so that $d_y(\psi) > 1$, but no left factor of $\sigma_y^{(n-1)}(A)$ has discrepancy larger than 1 or less than -1. Without loss of generality, we may assume that ψ terminates with an A (i.e. this ψ is a left-factor of minimal length with an excessively positive discrepancy for any ω_y across all y). Since the letter A appears exactly as the first letter of $\sigma_y(\star)$ for any letter $\star$, we conclude that ψ takes the form

$$\psi = \sigma_y(\psi')A,$$

where ψ' must be a left factor of $\sigma_{\gamma(y)}^{(n-1)}(A)$ (see (5.1)). We may then compute

$$\begin{aligned}
1 &< d_y(\psi) \\
1 &< d_y(\sigma_y(\psi')) + (1-y) \\
1 &< -yd_{\gamma(y)}(\psi') + 1 - y \\
0 &< -y(d_{\gamma(y)}(\psi') + 1) \\
-1 &> d_{\gamma(y)}(\psi')
\end{aligned}$$

which violates our inductive hypothesis that ψ was of minimal length to have such an excessive discrepancy. A similar proof holds if we assumed that $d_y(\psi) < -1$. $\square$

We have therefore completed the proof of Theorem 6.1.

A more complicated but similarly motivated example was developed in [**29**]. The goal is still to encode rotation by some irrational $\theta < 1/2$, but the substitutions are defined on three letters A, B, and C, with A corresponding to $[0, 1/2)$, B to $[1/2, 1-\theta)$, and C to $[1-\theta, 1)$. In contrast to self-similar systems, the symbolic discrepancy may be forced to be bounded below or above by a somewhat arbitrary choice of function, but cannot in general be bounded above and below by the same function. Note in particular that in contrast to 6.1 our coding includes an interval which was independent of the choice of θ.

For the sake of brevity we omit the particular substitutions σ_θ (see [**30**, Table 1]) and the particular map g (see [**30**], proof of ergodicity in [**30**, Theorem 1]). We mention, however, that the symbolic discrepancy $D_N(X_\theta, A)$ corresponds to the ergodic sums under rotation by θ of the function $\left(\chi_{[0,1/2)}(x) - 1/2\right)$.

THEOREM 6.2. *Let $f(x)$ be a continuous, positive, nondecreasing, unbounded, regularly varying function (i.e. $f(kx) \in O(f(x))$ and $f(x) \in O(f(kx))$ for any $k > 0$) which is defined for all sufficiently large $x \geq C$. Define the sequence $\{F(n)\}$ for $n \geq C$ by*

$$F(n) = \int_C^n f(x)dx.$$

Define the A-discrepancy by

$$D_n(X_\theta, A) = \sup_{\omega \in X_\theta} D_n(\omega, A),$$

where the word A is fixed but we vary over all elements of the space X_θ.

(1) [**29**, Theorem 1.4]: *If the partial quotients of θ are bounded, then $D_n(X_\theta, A) = (O \cap \Omega)(\log n)$.*
(2) [**29**, Theorem 1.2]: *One may construct θ so that $D_n(X_\theta, A) \notin o(f(n))$, or we may construct θ so that $D_n(X_\theta, A) \in o(f(n))$.*
(3) [**30**, Theorem 2]: *For Lebesgue almost every θ, either*

$$D_n(X_\theta, A) \in o(F(\log n))$$

or $D_n(X_\theta, A) \notin o(f(\log n))$ according to whether

$$\int_C^\infty \frac{1}{f(x)}dx < \infty \quad \text{or} \quad \int_C^\infty \frac{1}{f(x)}dx = \infty.$$

The notation and language varies slightly in the original references; here we have brought those results under a unified umbrella. We correct here a typo in the original reference in regards to the final entry; the integrand should be the reciprocal of $f(x)$, as is written here. The first simply states that when the dynamically defined substitution system is created from a θ with bounded partial quotients (for example: any quadratic surd or self-similar rotation) the discrepancy is precisely describable. However, one may find specific θ with discrepancy growing as fast as desired (as long as the growth is $o(n)$) or as slow as desired (as long as it is unbounded). If one wishes to search for generic behavior, however, the infrequent appearance of large partial quotients prohibits the possibility of a single rate of divergence acting as both an upper and lower bound on the discrepancy, an imprecision reflected in the last item.

The key observation in the proof of the above is the interplay between the substitutions and the discrepancy, just as in our proof of 6.1. For all of the substitutions and any word ω of length n for example, it holds that

$$\left(\sum_{i=0}^{n-1} \chi_A(\omega_i) - n/2\right) = \pm \left(\sum_{i=0}^{|\sigma(\omega)|-1} \chi_A(\sigma(\omega)_i) - |\sigma(\omega)|/2\right),$$

where the choice of $\pm$ is explicit according to the substitution σ, which is determined by the rotation parameter and a map similar to the Gauss map (see **[29]** for details), and the maximal discrepancy achieved by any left factor is bounded above and below by explicit quantities derived from the rotation. Furthermore, for all σ the images of every letter under σ are *heavy words* (their minimal intermediate ergodic sum is the final ergodic sum; 2), and the one-sided bound inherent in that definition, combined with the explicit interaction between the discrepancy and the substitutions σ outlined above, makes for easier computation of overall bounds on discrepancy.

7. Conclusion

The notion of the *dynamically defined substitution scheme* seems a natural generalization of the use of primitive substitutions to give codings of self-similar dynamical systems. Some properties of this process would be of general interest.

While the limiting word ω_y arises in a natural way, it is not immediately clear what properties the system (X_y, S) may have. For example, assume that every $\sigma_y^{(n)}$ is primitive, and therefore generates a uniquely ergodic symbolic dynamical system. So for any y, n, there is a well-defined measure $\mu_{y,n}$, the unique preserved measure on the uniquely ergodic system defined by $\sigma_y^{(n)}$, but some questions arise:

QUESTION 1. Under what circumstances does the measure

$$\mu_y = \lim_{n\to\infty} \mu_{y,n}$$

exist for almost every y? Under what circumstances is (X_y, S, μ_y) ergodic or uniquely ergodic?

When our substitutions are derived from induced maps on a known system (X, T, μ), we may use known properties of that system, but it is quite possible to create a dynamically defined substitution scheme with no obvious properties.

EXAMPLE 4. Let $\Sigma = [0,1)$ and γ be the Gauss map as before; set $\gamma(0) = 0$ for convenience. For $y \in \Sigma$, define the substitution

$$\sigma_y : \begin{array}{l} A \mapsto A^{a_1} B \\ B \mapsto A B^{a_1 - 1}, \end{array}$$

where a_1 is the first partial quotient of y, as before. Note the similarity to the substitutions which gave our Sturmian sequences, but now $\sigma_y(A)$ contains many initial As, rather than many terminal Bs. One may verify that every $\sigma_y^{(n)}$ is primitive y and $n \geq 2$.

Is the system (X_y, S) ergodic for almost every y with respect to some natural measure μ? Unique ergodicity seems unlikely, as for generic y there is a sparse subsequence of partial quotients a_n which are comparable in size to the sum $a_1 + a_2 + \cdots + a_{n-1}$ (see e.g. the final statements of [**19**], or [**7**] for a more in-depth investigation of these sums of a_i). This will result in very long stretches of both A and B which are then subjected to the substitution $\sigma_y^{(n-1)}$. It appears likely that this phenomenon will obstruct the uniform convergence required for unique ergodicity, even if it might not interfere with $\mu_{n,y}(\psi)$ converging. For those y which are quadratic surds, and therefore the sequence a_i is eventually periodic, we will end up with a fixed primitive substitution (corresponding to the initial partial quotients) applied to a self-similar system (corresponding to the eventually periodic tail), but generic y will not be so well-behaved. Even y which are not quadratic surds but are *badly approximable* (the a_i are bounded) may perhaps fail to generate uniquely ergodic systems (corresponding to the possible non-convergence of $\frac{1}{n}\sum_{i=1}^{n} a_i$ as $n \to \infty$).

EXAMPLE 5. Continue with $\Sigma = [0,1)$, but now consider $g(y) = 2y \mod 1$. Define the substitutions

$$\sigma_y : \begin{cases} \begin{array}{l} A \mapsto AB \\ B \mapsto A \end{array} & (y < 1/2) \\ \\ \begin{array}{l} A \mapsto B \\ B \mapsto BA \end{array} & (y \geq 1/2) \end{cases}$$

This situation models the following: we randomly choose substitutions according to a fair coin. Either A is 'favored' (by 'growing' $A \mapsto AB$ and replacing $B \mapsto A$), or B is 'favored.' We denote the first case as σ_A and the second as σ_B. Both are primitive substitutions.

Suppose we let the binary expansion of y begin

$$y = \frac{0}{2} + \frac{1}{4} + \frac{1}{8} + \frac{0}{16} + \frac{0}{32} + \cdots$$

Then we compute the words

$$\begin{aligned} \sigma_A(A) &= AB \\ \sigma_A \circ \sigma_B(A) &= A \\ \sigma_A \circ \sigma_B \circ \sigma_B(A) &= AAB \\ \sigma_A \circ \sigma_B \circ \sigma_B \circ \sigma_A(A) &= AABAABA \\ \sigma_A \circ \sigma_B \circ \sigma_B \circ \sigma_A \circ \sigma_A(A) &= AABAABAAAB \\ &\vdots \end{aligned}$$

Generic behavior of the resulting system (X_y, S), or even just one particular (X_y, S) for irrational y, would be of interest.

These examples and questions are posed to indicate that a wide variety of systems and problems may be described by these dynamically defined substitution systems, a way of coupling substitutions to a transformation on a base space. These objects embrace self-similar systems as the simplest example (the base transformation is a one-point space), give a mechanism for generating Sturmian sequences, describe the 1/2-discrepancy of circle rotations, and much more. It would be of general interest to study these objects further in any of the following directions:

- What other classes of systems may be represented by such substitutions? For example, the process of *Rauzy-Veech induction* in the study of *interval exchange transformations* is a way to use an induced map to generate a new, but similar transformation, with new parameters chosen from a parameter space. 5 shows the possibility of modeling 'random substitutions,' a very different interpretation.
- Under what circumstances can one derive useful results about the symbolic system (X_y, S) for generic y, such as ergodicity, spectral or discrepancy results, etc.?

Acknowledgments

We are grateful to an anonymous referee who not only found numerous typos in the original manuscript, but made several suggestions for improving the readability and organization of this paper.

References

[1] Boris Adamczewski, *Symbolic discrepancy and self-similar dynamics* (English, with English and French summaries), Ann. Inst. Fourier (Grenoble) **54** (2004), no. 7, 2201–2234 (2005). MR2139693

[2] M. D. Boshernitzan and C. R. Carroll, *An extension of Lagrange's theorem to interval exchange transformations over quadratic fields*, J. Anal. Math. **72** (1997), 21–44, DOI 10.1007/BF02843152. MR1482988

[3] Xavier Bressaud, Alexander I. Bufetov, and Pascal Hubert, *Deviation of ergodic averages for substitution dynamical systems with eigenvalues of modulus 1*, Proc. Lond. Math. Soc. (3) **109** (2014), no. 2, 483–522, DOI 10.1112/plms/pdu009. MR3254932

[4] Alexander I. Bufetov and Boris Solomyak, *Limit theorems for self-similar tilings*, Comm. Math. Phys. **319** (2013), no. 3, 761–789, DOI 10.1007/s00220-012-1624-7. MR3040375

[5] Vincent Canterini and Anne Siegel, *Automate des préfixes-suffixes associé à une substitution primitive* (French, with English and French summaries), J. Théor. Nombres Bordeaux **13** (2001), no. 2, 353–369. MR1879663

[6] David Damanik and Daniel Lenz, *Substitution dynamical systems: characterization of linear repetitivity and applications*, J. Math. Anal. Appl. **321** (2006), no. 2, 766–780, DOI 10.1016/j.jmaa.2005.09.004. MR2241154

[7] Harold G. Diamond and Jeffrey D. Vaaler, *Estimates for partial sums of continued fraction partial quotients*, Pacific J. Math. **122** (1986), no. 1, 73–82. MR825224

[8] Michael Drmota and Robert F. Tichy, *Sequences, discrepancies and applications*, Lecture Notes in Mathematics, vol. 1651, Springer-Verlag, Berlin, 1997. MR1470456

[9] Jean-Marie Dumont and Alain Thomas, *Systemes de numeration et fonctions fractales relatifs aux substitutions* (French, with English summary), Theoret. Comput. Sci. **65** (1989), no. 2, 153–169, DOI 10.1016/0304-3975(89)90041-8. MR1020484

[10] N. Pytheas Fogg, *Substitutions in dynamics, arithmetics and combinatorics*, Lecture Notes in Mathematics, vol. 1794, Springer-Verlag, Berlin, 2002. Edited by V. Berthé, S. Ferenczi, C. Mauduit and A. Siegel. MR1970385

[11] Gustav A. Hedlund, *On the metrical transitivity of the geodesics on closed surfaces of constant negative curvature*, Ann. of Math. (2) **35** (1934), no. 4, 787–808, DOI 10.2307/1968495. MR1503197

[12] C. Holton and L. Q. Zamboni, *Directed graphs and substitutions*, Theory Comput. Syst. **34** (2001), no. 6, 545–564. MR1865811

[13] Svetlana Katok, *Fuchsian groups, geodesic flows on surfaces of constant negative curvature and symbolic coding of geodesics*, Homogeneous flows, moduli spaces and arithmetic, Clay Math. Proc., vol. 10, Amer. Math. Soc., Providence, RI, 2010, pp. 243–320. MR2648696

[14] Svetlana Katok and Ilie Ugarcovici, *Symbolic dynamics for the modular surface and beyond*, Bull. Amer. Math. Soc. (N.S.) **44** (2007), no. 1, 87–132, DOI 10.1090/S0273-0979-06-01115-3. MR2265011

[15] Svetlana Katok and Ilie Ugarcovici, *Structure of attractors for (a,b)-continued fraction transformations*, J. Mod. Dyn. **4** (2010), no. 4, 637–691. MR2753948

[16] Svetlana Katok and Ilie Ugarcovici, *Theory of (a,b)-continued fraction transformations and applications*, Electron. Res. Announc. Math. Sci. **17** (2010), 20–33, DOI 10.3934/era.2010.17.20. MR2644834

[17] Svetlana Katok and Ilie Ugarcovici, *Applications of (a,b)-continued fraction transformations*, Ergodic Theory Dynam. Systems **32** (2012), no. 2, 755–777, DOI 10.1017/S0143385711000460. MR2901369

[18] Harry Kesten, *On a conjecture of Erdős and Szüsz related to uniform distribution* mod 1, Acta Arith. **12** (1966/1967), 193–212. MR0209253

[19] A. Ya. Khinchin, *Continued fractions*, Translated from the third (1961) Russian edition, Dover Publications, Inc., Mineola, NY, 1997. With a preface by B. V. Gnedenko; Reprint of the 1964 translation. MR1451873

[20] Nicolas Kryloff and Nicolas Bogoliouboff, *La théorie générale de la mesure dans son application à l'étude des systèmes dynamiques de la mécanique non linéaire* (French), Ann. of Math. (2) **38** (1937), no. 1, 65–113, DOI 10.2307/1968511. MR1503326

[21] L. Kuipers and H. Niederreiter, *Uniform distribution of sequences*, Wiley-Interscience [John Wiley & Sons], New York-London-Sydney, 1974. Pure and Applied Mathematics. MR0419394

[22] François Ledrappier and Omri Sarig, *Fluctuations of ergodic sums for horocycle flows on $\mathbb{Z}^d$-covers of finite volume surfaces*, Discrete Contin. Dyn. Syst. **22** (2008), no. 1-2, 247–325, DOI 10.3934/dcds.2008.22.247. MR2410958

[23] Daniel Lenz, *Uniform ergodic theorems on subshifts over a finite alphabet*, Ergodic Theory Dynam. Systems **22** (2002), no. 1, 245–255, DOI 10.1017/S0143385702000111. MR1889573

[24] M. Lothaire, *Combinatorics on words*, Cambridge Mathematical Library, Cambridge University Press, Cambridge, 1997. With a foreword by Roger Lyndon and a preface by Dominique Perrin; Corrected reprint of the 1983 original, with a new preface by Perrin. MR1475463

[25] Marston Morse and Gustav A. Hedlund, *Symbolic dynamics II. Sturmian trajectories*, Amer. J. Math. **62** (1940), 1–42. MR0000745

[26] Jakob Nielsen, *Untersuchungen zur Topologie der geschlossenen zweiseitigen Flächen* (German), Acta Math. **50** (1927), no. 1, 189–358, DOI 10.1007/BF02421324. MR1555256

[27] Martine Queffélec, *Substitution dynamical systems—spectral analysis*, 2nd ed., Lecture Notes in Mathematics, vol. 1294, Springer-Verlag, Berlin, 2010. MR2590264

[28] David Ralston, *Heaviness in symbolic dynamics: substitution and Sturmian systems*, Discrete Contin. Dyn. Syst. Ser. S **2** (2009), no. 2, 287–300, DOI 10.3934/dcdss.2009.2.287. MR2505639

[29] David Ralston, *Substitutions and 1/2-discrepancy of* $\{n\theta + x\}$, Acta Arith. **154** (2012), no. 1, 1–28, DOI 10.4064/aa154-1-1. MR2943672

[30] David Ralston, *Generic* $\frac{1}{2}$*-discrepancy of* $\{n\theta + x\}$, New York J. Math. **20** (2014), 195–208. MR3177171

[31] G. Rauzy, *Sequences defined by iterated morphisms*, Sequences (Naples/Positano, 1988), Springer, New York, 1990, pp. 275–286. MR1040317

[32] Caroline Series, *The geometry of Markoff numbers*, Math. Intelligencer **7** (1985), no. 3, 20–29, DOI 10.1007/BF03025802. MR795536

[33] Caroline Series, *The modular surface and continued fractions*, J. London Math. Soc. (2) **31** (1985), no. 1, 69–80, DOI 10.1112/jlms/s2-31.1.69. MR810563

[34] Caroline Series, *Geometrical Markov coding of geodesics on surfaces of constant negative curvature*, Ergodic Theory Dynam. Systems **6** (1986), no. 4, 601–625, DOI 10.1017/S0143385700003722. MR873435

[35] Caroline Series, *Symbolic dynamics for geodesic flows*, Proceedings of the International Congress of Mathematicians, Vol. 1, 2 (Berkeley, Calif., 1986), Amer. Math. Soc., Providence, RI, 1987, pp. 1210–1215. MR934325

[36] Caroline Series, *Geometrical methods of symbolic coding*, Ergodic theory, symbolic dynamics, and hyperbolic spaces (Trieste, 1989), Oxford Sci. Publ., Oxford Univ. Press, New York, 1991, pp. 125–151. MR1130175

Department of Mathematics, SUNY College at Old Westbury, Old Westbury, New York 11568

E-mail address: `ralstond@oldwestbury.edu`

Contemporary Mathematics
Volume **678**, 2016
http://dx.doi.org/10.1090/conm/678/13653

Nearly continuous even Kakutani equivalence of nearly continuously rank-one transformations

Bethany D. Springer

Dedicated to Dan Rudolph and Andrés del Junco.
The former's vision and the latter's encouragement made this work possible.

Abstract. Nearly continuous dynamical systems, a relatively new field of study, blends together topological dynamics and measurable dynamics/ergodic theory by asking that properties hold modulo sets both meager and of measure zero. In the measure theoretic category, two dynamical systems (X, T) and (Y, S) are called Kakutani equivalent if there exist measurable subsets $A \subset X$ and $B \subset Y$ such that the induced transformations T_A and S_B are measurably conjugate. We say that a set $A \subset X$ is nearly clopen if it is clopen in the relative topology of a dense G_δ subset of full measure. Nearly continuous Kakutani equivalence refines the measure-theoretic notion by requiring the sets A and B to be nearly clopen and T_A and S_B to be nearly continuously conjugate. If A and B have the same measure, then we say that the systems are nearly continuously evenly Kakutani equivalent. All irrational rotations of the circle and all odometers belong to the same equivalence class for nearly continuous even Kakutani equivalence. For our first main result, we prove that if A and B are nearly clopen subsets of the same measure of X and Y respectively, and if ϕ is a nearly continuous conjugacy between T_A and S_B, then ϕ extends to a nearly continuous orbit equivalence between T and S. We also prove that if $A \subset X$ and $B \subset Y$ are nearly clopen sets such that the measure of A is larger than the measure of B, and if T is a nearly uniquely ergodic transformation and T_A is nearly continuously conjugate to S_B, then there exists $B' \subset Y$ such that T is nearly continuously conjugate to $S_{B'}$. We then introduce the natural topological analog of rank-one transformations, called nearly continuously rank-one transformations, and show that all nearly continuously rank-one transformations are nearly continuously evenly Kakutani equivalent to the class containing all adding machines.

1. Introduction

In ergodic theory, we ask properties to hold modulo sets of measure zero. In the following theory, dubbed nearly continuous dynamics, we add a topological component by asking properties to hold modulo F_σ sets of measure zero. Denker and Keane [**5**] set down the original foundations for the theory, which then lay dormant until Hamachi and Keane [**7**] revived interest and helped motivate del Junco,

2010 *Mathematics Subject Classification.* Primary 37A20; Secondary 37B99.
Key words and phrases. Nearly continuous, Kakutani equivalence, rank-one.
Author supported in part by NSF Grant #DMS-0700874 AMD006.

Rudolph, and Weiss [**1**] to further investigate the theory. For a more complete understanding of the background of this theory, we refer the readers to [**5**] and [**1**]. Transformations must be homeomorphisms in the relative topology of some G_δ set of full measure. We ask sets to be nearly clopen instead of measurable, i.e. clopen in the relative topology when restricted to a G_δ subset of full measure. A nearly continuous dynamical system consists of a Polish probability space and a transformation which is an ergodic measure preserving homeomorphism when restricted to an invariant G_δ subset of full measure. We consider two transformations to be equivalent if they agree upon a G_δ subset of full measure.

In the 1940s, S. Kakutani introduced a new type of equivalence between ergodic measure preserving transformations of finite measure spaces relevant to studying the relationships between measurable cross sections of measurable flows. Two ergodic measure preserving transformations S and T are called Kakutani equivalent if they induce isomorphic transformations.

In 1984, A. del Junco and D. Rudolph extended the definition of even Kakutani equivalence to ergodic $\mathbb{Z}^n$ actions in [**3**], first by taking Katok cross-sections of a flow, and second, by determining the existence of an orbit preserving injection with an extra asymptotic linearity condition. In one dimension, this translated to the classical theory of Kakutani equivalence with an addition of an orbit equivalence between the two systems. When A. del Junco, D. Rudolph, and B. Weiss explored Kakutani equivalence in the context of what we are calling nearly continuous dynamical systems [**1**], this element of orbit equivalence persisted in the definition. The authors defined even Kakutani equivalence to be an orbit equivalence on sets of full measure which gave a conjugacy when restricted to induced transformations of nearly clopen subsets of the same measure. We give a slightly different definition for nearly continuous even Kakutani equivalence, asking only for the conjugacy between induced transformations on nearly clopen subsets of the same measure. The first aim of the paper is to prove the following theorem which demonstrates that our definition of nearly continuous even Kakutani equivalence is synonymous with the definition given by del Junco, Rudolph and Weiss:

THEOREM 1.1. *Let (X, τ, μ, T) and (Y, τ, ν, S) be two nearly continuous dynamical systems. If $A \subset X$ and $B \subset Y$ are nearly clopen sets with $\mu(A) = \nu(B)$ such that there exists a nearly continuous conjugacy ϕ between the induced maps T_A and S_B, then ϕ extends to a nearly continuous orbit equivalence between T and S.*

Naturally, one would ask about the consequences of starting with a nearly continuous conjugacy between nearly clopen sets of different measure. Answering this question in a satisfactory manner requires the use of nearly unique ergodicity, and we come to the following theorem.

THEOREM 1.2. *Let (X, τ, μ, T) and (Y, τ, ν, S) be two nearly continuous dynamical systems such that $A \subset X$ and $B \subset Y$ are nearly clopen with $\mu(A) > \nu(B) > 0$ and with T_A nearly continuously conjugate to S_B. If (X, τ, μ, T) is nearly uniquely ergodic, then there exists a nearly clopen set $\bar{B} \subset Y$ such that $S_{\bar{B}}$ is nearly continuously conjugate to T.*

This theorem leads to interesting results. Roychowdhury and Rudolph [**8**] proved that all adding machines are nearly continuously evenly Kakutani equivalent, and Dykstra and Rudolph [**6**] added irrational rotations to the class. We apply the above theorem to show that a class of transformations, which is here-in named nearly continuously rank-one, belongs to the equivalence class of irrational rotations.

This work was begun jointly with Daniel Rudolph. Rudolph guided the direction of the paper by selecting the various conjectures herein. Unfortunately, Daniel Rudolph passed away shortly after we completed a proof of Theorem 1.2, and the rest of the exposition was finished without him.

The structure of the paper is as follows: section 2 gives basic definitions and facts. Section 3 provides a proof of Theorem 1.1. Section 4 defines nearly unique ergodicty and provides a proof of Theorem 1.2. Section 5 defines nearly continuously rank-one transformations and shows that such transformations are nearly continuously evenly Kakutani equivalent to irrational rotations of the unit circle.

2. Definitions and Basic Facts

The state space X is always an uncountable Polish space: a separable topological space neither finite nor homeomorphic to $\mathbb{N}$ permitting at least one metric for which the space is complete. The space is endowed with a non-atomic Borel probability measure, μ, of full support. Recall that Polish spaces are G_δ subspaces of compact metric spaces. The topology τ is the Polish topology, and whenever we write (X, τ, μ), we mean precisely such a space, referred to as a *Polish probability space.*

DEFINITION 2.1. A G_δ subset of full measure $X_0 \subseteq X$ is called a **nearly full** subset.

Because the support of μ is X, any nearly full subset is dense in X. In general, we say that property R of an object holds *nearly* on the space X if there exists a nearly full subset $X_0 \subseteq X$ such that R holds in the relative topology of X_0.

DEFINITION 2.2. A set $C \subseteq X$ is said to be **nearly clopen** if there exists a nearly full subset $X_0 \subseteq X$ such that $C \cap X_0$ is clopen in the relative topology of X_0.

For a more complete development of nearly clopen sets, as well as assurance of their utility in this theory, we refer the reader to [**1**]. A **nearly clopen partition** $\mathcal{P}$ of a Polish probability space (X, τ, μ) is a set of pairwise disjoint nearly clopen sets whose union $\bigcup_{p\in\mathcal{P}} p$ is within a meager null set of X.

LEMMA 2.3. *If $\mathcal{P}$ is a countable nearly clopen partition of (X, τ, μ), and $\mathcal{Q}$ is a countable collection of partition elements of $\mathcal{P}$, then the union of all $q \in \mathcal{Q}$ is nearly clopen.*

PROOF. Let X_0 be the nearly full subset for which $\mathcal{P}$ is a partition consisting of clopen sets. Then $\bigcup_{q\in\mathcal{Q}} q$ is open in X_0. As $\bigcup_{p\in\mathcal{P}\setminus\mathcal{Q}} p$ is open in X_0, its complement $\bigcup_{q\in\mathcal{Q}} q$ is closed in X_0 and hence nearly clopen.

□

DEFINITION 2.4. Let $\phi : X \to Y$ be a transformation of Polish probability spaces (X, τ, μ) and (Y, τ, ν). If there exist nearly full subsets $X_0 \subseteq X$ and $Y_0 \subseteq Y$ such that $\phi : X_0 \to Y_0$ is a homeomorphism, then ϕ is a **near homeomorphism** (or **nearly continuous**) and the spaces are called **nearly homeomorphic**.

Observe that any two Polish probability spaces are homeomorphic. We use the abbreviation n.c. for nearly continuous or nearly continuously.

DEFINITION 2.5. Let (X, τ, μ) be a Polish probability space. We call (X, τ, μ, T) such that T is an ergodic, measure-preserving homeomorphism of X a **nearly continuous dynamical system**.

In measurable dynamics, if two functions agree on all but a null set, then we consider those functions to be equivalent. Naturally then, in nearly continuous dynamics, two functions are considered equivalent if they agree on all but a meager null set. Likewise, nearly clopen sets are equivalent if they differ only by a meager null set. A T-invariant nearly full subset of X is called a **carrier**. The intersection of two carriers is a carrier. In proofs to follow in later chapters, it is often necessary to restrict T to a carrier $X_0 \subseteq X$. The subspace X_0 is then again a Polish space, $T : X_0 \to X_0$ is a homeomorphism, and studying (X_0, τ, μ, T) is equivalent to studying (X, τ, μ, T).

Nearly clopen sets play an important role in the study of induced transformations. Let (X, τ, μ, T) be a nearly continuous dynamical system, and $A \subseteq X$ be nearly clopen. There exists a nearly full set $X' \subset X$ such that $A \cap X'$ is clopen in X'. Let $X_1 = \bigcap_{i=-\infty}^{\infty} T^{-i}(X')$, giving a carrier of T such that $A_1 = A \cap X_1$ is clopen in X_1. The set A_1 is a nearly full subset of A. Then, define the return time function $r_{A_1} : A_1 \to \mathbb{N}$ by

$$r_{A_1}(x) = \inf\{r > 0 : T^r(x) \in A_1\}.$$

The induced map $T_{A_1} : A_1 \to A_1$ is defined as $T_{A_1} = T^{r_{A_1}}$.

LEMMA 2.6. *Let (X, τ, μ, T) be a n.c. dynamical system and let $A \subset X$ be nearly clopen. Then, the return time function to A is nearly continuous and T_A is nearly a homeomorphism on A.*

PROOF. Define X_1 be a carrier of T such that $A_1 = A \cap X_1$ is clopen in X_1. For $r = 1, 2, \ldots$, define

$$B_r = \{x \in A_1 : r_{A_1}(x) = r\}.$$

Each B_r is clopen in X_1 and $r_{A_1}(x)$ is constant on each B_r and thus nearly continuous on A_1. For $x \in B_r$, $r_{A_1}(x) = r$ so that $T_{A_1} = T^r$ on B_r. Thus, the induced map is a function defined piecewise by homeomorphisms on each set in a clopen decomposition of A_1.

□

DEFINITION 2.7. Given a n.c. dynamical system (X, τ, μ, T) and a nearly clopen set $A \subset X$, the **skyscraper** over A is the ordered list of sets

$$A, \quad T(A)\backslash A, \quad T^2(A)\backslash(A \cup T(A)), \quad T^3(A)\backslash(\textstyle\bigcup_{k=0}^{2} T^i(A)), \ldots$$

We refer to these sets as **levels** of the skyscraper, and A is called the **base**.

The levels are commonly visualized as intervals, with A on the bottom, $T(A)\backslash A$ above, etc. The transformation T moves points in a level to the level above. If the image of a point is not in the next level, then the image appears in the base. For an $x \in A$, T moves the point x up the levels of the skyscraper to height $r_A(x)$, then to the point $T_A(x)$ in the base.

LEMMA 2.8. *Let (X, τ, μ, T) be a nearly continuous dynamical system. Given a positive integer N, there exists a nearly clopen set $B \subset X$ such that the skyscraper over B has height at least N.*

PROOF. Select a nearly clopen set $A \subset X$ such that $0 < \mu(A) < \frac{1}{N}$. Decompose A into nearly clopen sets A_i such that $r_A(x) = i$ for all $x \in A_i$. Note that the images of all A_i are nearly clopen. The collection of sets $\{T^j(A_i) : i = 1, 2, \dots,$ and $0 \leq j =\leq i-1\}$ gives a nearly clopen partition of X. Let $B = \bigcup_{i=N}^{\infty} A_i$. By Lemma 2.3, B is a nearly clopen set such that the skyscraper over B has height at least N. □

DEFINITION 2.9. Given a skyscraper over a nearly clopen set A, and a nearly clopen set $B \subset A$ such that $r_A(x) = r$ for all $x \in B$, we call the ordered list of sets

$$B,\, T(B), \dots, T^{r-1}(B)$$

a **column** of the skyscraper, and B is called the base.

As the return time function is nearly continuous, any skyscraper may be decomposed into columns with constant return times for each base such that the levels of the columns are nearly clopen.

DEFINITION 2.10. Given a n.c dynamical system (X, τ, μ, T) and a nearly clopen set A, if there exists an ordered list of sets

$$A, T(A), T^2(A), \dots, T^{n-1}(A)$$

such that the sets are pairwise disjoint, it is called the **nearly clopen tower** of height n over A. The sets are called levels of the tower and A is called the base.

The proof of the existence of Rohklin towers with nearly clopen levels is quite similar to the proof for Rohklin towers with measurable levels, and is thus omitted. However, the analogy of the Alpern multi-tower lemma, necessary for a later theorem, calls for a different proof. To simplify the proof, we first define a tiling, then give the equivalent of Alpern's multi-Tower lemma in the nearly continuous category.

DEFINITION 2.11. Given an interval $I = [m, M-1]$ of $\mathbb{N}$, and a finite set of integers $J \subset \mathbb{N}$, we say that the set of intervals of $\mathbb{N}$ whose lengths come from J **tile** I if I may be written as a disjoint union of intervals whose lengths come from J. We call an ordered list of intervals $[m, m+j_1-1]$, $[m+j_1, m+j_1+j_2-1]$,...,$[M-j_k, M]$ with $j_i \in J$ a **tiling** of I. The sub-intervals of length j_i are called **tiles**.

LEMMA 2.12. *Let (X, τ, μ, T) be a nearly continuous dynamical system. For every list of integers $n_1, \dots, n_k$ with $k \geq 2$ and gcd $\{n_i\} = 1$, and a probability vector $(p_1, \dots, p_k)$, there exists a decomposition of X into k disjoint, nearly clopen towers such that the ith tower has height n_i and mass p_i.*

PROOF. Given a set of numbers $n_1, \dots, n_k$ with $k \geq 2$ such that $gcd\{n_i\} = 1$, there exists a positive integer N such that all $m \geq N$ may be written as a positive linear combination of the n_i. Let $P = \Pi_{i=1}^{k} n_i$.

Select a nearly clopen set A such that the skyscraper over A has height greater than $\frac{N+P}{\min_i\{p_i\}}$. For each $j > \frac{N+P}{\min_i\{p_i\}}$ such that there exists $x \in A$ with $r_A(x) = j$, let $A_j = \{x \in A : r_A(x) = j\}$, giving a nearly clopen decomposition of the skyscraper into columns over the sets A_j with distinct heights.

Write $j = lP + m$ with $N \leq m < N + P$. The integers in the interval $[0, lP+m-1]$ of $\mathbb{N}$ index the levels of the column over A_j by height. Tile $[0, lP-1]$ using l tiles of length P. Tile $[lP, lP+m-1]$ using tiles of length n_1,...,n_k, which is possible as $m \geq N$. If a tile has length n_r, then label the level indexed by the integer at the beginning of the tile with $\tilde{n}_r$. If a tile has length P, then label the level indexed by the integer at the beginning of the tile with $\tilde{P}$. Repeat for each column. For $r = 1, \ldots, k$, define $\tilde{B}_r$ to be the union of levels labeled by $\tilde{n}_r$, and define $\tilde{B}_P$ to be the union of levels labeled by $\tilde{P}$. There are now $k+1$ nearly clopen sets $\tilde{B}_1$,...,$\tilde{B}_k$ and $\tilde{B}_P$ which are bases of disjoint towers of heights $n_1, \ldots, n_k$ and P. As $N \leq m \leq N+P$,

$$\frac{m}{\min_i\{p_i\}} < \frac{N+P}{\min_i\{p_i\}} < j = Pl + m$$

so that $\frac{m}{Pl+m} < \min_i\{p_i\}$ and

$$\mu\left(\bigcup_{i=0}^{n_r-1} T^i\left(\tilde{B}_r\right)\right) < p_r \text{ for each } r = 1, \ldots, k.$$

Let $q_r = p_r - \mu(\bigcup_{i=0}^{n_r-1} T^i(\tilde{B}_r)) < p_r$ for each $r = 1, \ldots, k$. Decompose $\tilde{B}_P$ into nearly clopen sets $C_1, \ldots C_k$ such that $\mu(C_r) = \frac{q_r}{P}$. For $r = 1, \ldots, k$, define

$$B_r = \tilde{B}_r \cup \bigcup_{i=0}^{\frac{P}{n_r}-1} T^{in_r} C_r.$$

Then, the tower over B_r has height n_r and mass p_r.

□

We now turn our attention to the various notions of equivalence between two nearly continuous dynamical systems.

2.1. Nearly Continuous Equivalences.

DEFINITION 2.13. Two systems (X, τ, μ, T) and (Y, τ, ν, S) are said to be **nearly continuously conjugate** if there exist carriers $X_0 \subset X$ and $Y_0 \subset Y$ and a homeomorphism $\phi : X_0 \to Y_0$ such that $\phi \circ T|_{X_0} = S|_{Y_0} \circ \phi$.

DEFINITION 2.14. For a nearly continuous dynamical system (X, τ, μ, T), the **orbit** of a point $x \in X$ is

$$Orb_T(x) = \{T^i(x) : i \in \mathbb{Z}\}.$$

DEFINITION 2.15. Two nearly continuous dynamical systems (X, τ, μ, T) and (Y, τ, ν, S) are **nearly continuously orbit equivalent** if there exist carriers $X_0 \subseteq X$ and $Y_0 \subseteq Y$, a homeomorphism $\phi : X_0 \to Y_0$, and continuous maps $p : X_0 \to \mathbb{Z}$ and $q : Y_0 \to \mathbb{Z}$ such that $\phi \circ T^p = S^q \circ \phi$ and $\phi(Orb_T(x)) = Orb_S(\phi(x))$ for $x \in X_0$.

In the nearly continuous category, del Junco and Şahin proved [**2**] proved the following analogue of Dye's theorem:

THEOREM 2.16. *(J.Ş.) Suppose (X, τ, μ, T) and (Y, τ, ν, S) are nearly continuous dynamical systems. Then, the systems are nearly continuously orbit equivalent.*

The following definition drives the rest of this paper. We note again that this definition differs from the one given in [**1**] and used in [**6**] and [**8**] in that it does not require an orbit equivalence between the systems.

DEFINITION 2.17. Two nearly continuous dynamical systems (X, τ, μ, T) and (Y, τ, ν, S) are **nearly continuously Kakutani equivalent** if there exist nearly clopen sets $A \subseteq X$ and $B \subseteq Y$ with $\mu(A) > 0$ and $\nu(B) > 0$ such that T_A and S_B are nearly continuously conjugate.

If $\mu(A) = \nu(B)$, then the systems are **nearly continuously evenly Kakutani equivalent** .

We have not found an example of two systems which are n.c Kakutani equivalent but are not n.c. evenly Kakutani equivalent. This paper is mostly concerned with nearly continuous even Kakutani equivalence in continuation of the work pursued thus far in the category of nearly continuous dynamical systems. However, we are still concerned with the non-even case as, historically, Kakutani equivalence was not restricted to interest in isomorphisms between induced transformations on sets of the same measure. The even case has been of more interest due to the original definition of nearly continuous even Kakutani equivalence as described by del Junco, Rudolph, and Weiss, which required an orbit equivalence between the full transformations. As we will see, the orbit equivalence is not necessary for the definition.

3. Theorem 1.1

3.1. Overview of the Proof. As $A \subset X$ and $B \subset Y$ are nearly clopen sets, there exist carriers $X_0 \subseteq X$ and $Y_0 \subseteq Y$ of T and S respectively such that $A_0 = A \cap X_0$ and $B_0 = B \cap Y_0$ are clopen in X and Y respectively, and $\phi : A_0 \to B_0$ is a homeomorphism and $\phi \circ T_{A_0} = S_{B_0} \circ \phi$. The system (X, τ, μ, T) is equivalent to (X_0, τ, μ, T), so to simplify notation, and without loss of generality, we let $X = X_0$, $A = A_0$, $Y = Y_0$, and $B = B_0$. Our goal is to extend ϕ to a carrier of T in such a way so as to establish an orbit equivalence, $\hat{\phi}$, while (nearly) preserving the conjugacy between the induced maps. We do so by defining a point map constructed piecewise on a clopen decomposition of X (based on return times to A and B), where $\hat{\phi}$ is a composition of the homeomorphisms T, T_A, ϕ, and S on each subset of the decomposition. To cast light on how the extension is defined, we begin with a visualization of the machinery used to create $\hat{\phi}$.

Construct the nearly clopen skyscraper of T over the set A. The height of the tower over a point $x \in A$ is

$$r_A(x) := \min\{n > 0 : T^n(x) \in A\}.$$

For any $x \in X$, define

$$h(x) := \min\{h \geq 0 : T^{-h}(x) \in A\}.$$

Let $\tilde{x} = T^{-h(x)}(x)$. The fiber of the tower containing the point x is the list of points

$$\left\{ \begin{matrix} T^{r_A(\tilde{x})-1}(\tilde{x}) \\ \vdots \\ x \\ \vdots \\ T(\tilde{x}) \\ \tilde{x} \end{matrix} \right\}.$$

We refer to fibers for points in X as *positive fibers*. To the right of the fiber containing x, list the fiber above $T_A(\tilde{x})$. Continue listing the positive fibers so that to the right of each base point $\tilde{x} \in A$, one sees the fiber above $T_A(\tilde{x})$, and, to the left of $\tilde{x} \in A$, one sees the fiber above $T_A^{-1}(\tilde{x})$. This ordered list of fibers gives what we call the positive frame for the point x.

For each point x in A, consider $\phi(x) \in B$. For each of these points, we extract a fiber of the skyscraper built via S over the set B. Define for $\tilde{y} \in B$,

$$r_B(\tilde{y}) := \min\{n > 0 : S^n(\tilde{y}) \in B\}.$$

Simply arrange the fibers from the skyscraper over B in the same manner as for the fibers above points in A, with the fiber for $S_B(\tilde{y})$ listed immediately to the right of the fiber for $\tilde{y} \in B$, and the fiber for $S_B^{-1}(\tilde{y})$ listed immediately to the left of the fiber for $\tilde{y} \in B$. Now, "flip" the fibers over the points in B so the point at the top of each list is in B. We refer to fibers for points in Y as negative fibers.

For the last step in this visualization, line up the base of the positive frame (points in A) across from the base of the negative frame (points in B) via the conjugacy. One traverses the diagram in the following manner: applying the induced map T_A to the base of the positive frame shifts the points to the right along the base, and applying the induced map S_B to the top of the lower frame shifts the points to the right across the negative frame. Apply T to move away from the base in the positive fibers and apply S to move away from the base in the negative fibers. Cross between the positive frame and the negative frame by applying ϕ or ϕ^{-1}. In Fig. 1, dots above or below represent points in A^C and B^C.

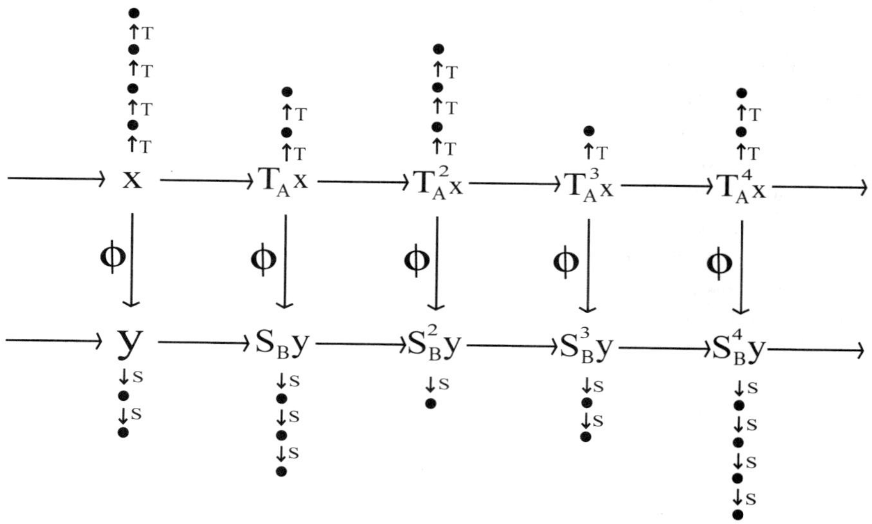

FIGURE 1. A segment of the full frame used for defining the extension of the conjugacy.

We create an orbit equivalence by developing a method for pairing up points from the positive fibers with points in the negative fibers. This method pairs up items in the positive fibers with items in the negative fibers directly across until either the positive fiber is neutral (to all points in the positive fiber, a map has been assigned) or the negative fiber is neutral (each point of the negative fiber is the image of a point in X). Then, it merely shifts the upper frame one to the right and pairs up as much of the remainder of the positive fibers with the remainder of the negative fibers, shifts again and pairs again, etc. Here, we merely "shift" the positive frame across the negative frame by utilizing the induced map on the base of the frames. We remark that this method is similar to the filling-scheme used by Chacon and Ornstein in [**4**].

A point $x \in A^C$ maps to a point $y \in B^C$ if, after a shift, the fiber across from x still has unused space to receive the point. A point in a positive fiber cannot be mapped to a negative fiber if, after a shift, the fiber across from it is already neutral.

3.2. Proof of Theorem 1.1. Define

$$\theta_k^n(x) := \sum_{i=k}^{n} r_A(T_A^i \circ T^{-h(x)}(x))$$

the sum of the lengths of the kth fiber through the nth fiber to the right of the positive fiber containing x, and

$$\psi_k^n(x) := \sum_{i=k}^{n} r_B(S_B^i \circ \phi \circ T^{-h(x)}(x)),$$

the sum of the lengths of the kth fiber through the nth fiber to the right of the negative fiber below the positive fiber containing x.

Let $n(x) = 0$ if $h(x) \leq r_B(\phi \circ T^{-h(x)}(x))$. Otherwise, let

$$n(x) = \min\{n > 0 : h(x) + \theta_1^n(x) \leq \psi_0^n(x)\}.$$

In essence, this is the smallest number of shifts required for there to be a point available in a negative fiber for the point x in a positive fiber. Let

$$d(x) := h(x) + \theta_1^{n(x)}(x) - \psi_0^{n(x)-1}(x).$$

Define $\hat{\phi}$ as:

$$\hat{\phi}(x) := \begin{cases} \phi(x) & \text{for} \quad x \in A \\ S^{d(x)} \circ \phi \circ T_A^{n(x)} \circ T^{-h(x)}(x) & \text{for} \quad x \notin A \end{cases}$$

In a similar manner, we define $\hat{\phi}^{-1}$. Define first

$$D(y) := \min\{D > 0 : S^{-D}(y) \in B\}.$$

$$\text{Let } \gamma_k^n(y) := \sum_{i=k}^{n} r_B(S_B^{-i} \circ S^{-D(y)}(y)),$$

the sum of the lengths of the kth through the nth fibers to the left of the negative fiber holding y. Let

$$\Psi_k^n(y) := \sum_{i=k}^{n} r_A(T_A^{-i} \circ \phi^{-1} \circ S^{-D(y)}(y)),$$

the sum of the lengths of the kth through the nth positive fibers to the left of the positive fiber across from the negative fiber containing y. The inverse map pairs a point in a negative fiber with a point in a positive fiber by sliding the lower frame backwards along the upper frame, in reverse fashion to what we described before.

The transformation is defined at the point y when, after a certain number of shifts, the positive fiber across from y has available space. If $D(y) \leq r_A(\phi^{-1} \circ S^{-D(y)}(y))$, then $m(y) = 0$. Otherwise,

$$\begin{aligned} m(y) &:= \min\{m > 0 : D(y) + \gamma_1^m(y) \leq \Psi_0^m(y)\}, \\ H(y) &:= D(y) + \gamma_1^m(y) - \Psi_0^m(y), \text{ and} \\ \hat{\phi}^{-1}(y) &:= \begin{cases} \phi^{-1}(y) & \text{for} \quad y \in B \\ T^{H(y)} \circ \phi^{-1} \circ S_B^{-m(y)} \circ S^{-D(y)}(y) & \text{for} \quad y \notin B \end{cases} \end{aligned}$$

LEMMA 3.1. *$\hat{\phi}^{-1}$ is the inverse of $\hat{\phi}$.*

The diagram of the $\hat{\phi}^{-1}$ construction is just the diagram of the $\hat{\phi}$ construction rotated 180°. We pair points starting at the base of the frames, and then work our way outward, shifting when necessary. Naturally, $D(\hat{\phi}(x)) = d(x)$, and the same number of shifts are required. Then, it is easily follows that $H(\hat{\phi}(x)) = h(x)$.

The maps $\hat{\phi}$ and $\hat{\phi}^{-1}$ need to be well-defined on nearly full sets. In order for $\hat{\phi}$ to be well-defined at a point, the values $h(x)$, $n(x)$, and $d(x)$ must exist and be finite. We may assume that $h(x)$ (and $D(y)$) are finite. Otherwise, we remove the entire orbit of x from X (or y from Y). To see that $n(x)$ (and $m(y)$) are finite on a subset of full measure, we begin with the following standard result from ergodic theory:

LEMMA 3.2. *Suppose $T : X \to X$ is an ergodic measure-preserving transformation, and $f : X \to \mathbb{R}$ is an integrable function such that $\int_X f(x) = 0$. Then, for almost every $x \in X$, there exists $n > 0$ such that $f(x) + f(T(x)) + f(T^2(x)) + \cdots + f(T^{n-1}(x)) \leq 0$.*

LEMMA 3.3. *The value $n(x)$ is finite on a set of full measure.*

PROOF. Let $X = A$, identified with B, and $T = T_A \cong S_B$. And, let $f(x) = r_A(x) - r_B(\phi(x))$. As $\mu(A) = \nu(B)$, $\int f(x) = 0$. Thus by Lemma 3.2, $n(x)$ is finite on a set of full measure. Fix n and let A_n be the set of points such that $\sum_{i=0}^{n-1} f(T^i(x)) \leq 0$ for the first time. As $f(x)$ takes only values in $\mathbb{Z}$, the set of points so that $\sum_{i=0}^{n-1} f(T^i(x)) \leq 0$ is nearly clopen. The union $\bigcup_n^\infty A_n$ gives a G_δ subset.

□

The proof that $\hat{\phi}^{-1}$ is defined on a set of full measure is essentially identical. We now define the sets $\hat{A}$ and $\hat{B}$ upon which $\hat{\phi}$ and $\hat{\phi}^{-1}$ are well-defined. First, let $A_{n,h,d}$ be the set of points for which $\hat{\phi} = S^d \circ \phi \circ T_A^n \circ T^{-h}$ for fixed n, d, and h. $A_{n,h,d}$ may tediously be written as a finite union of finite intersections of images of A and B via the homeomorphisms S, ϕ, and T. Thus $A_{n,h,d}$ is clopen in X. Let $B_{m,D,H} = \{y \in Y : \hat{\phi}^{-1}(y) = T^H \circ \phi^{-1} \circ S_B^{-m} \circ S^{-D}(y)\} = \hat{\phi}(A_{m,H,D})$. Define

$$\hat{A} := \bigcup_{n=0}^{\infty} \bigcup_{h=0}^{\infty} \bigcup_{d=0}^{\infty} A_{n,h,d}\,, \text{ and}$$

$$\hat{B} := \bigcup_{m=0}^{\infty} \bigcup_{D=0}^{\infty} \bigcup_{H=0}^{\infty} B_{m,D,H}\,.$$

LEMMA 3.4. *The sets $\hat{A}$ and $\hat{B}$ are nearly full sets.*

PROOF. Both $\hat{A}$ and $\hat{B}$ are open in X as they are the countable union of clopen sets in X, and of full measure by the previous lemma. □

LEMMA 3.5. *The map $\hat{\phi}$ is continuous on $\hat{A}$ and $\hat{\phi}^{-1}$ is continuous on $\hat{B}$.*

PROOF. Decompose $\hat{A}$ into the countable union of clopen sets $A_{n,h,d}$. On each of these clopen sets, $\hat{\phi}$ is the composition $S^d \circ \phi \circ T_A^n \circ T^{-h}$ of functions continuous in the relative topology of X. Decomposing $\hat{B}$ in a similar manner shows that $\hat{\phi}^{-1}$ is also continuous in the relative topology of $\hat{B}$. □

LEMMA 3.6. *$\hat{\phi}$ is measure preserving on $\hat{A}$ and $\hat{\phi}^{-1}$ is measure preserving on $\hat{B}$.*

PROOF. Identical to the proof of Lemma 3.5. □

We define the final carriers. Let

$$X^* := \bigcap_{i=-\infty}^{\infty} T^{-i}(\hat{A}) \cap \hat{\phi}^{-1}\left(\bigcap_{i=-\infty}^{\infty} S^{-i}(\hat{B})\right)$$

and

$$Y^* := \hat{\phi}(X^*).$$

LEMMA 3.7. *The sets X^* and Y^* on which $\hat{\phi}$ and $\hat{\phi}^{-1}$ are defined are G_δ sets of full measure.*

PROOF. Note that $\hat{A}$ and $\hat{B}$ are both G_δ sets of measure 1 relative to X and Y.

□

In conclusion of the proof of Theorem 1.1, we have the existence of carriers $X^* \subseteq X$, $Y^* \subseteq Y$, and a nearly continuous orbit equivalence $\hat{\phi} : X^* \to Y^*$.

4. Nearly Unique Ergodicity and Non-even n.c. Kakutani Equivalence

In this section, we begin by defining nearly unique ergodicity, a concept introduced by Denker and Keane in [**5**] as strict ergodicity and later re-named nearly unique ergodicity by del Junco, Rudolph, and Weiss in [**1**]. The authors used the concept of nearly unique ergodicity to demonstrate that n.c. Kakutani equivalence is strictly stronger than measured Kakutani equivalence (section 6 of [**1**] contains the example). While we include the necessary facts, for more detail and discussion, see [**1**] and [**5**].

4.1. Nearly Unique Ergodicity.

DEFINITION 4.1. A sequence of functions $\{f_n\}_{n\in\mathbb{N}}$ on a Polish probability space (X, τ, μ) is said to converge **nearly uniformly** if there exists a nearly full subset $X_0 \subseteq X$ on which the sequence converges uniformly.

DEFINITION 4.2. Suppose (X, τ, μ, T) is a nearly continuous dynamical system and suppose $f \in L_1(\mu)$. If the ergodic averages of f

$$A_n(f) = \frac{1}{n}\sum_{i=0}^{n-1} f(T^i)$$

converge nearly uniformly to $\int f d\mu$ for all nearly bounded and nearly continuous functions f, we say T is **nearly uniquely ergodic**.

We point out that our definition differs slightly from that of [**1**] in that we only require f to be nearly continuous and nearly bounded. Note that one could always restrict the dynamics to a carrier $X_0 \subseteq X$ so that f is continuous and bounded in X_0, and that studying the dynamics of (X_0, τ, μ, T) is equivalent to studying (X, τ, μ, T).

THEOREM 4.3. *If (X, μ, T) is a uniquely ergodic system on a compact metric space, then it is nearly uniquely ergodic.*

PROOF. Let f be a continuous function on X. The ergodic averages of f converge uniformly, hence nearly uniformly. Note that if the ergodic averages of f converge nearly uniformly and f' is nearly equal to f, then the ergodic averages of f' converge nearly uniformly. It now follows from the proof of Theorem 5.2 in [**1**] that for any nearly clopen set $A \subset X$, the ergodic averages of 1_A converge nearly uniformly to $\mu(A)$. By Lemma 5.3 of [**1**], T is nearly uniquely ergodic.

□

Hence, well known systems such as irrational rotations of the unit circle and adding machines are nearly uniquely ergodic. If a system (X, τ, μ, T) is nearly continuously conjugate to a uniquely ergodic system, then it is nearly uniquely ergodic. The converse is true as well.

4.2. The Proof of Theorem 1.2. In the measure theoretic category, it is not difficult to show that if T and S are two ergodic measure preserving transformations of finite measure spaces (X, μ) and (Y, ν) respectively, and if $A \subset X$ and $B \subset Y$ are measurable sets with $\mu(A) > \nu(B) > 0$, and with T_A and S_B measurably conjugate, then one can find a measurable set $B' \subset Y$ such that T is measurably conjugate to $S_{B'}$. One may easily find an example that the analogous result does not hold in the nearly continuous category without the added restriction of nearly unique ergodicity.

PROOF. Let X_0 and Y_0 be carriers for T and S, respectively, such that $A_0 = A \cap X_0$ and $B_0 = B \cap Y_0$ are clopen in their relative topologies, and let $\phi : A_0 \to B_0$ be the homeomorphism giving the conjugacy between T_{A_0} and S_{B_0}. As before, we assume that $X_0 = X$, $A_0 = A$, $Y_0 = Y$, and $B_0 = B$. The idea for the proof is to select $A' \subseteq A$ such that the return times for points in A' are smaller than the return times for points in $\phi(A')$.

Let $\epsilon > 0$ be given such that

$$\epsilon < \frac{1}{4}\left(\frac{1}{\nu(B)} - \frac{1}{\mu(A)}\right). \tag{4.1}$$

By Theorem 5.4 of [**1**], as (X, τ, μ, T) is nearly uniquely ergodic and A is a nearly clopen set of positive measure, T_A is nearly uniquely ergodic. We use the nearly unique ergodicity of T_A to find a set with "nice" return times. For $x \in A$, let

$$r_A(x) := \min\{r > 0 : T^r(x) \in A\}.$$

Note that $r_A(x)$ is bounded. There exists an $N_1 = N_1(\epsilon)$ such that

$$\left|\frac{1}{n}\sum_{i=1}^{n} r_A\left(T_A^i\right)(x) - \int_A r_A d\mu_A\right| < \epsilon \tag{4.2}$$

for all $n > N_1$ and nearly all x.

As T_A is n.c. conjugate to S_B, S_B is nearly uniquely ergodic on B. For $y \in B$, let

$$r_B(y) := \min\{r > 0 : S^r(y) \in B\}.$$

Note that the ergodic averages for $r_B(y)$ converge to $\int_B r_B(y)d\nu_B = \frac{1}{\nu(B)}$, but may not converge nearly uniformly as $r_B(y)$ may not be bounded. For an $M \in \mathbb{N}$, define $f : B \to \mathbb{N}$ by $f = \min\{r_B(y), M\}$. Note that f is a bounded and nearly continuous function. Select M so that

$$\int_B f d\nu_B > \frac{1}{\nu(B)} - \frac{\epsilon}{2}. \tag{4.3}$$

Note that

$$\frac{1}{n}\sum_{k=0}^{n-1} f(S_B^k(y)) \leq \frac{1}{n}\sum_{k=0}^{n-1} r_B(S_B^k(y)). \tag{4.4}$$

There exists $N_2(\epsilon)$ s.t.

$$\left|\frac{1}{n}\sum_{k=0}^{n-1} f(S_B^k(y)) - \int_B f(y)d\nu_B\right| < \frac{\epsilon}{2} \tag{4.5}$$

for all $n > N_2$ and nearly all $y \in B$. Then

$$\begin{aligned}
\int_B f(y)d\nu_B - \frac{\epsilon}{2} &< \frac{1}{n}\sum_{k=0}^{n-1} f(S_B^k(y)) && \text{by (4.3)}\\
\Rightarrow \frac{1}{\nu(B)} - \frac{\epsilon}{2} - \frac{\epsilon}{2} &< \frac{1}{n}\sum_{k=0}^{n-1} f(S_B^k(y)) && \text{by (4.5)}\\
\Rightarrow \frac{1}{\nu(B)} - \epsilon &< \frac{1}{n}\sum_{k=0}^{n-1} f(S_B^k(y)) \\
\Rightarrow \frac{1}{\nu(B)} - \epsilon &< \frac{1}{n}\sum_{k=0}^{n-1} r_B(S_B^k(y)). && \text{by (4.4)} \qquad (4.6)
\end{aligned}$$

Let $N = \max\{N_1, N_2, \frac{\mu(A)}{\mu(A)-\nu(B)}\}$. By Lemma 2.12, we may select a nearly clopen set $A' \subset A$ such that the multi-tower built over A' using T_A has height $2N$ on a set of positive measure and height $2N+1$ on a set of positive measure. Let $B' = \phi(A')$. Define

$$\begin{aligned}
r_{A'}(x) &:= \min\{r > 0 : (T_A)^r(x) \in A' \text{ for } x \in A'\}\\
r_{B'}(y) &:= \min\{r > 0 : (S_B)^r(y) \in B' \text{ for } y \in B'\}.
\end{aligned}$$

The return time $r_{A'}$ to A' via T_A is either $2N$ or $2N+1$.

$$\sum_{k=0}^{2N-1} r_A(T_A^k(x)) \leq r_{A'}(x) \leq \sum_{k=0}^{2N} r_A(T_A^k(x)). \tag{4.7}$$

As T_A and S_B are conjugate, B' also has $2N$ and $2N+1$ as the only return times via S_B, and

$$\sum_{k=0}^{2N-1} r_B(S_B^k(y)) \leq r_{B'}(y) \leq \sum_{k=0}^{2N} r_B(S_B^k(y)). \tag{4.8}$$

It only remains to check that the tower over B' via S is taller than the tower over A' via T.

By (4.2) and (4.7),

$$\begin{aligned} \frac{1}{2N+1}\sum_{i=0}^{2N} r_A(T_A^i(x)) &< \frac{1}{\mu(A)} + \epsilon \\ \Rightarrow r_{A'} &< (2N+1)\left(\frac{1}{\mu(A)} + \epsilon\right). \end{aligned} \tag{4.9}$$

By (4.6) and (4.8),

$$\begin{aligned} \frac{1}{2N}\sum_{i=0}^{2N} r_B(S_B^i(y)) &\geq \frac{1}{\nu(B)} - \epsilon \\ \Rightarrow r_{B'} &> 2N\left(\frac{1}{\nu(B)} - \epsilon\right). \end{aligned} \tag{4.10}$$

Combining (4.9) and (4.10), we have

$$\begin{aligned} (2N+1)\left(\frac{1}{\mu(A)} + \epsilon\right) &< 2N\left(\frac{1}{\nu(B)} - \epsilon\right) \\ \Rightarrow (2N+1)\left(\frac{3}{4}\frac{1}{\mu(A)} + \frac{1}{4}\frac{1}{\nu(B)}\right) &< 2N\left(\frac{3}{4}\frac{1}{\nu(B)} + \frac{1}{4}\frac{1}{\mu(A)}\right) \qquad \text{by (4.1)} \\ \Rightarrow \frac{3}{4}\frac{1}{\mu(A)} + \frac{1}{4}\frac{1}{\nu(B)} &< N\left(\frac{1}{\nu(B)} - \frac{1}{\mu(A)}\right) \end{aligned}$$

which holds as $N > \frac{\mu(A)}{\mu(A)-\nu(B)}$.

Let

$$h(x) := \min\{h \geq 0 : T^{-h}(x) \in A'\}.$$

Define

$$\hat{\phi}(x) := S^{h(x)} \circ \phi \circ T^{-h(x)}(x) \text{ for } x \in X.$$

Let

$$\bar{B} := \hat{\phi}(X).$$

It is easy to see that $\bar{B}$ is nearly clopen and that $S_{\bar{B}}$ is nearly continuously conjugate to T. $\square$

We give interesting applications of Theorem 1.2, some of which come into play in the following section.

Recall from [**8**] and [**6**] that all odometers and all irrational rotations of the circle are nearly continuously evenly Kakutani equivalent to the binary odometer.

LEMMA 4.4. *Let (Y, τ, ν, S) be any irrational rotation of the unit circle or adding machine. Given any value α, $0 < \alpha < 1$, there exists a nearly clopen set $D \subset Y$ with $\nu(D) = \alpha$ such that S_D is nearly continuously conjugate to either an irrational rotation of the unit circle (α irrational) or an adding machine (α rational).*

PROOF. First, we must find a nearly continuous dynamical system (X, τ, μ, T) and a nearly clopen set $A \subset X$ with $\mu(A) = \alpha$ such that T_A is either an irrational rotation (α irrational) or an adding machine (α rational).

For α irrational, let (X, τ, μ, T) be the irrational rotation of the unit circle by α. Let A be the interval $[0, \alpha]$, and induce on this set. One sees that this induced transformation may be represented by the interval exchange where $[0, 1 - n\alpha]$ maps to $[(n+1)\alpha - 1, \alpha]$ and $[1 - n\alpha, \alpha]$ maps to $[0, (n+1)\alpha - 1]$ where $n = \lfloor \frac{1}{\alpha} \rfloor$, the fractional part of $\frac{1}{\alpha}$. Thus, T_A is an irrational rotation and $\mu(A) = \alpha$

For α rational, write α as $\frac{p}{q}$ where $p, q \in \mathbb{N}$. Let (X, τ, μ, T) be the adding machine produced by cutting the stage i stack/tower into q columns of equal width before stacking. Let the set A be the first p levels of the first tower of height q. Note that $\mu(A) = \alpha$. The induced map T_A on A is an adding machine produced by cutting an interval into p sub-intervals of equal width to create the first stack, then cutting the subsequent stacks into q columns of equal width.

As S and T are n.c. evenly Kakutani equivalent, there exist nearly clopen subsets $B \subset X$ and $C \subset Y$ and a n.c. map $\phi : B \to C$ such that T_B and S_C are n.c. conjugate. There exists k such that $\mu(T^k(B) \cap A) \neq 0$. Let $A' = T^k(B) \cap A$ and $C' = \phi(T^{-k}(A'))$. The induced maps $T_{A'}$ and $S_{C'}$ are n.c. conjugate. As $\mu_A(A') > \nu(C')$, and T_A is nearly uniquely ergodic, there exists $D \subset Y$, nearly clopen, such that T_A is nearly continuously conjugate to S_D. Then $\nu(D) = \frac{\nu(C')}{\mu_A(A')} = \frac{\nu(C')}{\mu(A')}\mu(A) = \mu(A)$. Thus, given an α, there exists a nearly clopen set D such that S_D is n.c. evenly Kakutani equivalent to an irrational rotation.

□

In the measure-theoretic category, given a dynamical system (X, μ, T), and any measurable subset $A \subset X$, T_A is evenly Kakutani equivalent to T. For the n.c. even Kakutani equivalence class of transformations containing irrational rotations of the unit circle, we obtain a similar result.

COROLLARY 4.2.1. Let (X, τ, μ, T) be any irrational rotation of the unit circle or adding machine. If $A \subset X$ is any nearly clopen set with $\mu(A) > 0$, then T_A is n.c. evenly Kakutani equivalent to T.

PROOF. Let $\mu(A) = \alpha$. By Lemma 4.4, there exists a nearly clopen set $B \subset X$ with $\mu(B) = \alpha$ such that T_B is n.c. conjugate to an irrational rotation of the circle or an adding machine. For some $k \in \mathbb{Z}$, $\mu(T^k(A) \cap B) \neq 0$. Let $B' = T^k(A) \cap B$ and $A' = A \cap T^{-k}(B)$. The induced maps $T_{B'}$ and $T_{A'}$ are n.c. conjugate via T^k and $\mu_A(A') = \mu_B(B')$. Thus T_B and T_A are n.c. evenly Kakutani equivalent □

5. Nearly Continuously Rank-One

The cutting and stacking of an interval to define a transformation is a well known process for creating examples of dynamical systems. For notation's sake, we describe the process. By a stack of intervals, we mean a collection of intervals of equal width which are placed one above another. To build a simple cutting and stacking construction using intervals of equal width at each stage, we need only to specify the number of columns into which to cut the stack during each stage and the number and placement of spacers (extra intervals) used. At stage i, we have a stack $\mathcal{S}_i$ of subintervals of equal width from $[0, 1]$, and one left over subinterval from which to cut spacers. The transformation defined at stage i, S_i, translates each level to the level above, and the domain includes all but the topmost level and leftover interval . Let c_i be the number of columns into which to cut the stack. If B_i represents the base of $\mathcal{S}_i$, and n_i represents the height of the stack, cut B_i into c_i intervals of equal width, forming $B_{i,1}, \ldots, B_{i,c_i}$, indexed from left to right. Cutting B_i thus cuts $\mathcal{S}_i$ into c_i columns if height n_i. For fixed i and j, the corresponding column is $\{B_{i,j}, S_i B_{i,j}, S_i^2 B_{i,j}, \ldots, S_i^{n_i-1} B_{i,j}\}$. Next, let $s_{i,j}^+$ be the number of spacers, subintervals cut from the leftover interval, to place above the jth column $1 \leq j \leq c_i$ during stage i, and let s_i^- denote the number of spacers to be placed below the first column. To move from stage i to $i+1$, cut the stack of intervals into c_i columns of equal width, w_{i+1}, place $s_{i,j}^+$ spacers of width w_{i+1} above the jth column and s_i^- below the first column, and re-stack by placing column $i+1$ (reading left to right) on top of column i for $1 \leq i < c_i$. The space Y is given by the collection of all intervals used–$[0,1]$. The transformation, S, is the limit of the transformations S_i. The collection of all intervals used at each stage generates the topology.

Adding machines and Chacon's map are standard examples of transformations built by such a process. Roychowdhury and Rudolph [**8**] proved that all adding machines are evenly nearly continuously Kakutani equivalent. Dykstra and Rudolph [**6**] proved that irrational rotations of the circle belong to the same equivalence class. Using these works as stepping stones, we define a class of transformations consisting of systems of a similar nature to Chacon's map, which we call nearly continuously rank-one, then prove they belong to the same equivalence class as irrational rotations.

DEFINITION 5.1. A system is **nearly continuously rank-one** if there exists a refining sequence of nearly clopen towers $\mathcal{T}_i$ whose heights increase to infinity, and whose levels generate the topology of a nearly full subset.

Before showing that nearly continuously rank-one transformations may be built from a cutting and stacking process, we include an essential theorem from Zhuravlev's thesis [**9**].

LEMMA 5.2. *Let (X, τ, μ) and (Y, τ, ν) be two Polish probability spaces, and let $\mathcal{A}$ and $\mathcal{B}$ denote Boolean algebras of equivalence classes of nearly clopen sets for X and Y, respectively. If $\Phi : \mathcal{A} \to \mathcal{B}$ is a measure preserving isomorphism, then Φ is induced by a nearly continuous point map, i.e. there exist nearly full subsets $X_0 \subseteq X$ and $Y_0 \subseteq Y$ and a measure preserving map $\phi : X_0 \to Y_0$ and $\phi(A) = \Phi(A)$ for each $A \in \mathcal{A}$.*

THEOREM 5.3. *Two nearly continuous dynamical systems (X, τ, μ, T) and (Y, τ, ν, S) are nearly continuously conjugate iff there exists a measure preserving isomorphism Φ of Boolean algebras of equivalence classes of nearly clopen sets $\mathcal{A}$ for X and $\mathcal{B}$ for Y such that $\Phi(T(A)) = S(\Phi(A))$ for all $A \in \mathcal{A}$.*

PROOF. The forward direction is obvious. For the reverse direction, by Theorem 5.2, Φ is induced by a nearly continuous point map ϕ such that for any $A \in \mathcal{A}$, $\phi(A) = \Phi(A)$. It only remains to show that $\phi(T(x)) = S(\phi(x))$ on a nearly full set. Let X_0 be the nearly full set on which ϕ is defined, and let $x \in X_0$. Then, there exists a nested sequence of nearly clopen sets $A_1 \supset A_2 \supset A_3 \supset \cdots$ such that

$x \in A_i$ for all $i \in \mathbb{N}$, so $x = \bigcap_{i=1}^{\infty} A_i$. Note also that $T(A_i)$ is a nearly clopen set in X_0 so that $\phi(T(A_i)) = \Phi(T(A_i))$. Then

$$\phi(T(x)) = \phi(T(\bigcap_{i=1}^{\infty} A_i)) = \bigcap_{i=1}^{\infty} \phi(T(A_i)) = \bigcap_{i=1}^{\infty} \Phi(T(A_i)).$$

For each A_i, $\Phi(T(A_i)) = S(\Phi(A_i))$, meaning

$$\phi(T(x)) = \bigcap_{i=1}^{\infty} S(\Phi(A_i)) = \bigcap_{i=1}^{\infty} S(\phi(A_i)) = S(\phi(\bigcap_{i=1}^{\infty} A_i)) = S(\phi(x))$$

□

LEMMA 5.4. *A nearly continuous dynamical system (X, τ, μ, T) is nearly continuously rank-one if and only if it is nearly continuously conjugate to a cutting and stacking of intervals consisting of one stack at each stage.*

PROOF. The backward direction is obvious. For the forward direction, suppose (X, τ, μ, T) is nearly continuously rank-one. Consider a refining sequence $\{\mathcal{T}_i\}_{i=1}^{\infty}$ of nearly clopen towers. Denote by B_i the base of tower $\mathcal{T}_i$, let n_i be its height, and $\mathcal{P}_i^j$ for $1 \leq j \leq n_i$ be the levels of the tower as well as the name of the partition element given by the level. At stage i, $\mathcal{P}_i^{n_i+1} = X \setminus \bigcup_{j=1}^{n_i} \mathcal{P}_i^j$ represents the residual set and final partition element with $\mu(\mathcal{P}_i^{n_i+1}) = \epsilon_i$. As we move from $\mathcal{T}_i$ to $\mathcal{T}_{i+1}$, each $\mathcal{P}_i^j = \bigcup_{k \in I_i^j} \mathcal{P}_{i+1}^k$ for some collection of indices $I_i^j \subset \mathbb{N}$ and $1 \leq j \leq n_i$. Let $m_i = \mu(B_i)$.

To define a cutting and stacking of intervals producing a transformation which is n.c. conjugate to (X, τ, μ, T), observe the transition from tower $\mathcal{T}_i$ to $\mathcal{T}_{i+1}$. Let $c_i = |I_i^1| = \frac{\mu(B_i)}{\mu(B_{i+1})}$. Starting from the base of $\mathcal{T}_{i+1}$, read off the names of levels according to the partition $\mathcal{P}_i$ given by $\mathcal{T}_i$. We see complete names $(\mathcal{P}_i^1, \mathcal{P}_i^2, \dots, \mathcal{P}_i^{n_i})$ of the tower $\mathcal{T}_i$ possibly with the name of the residual set $\mathcal{P}_i^{n_i+1}$ inserted between the complete tower names. Let $s_{i,j}^+$ be the number of $\mathcal{P}_i^{n_i+1}$ which occur after the jth appearance of the complete name of $\mathcal{T}_i$ reading up the tower $\mathcal{T}_{i+1}$, and let s_i^- be the number of $\mathcal{P}_i^{n_i+1}$ which occur before the first full tower name.

Begin with n_1 intervals of width $w_1 = m_1$, and, for each i and $1 \leq j \leq c_i$, with $s_{i,j}^+ + s_i^-$ intervals of width $w_i = \frac{w_1}{\Pi_{k=1}^i c_k}$. Follow the procedure to build the cutting and stacking construction. Let S be the transformation which arises in the limit from $S_i(\mathcal{P}_i^j) = \mathcal{P}_i^{j+1}$ for $1 \leq j < n_i$.

Let Φ be the map which arises naturally in the identification of levels of the towers $\mathcal{T}_i$ with the sub-intervals of $[0, 1]$ which form the levels of the stacks $\mathcal{S}_i$. This is obviously a measure preserving isomorphism between the Boolean algebra of nearly clopen sets generated by the levels from the towers for (X, τ, μ, T) and the Boolean algebra of intervals generated by the levels of the cutting and stacking for (Y, τ, λ, S). By Theorem 5.3, Φ is induced by a nearly continuous point map ϕ and ϕ is a nearly continuous conjugacy between T and S. □

THEOREM 5.5. *All nearly continuously rank-one transformations are nearly continuously evenly Kakutani equivalent to any irrational rotation of the unit circle.*

PROOF. Let (X, τ, μ, T) be a nearly continuously rank-one transformation. By the previous theorem, (X, τ, μ, T) is n.c. conjugate to a cutting and stacking.

At stage i, suppose we cut the stack into c_i columns of equal width and place $s_{i,j}$ spacers above column j for $j = 1, \dots, c_i$. In moving to stage $i+1$ we then stack the columns and spacers in a specified order.

Let I be the union of the levels of the very first tower, before cutting and stacking and adding any spacers, i.e., I will be cut into c_1 sets of equal size. I is nearly clopen, and inducing T_I on I ignores the addition of all spacers. T_I is thus n.c. conjugate to an adding machine.

Suppose $\mu(I) = \alpha$. Let (Y, τ, ν, S) be any adding machine. Whether α is irrational or rational, by Lemma 4.4, there exists a nearly clopen set $J \subset Y$ with $\nu(J) = \alpha$ such that S_J is n.c. conjugate to an adding machine or an irrational rotation of the circle. As T_I is n.c. evenly Kakutani equivalent to S_J and $\nu(J) = \mu(I)$, T is n.c. evenly Kakutani equivalent to S.

□

6. Conclusion

It is known that irrational rotations of the unit circle are rank-one in the measure theoretic sense. The standard proof that irrational rotations are measure-theoretic rank-one does not extend into the nearly continuous category. It is also known that irrational rotations of the unit circle are not interval rank-one; there does not exist a refining sequence of towers whose levels are subintervals of the unit circle. This leads to an interesting question: are irrational rotations nearly continuously rank-one? Can one find a refining sequence of towers whose levels are nearly clopen for irrational rotations of the unit circle?

In the measure-theoretic category, if A is a measurable subset, then T_A is evenly Kakutani equivalent to T. In the nearly continuous category, for the n.c. even Kakutani equivalence class of transformations containing irrational rotations, we showed that if A is a nearly clopen set of positive measure, then T_A is n.c evenly Kakutani equivalent to T. The proof relied upon the nearly unique ergodicity of irrational rotations of the unit circle and adding machines. For any n.c. dynamical system, (X, τ, μ, T), and nearly clopen subset $A \subset X$, is it possible that T_A is n.c. evenly Kakutani equivalent to T?

References

[1] Andres del Junco, Daniel J. Rudolph, and Benjamin Weiss, *Measured topological orbit and Kakutani equivalence*, Discrete Contin. Dyn. Syst. Ser. S **2** (2009), no. 2, 221–238, DOI 10.3934/dcdss.2009.2.221. MR2505635

[2] Andrés del Junco and Ayşe Şahin, *Dye's theorem in the almost continuous category*, Israel J. Math. **173** (2009), 235–251, DOI 10.1007/s11856-009-0090-1. MR2570667

[3] Andrés del Junco and Daniel J. Rudolph, *Kakutani equivalence of ergodic* $\mathbf{Z}^n$ *actions*, Ergodic Theory Dynam. Systems **4** (1984), no. 1, 89–104. MR758896

[4] R. V. Chacon and D. S. Ornstein, *A general ergodic theorem*, Illinois J. Math. **4** (1960), 153–160. MR0110954

[5] Manfred Denker and Michael Keane, *Almost topological dynamical systems*, Israel J. Math. **34** (1979), no. 1-2, 139–160 (1980), DOI 10.1007/BF02761830. MR571401

[6] Andrew Dykstra and Daniel J. Rudolph, *Any two irrational rotations are nearly continuously Kakutani equivalent*, J. Anal. Math. **110** (2010), 339–384, DOI 10.1007/s11854-010-0009-0. MR2753297

[7] Toshihiro Hamachi and Michael S. Keane, *Finitary orbit equivalence of odometers*, Bull. London Math. Soc. **38** (2006), no. 3, 450–458, DOI 10.1112/S0024609306018455. MR2239040

[8] Mrinal Kanti Roychowdhury and Daniel J. Rudolph, *Nearly continuous Kakutani equivalence of adding machines*, J. Mod. Dyn. **3** (2009), no. 1, 103–119, DOI 10.3934/jmd.2009.3.103. MR2481334

[9] Vladimir Zhuravlev, *Two theorems of Dye in the almost continuous category*, ProQuest LLC, Ann Arbor, MI, 2009. Thesis (Ph.D.)–University of Toronto (Canada). MR2753148

Department of Mathematics, Colorado State University, Fort Collins, Colorado 80523

Current address: Department of Mathematics, Regis University, Denver, Colorado, 80221

E-mail address: bspringer@regis.edu

Contemporary Mathematics
Volume **678**, 2016
http://dx.doi.org/10.1090/conm/678/13654

Dynamics of self-similar interval exchange transformations on three intervals

Kelly Yancey

ABSTRACT. It is well known that almost all interval exchange transformations are rigid and weakly mixing. Jon Chaika proposed the following question: Is it true that self-similar IETs cannot be both rigid and weakly mixing? The set of self-similar IETs are a set of measure zero and are the ones that are periodic under Rauzy induction. This paper answers Chaika's question in the affirmative for the case where three intervals are exchanged.

1. Introduction

Interval exchange transformations have garnered a lot of attention in the recent years as there has been substantial progress made in regards to an old question of Veech. Given a normalized length vector with d components and a permutation on d letters, a d-interval exchange transformation is an exchange of the d intervals according to the permutation. The question of Veech is the following: Are almost all interval exchange transformations simple, rigid, and weakly mixing? For a fixed permutation, almost all refers to Lebesgue measure on the unit simplex. Veech himself proved that almost all IETs (interval exchange transformations) are rigid [**15**] and Katok and Stepin proved that almost all 3-IETs are weakly mixing [**9**]. Later, in 2007, Avila and Forni showed that almost all IETs that are not rotations are weakly mixing [**2**]. The result of Avila and Forni was a major breakthrough in the theory of IETs. The simplicity part of Veech's question remains open.

Even though almost all IETs are weakly mixing and rigid, these results do not shed any light on the behavior of classes of IETs that have measure zero. For example, a special set of measure zero IETs are the self-similar ones, also referred to as pseudo-Anosov IETs. In the particular case of 3-IETs the results in [**9**] show that self-similar IETs are weakly mixing. The class of self-similar IETs is periodic under the scheme of Rauzy induction and can be thought of as arising from substitutions. This paper explores the following question of Chaika:

QUESTION 1.1. *Is it true that self-similar IETs cannot be both weakly mixing and rigid?*

The goal is to prove that while almost all IETs are weakly mixing and rigid, these behaviors are incompatible notions in the realm of self-similar IETs.

This work was partially supported by NSF Grant DMS 1201534.

This paper explores the dynamics of self-similar 3-IETs and answers the question in the affirmative for this case.

THEOREM 1.2. *Self-similar 3-IETs are not rigid.*

To prove this theorem we view the self-similar IET first as a substitution system and then as a cutting and stacking transformation. We then proceed in a combinatorial way to show the absence of rigidity.

A result of Ferenczi, Holton, and Zamboni [**6**], which states that 3-IETs either have minimal self-joinings or are rigid, allows us to conclude immediately that self-similar 3-IETs have minimal self-joinings. Also, this class of 3-IETs arrise from primitive substitutions and as such are uniquely ergodic. Thus, self-similar 3-IETs satisfy Sarnak's conjecture, that is, this class of transformations is disjoint from the Mobius function. For more information on Sarnak's conjecture see [**4**] and [**13**].

The structure of this paper is as follows: In Section 2 we give some background on substitutions, cutting and stacking transformations, and IETs. We prove Theorem 1.2 in Section 3 and in Section 4 we prove the minimal self-joinings property for a specific 3-IET directly.

2. Preliminaries

Suppose (X, β, μ) is a Lebesgue probability space. Let $T : X \to X$ be an invertible measure-preserving transformation. In this case we will call (X, β, μ, T) a *dynamical system.* We will begin with a few standard definitions from ergodic theory.

DEFINITION 2.1. *The transformation T is weakly mixing if $f \in L_2$ and $f \circ T = \lambda f$ for some $\lambda \in \mathbb{C}$ implies that f is constant almost everywhere.*

DEFINITION 2.2. *The transformation T is mildly mixing if there are no non-constant rigid L_2 functions. (A function $f \in L_2$ is rigid if there exists an increasing sequence of natural numbers (n_m) such that $f \circ T^{n_m} \to f$ in L_2.)*

An equivalent formulation of mildly mixing is that there exists no measurable set A with $0 < \mu(A) < 1$ such that $\liminf_{n\to\infty} \mu(T^n A \triangle A) = 0$.

DEFINITION 2.3. *The transformation T is strongly mixing if*

$$\mu(T^n A \cap B) \to \mu(A)\mu(B)$$

for every $A, B \in \beta$.

DEFINITION 2.4. *The transformation T is rigid if there exists an increasing sequence of natural numbers (n_m) such that*

$$\lim_{m\to\infty} \mu(T^{n_m} A \triangle A) = 0$$

for every set A of positive measure.

With these definitions it is easy to see that rigidity and mild mixing are incompatible notions. Rigidity is also incompatible with strong mixing. Now we will give some definitions from the theory of joinings. For more information on joinings see [**7**] and [**11**] and for joinings related to substitutions see [**14**].

DEFINITION 2.5. *For any integer $k \geq 2$, a k-fold self-joining of (X, β, μ, T) is a measure defined on the Cartesian product X^k that is T^k-invariant and whose marginals are μ.*

Let $J(X)$ denote the set of joinings of (X, β, μ).

DEFINITION 2.6. *A dynamical system (X, β, μ, T) has minimal self-joinings of order k, k-fold MSJ for short, if every ergodic k-fold self-joining of the system $\nu \in J(X)$ is either the product measure μ^k or an off diagonal measure (i.e. $\nu(A_1 \times \cdots \times A_k) = \mu(A_1 \cap T^{i_2} A_2 \cap \cdots \cap T^{i_k} A_k)$).*

DEFINITION 2.7. *A dynamical system has minimal self-joinings, MSJ, if it has minimal self-joinings of every order.*

In Rudolph's book [**11**] it is shown that transformations with MSJ have trivial centralizer (only the powers commute with the transformation) and are prime (there are no factors). With these observations and the fact that rigid transformations have uncountable centralizer [**8**], we have that a transformation with MSJ is mildly mixing. Thus if you can show that a transformation has MSJ then it cannot be rigid.

The following theorem is helpful in proving that self-similar IETs have MSJ.

THEOREM 2.8. [**12**] *Suppose T has 2-fold MSJ, is weakly mixing, and not strongly mixing, then T has MSJ.*

2.1. Substitution Systems. Suppose $\mathcal{A}$ is a finite *alphabet* of d letters. A *word* is a finite sequence of letters from the alphabet, also referred to as a *block*. The set $\mathcal{A}^*$ refers to the set of all finite words over the alphabet $\mathcal{A}$. The natural metric on $\mathcal{A}^{\mathbb{Z}}$ is defined by $d(x,y) = \sum_{-\infty}^{\infty} \frac{|x_i - y_i|}{2^{|i|}}$. With this metric $\mathcal{A}^{\mathbb{Z}}$ is a compact metric space and the shift map $T : \mathcal{A}^{\mathbb{Z}} \to \mathcal{A}^{\mathbb{Z}}$ defined by $T(x)_i = x_{i+1}$ is a homeomorphism. A set that is closed and shift-invariant is called a *subshift* or *shift space.*

A special class of subshifts that we will study in this paper are substitutions. A map $\theta : \mathcal{A} \to \mathcal{A}^*$ which induces a map from $\mathcal{A}^*$ to $\mathcal{A}^*$ by $\theta(ab) = \theta(a)\theta(b)$ where $a, b \in \mathcal{A}$ is called a *substitution.* A word is *admissible* if it occurs as a subword of $\theta^n(a)$ for some $n \in \mathbb{N}$ and $a \in \mathcal{A}$. The *substitution space* is then defined as the set of all sequences in $\mathcal{A}^{\mathbb{Z}}$ such that every finite word is admissible. We will denote the substitution space associated to the substitution θ by X_θ.

The incidence matrix M associated to a substitution θ has entries m_{ij} where m_{ij} is the number of occurrences of i in $\theta(j)$. The substitution is said to be *primitive* if the incidence matrix is primitive. Substitution spaces arising from primitive substitutions are minimal and uniquely ergodic [**3**].

2.2. Cutting and Stacking. In this paper we will be viewing substitutions as finite rank transformations and in particular will be viewing these transformation through the lens of cutting and stacking. We begin by describing the procedure of cutting and stacking for rank-one maps.

We inductively define a sequence of towers, X_n, each of height h_n. Each X_n is a column of h_n disjoint intervals with equal measure denoted by $\{I_{n,0}, ..., I_{n,h_n-1}\}$. The elements of X_n are called *levels.* We often refer to $I_{n,0}$ as the *bottom level* and I_{n,h_n-1} as the *top level* of X_n. A transformation, T_n, is defined on $\{I_{n,0}, ..., I_{n,h_n-2}\}$ by moving up one level. That is, $T_n(I_{n,i}) = I_{n,(i+1)}$ for all $0 \le i < h_n - 1$. Note that T_n is not defined on the top level of X_n. Thus, we must define X_{n+1} by first cutting X_n into q_n subcolumns of equal width. We may then add any number of new levels (called *spacers*) above each subcolumn. Now, we stack every subcolumn of X_n above the subcolumn to its left to form X_{n+1}. Thus, X_{n+1} consists of q_n copies of

X_n which may be separated by spacers. Finally, we define $T = \lim_{n\to\infty} T_n$. The transformation T is called a *rank-one* map. For example, Chacon's transformation was the first example of a map that is weakly mixing, but not strongly mixing and is a classic example of a rank-one map.

In this paper we will be using the technique of cutting and stacking with 3 towers instead of 1 tower as in the rank-one case. There will be no spacers, only the subcolumns of each tower will be rearranged. That is, each of the 3 towers of X_n, call them $A^{(n)}, B^{(n)}, C^{(n)}$, will be cut into a given number of subcolumns, and restacked according to a given substitution to form the new columns of X_{n+1}. The transformation acts in the same way as the rank-one case. We will denote by $h_{n,A}$, $h_{n,B}$ and $h_{n,C}$ respectively the heights of the towers $A^{(n)}$, $B^{(n)}$ and $C^{(n)}$.

2.3. Interval Exchange Transformations and Rauzy Induction. In this section we will define interval exchange transformations (IETs for short), review the basics of Rauzy induction, and discuss a special class of IETs, the self-similar ones. For more information on IETs see Viana's notes [**16**].

A d-IET is a piecewise continuous map of the unit interval in which d subintervals are exchanged according to a permutation π.

DEFINITION 2.9. *Suppose $d \geq 2$ and let $\mathcal{A}$ be an alphabet with d letters. Given a partition $\{I_\alpha : \alpha \in \mathcal{A}\}$ of $[0,1)$ into d subintervals each of length λ_α and a permutation $\pi = (\pi_t, \pi_b)$ of the d letters, define the d-IET $T : [0,1) \to [0,1)$ by*

$$T(x) = x + \sum_{\pi_b(\beta)<\pi_b(\alpha)} \lambda_\beta - \sum_{\pi_t(\beta)<\pi_t(\alpha)} \lambda_\beta \qquad \textit{for } x \in I_\alpha.$$

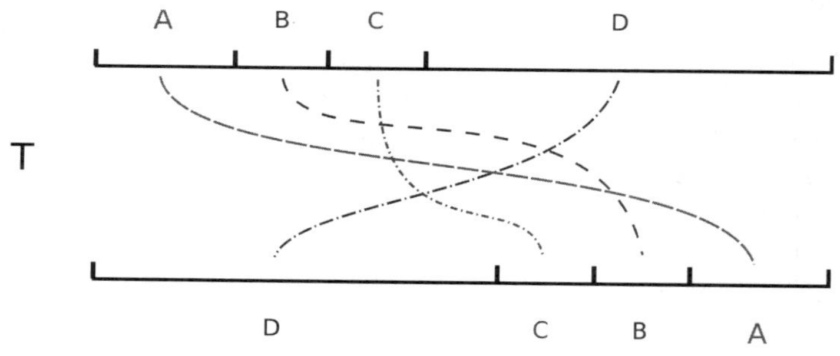

FIGURE 1. 4-IET

An example of a 4-IET with permutation

$$\pi = \begin{pmatrix} \pi_t \\ \pi_b \end{pmatrix} = \begin{pmatrix} A & B & C & D \\ D & C & B & A \end{pmatrix}$$

is given in Figure 1.

Rauzy induction is a specific procedure for inducing on a subinterval of the IET. If the original IET is an exchange of d intervals, the induced map will also be a d-IET. In Rauzy induction there are two types of inducing schemes, top and bottom, depending on the length of the last subinterval. First we will describe the top procedure.

Suppose that $\pi_t(\alpha) = d$ and $\pi_b(\beta) = d$, that is, I_α is the last subinterval of the partition of $[0,1)$ and I_β is the last subinterval of the partition of $T([0,1))$. In the *top produce* we assume that $\lambda_\alpha > \lambda_\beta$. Define the interval J by $J = [0, 1-\lambda_\beta)$ and under these assumptions induce on the interval J to produce a new d-IET, T'. That is, T' is the first return map to the interval J. In this case we call I_α the *winner* and I_β the *loser*. The new IET T' has permutation $\pi' = (\pi'_t, \pi'_b)$ where $\pi'_t = \pi_t$ and in π'_b, α is replaced by α followed by β and β is deleted at the right end (in other terms we perform a cyclic permutation on symbols to the right of the winner symbol α).

For the *bottom procedure* assume that $\lambda_\alpha < \lambda_\beta$. Define the interval J by $J = [0, 1-\lambda_\alpha)$. Produce a new d-IET by inducing on the interval J. In this case we call I_β the winner. The new IET T' has permutation $\pi' = (\pi'_t, \pi'_b)$ where in π'_t, β is replaced by β followed by α and α is deleted at the right end (in other terms we perform a cyclic permutation on symbols to the right of the winner symbol β) and $\pi'_b = \pi_b$.

We will assume that all IETs satisfy the infinite distinct orbit condition (i.d.o.c) which says that the orbits of discontinuities are infinite and distinct. This is also known as the Keane condition. It ensures that Rauzy induction may be repeatedly applied [**16**].

The following figure displays an example of Rauzy induction applied to a 4-IET. See Example 2.10 for the permutation.

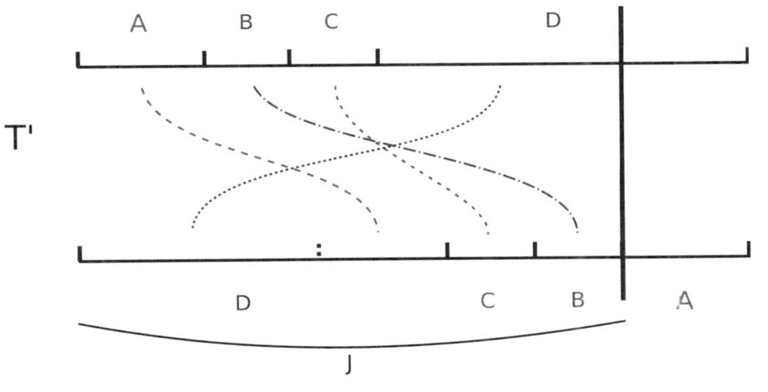

FIGURE 2. Rauzy Induction for Example 2.10

EXAMPLE 2.10. *Consider the IET from Figure 1. As you can see from Figure 2 this is a top procedure and D is the winner. Thus the new permutation π' is given by*

$$\pi = \begin{pmatrix} A & B & C & D \\ D & A & C & B \end{pmatrix}.$$

2.4. Self-Similar IETs. The information obtained from Rauzy induction can be put into a diagram called a *Rauzy diagram*. Fix a permutation of d letters, π. Consider the graph whose vertices are permutations of d letters and put an edge from π_1 to π_2 if you can obtain the permutation π_2 from π_1 through Rauzy induction. The edges of this directed graph are labeled by the winner symbols from

Rauzy induction. The Rauzy diagram associated to the given permutation π is the connected component of this graph that contains π.

For the rest of this paper we will be considering 3-IETs. The only irreducible 3-IET is given by the permutation $\pi = \begin{pmatrix} A & B & C \\ C & B & A \end{pmatrix}$. The following figure displays the Rauzy diagram for this permutation.

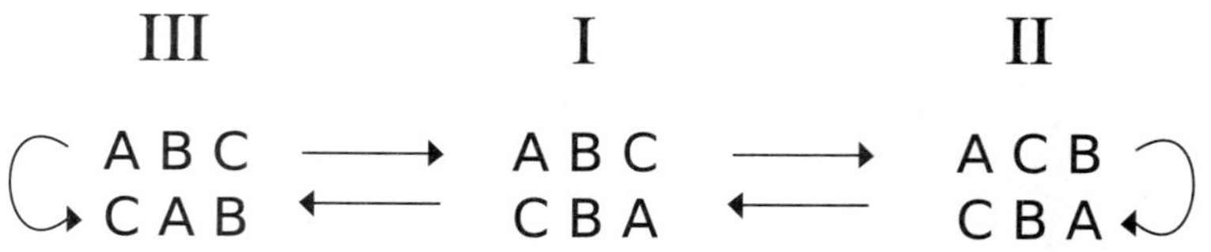

FIGURE 3. Rauzy Diagram on 3 Intervals

Associated to each edge in the Rauzy diagram is a substitution on three letters. Recall that J was the interval that we induced on in the induction scheme. The substitution is obtained by following the images of elements of J in $[0,1)$ under T until they return to J. For example, the substitution from the top procedure induction in Example 2.10 is the following:

$$A \mapsto AD; B \mapsto B; C \mapsto C; D \mapsto D.$$

The substitutions corresponding to the edges of the Rauzy diagram from Figure 3 are below.

$$\begin{aligned}
I \to II &: \quad A \mapsto A; B \mapsto B; C \mapsto AC \\
II \to II &: \quad A \mapsto AB; B \mapsto B; C \mapsto C \\
II \to I &: \quad A \mapsto A; B \mapsto AB; C \mapsto C \\
I \to III &: \quad A \mapsto AC; B \mapsto B; C \mapsto C \\
III \to III &: \quad A \mapsto A; B \mapsto B; C \mapsto BC \\
III \to I &: \quad A \mapsto A; B \mapsto BC; C \mapsto C
\end{aligned}$$

Associated to each above substitution is a 3×3 incidence matrix M. To obtain the incidence matrix of the substitution that is derived from moving from permutation I to permutation II and then back to I, you simply multiply the individual incidence matrices from each substitution together. In this case you would obtain

$$\begin{pmatrix} 1 & 0 & 1 \\ 0 & 1 & 0 \\ 0 & 0 & 1 \end{pmatrix} \begin{pmatrix} 1 & 1 & 0 \\ 0 & 1 & 0 \\ 0 & 0 & 1 \end{pmatrix} = \begin{pmatrix} 1 & 1 & 1 \\ 0 & 1 & 0 \\ 0 & 0 & 1 \end{pmatrix}.$$

We are now ready to define the class of self-similar IETs. Intuitively, a *self-similar IET* is one that after inducing a finite number of times (using Rauzy induction) and rescaling so that you have a map of the unit interval, you arrive back to your original IET.

DEFINITION 2.11. *The IET defined by permutation π and length vector λ is self-similar if there exists a loop in the Rauzy diagram beginning at π and an associated Perron-Frobenius matrix M such that $M\lambda = \xi\lambda$ where ξ is the dominant eigenvalue of the matrix M.*

Thus, self-similar IETs correspond to loops in the Rauzy diagram that see every label at least once, where the labels are given by the winners in the induction procedure.

It was shown in [**9**] that self-similar 3-IETs are weakly mixing. Weak mixing in this case also follows directly from the Veech criterion in [**15**].

3. 3-IETs are not Rigid

In this section we will show that self-similar 3-IETs are not rigid. One way to accomplish this goal is to view a self-similar 3-IET as a cutting and stacking transformation and proceed in a combinatorial way to obtain the absence of rigidity.

Consider the loop $CBCAA$ in the Rauzy diagram from Figure 3. This loop gives rise to a self-similar 3-IET. To study self-similar IETS we will study such loops in the Rauzy diagram and their corresponding substitutions. The substitutions will be defined over the alphabet $\mathcal{A} = \{A, B, C\}$.

We will always begin our loops in the Rauzy diagram from the permutation labeled I. Starting from this permutation there are two possible directions to follow in the directed graph. We begin by considering loops that first move from permutation I to permutation II.

Consider the path generated by the loop $ABACC$ in the Rauzy diagram. This loop gives rise to a self-similar IET on three intervals determined by the substitution θ defined by

$$\begin{aligned} \theta(A) &= ABAC \\ \theta(B) &= ABBAC \\ \theta(C) &= AC. \end{aligned}$$

Associated to the above substitution is a 3×3 incidence matrix M where

$$M = \begin{pmatrix} 2 & 2 & 1 \\ 1 & 2 & 0 \\ 1 & 1 & 1 \end{pmatrix}.$$

The matrix M has eigenvector $(1, -1, 1)^T$ with eigenvalue 1 in the neutral direction, which in the language of cutting and stacking means that the sum of the heights of the first and third columns is always equal to the height of the second column plus one. That is,

$$h_{n,A} + h_{n,C} = h_{n,B} + 1.$$

Let T denote the transformation associated (via cutting and stacking) to the IET generated by the loop $ABACC$ in the Rauzy diagram. The proposition below says that this transformation is not rigid. To show that T is not rigid, it suffices to find a set of positive measure E such that $\liminf_{n\to\infty} \mu(T^n E \triangle E) > 0$. For our purposes, let E be the bottom level of the A tower at some stage m.

PROPOSITION 3.1. *The self-similar IET generated by the loop $ABACC$ in the Rauzy diagram is not rigid.*

PROOF. Suppose for a contradiction, that $\liminf_{n\to\infty} \mu(T^n E \triangle E) = 0$. That means that there exists a subsequence of natural numbers (n_m) such that $\mu(T^{n_m} E \triangle E) \to 0$. Our goal is to follow iterates of the set E and show that if our assumption is true then it must be the case that $TE = E$. To do this we will rely heavily on the fact that $h_{n,A} + h_{n,C} = h_{n,B} + 1$. This will produce a contradiction since E has positive measure strictly less than one and T is ergodic. In [**1**] Aaronson and Weiss use this approach to show that the classic Chacon's map is not rigid. We are adapting their approach to the setting of finite rank transformations.

Recall that we denote by $h_{n,A}$, $h_{n,B}$ and $h_{n,C}$ respectively the heights of the towers $A^{(n)}$, $B^{(n)}$ and $C^{(n)}$ at stage n.

Let $\epsilon > 0$, and let N be such that $\mu(T^N E \triangle E) < \epsilon$. Fix n such that $N \in [h_{n,A}, h_{n+1,A} - 1]$ and assume that $n \geq m$.

First consider the case when $n = m$. That is, E is the bottom level of the A tower at stage n and $\mu(T^N E \triangle E) < \epsilon$ where $N \in [h_{n,A}, h_{n+1,A} - 1]$. This case will display the main mechanism in the proof.

Before we proceed, we need some notation for how levels in X_n appear in X_{n+2}. This will help us keep track of where the set E travels under iterates of T. It may be helpful to view the substitution after two steps to help visualize how the towers appear.

$$A \mapsto ABACABBACABACAC \tag{1}$$

$$B \mapsto ABACABBACABBACABACAC \tag{2}$$

$$C \mapsto ABACAC. \tag{3}$$

Let us now consider what happens during stage n of the construction to produce the next stage $n+1$. At the n-th stage, the tower $A^{(n)}$ is cut into five pieces, the tower $B^{(n)}$ is cut into three pieces, and the tower $C^{(n)}$ is cut into three pieces before the subcolumns are stacked together. Let A_n be the bottom level of $A^{(n)}$, B_n the bottom level of $B^{(n)}$, and C_n the bottom level of $C^{(n)}$. Thus $A^{(n)}$ is comprised of levels $A_n, TA_n, \ldots, T^{h_{n,A}-1}A_n$, $B^{(n)}$ is comprised of levels $B_n, TB_n, \ldots, T^{h_{n,B}-1}B_n$, and $C^{(n)}$ is comprised of levels $C_n, TC_n, \ldots, T^{h_{n,C}-1}C_n$. As we described before, A_n is cut into five pieces, call them $A_{n,0}, \ldots, A_{n,4}$. Taking into account the way the columns are stacked together to form X_{n+1}, we see that $A_{n+1} = A_{n,0}$, $B_{n+1} = A_{n,2}$, and $C_{n+1} = A_{n,4}$.

Figure 4 displays how X_1 is constructed from X_0.

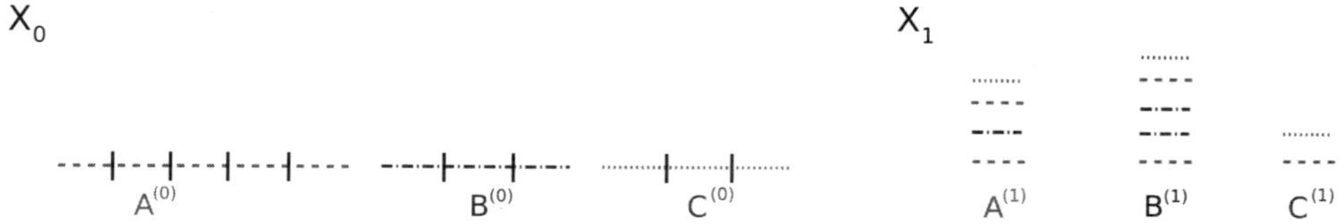

FIGURE 4. Construction of X_1 from X_0

As you can see from the figure, $h_{1,A} + h_{1,C} = 4 + 2 = 6 = 5 + 1 = h_{1,B} + 1$.

At this point it is helpful to define the size of the gaps between consecutive occurrences of $A^{(n)}$ inside the towers of X_{n+2}. To that end define

$$e_j = \begin{cases} h_{n,B}, & \text{if } j = 0, 4, 7, 13, 16 \\ 2h_{n,B}, & \text{if } j = 2, 9, 11 \\ h_{n,C}, & \text{if } j = 1, 3, 5, 8, 10, 12, 14, 17. \end{cases}$$

Notice that e_j is not defined when $j = 6, 15, 18$. Figure 5 displays how the e_j terms were determined for $0 \le j \le 5$ which corresponds to gaps between appearances of $A^{(n)}$ inside $A^{(n+2)}$.

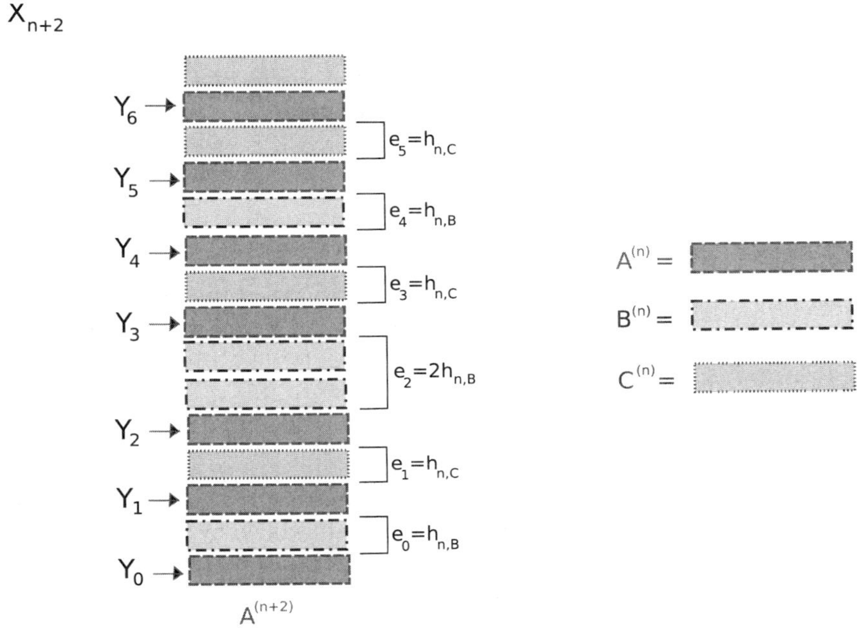

FIGURE 5. Tower $A^{(n+2)}$

Define k_j for $j = 0, 1, \ldots, 18$ by the recursive formula

$$k_j = 0 \quad \text{for } j = 0, 7, 16 \qquad \text{and} \qquad k_{j+1} = k_j + h_{n,A} + e_j \quad \text{otherwise.}$$

These variables allow us to easily describe the appearance of $A^{(n)}$ blocks inside the three towers at stage $n+2$. In that vein, let

$$Y_j = \begin{cases} \bigcup_{i=0}^{h_{n,A}-1} T^{i+k_j} A_{n+2}, & \text{if } j = 0, 1, \ldots, 6 \\ \\ \bigcup_{i=0}^{h_{n,A}-1} T^{i+k_j} B_{n+2}, & \text{if } j = 7, 8, \ldots, 15 \\ \\ \bigcup_{i=0}^{h_{n,A}-1} T^{i+k_j} C_{n+2}, & \text{if } j = 16, 17, 18. \end{cases}$$

You can view Y_j for $0 \le j \le 6$ in Figure 5. Analyzing the Y_j we see that $Y_{j+1} = T^{h_{n,A}+e_j} Y_j$ for all $j \ne 6, 15, 18$.

Recall that $N \in [h_{n,A}, h_{n+1,A} - 1]$. If we write N as $N = ah_{n,A} + b$ then we need to factor $h_{n+1,A}$ in terms of $h_{n,A}$ to determine the potential values for a and b. Now, notice that

$$h_{n+1,A} = 2h_{n,A} + h_{n,B} + h_{n,C} = 2h_{n,A} + (h_{n,A} + h_{n,C} - 1) + h_{n,C} = 3h_{n,A} + 2h_{n,C} - 1.$$

Thus if we write $N = ah_{n,A} + b$ then we have $a = 1, 2, 3$ and $0 \le b < h_{n,A}$ (except when $a = 3$, in which case $b \le 2h_{n,C} - 2$).

For fixed a (determined by N above) let $e_{j,a} = e_j + e_{j+1} + \cdots + e_{j+a-1}$. Recall that our set E belongs to $A^{(n)}$ which has been partitioned into pieces Y_j for $0 \le j \le 18$. To analyze $T^N E$ we can analyze $T^N Y_j$. Observe

$$T^N Y_j = T^{ah_{n,A}+b} Y_j = T^{b-e_{j,a}} Y_{j+a}.$$

Suppose S_1, S_2 are two sets of positive measure. By $S_1 \overset{\epsilon}{\approx} S_2$ we will mean $\mu(S_1 \triangle S_2) < \epsilon$. First we will concentrate on the pieces of E that are inside the tower $A^{(n+2)}$, that is, we will focus on $E \cap Y_j$ where $0 \le j \le 6$. Here we have

$$T^N(E \cap Y_j) = T^N E \cap T^N Y_j \overset{\epsilon}{\approx} E \cap T^{b-e_{j,a}} Y_{j+a}$$

for $0 \le j < 6 - a$. Also notice that

$$T^N(E \cap Y_j) = T^{b-e_{j,a}}(E \cap Y_{j+a})$$

for $0 \le j < 6 - a$. This is true because E appears in the same position inside each Y_j. In particular, E is the bottom level of each Y_j. Putting these together we see that

$$E \cap Y_{j+a} \overset{\epsilon}{\approx} T^{-b+e_{j,a}} E \cap Y_{j+a}.$$

A similar calculation can be carried out for the rest of the j values. Thus $E \overset{\epsilon}{\approx} T^{-b+e_{j,a}} E$.

Now, if we analyze the formulas for e_j we see that for any a we can find j, j' such that $|e_{j,a} - e_{j',a}| = h_{n,B}$. Recall that e_j is determined by the length of the gap between consecutive appearances of Y_j, which is really determined by the length of the gap between consecutive A blocks in Equations (1) – (3). The difference of $h_{n,B}$ comes from the fact that there is an A block followed by a B block and a B block followed by a B block. Hence, $E \overset{2\epsilon}{\approx} T^{h_{n,B}} E$. To be more clear, for $a = 1, 2, 3$ choose $j = 0$, $j' = 9$, that is pick j so that Y_j is the first appearance of a Y_i inside $A^{(n+1)}$ in $A^{(n+2)}$ and j' so that $Y_{j'}$ is the first appearance of a Y_i inside $B^{(n+1)}$ in $B^{(n+2)}$.

Now we will use the fact that $h_{n,A} + h_{n,C} = h_{n,B} + 1$. Recall that E is the bottom level of $A^{(m)} = A^{(n)}$ which appears in X_n. Thus E is the bottom level in each Y_j. In particular, E is in the same position inside each Y_j as we noted earlier. Observe the following about Y_1:

$$E \cap Y_2 = T^{h_{n,A}+h_{n,C}}(E \cap Y_1) = T^{h_{n,A}+h_{n,C}} E \cap Y_2 = T^{h_{n,B}+1} E \cap Y_2 \overset{2\epsilon}{\approx} TE \cap Y_2.$$

Since E occupies the same position inside each block Y_j, the same estimate holds for every j. Thus, the above says that

$$E \overset{2\epsilon}{\approx} TE.$$

Next consider the case when $n = m + 1$. Thus our set E is the bottom level of $A^{(m)}$ and our rigidity time N is between $h_{m+1,A}$ and $h_{m+2,A} - 1$. This time around we want to keep track of how X_m appears inside $X_{n+2} = X_{m+3}$. The substitution is longer, but the same variables can be defined. That is, the set $A^{(m)}$ has been broken up into sets Y_j where the Y_j blocks belong to $X_{m+3} = X_{n+2}$ and e_j represents the length of consecutive gaps between Y_j. This time around if we write $N = ah_{m,A} + b$ (and remember that $N \in [h_{n,A}, h_{n+1,A} - 1] = [h_{m+1,A}, h_{m+2,A} - 1]$) then we need to factor $h_{m+1,A}$ and $h_{m+2,A}$ in terms of $h_{m,A}$. Using the relationship between the heights of the towers we see that the possible values of a and b are $a = 3, 4, \ldots, 13$

and $0 \leq b < h_{m,A}$ (except when $a = 3$, in which case $2h_{m,C} - 1 \leq b < h_{m,A}$ and when $a = 13$, in which case $0 \leq b < 2h_{m,C} + 2h_{m-1,C} - 3$).

Fix a (which depends on N). As before, the key step is finding j, j' such that $|e_{j,a} - e_{j',a}| = h_{m,B}$. The structure of the substitution allows you to find such j, j' (as in the previous case). We will use the previously constructed j and j' to form the new ones for this case. Notice that the useful blocks to analyze from the previous case were Y_0 and Y_9. Those blocks were then subdivided and restacked. We want to consider the restacked sub-blocks that appear inside $A^{(n+2)} = A^{(m+3)}$ and $B^{(n+2)} = B^{(m+3)}$. That means that we will analyze Y_0 (the first appearance of a Y_i inside $A^{(n+1)}$ in $A^{(n+2)}$) and Y_{35} (the third appearance of a Y_i inside $B^{(n+1)}$ in $B^{(n+2)}$). Now, $|e_{j,a} - e_{j',a}| = h_{m,B}$ for any a in the desired range because the difference of $h_{m,B}$ comes from the a values in the previous case ($n = m$) and then the blocks agree after that initial disparity for the desired a values.

The remainder of the proof for this case follows in exactly the same way as when $n = m$. The same analysis can be done for all n where $n \geq m$. In each case the same contradiction is reached and T is not rigid.

□

Above we considered a specific loop in the Rauzy diagram and proved that the corresponding IET is not rigid. Generalized loops in the Rauzy diagram that begin at the same permutation (permutation I) and have A as the first letter in the loop act in a similar manner. Consider the path given by $AB^kACB^\ell C$ where $k, \ell \in \mathbb{Z}$ with at least one of k, ℓ nonzero where the notation B^i means that the letter B is repeated i-times. This path is a loop that covers every letter in the alphabet and generates a self-similar IET. We will consider two separate components in the above path, $CB^\ell C$ and AB^kA. Notice that $CB^\ell C$ corresponds to the following substitution:

$$\begin{aligned}\theta_1(A) &= AC\\ \theta_1(B) &= B^{\ell+1}C\\ \theta_1(C) &= B^\ell C\end{aligned}$$

Likewise, AB^kA corresponds to the following substitution:

$$\begin{aligned}\theta_2(A) &= AB^k\\ \theta_2(B) &= AB^{k+1}\\ \theta_2(C) &= AC.\end{aligned}$$

Putting these two together gives the following substitution, θ, associated to the complete loop:

$$\begin{aligned}\theta(A) &= AB^kAC\\ \theta(B) &= (AB^{k+1})^{\ell+1}AC\\ \theta(C) &= (AB^{k+1})^\ell AC.\end{aligned}$$

For this loop, the incidence matrix associated to the substitution is

$$M = \begin{pmatrix} 2 & \ell+2 & \ell+1 \\ k & (k+1)(\ell+1) & (k+1)\ell \\ 1 & 1 & 1 \end{pmatrix}$$

and has eigenvector $(1,-1,1)^T$ with eigenvalue 1 in the neutral direction. This allows us to conclude that we have the same relationship between the heights of the towers $A^{(n)}, B^{(n)}, C^{(n)}$ as before. Namely

$$h_{n,A} + h_{n,C} = h_{n,B} + 1.$$

PROPOSITION 3.2. *The self-similar IET generated by the loop $AB^kACB^\ell C$ in the Rauzy diagram is not rigid.*

The proof is the same as the proof of the previous proposition. Instead of supplying all of the details we will describe the main components of the proof. To begin, let E be the bottom level of the A tower in the m-th stage. Let $\epsilon > 0$ and suppose for a contradiction that N is such that $\mu(T^N E \triangle E) < \epsilon$ and $N \in [h_{n,A}, h_{n+1,A} - 1]$ where $n \geq m$. Recall that the first step is to analyze how X_n appears in X_{n+2}. To do that, consider two iterates of the substitution:

$$\begin{aligned} A &\mapsto AB^kAC[(AB^{k+1})^{\ell+1}AC]^k AB^kAC(AB^{k+1})^\ell AC \\ B &\mapsto [AB^kAC[(AB^{k+1})^{\ell+1}AC]^{k+1}]^{\ell+1} AB^kAC(AB^{k+1})^\ell AC \\ C &\mapsto [AB^kAC[(AB^{k+1})^{\ell+1}AC]^{k+1}]^{\ell} AB^kAC(AB^{k+1})^\ell AC. \end{aligned}$$

The two key components of the previous proof remain present in this more complicated version. Let $N = ah_{m,A} + b$. The first component is that you can always find j, j' such that $|e_{j,a} - e_{j',a}| = h_{m,B}$. This allows you to conclude that $T^{h_{m,B}}E \approx E$. The second component is that the sequence of levels $A^{(n)}C^{(n)}A^{(n)}$ appears in each of the three towers in X_{n+2}. Using this information and the fact that $h_{m,A} + h_{m,C} = h_{m,B} + 1$ you can conclude that $TE \approx E$, which is a contradiction.

Completely generalized loops that start at permutation I and first travel to permutation II are a countable number of combinations of the above loops, and have the same structure and neutral eigenvector. Thus we have the following theorem:

THEOREM 3.3. *Any self-similar IET generated by a loop in the Rauzy diagram beginning at permutation I and first traveling to permutation II is not rigid.*

Since the Rauzy diagram is symmetric about permutation I, we have proved Theorem 1.2.

4. Minimal Self-Joinings of 3-IETs

We are now ready to discuss the MSJ property of self-similar 3-IETs. In 2005 Ferenczi, Holton, and Zamboni proved the following theorem:

THEOREM 4.1. [**6**] *Every 3-IET either has MSJ or is rigid.*

In light of Theorem 1.2 we obtain a corollary about self-similar 3-IETs.

COROLLARY 4.2. *Self-similar 3-IETs have MSJ.*

As stated in the introduction, these are an important class of 3-IETs, as they satisfy Sarnak's conjecture. Let $\phi(n)$ denote the Mobius function. Sarnak's conjecture says that the Mobius function is disjoint from deterministic dynamical systems. Specifically it states that for any deterministic topological dynamical system (X,T) as $N \to \infty$ we have

$$\sum_{n \leq N} \phi(n) f(T^n x) = O(n)$$

for $x \in X$ and $f \in C(X)$.

In [**4**] it is stated that any uniquely ergodic topological dynamical system (X,T) such that T^p and T^q are disjoint for $p \neq q$ satisfies Sarnak's conjecture. In [**5**] del Junco and Rudolph showed that if T is weakly mixing and has minimal self-joinings then T^p and T^q are disjoint for $p \neq q$. Thus self-similar 3-IETs are disjoint from the Mobius function.

Currently there are no known examples of d-IETs where $d > 3$ that have MSJ.

4.1. Example. Proving that a transformation has MSJ is not an easy task. In this section we concentrate on one self-similar IET and show that it has MSJ. We do this directly, by viewing the IET as a substitution system. Unlike the result from [**6**], our substitution arises from Rauzy induction.

Consider the substitution defined by the loop $ABACC$ in the Rauzy diagram, call this substitution θ. This is the substitution that was discussed in detail in the previous section. Let T denote the shift map on the substitution space X_θ.

THEOREM 4.3. *The map T has MSJ.*

Self-similar IETs on 3 intervals are weakly mixing ([**9**]) and never strongly mixing ([**10**]). Thus to prove the above theorem, we only need to show 2-fold MSJ. The main tools to prove 2-fold MSJ are the following two lemmas by Rudolph.

LEMMA 4.4. *(Lemma 6.14 from [**11**]) If $(X.\beta,\mu,T)$ is ergodic and $\nu \in J(X)$ is a 2-fold ergodic joining that is $(Id \times T)$ invariant, then $\nu = \mu \times \mu$.*

LEMMA 4.5. *(Lemma 6.15 from [**11**]) Let (X,β,μ,T) be an ergodic dynamical system and $\{P_i\}$ a countable set of cylinders generating β. Let $\overline{\beta} = \{P_\ell \times P_m\}$ be a countable generating algebra of $X \times X$. Assume that*

(1) *$\nu \in J(X)$ is a 2-fold ergodic joining*

(2) *$(x,y) \in X \times X$ satisfies*

$$\lim_{N\to\infty} \frac{1}{N}\sum_{n=0}^{N-1} \chi_A(T^{-n}x,T^{-n}y) = \lim_{N\to\infty} \frac{1}{N}\sum_{n=0}^{N-1} \chi_A(T^{n}x,T^{n}y) = \nu(A)$$

for all $A \in \overline{\beta}$

(3) *there exists intervals $L_k = [i_k, j_k] \subset \mathbb{Z}$, intervals $M_k \subset \mathbb{Z}$, $t_k \in \mathbb{Z}$, and $\gamma > 0$ such that $i_k \leq 0 \leq j_k$, $j_k - i_k \to \infty$, $M_k \subset L_k$, $M_k + t_k \subset L_k$, and $|M_k| \geq \gamma |L_k|$*

(4) *for any cylinder sets P_ℓ and P_m there exists K such that if $k \geq K$ then for all $i \in M_k$*

$$T^i x \in P_\ell \text{ if and only if } T^{i+t_k}x \in P_\ell$$
$$T^i y \in P_m \text{ if and only if } T^{i+t_k+1}y \in P_m.$$

Then, ν is $(Id \times T)$ invariant and hence $\nu = \mu \times \mu$.

The first thing that we need to do is determine the structure of long words. Our substitution θ is defined on three letters, A, B, C, and given by

$$\begin{aligned}\theta(A) &= ABAC\\ \theta(B) &= ABBAC\\ \theta(C) &= AC.\end{aligned}$$

Define D_1 to be the word $\theta(A)$ with the first letter removed and an A appended to the end. Thus, $D_1 = BACA$. Similarly, define D_n to be the word $\theta^n(A)$ with the first letter removed and an A appended to the end. This means that the length of D_n is the length of $\theta^n(A)$. Let $\overline{D_n} = D_n D_{n-1} \cdots D_1$ and define $H_n = |\theta^n(A)| + \cdots + |\theta(A)|$ to be the length of $\overline{D_n}$.

LEMMA 4.6. *Iterates of the substitution θ have the following form for $n \geq 2$:*

$$\begin{aligned} \theta^n(A) &= A\overline{D_{n-1}}BD_{n-1}\overline{D_{n-1}}C \\ \theta^n(B) &= A\overline{D_{n-1}}B\overline{D_{n-1}}BD_{n-1}\overline{D_{n-1}}C \\ \theta^n(C) &= A\overline{D_{n-1}}C. \end{aligned}$$

PROOF. This will be a proof by induction. For the case $n = 2$ consider

$$\begin{aligned} \theta^2(A) &= \theta(ABAC) = ABACABBACABACAC = AD_1BD_1D_1C \\ \theta^2(B) &= \theta(ABBAC) = ABACABBACABBACABACAC = AD_1BD_1BD_1D_1C \\ \theta^2(C) &= \theta(AC) = ABACAC = AD_1C. \end{aligned}$$

Now suppose that the formula holds for an arbitrary n. Thus, for instance $\theta^n(A) = A\overline{D_{n-1}}BD_{n-1}\overline{D_{n-1}}C$. From the definition of D_n we see that $D_n = \overline{D_{n-1}}BD_{n-1}\overline{D_{n-1}}CA$. Observe,

$$\begin{aligned} \theta^{n+1}(A) &= \theta^n(ABAC) = \theta^n(A)\theta^n(B)\theta^n(A)\theta^n(C) \\ &= \left(A\overline{D_{n-1}}BD_{n-1}\overline{D_{n-1}}C\right)\left(A\overline{D_{n-1}}B\overline{D_{n-1}}BD_{n-1}\overline{D_{n-1}}C\right) \\ &\quad \times \left(A\overline{D_{n-1}}BD_{n-1}\overline{D_{n-1}}C\right)\left(A\overline{D_{n-1}}C\right) \\ &= A\left(\overline{D_{n-1}}BD_{n-1}\overline{D_{n-1}}CA\right)\overline{D_{n-1}}B\left(\overline{D_{n-1}}BD_{n-1}\overline{D_{n-1}}CA\right) \\ &\quad \times \left(\overline{D_{n-1}}BD_{n-1}\overline{D_{n-1}}CA\right)\overline{D_{n-1}}C \\ &= AD_n\overline{D_{n-1}}BD_nD_n\overline{D_{n-1}}C \\ &= A\overline{D_n}BD_n\overline{D_n}C. \end{aligned}$$

In a similar fashion we have

$$\begin{aligned} \theta^{n+1}(B) &= \theta^n(ABBAC) = \theta^n(A)\theta^n(B)\theta^n(B)\theta^n(A)\theta^n(C) \\ &= A\overline{D_{n-1}}BD_{n-1}\overline{D_{n-1}}CA\overline{D_{n-1}}B\overline{D_{n-1}}BD_{n-1}\overline{D_{n-1}}C \\ &\quad A\overline{D_{n-1}}B\overline{D_{n-1}}BD_{n-1}\overline{D_{n-1}}CA\overline{D_{n-1}}BD_{n-1}\overline{D_{n-1}}CA\overline{D_{n-1}}C \\ &= AD_n\overline{D_{n-1}}BD_n\overline{D_{n-1}}BD_nD_n\overline{D_{n-1}}C \\ &= A\overline{D_n}B\overline{D_n}BD_n\overline{D_n}C \end{aligned}$$

and

$$\begin{aligned} \theta^{n+1}(C) &= \theta^n(AC) = \theta^n(A)\theta^n(C) \\ &= A\overline{D_{n-1}}BD_{n-1}\overline{D_{n-1}}CA\overline{D_{n-1}}C \\ &= AD_n\overline{D_{n-1}}C \\ &= A\overline{D_n}C. \end{aligned}$$

□

A point in X_θ can be written in terms of $\theta^n(A), \theta^n(B)$, and $\theta^n(C)$. The above lemma tells us what form these blocks take when $n \geq 2$. Thus if the length of an admissible word is greater than or equal to 20 then it can be written in terms of the previous lemma.

Proof of Theorem 4.3. Let $\{P_i\}$ be a countable set of cylinders generating the sigma-algebra β. Suppose that $\overline{\beta} = \{P_\ell \times P_m\}$ is a countable set generating the algebra for $X \times X$. Suppose that $\nu \in J(X)$ is an ergodic joining. Let $(x, y) \in X \times X$ be a ν generic point satisfying condition (2) of Lemma 4.5.

If x and y are in the same orbit then there exists $k \in \mathbb{Z}$ such that $T^k x = y$. Then ν is an off diagonal measure, that is, it is the image of μ under the map $z \mapsto (z, T^k z)$.

Now suppose x and y are in different orbits. Write x and y in terms of $\theta^n(A), \theta^n(B)$, and $\theta^n(C)$ blocks. Let $k_n \in \mathbb{Z}$ be such that the block containing the zeroth place of x, denoted by x_0, and the block containing the zeroth place of $T^{k_n} y$, denoted by $(T^{k_n} y)_0$, start at the same location. Then k_n is bounded by half the length of the longest block. Thus, $|k_n| \leq \frac{|\theta^n(B)|}{2}$. Since x and y are in different orbits, there exists a block where they differ, call it the s_nth block. Note that s_n negative means that x and y differ at a block that occurs in the negative direction (i.e. to the left of the zeroth position). Choose s_n such that $|s_n|$ is minimal, that is, s_n should be the first block where they disagree.

Suppose $|s_n| > 20$. In this case, consider a coding of x and $T^{k_n} y$ by $\theta^{n+1}(A)$, $\theta^{n+1}(B)$, and $\theta^{n+1}(C)$ blocks. Since $|s_n| > 20$, the coding in terms of θ^n blocks agree in a large neighborhood of the block that contains x_0. To be precise, the blocks agree in at least 19 places in each direction. Since the longest component of the substitution is $\theta(B)$ (which has length 5), the θ^{n+1} blocks from x and $T^{k_n} y$ agree and are aligned around the block that contains x_0. Thus, there is no need to shift to realign the blocks in this case. Since the block lengths are increasing in length as n increases, we have that $|s_{n+1}| < |s_n|$. Note that we can continue this process until we find n' such that $|s_{n'}| \leq 20$. For notational purposes, we will assume that n satisfies this requirement and that y is shifted by k (instead of k_n). Let ω_{s_n} be the position in x where the s_nth block begins.

The remainder of the proof is split into cases depending on the structure of the first block where x and $T^k y$ differ. Suppose that $s_n > 0$.

Case 1. Suppose that $x_{\omega_{s_n}}$ is the beginning of $\theta^n(A)$ and $(T^k y)_{\omega_{s_n}}$ is the beginning of $\theta^n(B)$. Since every block ends in the same way, namely $\overline{D_{n-1}}C$, and begins in the same way, namely $A\overline{D_{n-1}}$, x and $T^k y$ have the following form around the ω_{s_n}th place:

$$x \; : \; \overline{D_{n-1}}CA\overline{D_{n-1}}BD_{n-1}\overline{D_{n-1}}CA\overline{D_{n-1}} \tag{4}$$

$$T^k y \; : \; \overline{D_{n-1}}CA\overline{D_{n-1}}B\overline{D_{n-1}}BD_{n-1}\overline{D_{n-1}}CA\overline{D_{n-1}} \tag{5}$$

If $\frac{1}{2}H_{n-1} \leq k \leq \frac{3}{2}H_{n-1}$ then define M_n to be the interval where the first $\overline{D_{n-1}}$ block from x and the second $\overline{D_{n-1}}$ block from y overlap (see Figure 6). The most you are shifting y is by $\frac{3}{2}H_{n-1}$. In which case the length of M_n is $\frac{1}{2}H_{n-1}+2$. When y is shifted by $\frac{1}{2}H_{n-1}$ the length of M_n is $\frac{1}{2}H_{n-1} - 2$.

Thus, in this case

$$|M_n| \geq \frac{1}{2}H_{n-1} - 2.$$

FIGURE 6. The case where $\frac{1}{2}H_{n-1} \leq k \leq \frac{3}{2}H_{n-1}$. In this figure, the long blocks have length H_{n-1}, the medium blocks have length $|\theta^{n-1}(A)|$ and the short blocks have length 1.

Figure 6 shows a shift close to $\frac{1}{2}H_{n-1}$. Define $t_n = H_{n-1} + 1$. Then for cylinders P_ℓ and P_m you can choose N large enough so that for $n \geq N$ we have

$$T^i x \in P_\ell \text{ if and only if } T^{i+t_n+1} x \in P_\ell$$

$$T^i y \in P_m \text{ if and only if } T^{i+t_n} y \in P_m$$

for all $i \in M_n - \max(|P_\ell|, |P_m|)$.

Define the interval L_n as the symmetric interval that contains the first block where x and y differ. Thus, the length of L_n is at most

$$2\left(|\omega_{s_n}| + |\theta^n(B)| + |k|\right) \leq 2\left(20|\theta^n(B)| + |\theta^n(B)| + \frac{1}{2}|\theta^n(B)|\right) \leq 44|\theta^n(B)|.$$

We can use the relationship between the length of the different letters iterated under θ to see that $|\theta^n(B)| \leq 6|\theta^{n-1}(A)|$. Thus,

$$\begin{aligned}
\liminf_{n\to\infty} \frac{|M_n|}{|L_n|} &\geq \liminf_{n\to\infty} \frac{\frac{1}{2}H_{n-1} - 2}{44|\theta^n(B)|} \\
&= \liminf_{n\to\infty} \frac{|\theta(A)| + \cdots |\theta^{n-1}(A)| - 4}{88\theta^n(B)} \\
&\geq \liminf_{n\to\infty} \frac{|\theta^{n-1}(A)|}{88 \cdot 6|\theta^{n-1}(A)|} > 0.
\end{aligned}$$

If $\frac{3}{2}H_{n-1} \leq k \leq \frac{1}{2}|\theta^n(B)|$ we can use the same M_n as before. It is split up for ease of reading. Now, the length of M_n is at least $H_{n-1} - \left(\frac{1}{2}H_{n-1} + \frac{1}{2}\left|\theta^{n-1}(A)\right|\right) = \frac{1}{2}\left(H_{n-1} - \left|\theta^{n-1}(A)\right|\right)$. Define $t_n = H_{n-1} + 1$. Then for cylinders P_ℓ and P_m you can choose N large enough so that for $n \geq N$ we have

$$T^i x \in P_\ell \text{ if and only if } T^{i+t_n+1} x \in P_\ell$$

$$T^i y \in P_m \text{ if and only if } T^{i+t_n} y \in P_m$$

for all $i \in M_n - \max(|P_\ell|, |P_m|)$.

We can again use the relationship between the heights to obtain $|\theta^n(B)| \leq 20|\theta^{n-2}(A)|$. Define L_n in the same way as before. Thus,

$$\begin{aligned}\liminf_{n\to\infty}\frac{|M_n|}{|L_n|} &\geq \liminf_{n\to\infty}\frac{\frac{1}{2}\left(H_{n-1}-|\theta^{n-1}(A)|\right)}{44|\theta^n(B)|}\\ &= \liminf_{n\to\infty}\frac{|\theta(A)|+\cdots|\theta^{n-2}(A)|}{88\theta^n(B)}\\ &\geq \liminf_{n\to\infty}\frac{|\theta^{n-2}(A)|}{88\cdot 20|\theta^{n-2}(A)|}>0.\end{aligned}$$

If $0 \leq k \leq \frac{1}{2}H_{n-1}$ then define M_n to be the interval where the second $\overline{D_{n-1}}$ block from x and the second $\overline{D_{n-1}}$ block from y overlap. Then $|M_n| \geq \frac{1}{2}H_{n-1}$. Define $t_n = H_{n-1} + |\theta^{n-1}(A)| + 2$. The remaining part of the argument from the above paragraphs hold here as well.

Therefore in this case the assumptions of Lemma 4.5 hold and $\nu = \mu \times \mu$.

Case 2. Suppose that $x_{\omega_{s_n}}$ is the beginning of $\theta^n(A)$ and $(T^k y)_{\omega_{s_n}}$ is the beginning of $\theta^n(C)$. The block that must precede $\theta^n(C)$ is $\theta^n(A)$ and the possible blocks that can precede $\theta^n(A)$ are $\theta^n(B)$ or $\theta^n(C)$. Since x and y are the same up until the ω_{s_n}-th place, this case isn't possible.

Case 3. Suppose that $x_{\omega_{s_n}}$ is the beginning of $\theta^n(B)$ and $(T^k y)_{\omega_{s_n}}$ is the beginning of $\theta^n(C)$. Since $\theta^n(A)$ must precede a $\theta^n(C)$ block and x and y are the same up until the ω_{s_n}-th place, the block that appears to the left of $\theta^n(B)$ in x must be $\theta^n(A)$. Also, $\theta^n(A)$ must be followed by a $\theta^n(C)$ block. Thus, x and $T^k y$ have the following form around the ω_{s_n}th place:

$$\begin{aligned} x &: A\overline{D_{n-1}}BD_{n-1}\overline{D_{n-1}}CA\overline{D_{n-1}}B\overline{D_{n-1}}BD_{n-1}\overline{D_{n-1}}C\\ T^k y &: A\overline{D_{n-1}}BD_{n-1}\overline{D_{n-1}}CA\overline{D_{n-1}}CA\overline{D_{n-1}}BD_{n-1}\overline{D_{n-1}}C\end{aligned}$$

If $H_{n-1} \leq k \leq \frac{1}{2}|\theta^n(B)|$ then define M_n to be the interval where the first $\overline{D_{n-1}}$ block from x and the second $\overline{D_{n-1}}$ block from y overlap. If y is shifted by H_{n-1} then $|M_n| = H_{n-1} - |\theta^n(A)| - 2$ and if y is shifted by $\frac{1}{2}|\theta^n(B)|$ then $|M_n| = \frac{1}{2}H_{n-1} + \frac{1}{2}|\theta^n(A)| - 1$. Thus, $|M_n| \geq H_{n-1} - |\theta^n(A)| - 2$.

Define $t_n = 3H_{n-1} + |\theta^n(A)| + 4$. Then for cylinders P_ℓ and P_m there is an N large enough so that for $n \geq N$ we have

$$T^i x \in P_\ell \text{ if and only if } T^{i+t_n+1}x \in P_\ell$$

$$T^i y \in P_m \text{ if and only if } T^{i+t_n}y \in P_m$$

for all $i \in M_n - \max(|P_\ell|, |P_m|)$.

Define the interval L_n as the symmetric interval that contains the first block where x and y differ. Thus, the length of L_n is at most $44|\theta^n(B)|$. We can use the relationship between the length of the different letters iterated under θ to see that

$|\theta^n(B)| \le 20|\theta^{n-2}(A)|$. Thus,

$$\begin{aligned}\liminf_{n\to\infty}\frac{|M_n|}{|L_n|} &\ge \liminf_{n\to\infty}\frac{H_{n-1}-|\theta^{n-1}(A)|-2}{44|\theta^n(B)|}\\ &= \liminf_{n\to\infty}\frac{|\theta(A)|+\cdots|\theta^{n-2}(A)|-2}{44\theta^n(B)}\\ &\ge \liminf_{n\to\infty}\frac{|\theta^{n-2}(A)|-2}{44\cdot 20|\theta^{n-2}(A)|}>0.\end{aligned}$$

If $0\le k\le \frac{1}{2}H_{n-1}$ then define M_n to be the interval where the third $\overline{D_{n-1}}$ block from x and the third $\overline{D_{n-1}}$ block from y overlap. Then $|M_n|\ge \frac{1}{2}H_{n-1}$. Define $t_n=H_{n-1}+1$.

For cylinders P_ℓ and P_m there is an N large enough so that for $n\ge N$ we have

$$T^ix\in P_\ell \text{ if and only if } T^{i+t_n+1}x\in P_\ell$$

$$T^iy\in P_m \text{ if and only if } T^{i+t_n}y\in P_m$$

for all $i\in M_n-\max(|P_\ell|,|P_m|)$.

Define the interval L_n as before. Thus,

$$\begin{aligned}\liminf_{n\to\infty}\frac{|M_n|}{|L_n|} &\ge \liminf_{n\to\infty}\frac{\frac{1}{2}H_{n-1}}{44|\theta^n(B)|}\\ &= \liminf_{n\to\infty}\frac{|\theta(A)|+\cdots|\theta^{n-1}(A)|}{88\theta^n(B)}\\ &\ge \liminf_{n\to\infty}\frac{|\theta^{n-1}(A)|}{88\cdot 6|\theta^{n-1}(A)|}>0.\end{aligned}$$

The last interval of k values to consider is $\frac{1}{2}H_{n-1}<k<H_{n-1}$. For these values we need two subcases, depending on what follows $\theta^n(B)$ from x and $\theta^n(A)$ from y.

Subcase 3.1. We still have all of the structure assumptions on x and y as in the beginning of Case 3. Additionally, suppose that $\theta^n(A)$ at the "end" of y is followed by $\theta^n(C)$, which in turn must be followed by the block $A\overline{D_{n-1}}$ since all θ^n blocks begin in that manner. Also, notice that $\theta^n(B)$ in x must be followed by $A\overline{D_{n-1}}$ for a similar reason. Hence, x and T^ky have the following form around the ω_{s_n}th place:

$$\begin{aligned} x &: A\overline{D_{n-1}}BD_{n-1}\overline{D_{n-1}}CA\overline{D_{n-1}}B\overline{D_{n-1}}BD_{n-1}\overline{D_{n-1}}CA\overline{D_{n-1}}\\ T^ky &: A\overline{D_{n-1}}BD_{n-1}\overline{D_{n-1}}CA\overline{D_{n-1}}CA\overline{D_{n-1}}BD_{n-1}\overline{D_{n-1}}CA\overline{D_{n-1}}CA\overline{D_{n-1}}\end{aligned}$$

Recall that we are assuming $\frac{1}{2}H_{n-1}<k<H_{n-1}$. Define M_n to be the interval where the first $\overline{D_{n-1}}$ block from x and the second $\overline{D_{n-1}}$ block from y overlap. Thus, $|M_n|\ge\frac{1}{2}H_{n-1}-2$. Define t_n to be $t_n=4H_{n-1}+|\theta^n(A)+6$. Then for cylinders P_ℓ and P_m there is N large enough so that for $n\ge N$ we have

$$T^ix\in P_\ell \text{ if and only if } T^{i+t_n}x\in P_\ell$$

$$T^iy\in P_m \text{ if and only if } T^{i+t_n+1}y\in P_m$$

for all $i \in M_n - \max(|P_\ell|, |P_m|)$. The interval L_n is defined in the same way and the length is at most $2(|\omega_{s_n}| + 2|\theta^n(B)| + |k|) \leq 46|\theta^n(B)|$. Thus,

$$\begin{aligned}\liminf_{n\to\infty} \frac{|M_n|}{|L_n|} &\geq \liminf_{n\to\infty} \frac{\frac{1}{2}H_{n-1} - 2}{46|\theta^n(B)|} \\ &= \liminf_{n\to\infty} \frac{|\theta(A)| + \cdots |\theta^{n-1}(A)| - 2}{92\theta^n(B)} \\ &\geq \liminf_{n\to\infty} \frac{|\theta^{n-1}(A)| - 2}{92 \cdot 6|\theta^{n-1}(A)|} > 0.\end{aligned}$$

Subcase 3.2. In this subcase we suppose that, in addition to the structure of x and y defined at the beginning of Case 3, we have that $\theta^n(A)$ at the "end" of y is followed by $\theta^n(B)$, which in turn must be followed by the block $A\overline{D_{n-1}}$. We also assume that $\theta^n(B)$ in x is followed by another $\theta^n(B)$. Notice that the case where $\theta^n(B)$ from x is followed by $\theta^n(A)$ reduces to Case 1. Under these assumptions x and $T^k y$ have the following form around the ω_{s_n}th place:

$$\begin{aligned}x \;:&\; A\overline{D_{n-1}}BD_{n-1}\overline{D_{n-1}}CA\overline{D_{n-1}}B\overline{D_{n-1}}BD_{n-1}\overline{D_{n-1}}C \\ &\; A\overline{D_{n-1}}B\overline{D_{n-1}}BD_{n-1}\overline{D_{n-1}}C \\ T^k y \;:&\; A\overline{D_{n-1}}BD_{n-1}\overline{D_{n-1}}CA\overline{D_{n-1}}CA\overline{D_{n-1}}BD_{n-1}\overline{D_{n-1}}C \\ &\; A\overline{D_{n-1}}B\overline{D_{n-1}}BD_{n-1}\overline{D_{n-1}}C\end{aligned}$$

Define M_n to be the interval where the first $\overline{D_{n-1}}$ block from x and the second $\overline{D_{n-1}}$ block from y overlap. Thus, $|M_n| \geq \frac{1}{2}H_{n-1} - 2$. Define t_n to be $t_n = 6H_{n-1} + 2|\theta^n(A)| + 8$. The rest of the argument is the same as Subcase 3.1.

Therefore in this case the assumptions of Lemma 4.5 hold and $\nu = \mu \times \mu$. □

Acknowledgements

The author would like to thank Giovanni Forni for being a valuable mentor and listening to these ideas; Matthew Yancey for help with the figures that appear in this paper; and the referee for valuable comments that made this paper better.

References

[1] Jon Aaronson and Benjamin Weiss, *Remarks on the tightness of cocycles.* part 2, Colloq. Math. **84/85** (2000), no. part 2, 363–376. Dedicated to the memory of Anzelm Iwanik. MR1784202

[2] Artur Avila and Giovanni Forni, *Weak mixing for interval exchange transformations and translation flows*, Ann. of Math. (2) **165** (2007), no. 2, 637–664, DOI 10.4007/annals.2007.165.637. MR2299743

[3] *Topics in symbolic dynamics and applications*, London Mathematical Society Lecture Note Series, vol. 279, Cambridge University Press, Cambridge, 2000. Papers from the CIMPA-UNESCO Summer School held at Universidad de la Frontera, Temuco, January 6–24, 1997; Edited by F. Blanchard, A. Maass and A. Nogueira. MR1776753

[4] J. Bourgain, P. Sarnak, and T. Ziegler, *Disjointness of Moebius from horocycle flows*, From Fourier analysis and number theory to Radon transforms and geometry, Dev. Math., vol. 28, Springer, New York, 2013, pp. 67–83, DOI 10.1007/978-1-4614-4075-8_5. MR2986954

[5] A. del Junco and D. Rudolph, *On ergodic actions whose self-joinings are graphs*, Ergodic Theory Dynam. Systems **7** (1987), no. 4, 531–557, DOI 10.1017/S0143385700004193. MR922364

[6] Sébastien Ferenczi, Charles Holton, and Luca Q. Zamboni, *Joinings of three-interval exchange transformations*, Ergodic Theory Dynam. Systems **25** (2005), no. 2, 483–502, DOI 10.1017/S0143385704000811. MR2129107

[7] Eli Glasner, *Ergodic theory via joinings*, Mathematical Surveys and Monographs, vol. 101, American Mathematical Society, Providence, RI, 2003. MR1958753

[8] A. B. Katok, Ja. G. Sinaĭ, and A. M. Stepin, *The theory of dynamical systems and general transformation groups with invariant measure* (Russian), Mathematical analysis, Vol. 13 (Russian), Akad. Nauk SSSR Vsesojuz. Inst. Naučn. i Tehn. Informacii, Moscow, 1975, pp. 129–262. (errata insert). MR0584389

[9] A. B. Katok and A. M. Stepin, *Approximations in ergodic theory* (Russian), Uspehi Mat. Nauk **22** (1967), no. 5 (137), 81–106. MR0219697

[10] Anatole Katok, *Interval exchange transformations and some special flows are not mixing*, Israel J. Math. **35** (1980), no. 4, 301–310, DOI 10.1007/BF02760655. MR594335

[11] Daniel J. Rudolph, *Fundamentals of measurable dynamics*, Oxford Science Publications, The Clarendon Press, Oxford University Press, New York, 1990. Ergodic theory on Lebesgue spaces. MR1086631

[12] V. V. Ryzhikov, *Self-joins of commutative actions with an invariant measure* (Russian), Mat. Zametki **83** (2008), no. 5, 792–795, DOI 10.1134/S0001434608050179; English transl., Math. Notes **83** (2008), no. 5-6, 723–726. MR2451369

[13] P. Sarnak, *Three lectures on the mobius function randomness and dynamics*, publications/ias.edu/sarnak/ (2011).

[14] Younghwan Son, *Substitutions, tiling dynamical systems and minimal self-joinings*, Discrete Contin. Dyn. Syst. **34** (2014), no. 11, 4855–4874, DOI 10.3934/dcds.2014.34.4855. MR3223833

[15] William A. Veech, *The metric theory of interval exchange transformations. I. Generic spectral properties*, Amer. J. Math. **106** (1984), no. 6, 1331–1359, DOI 10.2307/2374396. MR765582

[16] M. Viana, *Dynamics of interval exchange transformations and teichmuller flows*, http://w3.impa.br/ viana/out/ietf.pdf.

Department of Mathematics, University of Maryland, Mathematics Building, College Park, Maryland 20742-4015

E-mail address: kbyancey1@gmail.com

Published Titles in This Series

678 **Joseph Auslander, Aimee Johnson, and Cesar E. Silva, Editors,** Ergodic Theory, Dynamical Systems, and the Continuing Influence of John C. Oxtoby, 2016

674 **Bogdan D. Suceavă, Alfonso Carriazo, Yun Myung Oh, and Joeri Van der Veken, Editors,** Recent Advances in the Geometry of Submanifolds, 2016

673 **Alex Martsinkovsky, Gordana Todorov, and Kiyoshi Igusa, Editors,** Recent Developments in Representation Theory, 2016

672 **Bernard Russo, Asuman Güven Aksoy, Ravshan Ashurov, and Shavkat Ayupov, Editors,** Topics in Functional Analysis and Algebra, 2016

671 **Robert S. Doran and Efton Park, Editors,** Operator Algebras and Their Applications, 2016

670 **Krishnendu Gongopadhyay and Rama Mishra, Editors,** Knot Theory and Its Applications, 2016

669 **Sergiĭ Kolyada, Martin Möller, Pieter Moree, and Thomas Ward, Editors,** Dynamics and Numbers, 2016

668 **Gregory Budzban, Harry Randolph Hughes, and Henri Schurz, Editors,** Probability on Algebraic and Geometric Structures, 2016

667 **Mark L. Agranovsky, Matania Ben-Artzi, Greg Galloway, Lavi Karp, Dmitry Khavinson, Simeon Reich, Gilbert Weinstein, and Lawrence Zalcman, Editors,** Complex Analysis and Dynamical Systems VI: Part 2: Complex Analysis, Quasiconformal Mappings, Complex Dynamics, 2016

666 **Vicenţiu D. Rădulescu, Adélia Sequeira, and Vsevolod A. Solonnikov, Editors,** Recent Advances in Partial Differential Equations and Applications, 2016

665 **Helge Glöckner, Alain Escassut, and Khodr Shamseddine, Editors,** Advances in Non-Archimedean Analysis, 2016

664 **Dihua Jiang, Freydoon Shahidi, and David Soudry, Editors,** Advances in the Theory of Automorphic Forms and Their L-functions, 2016

663 **David Kohel and Igor Shparlinski, Editors,** Frobenius Distributions: Lang-Trotter and Sato-Tate Conjectures, 2016

662 **Zair Ibragimov, Norman Levenberg, Sergey Pinchuk, and Azimbay Sadullaev, Editors,** Topics in Several Complex Variables, 2016

661 **Douglas P. Hardin, Doron S. Lubinsky, and Brian Z. Simanek, Editors,** Modern Trends in Constructive Function Theory, 2016

660 **Habib Ammari, Yves Capdeboscq, Hyeonbae Kang, and Imbo Sim, Editors,** Imaging, Multi-scale and High Contrast Partial Differential Equations, 2016

659 **Boris S. Mordukhovich, Simeon Reich, and Alexander J. Zaslavski, Editors,** Nonlinear Analysis and Optimization, 2016

658 **Carlos M. da Fonseca, Dinh Van Huynh, Steve Kirkland, and Vu Kim Tuan, Editors,** A Panorama of Mathematics: Pure and Applied, 2016

657 **Noé Bárcenas, Fernando Galaz-García, and Mónica Moreno Rocha, Editors,** Mexican Mathematicians Abroad, 2016

656 **José A. de la Peña, J. Alfredo López-Mimbela, Miguel Nakamura, and Jimmy Petean, Editors,** Mathematical Congress of the Americas, 2016

655 **A. C. Cojocaru, C. David, and F. Pappalardi, Editors,** SCHOLAR—a Scientific Celebration Highlighting Open Lines of Arithmetic Research, 2015

654 **Carlo Gasbarri, Steven Lu, Mike Roth, and Yuri Tschinkel, Editors,** Rational Points, Rational Curves, and Entire Holomorphic Curves on Projective Varieties, 2015

653 **Mark L. Agranovsky, Matania Ben-Artzi, Greg Galloway, Lavi Karp, Dmitry Khavinson, Simeon Reich, Gilbert Weinstein, and Lawrence Zalcman, Editors,** Complex Analysis and Dynamical Systems VI: Part 1: PDE, Differential Geometry, Radon Transform, 2015

652 **Marina Avitabile, Jörg Feldvoss, and Thomas Weigel, Editors,** Lie Algebras and Related Topics, 2015

651 **Anton Dzhamay, Kenichi Maruno, and Christopher M. Ormerod, Editors,** Algebraic and Analytic Aspects of Integrable Systems and Painlevé Equations, 2015

650 **Jens G. Christensen, Susanna Dann, Azita Mayeli, and Gestur Ólafsson, Editors,** Trends in Harmonic Analysis and Its Applications, 2015

649 **Fernando Chamizo, Jordi Guàrdia, Antonio Rojas-León, and José María Tornero, Editors,** Trends in Number Theory, 2015

648 **Luis Álvarez-Cónsul, José Ignacio Burgos-Gil, and Kurusch Ebrahimi-Fard, Editors,** Feynman Amplitudes, Periods and Motives, 2015

647 **Gary Kennedy, Mirel Caibăr, Ana-Maria Castravet, and Emanuele Macrì, Editors,** Hodge Theory and Classical Algebraic Geometry, 2015

646 **Weiping Li and Shihshu Walter Wei, Editors,** Geometry and Topology of Submanifolds and Currents, 2015

645 **Krzysztof Jarosz, Editor,** Function Spaces in Analysis, 2015

644 **Paul M. N. Feehan, Jian Song, Ben Weinkove, and Richard A. Wentworth, Editors,** Analysis, Complex Geometry, and Mathematical Physics, 2015

643 **Tony Pantev, Carlos Simpson, Bertrand Toën, Michel Vaquié, and Gabriele Vezzosi, Editors,** Stacks and Categories in Geometry, Topology, and Algebra, 2015

642 **Mustapha Lahyane and Edgar Martínez-Moro, Editors,** Algebra for Secure and Reliable Communication Modeling, 2015

641 **Maria Basterra, Kristine Bauer, Kathryn Hess, and Brenda Johnson, Editors,** Women in Topology, 2015

640 **Gregory Eskin, Leonid Friedlander, and John Garnett, Editors,** Spectral Theory and Partial Differential Equations, 2015

639 **C. S. Aravinda, William M. Goldman, Krishnendu Gongopadhyay, Alexander Lubotzky, Mahan Mj, and Anthony Weaver, Editors,** Geometry, Groups and Dynamics, 2015

638 **Javad Mashreghi, Emmanuel Fricain, and William Ross, Editors,** Invariant Subspaces of the Shift Operator, 2015

637 **Stéphane Ballet, Marc Perret, and Alexey Zaytsev, Editors,** Algorithmic Arithmetic, Geometry, and Coding Theory, 2015

636 **Simeon Reich and Alexander J. Zaslavski, Editors,** Infinite Products of Operators and Their Applications, 2015

635 **Christopher W. Curtis, Anton Dzhamay, Willy A. Hereman, and Barbara Prinari, Editors,** Nonlinear Wave Equations, 2015

634 **Steven Dougherty, Alberto Facchini, André Leroy, Edmund Puczyłowski, and Patrick Solé, Editors,** Noncommutative Rings and Their Applications, 2015

633 **Delaram Kahrobaei and Vladimir Shpilrain, Editors,** Algorithmic Problems of Group Theory, Their Complexity, and Applications to Cryptography, 2015

632 **Gohar Kyureghyan, Gary L. Mullen, and Alexander Pott, Editors,** Topics in Finite Fields, 2015